OPTIMUM DESIGN OF MECHANICAL ELEMENTS

A unique book of basic techniques and practical applications for the optimal design of mechanical elements in realistic design settings—

The first edition of OPTIMUM DESIGN OF MECHANICAL ELEMENTS established Ray C. Johnson as a pioneer in the field of mechanical design optimization. Published in 1961, the book introduced Professor Johnson's practical and powerful method for handling a wide variety of critical design problems. This second edition, greatly expanded and updated, incorporates important new developments in basic techniques and applications.

The first five chapters review important background material, including such practical mathematical topics as curve fitting, finite differences and equation simplification; the effects of manufacturing errors on product performance; material properties; failure theories; and the important problem of selecting factors of safety for original design use. The book then presents an updated explanation of the method of optimum design (MOD) along with simple illustrative examples. The second edition also contains a discussion of automated optimal design (AOD) with high-speed digital computers along with illustrative examples and a correlation with MOD solutions. Optimization problems and solutions are then given for such simple mechanical elements as tensile bars, torsion bars, and shafts in combined loading and also for complex mechanical elements, including helical and spur gears, helical springs, and hydrostatic journal bearings. The solutions illustrate basic techniques that can be applied directly to countless practical problems in the design of machinery, mechanisms, mechanical assemblages and products of all types. Many practical exercises are included at the end of each chapter.

Proven in the successful, and sometimes spectacular, design of industrial products and machinery for more than twenty-five years, the material contained in this volume will be valuable to both professionals and students.

OPTIMUM DESIGN OF MECHANICAL ELEMENTS

Second Edition

RAY C. JOHNSON

Higgins Professor of Mechanical Engineering
Worcester Polytechnic Institute and
Consulting Engineer

A Wiley-Interscience Publication

JOHN WILEY & SONS, INC.
New York · Chichester · Brisbane · Toronto

Library of Congress Cataloging in Publication Data

Johnson, Ray C.
Optimum design of mechanical elements.

"A Wiley-Interscience publication."
Includes bibliographical references and index.
1. Machinery—Design—Mathematical models.
2. Machinery—Design—Data processing. I. Title.
TJ233.J63 1979 621.8′15 79-14363
ISBN 0-471-03894-6

Printed in the United States of America

10 9 8 7 6 5 4 3 2 1

原著者：Ray C. Johnson
發行所：北門出版社
發行人：沈 家 駿
地 址：台北市永吉路68巷19～3號
電 話：7680934
總經銷：北門書城有限公司
地 址：台北市忠孝西路一段209號二樓
電 話：3822115 • 3145032
郵撥帳號：107525號
新聞局出版登記局版台業字第2969號
中華民國75年3月 日
售價：NT$ 元整

Foreword

The first edition of *Optimum Design of Mechanical Elements* established Ray C. Johnson as an important pioneer in the field of mechanical design optimization. Professor Johnson's method of optimum design was conceived in the early 1950s to handle critical industrial design problems with high speed machinery. Readers of the second edition of this book have the benefits of the author's wide experience in the intervening years both in applying the method in consultancy work for industry and in teaching the method to senior and graduate engineering students and to practicing engineers.

The method of optimum design is a powerful technique for those confronted with practical mechanical design problems. Researchers and teachers with wider interests in design optimization methods should also find the approach of interest. A distinctive feature of the method is the insight which it gives into the nature of an optimization problem and its possible solutions.

Finally, the readers of this book, in addition to learning the procedures of the method of optimum design, can observe and hopefully gain from Professor Johnson's considerable skill in modeling real situations for optimization purposes to arrive at meaningful and practical solutions.

JACK ELLIS
Lecturer in Mechanical Engineering
University of Salford,
United Kingdom

Preface to the Second Edition

Since publication of the first edition much progress has been made in the development of basic techniques and applications for the optimal design of mechanical elements. Hence, the primary purpose for this second edition is to incorporate some of the developments in the original work. As related to practical applications of mechanical elements, emphasis will be on two basic optimization techniques which are particularly suitable to solution of problems in such cases. First, developments in the method of optimum design (MOD) are incorporated and illustrated with many practical examples. Incidentally, the MOD was originally published in the first edition. Secondly, with the widespread availability of digital computers since the first edition, applications of automated optimal design (AOD) are incorporated in this work.

The first five chapters of the second edition present background information which has been up-dated and expanded for later application. Chapter 2 on mathematics now includes some basic background on the gradient search technique used in development of the AOD algorithm OPTIGO in Chapter 6. Chapter 3 on manufacturing errors contains pertinent background for an understanding of the need for factor of safety in design work. Chapter 4 on material properties and failure theories reviews some information which is particularly vital to the optimal design of mechanical elements as applied in later chapters. Chapter 5 includes some techniques which can be applied to the selection of factor of safety for original design use. Again, this is a topic which is particularly vital in the optimal design of mechanical elements.

In Chapter 6, we start by considering some basic characteristics for mechanical elements and their relation to the general problem of design optimization. This is followed by an up-dated explanation of the MOD with simple illustrative examples. The chapter is concluded with a consideration of

AOD as particularly applied to the algorithm OPTIGO, including some simple illustrative examples and a correlation with MOD solutions.

Some optimization problems for simple mechanical elements are covered in Chapter 7. Solutions are obtained for these problems by the MOD technique, which is most appropriate for such problems. However, the reader might wish to solve some of the exercises by the AOD technique to gain experience in this respect.

Some optimization problems for complex mechanical elements are covered in Chapter 8. Solutions to the helical gearset and the helical spring problems are obtained by the MOD, which incorporates curve-fitting techniques and other practical approaches for handling complicated problems using three dimensional variation diagrams. However, the hydrostatic journal bearing problem of Chapter 8 involves a very complicated equation system, and it is solved by the AOD technique using the algorithm OPTIGO. A variety of other complex mechanical elements are presented as exercises which may be solved by either the MOD or AOD techniques.

In Appendix A, a brief summary of nomenclature is presented for the MOD and for the AOD algorithm OPTIGO. In Appendix B, a Fortran IV listing of OPTIGO is presented, including the printouts for a simple example from Chapter 6. Finally, in Appendix C another recently developed algorithm for AOD is described and illustrated, as contributed by G. E. Johnson of the University of Virginia and M. A. Townsend of Vanderbilt University. The numerical example used here is a problem from Chapter 7, and the solution correlates very well with the MOD solution presented in that chapter.

The second edition has been written either for college adoption or for self-study by engineers in professional practice. To facilitate the learning process, an ample number of exercises are presented at the ends of Chapters 2 through 8. In this way experience can be gained for application of the basic material to original problems encountered in practice.

In conclusion, I wish to thank the many people who have contributed so much to the development of this work. Comments and suggestions on the first edition received from professors and engineers around the world are too numerous to itemize. Likewise, the opportunity for teaching the subject material at Worcester Polytechnic Institute has been invaluable, including my contact with colleagues and students on both the undergraduate and graduate levels. Finally, the numerous opportunities to apply the techniques in many consulting engagements for industry has been an indispensable experience so necessary for the development of this work.

RAY C. JOHNSON

Holden, Massachusetts
August, 1979

Contents

OPTIMUM DESIGN OF MECHANICAL ELEMENTS

1

Introduction

.1 Mechanical Design

Mechanical *design* is one of the most important problems which can confront a practicing engineer. This critical task of synthesis *can* be both distasteful and elusive because of the complex relationship existing among the factors associated with the definition of design itself. We can define mechanical design as the *selection* of materials and geometry which satisfies specified and implied functional requirements while remaining within the confines of inherently unavoidable limitations. Hence the basic problem of mechanical *design* is considerably more difficult than the problem of mechanical *analysis*, since generally for the latter both geometry and materials are assumed to be constants and limitations such as space restrictions are of no significance.

The overall objective of mechanical design is stated in the definition above. For most specific real problems we would find an infinite number of *possible* design solutions for the overall objective. Hence any one of these possible design solutions might be designated an *adequate design*, because it represents a synthesis which merely *satisfies* the functional requirements within the confines of the existing limitations. Thus since the overall objective of mechanical design can often be satisfied by more than one possible design solution, we have a situation naturally conducive to cut-and-try techniques; hence such practices are inherently characteristic of *adequate design.* To obtain a more explicit method of design, the overall objective must be more clearly defined. Let us now consider how we might define the overall objective more explicitly.

1.2 Optimum Design

Associated with any mechanical element are certain inherently unavoidable *undesirable effects*. For instance, in any specific mechanical element such effects as stresses, deflections, vibrations, space occupancy, weight, and cost might be undesirable, and the degree of significance for each effect depends on the particular application. Hence a *tolerable* limit will exist for the permissible degrees of each undesirable effect. Similarly, associated with any mechanical element are certain *desirable effects*, such as power transmission *capability*, energy absorption *capability*, momentary overload *capacity*, varying force transmission *capability*, speed *capability*, usable length of life, and factor of safety.

Any *one* of the previously mentioned effects, among other possibilities, can be the basis upon which an *explicit* overall objective in mechanical design is obtained. Hence, depending on the particular problem at hand, we can so conduct our method of design as to *minimize* the *most significant* undesirable effect; or we might realize that it would be more beneficial so to conduct our method of design as to *maximize* the *most significant* desirable effect. Thus, in initially deciding on one of these goals, we introduce an additional design objective, and since it is clearly defined it results in an explicit design procedure for selecting materials and geometry. In doing this we arrive at a mechanical design which is the best possible one from the standpoint of the most significant effect, and we call it the *optimum design*.

A *method of optimum design* [1.9–1.16] derived from overall considerations of many specific optimum design studies has been formulated, and it will be presented in Chapter 6. From the previous discussions, it is obvious that the method of optimum design is an *explicit* method of design. Also, it should be emphasized that the method does *not* ignore any of the functional requirements or any of the existing limitations. Hence a satisfactory design is determined in an explicit manner, it is the best possible design from the standpoint of the most significant effect (which will be called the *basis* of optimum design), and all undesirable effects are *within* their permissible ranges of magnitude.

Generally, the experienced design engineer will be able to choose the appropriate objective for optimization depending on the requirements of the particular application at hand. The critical phenomena of significance should be summarized for the device to be designed. The appropriate optimization objective can then be selected and constraints can be imposed on the remaining phenomena to assure an acceptable design from all standpoints. However, the initial formulation for the optimization problem should not be more complicated than necessary at least to start, and this often requires the making of some simplifying assumptions. The strategy for simplification in this respect will be covered in Chapter 6, together with a consideration of validity for the assumptions made in formulating the problem.

1.3 Some Aspects of a Specific Design Problem

In this section we shall explore various aspects of design solutions to a relatively simple design problem. First, we shall consider a typical *adequate design* solution which is characterized by traditional influences and cut-and-try techniques. Such a design will be entirely *satisfactory* from the standpoint of the overall objectives in the statement of the problem. Second, we shall consider an *optimum design* solution to the same problem which will illustrate the possibility of reaping previously hidden benefits by such an approach. The conduction of design in this manner will explicitly determine values for geometry and material, and the effects of existing limitations on the optimum design will be illustrated. Finally, we shall illustrate the additional benefits which can be gained often by going beyond traditional bounds to an improvement in basic geometrical shape or to an improvement in manufacturing technique. It is hoped that what follows will inspire the reader by illustrating the value of the optimum design approach in providing an explicit method of synthesis and simultaneously resulting in the most desirable design solution.

The Basic Design Problem. Design a plastic tray capable of holding a specified volume of liquid, V, such that the liquid has a specified depth, H, and the wall thickness of the tray is to be a specified thickness, T. The tray is to be manufactured in large quantities.

Adequate Design Solution. By intuition we might select the basic geometrical shape to be rectangular, as depicted in figure 1.1. Hence, assuming the wall thickness T to be small compared with the dimensions b and l, we can write equation 1.1 for the liquid volume V.

$$V = blH \tag{1.1}$$

Therefore we can arbitrarily select *any* value for b and then use equation 1.1 to solve for the corresponding value for l. The geometrical design so obtained would be entirely satisfactory for the problem as stated, and obviously there would be an infinite number of possible solutions.

In selecting the material for the tray we would consider such factors as the available manufacturing techniques and possible chemical reactions of the tray material with the liquid which will be held. Therefore, from our experience, we might select vacuum forming as our manufacturing method, using an acrylic thermoplastic sheet as our material. Hence the design of our plastic tray would be complete, and the basic problem as specifically stated would be *satisfied* in its entirety.

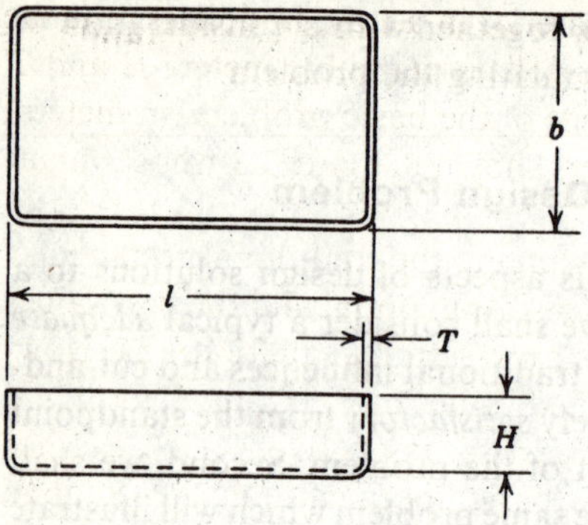

Figure 1.1 Typical tray with traditional rectangular shape.

Optimum Design Solution. The basic design problem as initially stated was purposely chosen to be a relatively simple one. Hence the undesirable effects of stresses, vibrations, deflections, and weight are insignificant considerations for this problem and we can ignore them. However, that the undesirable effect of cost is significant is *implied* by the statement of the basic problem, since the tray is to be manufactured in large quantities. Therefore the most significant undesirable effect for this problem is cost, and a worthwhile explicit objective would be to design the tray on the basis of minimizing the cost. Thus, from the standpoint of cost, we might question whether our adequate design solution is the best one with respect to geometry, material, and manufacturing method.

To arrive at the optimum design in an explicit manner, we must express mathematically the quantity on which the optimum design is based. Hence the cost C of a tray could be written as follows, and this would be known as the *primary design equation* (P.D.E.):

$$C = C_o + C_t + C_l + C_m \tag{1.2}$$

In equation 1.2, C is the total cost, C_o is the cost of overhead, C_t is the tooling cost, C_l is the labor cost, and C_m is the cost of the plastic material, all on a per tray basis.

To begin, let us assume that the vacuum-forming technique will be our *available* manufacturing method. Hence overhead, tooling, and labor costs will be essentially independent of reasonable geometrical shapes and feasible plastic materials, and we shall assume from the problem statement large quantities of production and specified values for V, H, and T. Therefore the optimum design of the tray would be determined only by the material cost term C_m in equation 1.2, which is [c (material volume)]. Thus for the traditional rectangular shape we have

$$C_m = c(bl + 2bH + 2lH)T \tag{1.3a}$$

In equation 1.3a, C_m is the cost of the material in a tray, in dollars; c is the unit volume cost of the tray material, in dollars per cubic meter; H and T are the specified values from the statement of the basic problem, in meters; and b and l are the geometrical parameters shown in figure 1.1 whose values are to be determined by *design*.

In determining the optimum design assuming the traditional rectangular shape, the objective in selecting the best feasible material and the best values for b and l will be to *minimize* cost C_m as expressed by equation 1.3a. However, we cannot ignore the initially specified values of V, H, and T for liquid volume, liquid depth, and tray wall thickness, respectively. Hence, since equation 1.1 *must* be satisfied from the basic specifications, we combine equation 1.1 with equation 1.3a by *eliminating* an *independent* design parameter such as l. Thus we obtain

$$C_m = c\left(\frac{V}{H} + 2bH + \frac{2V}{b}\right)T \tag{1.3b}$$

This equation expresses the quantity C_m upon which the optimum design is based, it takes into account *all* the originally stated specifications, and all terms on the right side of the equation are *independent* of each other. We might think of equation 1.3b as the *developed primary design equation*, since from it we can extract explicitly the optimum design of the tray.

With respect to the selection of the optimum feasible material, the term c in equation 1.3b is the only one which depends on the material. Hence, to minimize C_m we would select from a discrete list of available *feasible* materials the material having the lowest value for the unit volume cost, c. Therefore, material parameter term c might be thought of as the *material selection factor*, and from equation 1.3b it is clear that the material selection for this problem is *independent* of the particular values for V, H, T, b, and l.

For each of the *available* materials, which are *feasible* for vacuum forming, specific values of c can readily be obtained. For simplicity of illustration, we assume that only two feasible thermoplastics, such as acrylic and polystyrene sheets, are available in our locality at this time. Values of unit volume cost c of the materials can be obtained from the trade journals and vendors, and typically we illustrate as follows.

	c	
	$/in³	$/m³
Acrylic	0.030	1831
Polystyrene	0.012	732

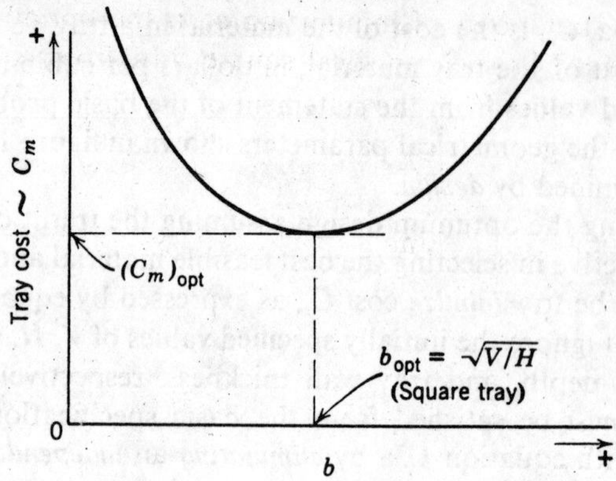

Figure 1.2 Cost versus geometrical parameter b for feasible rectangular tray proportions. (V, H, T, c are constant; see equation 1.3b and figure 1.1.)

Therefore, for this illustrative problem, the optimum available material is the polystyrene because its value of c is lower.

With respect to the selection of the optimum geometry for the rectangular tray, we would consider the variation of the cost C_m with respect to the independent variable b, from the developed primary design equation 1.3b. Obviously, this would be as shown in figure 1.2, with minimum cost occurring for an optimum value of b which we shall designate b_{opt}. This value is determined as follows, from equation 1.3b.

$$\frac{\partial C_m}{\partial b} = (c)(2T)\left(H - \frac{V}{b^2}\right) = 0$$

$$\therefore \quad b_{opt} = \sqrt{\frac{V}{H}}$$

Placing this value for b in equation 1.1, we can obtain the optimum value for l:

$$l = \frac{V}{bH}$$

Therefore,

$$l_{opt} = \frac{V}{H(b_{opt})} = \frac{V}{H\sqrt{V/H}} = \sqrt{\frac{V}{H}} = b_{opt}$$

Thus, from the standpoint of minimizing cost, the optimum rectangular shape is a square. For such an ideal square tray, from equation 1.3b its material cost C_m would be

$$(C_m)_{opt} = c\left[\frac{V}{H} + 2\left(\sqrt{\frac{V}{H}}\right)H + \frac{2V}{\sqrt{V/H}}\right]T$$

Therefore,

$$(C_m)_{opt} = c\left(\frac{V}{H} + 4\sqrt{VH}\right)T \qquad (1.3c)$$

Often space restrictions exist in real design problems. Hence, for the intended usage of the tray, it might be required to fit between two structures while resting on a shelf of specified depth. Therefore limit equations might exist expressing permissible values of b and l as follows:

$$b \leq b_{max} \qquad (1.4)$$

$$l \leq l_{max} \qquad (1.5)$$

In equations 1.4 and 1.5, b_{max} and l_{max} are the values of the space restrictions existing on feasible values for b and l respectively. Obviously, the particular values for b_{max} and l_{max} might affect the optimum geometry of the rectangular tray. If b_{max} and l_{max} are *both* equal to or greater than $\sqrt{V/H}$, the ideal square tray is still the optimum design. However, if either b_{max} or l_{max} is less than $\sqrt{V/H}$, the optimum design is the tray which barely fits within the space restriction. For the case where b_{max} is less than $\sqrt{V/H}$, the optimum design is depicted in figures 1.3 and 1.4. Also, it is possible that *no* geometrical design satisfying the specifications within the space restrictions is feasible. This possibility might be termed a case of *incompatible specifications*, which would exist if the following relationship existed (from equation 1.1).

$$b_{max} l_{max} < \frac{V}{H} \qquad (1.6)$$

Now we have completed our optimum design study for the tray, having arbitrarily *assumed* the traditional rectangular shape to begin with. However, we might rightfully question this arbitrary choice, since some other basic geometrical shape might possibly result in a lower cost while still satisfying the functional requirements of V, H, and T as specified in the original statement of the problem. In fact, we would find that a circular tray would result

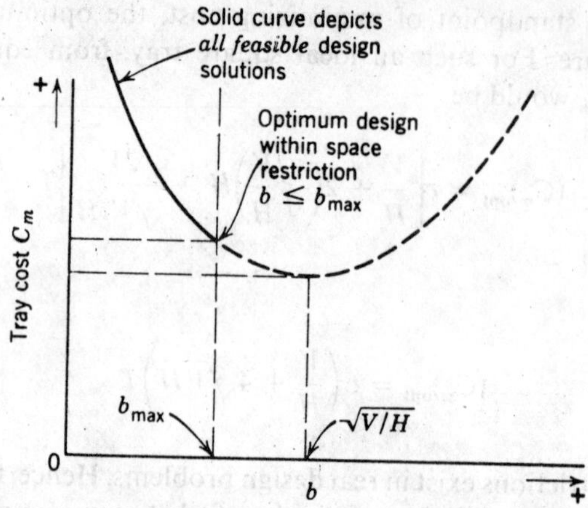

Figure 1.3 Cost versus geometrical parameter b for feasible rectangular tray proportions within existing space restriction $b \leq b_{max}$. (V, H, T, c are constant; see equation 1.3b and figure 1.4.)

in the lowest possible value for material cost C_m, assuming specified values for V, H, and T, and this *could* be determined explicitly by using the calculus of variations. Let us write an equation expressing the material cost C_m for a circular tray as shown in figure 1.5. Hence

$$C_m = c\left(\frac{\pi D^2}{4} + \pi DH\right)T \tag{1.7a}$$

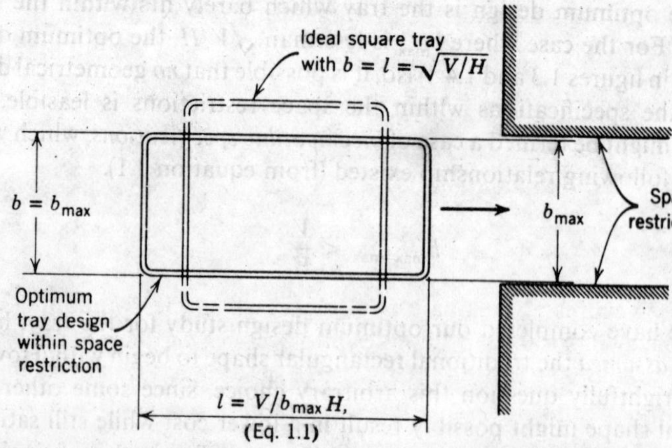

Figure 1.4 Optimum rectangular shape tray design within existing space restrictions. (V, H, T are specified values.)

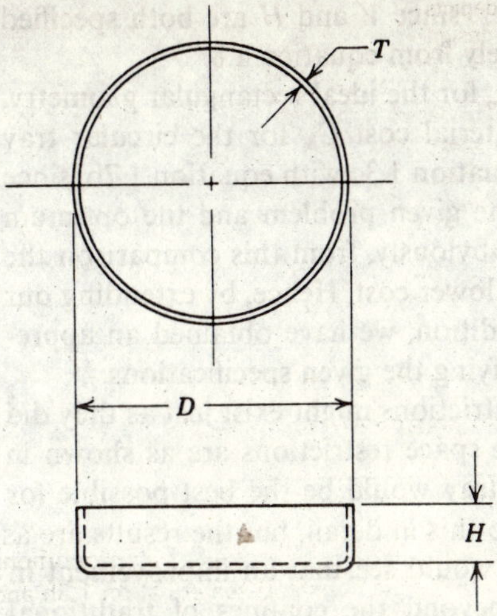

Figure 1.5 Circular tray design, representing a deviation from the traditional rectangular shape, resulting in a lower cost C_m. (Assume V, H, T are specified values.)

In this equation all terms are as previously defined except D, which is shown in figure 1.5. Incidentally, equation 1.7a assumes that T is much smaller than D, which is reasonable. For the circular tray, the specified liquid volume V can be expressed as

$$V = \left(\frac{\pi D^2}{4}\right) H \tag{1.8}$$

Combining equations 1.8 and 1.7a by eliminating D, we obtain

$$C_m = c\left(\frac{V}{H} + 2\sqrt{\pi V H}\right) T$$

Therefore

$$C_m = c\left(\frac{V}{H} + 3.54\sqrt{V H}\right) T \tag{1.7b}$$

Equation 1.7b expresses the quantity C_m upon which the optimum design is based for the cylindrical shape, it takes into account *all* the originally stated specifications, and all terms on the right side of the equation are *independent* of each other. Hence equation 1.7b is the *developed primary design equation*, since from it we can extract explicitly the optimum design. Thus the optimum material is selected from the list of available feasible ones on the basis of minimum value for the term c, which is the same material selection factor as that for the rectangular shape. However, since V and H are both specified values, diameter D is determined uniquely from equation 1.8.

We can compare the material cost C_m for the ideal rectangular geometry, which is the square tray, with the material cost C_m for the circular tray geometry of figure 1.5 by comparing equation 1.3c with equation 1.7b, since V, H, and T are specified values for the given problem and the optimum material for both shapes is the same. Obviously, from this comparison the circular tray geometry will result in the lower cost. Hence, by extending our considerations beyond the limits of tradition, we have obtained an appreciably better tray design while still satisfying the given specifications.

For the circular tray design, space restrictions might exist just as they did for the rectangular shape. Hence, if the space restrictions are as shown in figure 1.4, we would find that an oval tray would be the best possible for minimizing cost C_m. We will not go into this in detail, but the results are as depicted in figure 1.6. Hence again we would see that an improvement in design would be obtained by going beyond the confines of traditional practices such as the restriction of our considerations to the rectangular shape.

As a final point, we might question whether our choice of the manufacturing method is the best possible one. Perhaps another available feasible manufacturing technique, such as a low cost molding process, would result in a lower tray cost than vacuum forming. We will not pursue this consideration further except to say that the general procedure would be to compare the

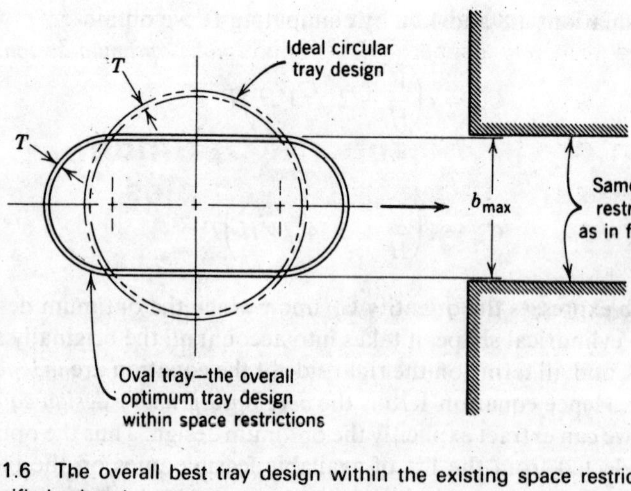

Figure 1.6 The overall best tray design within the existing space restrictions. (V, H, T are specified values.)

optimum tray designs for each manufacturing method from the standpoint of *total cost*, including overhead, tooling, labor, and material.

Some Salient Characteristics of the Preceding Design Studies. From the tray design solutions which have been studied, we can observe some salient points of interest which are characteristic for mechanical design in general. For one, although the adequate design solution *satisfied* the specified problem in its entirety, we reaped "hidden" benefits by introducing an explicit objective upon which the optimum design was based. Second, introduction of the explicit objective resulted in an explicit determination of both material and geometry. Last, by deviating from the imaginary confines of tradition, such as the rectangular tray shape, we were able to obtain an additional improvement in design, such as the circular tray.

The salient characteristics of this specific design study are typical for almost all mechanical elements. The basis for optimum design can be any one of innumerable possibilities, such as the most significant undesirable effect or the most significant desirable effect, the selection depending on its relative importance for the particular problem at hand. Hence, depending on the particular mechanical element and its intended usage, we might wish to minimize either stresses, deflections, vibrations, space occupancy, weight, or cost. On the other hand, we might desire to maximize either power transmission *capability*, energy absorption *capability*, overload *capacity*, varying force *capacity*, speed *capability*, usable length of life, or factor of safety. The important point to realize is that a general method exists for arriving at explicit determinations of both geometry and materials. This *method of optimum design* is presented in Chapter 6, and there are many illustrative examples in the following chapters. The simple tray design problem covered here has served as merely a superficial introduction to *optimum design*.

Evolution of the Method of Optimum Design—Some Industrial Examples [1.1–1.8]

1.4 Optimum Design by Necessity

The purpose of the remainder of this chapter is to present some physical evidence of some specific optimum design solutions to some actual industrial problems. The photographs presented in the following pages really represent the embryonic foundation from which this book was developed. The mechanisms shown in these photographs were designed for high speed machines, and hence significant dynamic stresses, forces, and vibrations were inherently unavoidable. Thus the author found it an absolute necessity to design certain mechanical elements on the basis of minimizing either dynamic stresses, forces, or vibrations. Therefore optimum design was initially studied by the author, in a relatively narrow field, because of the *need* for solving specific design problems with which he was confronted.

The initial design studies made on the mechanisms shown in the following pages resulted in the publication of some of the technical articles listed at the end of this chapter. However, in recent years the author has conducted optimum design studies for many other mechanical elements on many other bases for optimum design, including considerations of the effects of inherently unavoidable limitations which confront practicing engineers. From these studies, the general *method of optimum design* presented in Chapter 6 evolved. Also, many of these specific studies are presented in Chapters 7 and 8.

1.5 High Speed Geneva Mechanism

One of the first problems encountered by the author which required an optimum design solution was the Geneva mechanism for a Minicard* film record handling machine. Some of the typical mechanical components for such a machine are shown in figure 1.7. In such a machine a small piece of cut film is transported with intermittent motion from a feeding magazine to a receiving magazine. The method for extraction and insertion of the Minicard film record from and into the magazines is not shown in the photograph. The film record is transported on a continuous flat steel belt, shown in figures 1.7 and 1.8. The belt is driven with high speed intermittent motion by means of a sprocket shown in figures 1.7 and 1.8, and this sprocket is driven by means of a 10-station Geneva mechanism, which can be seen in figure 1.9. The desired speed of operation for the film record transport belt was 30 cps.

The inertial load for the Geneva mechanism is the drive sprocket, the film record transport belt, and the idler drum, all of which can be seen in figure 1.7. Hence, in designing the Geneva mechanism, the overall objective which must be satisfied is to drive a given inertial load with a specified high rate of intermittent motion. However, it was determined that dynamic stresses would be significant at the vicinity of contact between the drive pin and the Geneva wheel shot. If the Geneva wheel were chosen too small, the significant dynamic stresses would be exorbitant because of the small torque arm for the drive pin with respect to the Geneva wheel. On the other hand, if the Geneva wheel were chosen too large, the significant dynamic stresses would be

* Trademark of the Eastman Kodak Company.

Figure 1.7 A typical Minicard film record handling machine showing feeding and receiving magazines, film record transport belt, drive, and idler sprockets. (Courtesy of Eastman Kodak Company.)

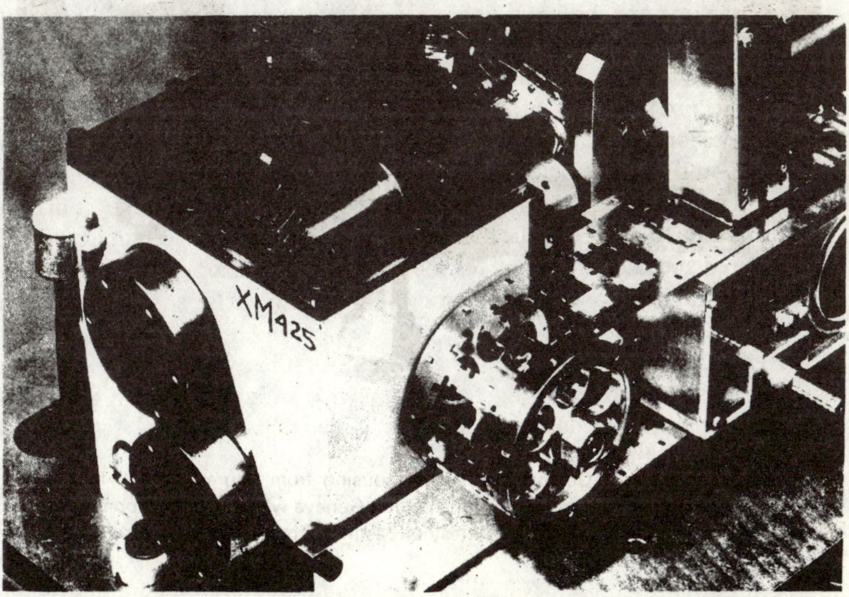

Figure 1.8 Close-up view of a Minicard film record handling machine showing feeding magazine, film record transport belt, drive sprocket, and housing for Geneva mechanism. (Courtesy of Eastman Kodak Company.)

Figure 1.9 Top view of Geneva mechanism housing from figure 1.8 with film record transport belt and drive sprocket removed, showing Geneva wheel in back. Top mechanism speed with inertial load = 30 cps. (Courtesy of Eastman Kodak Company.)

exorbitant because of the large added inertia of the Geneva wheel. In either event, we know that fatigue damage would occur near the surface of the drive pin and it would progress to a condition of noticeable wear and eventually the mechanism would malfunction. Hence the explicit objective introduced for designing the Geneva mechanism was minimization of the significant dynamic stresses, and this was the chosen basis for optimum design.

The solution to this problem is given in reference 1.3. For a broader viewpoint of the problems which might be encountered in the design of high speed Geneva mechanisms, the reader should also refer to Chapter 8 and to reference 1.9.

1.6 An Unconventional High Speed Intermittent Motion Mechanism

The use of a Geneva mechanism for very high speed applications often results in significant degrees of certain inherent undesirable characteristics. First, for a Geneva mechanism having a given number of wheel slots the ratio of dwell period to motion period is fixed. Hence, for a specified indexing angle, the designer does not have control over the percentage of the cycle during which the inertial load is standing still, and this might be undesirable for certain applications. Second, since the drive shaft makes one complete revolution for each cycle of indexing, for very high rates of operation the drive shaft carrying the Geneva pin must rotate at very high rates of speed, thereby introducing problems in the precision mechanism because of vibrations of the drive shaft. Third, wear in the Geneva mechanism is primarily concentrated at a highly localized region of one element, which is the Geneva drive pin. Fourth, the designer has no control over the general acceleration characteristics of the Geneva mechanism. Hence the characteristic abrupt changes in acceleration at the beginning and end of the Geneva wheel motion are conducive to vibratory components of motion for the inertial load. Finally, a small amount of backlash or clearance is always required between the drive pin and the wheel slot, and therefore a completely controlled motion for the Geneva wheel is not obtained. However, during a single cycle of motion the direction of acceleration reverses for the inertial load, causing the drive pin to transfer its contact from one side of the wheel slot to the other. Thus vibrations are likely to be introduced on the inertial load motion because of the unavoidable existence of backlash in the Geneva mechanism.

For a given inertial load and for a specified indexing angle, an upper limit exists on the permissible speed of operation for a Geneva mechanism because of the previously mentioned inherently unavoidable characteristics. For the optimum designed Geneva mechanism in the Minicard film record handling machine discussed in the previous section, the upper speed of operation was close to 30 cps. However, for an experimental Minicard film record handling machine it was desired to obtain a higher speed of operation for the film record transport belt. Thus the unconventional intermittent motion mechanism shown in figure 1.10 was designed; it illustrates the benefits which can be gained often by deviating from traditional practices.

The unconventional high speed intermittent motion mechanism shown in figure 1.10 was synthesized from the basic elements of figure 1.11, and it

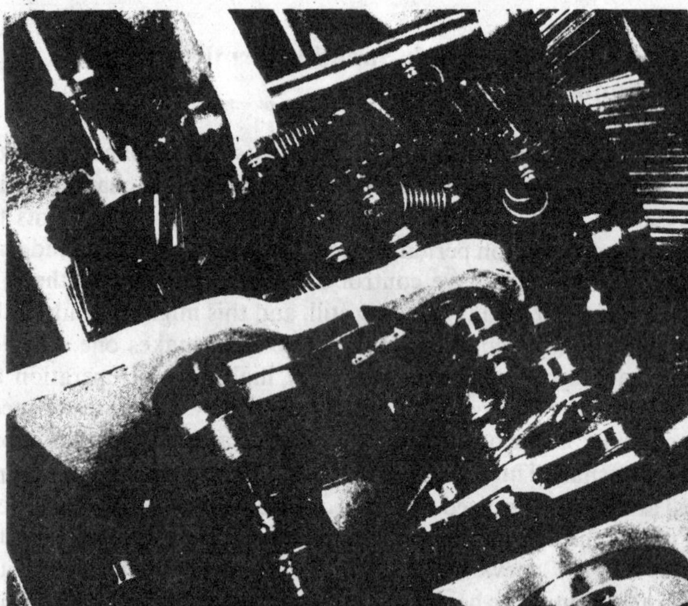

Figure 1.10 Experimental high speed intermittent motion mechanism shown with cover removed. Note cam drive shaft, cam follower arms, follower bearings, and helical extension springs. Inertial load normally on front shaft removed for photograph. (Courtesy of Eastman Kodak Company.)

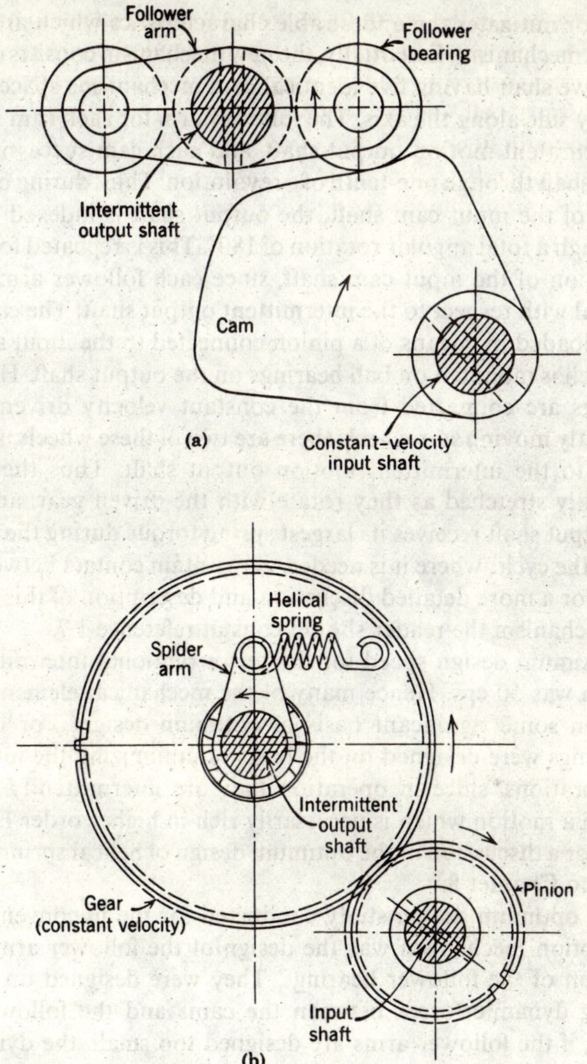

Figure 1.11 Basic elements used in synthesis of new indexing mechanism shown in figure 1.10. (*a*) Typical cam and follower linkage for indexing mechanism. (*b*) Method for supplying spring torque to intermittent motion shaft.

eliminates or mitigates the undesirable characteristics which are inherent in the Geneva mechanism. Essentially, the new mechanism consists of a constant velocity drive shaft having five identical cam mechanisms spaced equiangularly side by side along the axis. The follower arm for each cam is connected to the intermittent motion output shaft, and each cam successively indexes the output shaft through one-tenth of a revolution. Thus, during one complete revolution of the input cam shaft, the output shaft is indexed five discrete times through a total angular rotation of 180°. This is repeated for the following revolution of the input cam shaft, since each follower arm is perfectly symmetrical with respect to the intermittent output shaft. The cam followers are spring loaded by means of a pinion connected to the input shaft driving a gear, which is mounted on ball bearings on the output shaft. Helical extension springs are connected from the constant velocity driven gear to an intermittently moving star wheel; there are two of these wheels, and both are connected to the intermittent motion output shaft. Thus the springs are intermittently stretched as they rotate with the driven gear, and the intermittent output shaft receives its largest spring torque during the decelerating portion of the cycle, where it is needed to maintain contact between cam and follower. For a more detailed discussion and description of this intermittent motion mechanism the reader should consult reference 1.7.

The maximum design speed for the unconventional intermittent motion mechanism was 50 cps. Hence many of the mechanical elements had to be designed on some significant basis of optimum design. For instance, the helical springs were designed on the basis of minimizing the internal longitudinal vibrations, since in operation thay are intermittently stretched at 50 cps with a motion which is necessarily rich in higher order Fourier components. For a discussion of the optimum design of helical springs, the reader is referred to Chapter 8.

Another optimum design study conducted for the unconventional intermittent motion mechanism was the design of the follower arms, including the selection of the follower bearings. They were designed on the basis of minimizing dynamic forces between the cams and the follower bearings. Obviously, if the follower arms are designed too small, the dynamic forces will be large because of the small torque arm used for driving the given inertial load. On the other hand, if the follower arms are too large, the dynamic forces will be high because of the large total inertia being driven. Hence there is an optimum size for the follower arms, including the follower bearings, for which the inherently unavoidable significant dynamic forces will be as small in amplitude as possible.

The experimental high speed intermittent motion mechanism shown in figure 1.10 was successfully operated under loaded conditions at approximately 45 cps for over 153 million cycles of operation. After this long period of continued operation, the cam mechanisms, including the follower bearings, showed no appreciable signs of wear or deterioration. Also, none of the 10 helical extension springs had broken or deteriorated during that period of time, although they all had been extended and released at 45 cps for over 153 million cycles. Finally, high speed motion pictures taken at frequent intervals during the testing period showed no evidence of significant vibratory motions or improper behavior of any kind. For a discussion of the high speed motion pictures and accelerometer measurements taken during the testing period, the reader should consult reference 1.7.

1.7 Optimum Design of a High Speed Cam Mechanism

An exceptionally high speed oscillating follower type of cam mechanism is shown in figure 1.12. This cam mechanism was designed for driving a given inertial load (which cannot be seen in the photograph) with a special type of oscillating motion. The design speed for the cam mechanism was 120 complete stops and starts per second. Hence dynamic stresses, forces, and vibrations were of utmost importance and optimum design had to be used in several places. The helical load spring had to be designed on the basis of

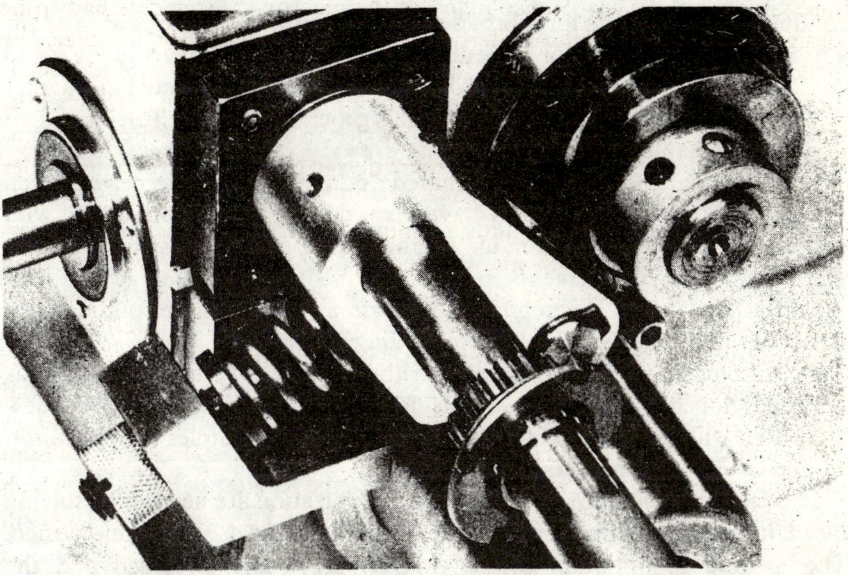

Figure 1.12 Experimental plunger-drive cam mechanism. Note size and location of load spring and size of follower bearing. Top speed of operation = 132 complete stops and starts per second. (Courtesy of Eastman Kodak Company.)

minimizing internal longitudinal vibrations, as presented in Chapter 8. Notice the following features, which are compatible with the optimum design conclusions of that chapter: the relatively large diameter for the spring in figure 1.12 and its reasonably close location with respect to the follower arm pivot axis, both of which are desirable from the standpoint of minimizing longitudinal vibrations in the spring. Also, the follower bearing in figure 1.12 was selected on the basis of minimizing the contact stress between cam and follower during dynamic operation, and this problem will be considered in Chapter 8. Incidentally, high speed motion pictures were taken of the cam mechanism in figure 1.12, including its inertial load, operating at 132 complete stops and starts per second, and no significant vibratory motions or misbehavior could be detected.

1.8 Optimal Design for Practical Mechanical Elements

In this chapter we have seen photographs of several industrial problems which were solved by using the results from various optimum design studies. Some of these initial optimization studies were conducted by *necessity* in order to mitigate inherently unavoidable dynamic stresses, forces, and vibrations to tolerable degrees in high speed mechanisms. Having been inspired by the results of these embryonic optimum design studies, the author conducted many others for a broader range of mechanical elements and for a broader range of bases for optimum design. These studies eventually led to the formulation of the general *method of optimum design* presented in Chapter 6. The extremely broad applicability of this general method will be illustrated in detail in Chapters 7 and 8.

In recent years, the field of automated optimal design has received much attention and many algorithms have been developed for application with digital computers. This technique of automated optimal design has much merit for mechanical elements, particularly in situations where design time is very critical and it is acceptable to obtain an approximate solution with but little insight into the total optimization picture.

An automated optimal design algorithm named OPTIGO recently has been developed with much success for mechanical elements. This new algorithm will be presented in Chapter 6. Practical examples of application will be covered in Chapters 6 and 8.

In conclusion, practical techniques of optimization are needed for solving mechanical element problems, particularly in the field of machine design. The method of optimum design (MOD) previously discussed and the algorithm OPTIGO are powerful optimization tools in this respect. Hence emphasis will be placed on these practical techniques in the Chapters that follow.

REFERENCES

The following publications, written by the author of this book, were based on design studies pertaining to some of the industrial examples shown in this chapter.

1.1 "Method of Finite Differences Provides Simple but Flexible Arithmetical Techniques for Cam Design," *Machine Design*, Nov. 1955, pp. 195–204.

1.2 "How To Design High-Speed Cam Mechanisms for Optimum Cam and Follower Proportions," *Machine Design*, Jan. 26, 1956, pp. 85–89.

1.3 "How To Design Geneva Mechanisms to Minimize Contact Stresses and Torsional Vibrations," *Machine Design*, March 22, 1956, pp. 107–111.

1.4 "A Rapid Method for Developing Cam Profiles Having Desired Acceleration Characteristics," *Machine Design*, Dec. 13, 1956, pp. 129–132.

1.5 "How Profile Errors Affect Cam Dynamics," *Machine Design*, Feb. 7, 1957, pp. 105–108.

1.6 "Method of Finite Differences in Cam Design—Accuracy—Applications," *Machine Design*, Nov. 14, 1957, pp. 159–161.

1.7 "The Development of a High Speed Indexing Mechanism," *Transactions of the Fourth Conference on Mechanisms*, Penton, Cleveland, Ohio, 1957, and *Machine Design*, Sept. 4, 1958, pp. 134–138.

1.8 "Optimum Design of Helical Springs for High Speed Operation," *Machine Design*, Sept. 18, 1958, pp. 155–160.

Some publications pertaining to the *Method of Optimum Design*, written subsequent to the first edition of this book, follow:

1.9 R. C. Johnson, *Mechanical Design Synthesis—Creative Design and Optimization*, 1st ed., Van Nostrand Reinhold, New York, 1971, 343 pp.; 2nd ed., Krieger, Huntingdon, N.Y., 11743, 1978, 448 pp.

1.10 J. Ellis, "Johnson's Method of Optimum Design Applied to a Problem with Simple Functional Relationships," *Computer Aided Design*, Vol. 8, No. 1, Jan. 1976, pp. 9–12.

1.11 J. Ellis, "An Introductory Problem in Mechanical Design Optimization," *International Journal of Mechanical Engineering Education*, Vol. 5, No. 2, 1977, pp. 121–126.

1.12 R. C. Johnson, "Design Synthesis," *Machine Design*, Penton, Cleveland, Ohio; Part 1, Oct. 18, 1973, pp. 149–155; Part 3, Dec. 13, 1973, pp. 164–169; Part 4, Dec. 27, 1973, pp. 52–57.

1.13 R. C. Johnson, "Three Dimensional Variation Diagrams for Control of Calculations in Optimum Design," *Transactions of the ASME, Journal of Engineering for Industry*, Aug. 1967, pp. 391–398.

1.14 M. A. Murray Lasso and E. Chicurel Uziel, *Applicaciones de Computacion a la Ingenieria*, Editorial Limusa, S. A., Mexico City, D. F., Mexico, 1975, pp. 167–174.

1.15 J. N. Siddall, *Analytical Decision-Making in Engineering Design*, Prentice-Hall, Englewood Cliffs, N.J., 1972, pp. 94, 192–195.

1.16 J. P. Vidosic, *Elements of Design Engineering*, Ronald, New York, 1969, pp. 223, 228–239.

2

Some Mathematical Background

The purpose of this chapter is to review some mathematical background to be used later in the book. This material includes two basic mathematical subjects employed in the optimization studies to follow. First, variation studies of optimization for mechanical element design are often facilitated by transforming given graphs, tabulated data, and complicated equations to suitable form by the application of approximation techniques to be covered here. This is sometimes necessary for executing the equation manipulation procedures in the method of optimum design (MOD). Second, computerized searching techniques are required for automated optimal design (AOD). In this respect the mathematical background for development of the algorithm OPTIGO in Chapter 6 will be covered here.

Some Approximations for Practical Functions

2.1 Simplicity and Accuracy Considerations

The use of approximations makes it possible to overcome seemingly formidable obstacles encountered in the conduct of explicit design studies. Hence, to implement the MOD, it will be necessary to convert graphical and tabular data to equation form; this will be accomplished by mathematical approximation techniques. Also, occasionally it will be necessary to simplify originally available complex equations in order to facilitate an optimum design solution. For either case the overall objective in obtaining the approximate mathematical equation will be twofold. First, the approximate equation must be *simple* enough that it can be manipulated appropriately in the conduct of the MOD as explained in Chapter 6. Second, the approximate equation must be reasonably *accurate* over the ranges of feasibility for the variables. Unfortunately, the attainment of simplicity is generally incompatible with the attainment of accuracy, and therefore in the conduct of explicit design studies the selection of an appropriate approximation equation is limited by *both* considerations. Thus the approximations we shall use in the conduct of the MOD generally will not be the same as those normally used in the conduct of *analytical* studies where accuracy alone is considered of paramount importance.

When selecting an appropriate approximation equation for implementing the MOD, it is generally advisable to strive to obtain as much simplicity as considerations of accuracy will permit. This poses a very difficult problem when appropriate approximations are being selected and a satisfactory selection based on knowledge and experience can generally be made. Hence in this chapter we shall emphasize *simple* approximations which have been found to be extremely powerful in the actual conduct of explicit *design* work, and considerations of accuracy will be presented briefly for each.

The amount of error which can be tolerated in approximation equations for the conduct of explicit *design* is generally much greater than what normally is expected in good *analysis* work, because, as many optimum design studies in later chapters will show, optimum numerical values for design parameters are generally determined by independent numerical *limitations* on materials or on geometry. Hence, using a reasonably simple approximation for expressing a function might enable us to prove that a certain design parameter should be placed at a certain numerical limit which is imposed only by independent material characteristics or by independent space limitations. Thus the optimum value for the design parameter would not be changed even if an exact mathematical equation had been used for expressing the function. However, using the exact mathematical equation might have eliminated the practical feasibility of explicitly proving that the design parameter should be placed at the certain limit value, which is independent of the design equation.

Thus approximations used in the explicit conduct of the MOD are merely powerful, indispensable means for obtaining an end, and the accuracy of the means should not be considered arbitrarily as synonymous with the accuracy of the end. Hence errors as great as 10 or 15% in approximation equations used in design studies often are not unreasonably large, and such errors generally will not affect the optimum values determined for the design parameters. Therefore, we shall find that the use of relatively simple approximations will enable us to determine explicitly optimum designs the determination of which otherwise would be out of the realm of practical feasibility. However, it should be emphasized that, as a final step after the determination of the optimum design, an accurate analysis should be conducted Thus the original complex equations, tabulated data, and graphical information should be used in a final step for analyzing the *designed* mechanical element in order to determine pertinent information, such as an accurate value for the load capacity of the particular element.

2.2 Mathematical Functions in Engineering

The design and analysis of the mechanical elements of interest in this book are primarily based on the proper use of mathematical functions. Hence, for a given helical spring, the axial deflection Δ is a function of the applied axial force F_S, expressed as $\Delta = f(F_S)$. In other words, for any spring of given design for each numerical value of applied force F_S there is a corresponding numerical value for the resulting deflection Δ. As another example, for a given ball bearing in a particular mode of operation, its radial load rating R_r is a function of its angular velocity ω, expressed as $R_r = f(\omega)$. In general, we say that y is a function of x, expressed as $y = f(x)$, if there exists a definite one-to-one correspondence between numerical values of the two variables.

In engineering work *several methods* for expressing mathematical functions between variables are commonly encountered. For one, the function between two variables can be expressed by means of a mathematical equation. The mathematical equation compactly and precisely relates the one-to-one correspondence between an infinite number of points over the useful range for the equation. In addition, it gives accurate information with respect to all derivatives and integrals of the function. Hence we might consider a mathematical equation the highest order method for function expression.

The second method for function expression is the graphical representation. The graph compactly and accurately relates the one-to-one correspondence between an infinite number of points over the useful range of the variables. The degree of accuracy depends on the size of the graph and the care exercised in drawing the graph. However, numerical methods have to be employed to determine derivatives and integrals, and in this respect accuracy is generally lost. Hence we might consider a graph a lower order method for function expression than a mathematical equation.

The third method is expressed in the form of tabulated data. The data accurately relate the one-to-one correspondence between a discrete finite number of points over the useful range of the variables. Good accuracy can exist for the given discrete points, but interpolation must be used to determine values between the given points. Numerical methods have to be used to determine derivatives and integrals, and, at best, values so determined are predicted on an anticipated or probable function variation between given points. Hence we might consider tabulated data the lowest order method for function expression.

2.3 Conversion Processes for Function Expression

In conducting explicit design studies on mechanical elements it is often necessary to convert the method for function expression from one type to another. For instance, if tabulated information is available in the literature for expressing a function, it will be necessary to convert this given information to mathematical equation form in order to conduct the explicit method of optimum design. Since this conversion process would be changing from a lower order of function expression to a higher order, we might think of it as a *regenerative* conversion process. The logical steps in performing the conversion in this illustration would be to convert the given tabulated data to a graphical representation, and then finally to express the function as a mathematical equation by converting from the graphical representation. The regenerative type of conversion process is often referred to as *curve fitting*.

Another example of a conversion process, which is occasionally necessary in explicit design studies, is the conversion of an unwieldy complex equation suitable only for analysis to a simplified equation which is acceptably accurate over the anticipated feasible ranges for the variables. In reality this type of conversion process would be from a higher order of function accuracy to a lower order, and hence we might think of it as a *degenerative* conversion process. The logical steps in performing the degenerative process in this illustration would be to convert the original complex equation to tabulated data, then to convert the tabulated data to a graphical representation, and finally to express the function in a simpler mathematical equation by curve fitting on the graphical representation. Essentially, this process involves a degenerative step and two regenerative steps. However, since the simplified equation will be less accurate than the original and since it will be valid over only limited ranges of anticipated variations, we must think of the overall conversion process as degenerative.

Many other examples of conversion processes are encountered in the explicit design of mechanical elements. For instance, in calculating moments of inertia of a complex body great simplification is obtained by dividing the body into simple parts. If the simple parts are approximations of the actual body, the conversion process might be considered degenerative. As a final example, the numerical solution of an equation with one unknown is really a conversion process, and if the numerical answer is irrational the conversion process is degenerative. Actually, the last illustration is merely the conversion from a mathematical equation to tabulated data, which expresses but one value, the numerical solution to the equation.

2.4 Errors in Conversion Processes

The goal of conversion processes for function expression will always be to arrive at a mathematical equation that can be handled efficiently in the design procedure. Conversion processes always involve the application of approximations, and, since design must be conducted with an acceptable degree of accuracy, the question of error is of paramount importance. The general definition of error e is given by equation 2.1, which expresses the error of an approximate value with respect to a true or reference value.

$$e = \frac{\text{approximate value} - \text{true value}}{\text{true value}} \tag{2.1}$$

It is clearly evident from equation 2.1 that the error of an approximate value for a function depends on what is meant by the true value for the function. The true value for a function might be given either by a mathematical equation, a graphical representation, or tabulated data. In fact, the true value of a function might be given only by the manufactured part itself, as when manufacturing errors have a significant influence on product performance.

2.5 Review of Continuous and Differentiable Functions [2.1]

The first kind of mathematical approximation which we will consider is the method of finite differences in section 2.6. A fundamental understanding of this method is closely allied with the basic mathematical definitions of continuity and differentiability for functions. Hence, for the benefit of practicing engineers, as an introduction we shall briefly review the basic concepts of mathematical continuity and differentiability. Also, this review should be helpful in creating a clear mental picture of the fundamental relationships among the various types of methods for function expression, emphasizing both the validity of curve fitting and the validity of the method of finite differences presented in section 2.6.

Let us consider a variable y as a function of the variable x, generally expressed by equation 2.2. For the sake of simplicity we shall consider only two variables and, if other independent variables exist, merely consider them as held constant.

$$y = f(x) \tag{2.2}$$

For ease of visualization, for equation 2.2 consider the general graphical representation of the function as depicted in figure 2.1. By definition we will say that f is a continuous function if the dependent variable y varies smoothly as the independent variable x is smoothly varied. In other words, continuity of f implies that there are no discrete jumps in y as x is smoothly varied.

The definition of continuity can be stated in more rigorous mathematical terms by referring to figure 2.1. Let the point in question be designated by the subscript 0; hence its y value is y_0 and its x value is x_0. Consider ϵ to be *any* real number greater than zero which determines a finite segment on the curve as shown in figure 2.1. Consider the existence of a real number δ greater than zero, which determines a slightly smaller segment on the curve as shown in figure 2.1. Now consider the point designated by 1 in figure 2.1 as any point on the curve within this smaller segment of the curve. *If* for *every* $\epsilon > 0$ there *exists* a $\delta > 0$ such that $|y_1 - y_0| < \epsilon$ for all points x_1 for which $|x_1 - x_0| < \delta$, *then* the function is considered to be continuous at

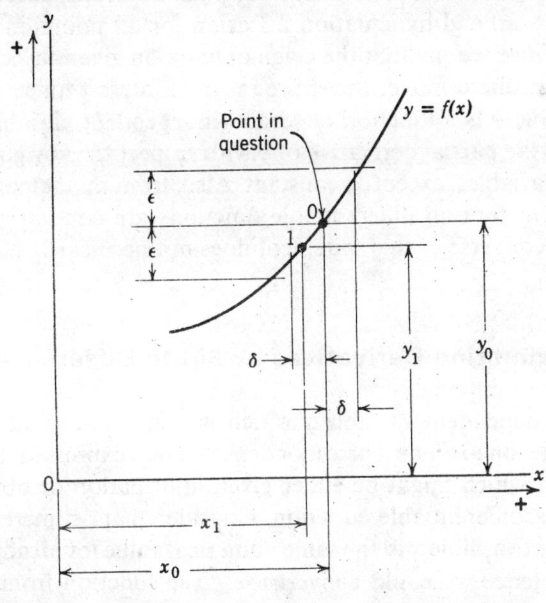

Figure 2.1 Graphical representation of a general function showing relationships used for mathematical definitions of continuity and differentiability at point 0.

point 0, the point in question. *If* point 0 is any general point on the curve within the range of interest, *then* the function is considered a continuous function within that entire range.

If a function is defined by a finite number of *tabulated* discrete values, it is always possible to perform a regenerative conversion to a continuous mathematical equation which precisely satisfies the tabulated data. Similarly, it is always possible to convert continuous graphical information to a continuous mathematical equation, with possibly some loss in accuracy. Hence, unless otherwise stated, all functions in this book will be considered expressible by continuous mathematical equations.

Again consider a variable y which is a function of the variable x and generally expressed by equation 2.2. The conversion of this equation to a graphical representation is typified in figure 2.1, and in that figure 0 designates the specific point under consideration. If in that figure, 1 designates a general neighboring point defined by the function, we define the derivative of y with respect to x at point 0 by equation 2.3 as follows:

$$\left(\frac{dy}{dx}\right)_0 = \lim_{x_1 \to x_0} \frac{(y_1 - y_0)}{(x_1 - x_0)} \tag{2.3}$$

where $x_1 \neq x_0$. *If* the limit defined by equation 2.3 *exists* at a point x_0, *then* we say that the function defined by equation 2.2 is differentiable at that point. *If* the function defined by equation 2.3 exists for all points in the range of interest for x, *then* we say that the original function given by equation 2.2 is a differentiable function over the entire range of interest for x.

If the variable y is a function of several independent variables, equation 2.3 will define the partial derivative of y with respect to x by considering all independent variables, except x, constant. Also, from mathematical analysis it can be shown that all differentiable functions are continuous functions. However, the converse of this statement does not necessarily hold true.

2.6 Approximating Derivatives by Finite Differences [2.2]

Consider a dependent variable y which is a function of an independent variable x, the one-to-one correspondence being expressed by means of tabulated data which might be either given information or obtained information from a differentiable equation. Consider that an increment Δx for the independent variable x is the same numerical value for all adjacent points in the table. Hence we could convert the given function from the tabular method of expression to a graphical representation, as shown in figure 2.2. In this figure 0 denotes the point under consideration, 1 denotes the point

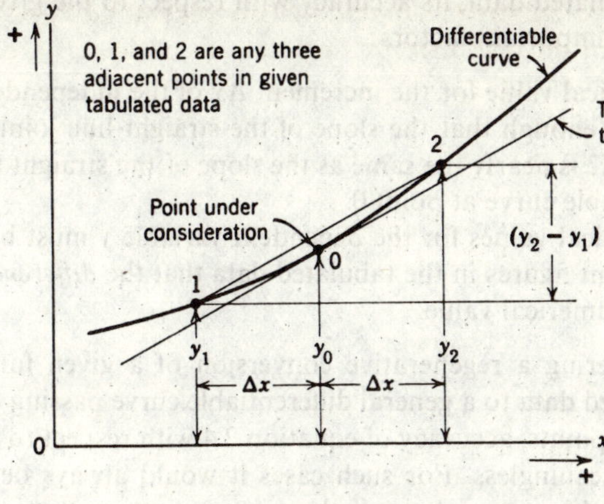

Figure 2.2 Graphical representation of given tabulated data showing a differentiable curve passing through three adjacent points Δx increments apart in the function data; also, pertinent relationships used in derivation of finite difference approximation for the first derivative of y with respect to x.

immediately preceding the one under consideration, and 2 designates the point immediately following the one under consideration. The three adjacent points of significance are separated by a numerical amount Δx with respect to the independent variable x, as shown in figure 2.2. Assume that a differentiable function has been passed through the three points of significance.

If we draw a straight line tangent to this differentiable function at the point in question, we realize from basic calculus that its slope would be the derivative of y with respect to x, from equation 2.3. However, if we draw a straight line between points 1 and 2, we notice that under certain conditions its slope is very nearly the same as that of the tangent line to the differentiable function curve at the point under consideration. In fact, in the limit, if we hypothetically considered Δx as tending to zero, the slopes of the two straight lines would be the same, as is obvious from a consideration of the definition of derivative. Hence an approximation for the derivative of y with respect to x would be

$$\lim_{\Delta x \to 0} \frac{(y_2 - y_1)}{2\Delta x} = \left(\frac{dy}{dx}\right)_0 \approx \frac{(y_2 - y_1)}{2\Delta x} \tag{2.4}$$

Equation 2.4 will be referred to as the basic finite difference first derivative approximation; in a degenerative conversion from a given differentiable equation to tabulated data, its accuracy with respect to the given equation depends on two important factors:

1. The numerical value for the increment Δx of the independent variable x must be small enough that the slope of the straight line joining points 1 and 2 in figure 2.2 is nearly the same as the slope of the straight tangent line to the differentiable curve at point 0.
2. The numerical values for the dependent variable y must be carried to enough significant figures in the tabulated data that the *difference* $(y_2 - y_1)$ is an accurate numerical value.

When considering a regenerative conversion of a given function from original tabulated data to a general differentiable curve passing through the tabulated value points, accuracy of equation 2.4 with respect to the original data is really meaningless. For such cases it would always be *possible* to pass a differentiable curve through the given points so that equation 2.4 would be *exact*. However, it would be more reasonable to consider passing an anticipated or probable curve through the given points. For such cases equation 2.4 will be reasonably accurate if the two previously mentioned factors are satisfied.

Again consider a dependent variable y which is a function of an independent variable x, the one-to-one correspondence being expressed by means of tabulated data which are either given information or obtained information from a differentiable equation. As before, consider that an increment Δx for the independent variable x has the same numerical value for all adjacent points in the table. Thus the tabulated information can be converted to a graphical representation as shown in the upper curve of figure 2.3, where again 0 designates the point under consideration, and 1 and 2 represent the adjacent points in the tabulated data. Consider that a differentiable curve has been passed through the three adjacent tabulated data points, 1, 0, and 2.

The method used for deriving equation 2.4 can be extended to the upper curve in figure 2.3. Consider A and B points on that differentiable curve, each located a numerical value $\Delta x/2$ from the tabulated data points 1, 0, and 2. Hence, by extending the method used in deriving equation 2.4, the first derivatives of the differentiable curve at points A and B are approximated as follows:

$$\lim_{\Delta x/2 \to 0} \frac{(y_0 - y_1)}{\Delta x} = \left(\frac{dy}{dx}\right)_A \approx \frac{(y_0 - y_1)}{\Delta x}$$

$$\lim_{\Delta x/2 \to 0} \frac{(y_2 - y_0)}{\Delta x} = \left(\frac{dy}{dx}\right)_B \approx \frac{(y_2 - y_0)}{\Delta x}$$

Hence approximate values for the first derivatives at A and B can be readily determined, and an approximate value for the first derivative at 0 can be

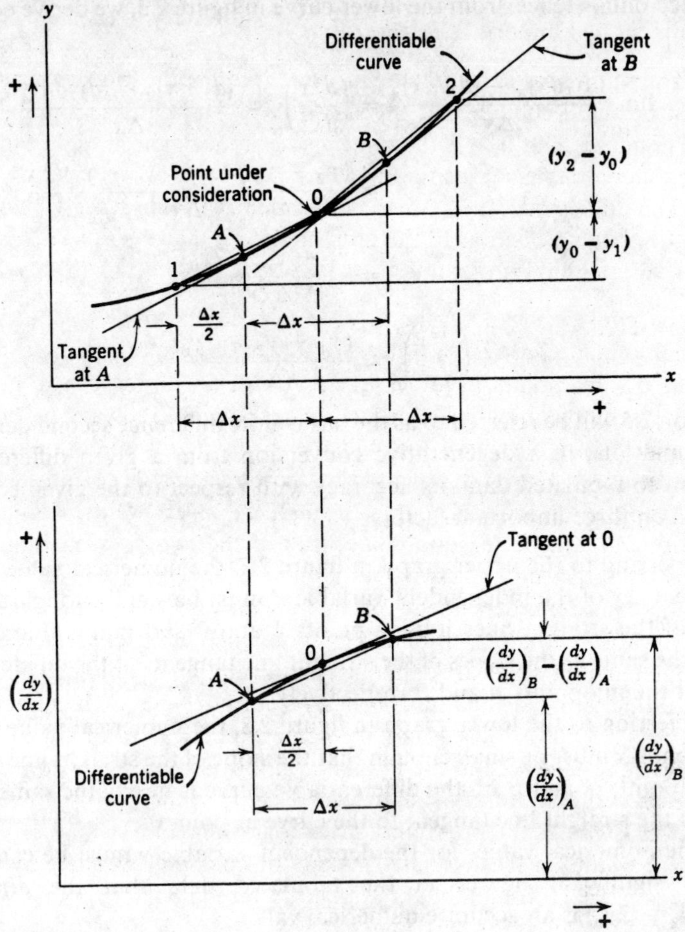

Figure 2.3 Graphical representation of given tabulated data for three adjacent points, 0, 1, and 2, Δx equal increments apart. Pertinent relationships used in derivation of finite difference approximation for the second derivative of y with respect to x are shown.

determined by equation 2.4. Now consider depicting the values for these three derivatives by the three points A, 0, and B in the lower graph of figure 2.3. Also consider passing a differentiable curve through these points as shown. We shall now extend the method used for deriving equation 2.4 to the lower curve in figure 2.3, and this will give us an approximation for the second derivative of y with respect to x at point 0 in terms of the original tabulated data. Hence, from the lower curve in figure 2.3, we derive equation 2.5

$$\lim_{\Delta x \to 0} \frac{(dy/dx)_B - (dy/dx)_A}{\Delta x} = \left(\frac{d^2y}{dx^2}\right)_0 \approx \frac{(dy/dx)_B - (dy/dx)_A}{\Delta x}$$

$$\approx \frac{(y_2 - y_0)/\Delta x - (y_0 - y_1)/\Delta x}{\Delta x}$$

$$= \frac{y_1 + y_2 - 2y_0}{(\Delta x)^2}$$

$$\therefore \quad \left(\frac{d^2y}{dx^2}\right)_0 \approx \frac{(y_1 + y_2 - 2y_0)}{(\Delta x)^2} \tag{2.5}$$

Equation 2.5 will be referred to as the basic finite difference second derivative approximation; in a degenerative conversion from a given differentiable equation to tabulated data, its accuracy with respect to the given equation depends on three important factors:

1. Referring to the upper graph in figure 2.3, the numerical value for the increment Δx of the independent variable x must be small enough that the slopes of the straight lines joining points 1 and 0 and points 0 and 2 are nearly the same as the slopes of the straight line tangents to the differentiable curve at the midpoints A and B respectively.
2. Referring to the lower graph in figure 2.3, the numerical value for the increment Δx must be small enough that the slope of the straight line joining the midpoints A and B on the differentiable curve is nearly the same as the slope of the straight line tangent to the curve at point 0.
3. The numerical values for the dependent variable y must be carried to enough significant figures in the tabulated data that the *difference* $(y_1 + y_2 - 2y_0)$ is an accurate numerical value.

When considering a regenerative conversion of the given function from original tabulated data to a general differentiable curve passing through the tabulated value points, accuracy of equation 2.5 with respect to the original data is really meaningless. For such cases it would always be *possible* to pass a differentiable curve through the given points so that equation 2.5 would be *exact*. However, it would be more reasonable to consider passing an anticipated or probable curve through the given tabulated value points. In that event, equation 2.5 will be reasonably accurate if the previously mentioned three factors are satisfied.

In considering the degenerative conversion of a function from a differentiable equation to tabulated data, equation 2.5 for approximating the second derivative is often very accurate for surprisingly large numerical values of the increment Δx. In fact, equation 2.5 is exact for *all* polynomials of the third degree or less, regardless of the numerical value of Δx. This can be proved by considering the general polynomial of the third degree,

$$y = ax^3 + bx^2 + cx + d \tag{2.6}$$

In this equation the coefficients a, b, c, and d are any real number values, either positive, negative, or zero. If we take the second derivative of y with respect to x exactly from equation 2.6, we obtain

$$\frac{d^2y}{dx^2} = 6ax + 2b \tag{2.7a}$$

Equation 2.7a is the exact second derivative of y with respect to x, and to calculate its value at any general point 0 we would merely substitute x_0 for x in that equation.

Let us now calculate the approximate value for the second derivative of y with respect to x for equation 2.6 at a general point 0 using equation 2.5, the finite difference approximation. We shall assume a general increment Δx for the independent variable, which might have any numerical value. Hence, referring to the derivation for equation 2.5, we have

$$x_1 = x_0 - \Delta x$$
$$x_2 = x_0 + \Delta x$$

Placing these relationships in equation 2.6, we obtain

$$y_0 = ax_0{}^3 + bx_0{}^2 + cx_0 + d$$
$$y_1 = a(x_0 - \Delta x)^3 + b(x_0 - \Delta x)^2 + c(x_0 - \Delta x) + d$$
$$y_2 = a(x_0 + \Delta x)^3 + b(x_0 + \Delta x)^2 + c(x_0 + \Delta x) + d$$

If we expand the right side of the last two of these equations, we are able to obtain exactly the following equation since most of the terms cancel out.

$$(y_1 + y_2 - 2y_0) = 6a(x_0)(\Delta x)^2 + 2b(\Delta x)^2$$

If we now use equation 2.5 for calculating the second derivative by the method of finite differences, we derive equation 2.7b

$$\left(\frac{d^2y}{dx^2}\right)_0 \approx \frac{(y_1 + y_2 - 2y_0)}{(\Delta x)^2} = \frac{6a(x_0)(\Delta x)^2 + 2b(\Delta x)^2}{(\Delta x)^2}$$
$$= 6a(x_0) + 2b \tag{2.7b}$$

Comparing equations 2.7b and 2.7a, we see that the basic finite difference second derivative approximation has *no* error for general polynomials of the third degree or less.

TABLE 2.1*

Random Checks of Accuracy for Equation 2.5 in Typical Degenerative Conversion Applications

Original Function	$x_0°$	Exact Value of $\left(\frac{d^2y}{dx^2}\right)_0$	Approximate Value of $\left(\frac{d^2y}{dx^2}\right)_0$ by Equation 2.5. $\Delta x = 5°$	% Error	$\Delta x = 10°$	% Error	$\Delta x = 30°$	% Error
$y = K \sin x$	30	−0.500K	−0.500K	0.0	−0.498K	0.4	−0.489K	2.2
$y = K \sin x$	60	−0.866K	−0.867K	0.1	−0.863K	0.4	−0.848K	2.1
$y = K \sin x$	90	−1.00K	−1.00K	0.0	−0.996K	0.4	−0.979K	2.1
$y = Kx^5$	30	540,000K	547,496K	1.4	570,000K	5.6	—	—

* Tabulated values are from reference 2.2.

In considering the degenerative conversions of other functions from exact equations to tabulated data, from random checks we obtain good accuracies with equation 2.5 for relatively large increments of the independent variables. The results of some of these checks from reference 2.2 for a sinusoidal function and for a simple polynomial of the fifth degree are summarized in table 2.1.

The method of finite differences is an extremely powerful mathematical tool in engineering calculations. It is particularly useful in solving difficult differential equations with known boundary conditions such as are encountered in stress analysis from the theory of elasticity, in dynamic analysis, and in the design of cam mechanisms. The simple relationship between second derivatives and displacements makes it possible to write a differential equation in terms of displacements only. As shown above the method is exact for degenerative conversions of all polynomial equations of the third degree or less. Also, the accuracy for other differentiable functions can be excellent even for large increments of the independent variable, as is evident from table 2.1.

Let us consider a final point of interest with respect to a characteristic of differentiable curves which is obvious from the method of finite differences. Suppose that the variable y is a function of many independent variables such as x, u, v, and w. Generally, we could express this function by

$$y = f(x, u, v, w) \tag{2.8}$$

Assume that this function f is differentiable with respect to each of the independent variables in the equation from the standpoint of partial derivatives.

If we consider the variation of y versus x, keeping u, v, and w constant, we will find that the variation will be as typically depicted in figures 2.1 and 2.2. We could apply the method of finite differences to approximate the first partial derivative of y with respect to x, which is the slope of the curve, at the point in question. For most functions encountered in practice, equation 2.4 enables us to approximate this slope quite accurately even for relatively large numerical values of the increment Δx of the independent variable x. Hence, considering the function continuous with respect to all independent variables, from the definition of continuity it would always be possible for us to add an increment δ to either u, v, or w without appreciably affecting the value of $(y_2 - y_1)$ for the relatively large increment Δx which is permitted for accuracy of equation 2.4. Hence, within a *limited neighborhood*, the variation of y with respect to x would be a family of parallel curves assuming discrete small changes in u, v, or w. This characteristic of parallel curves for differentiable functions of many variables within a limited range is a very important one, and it is especially useful in curve fitting for functions of many variables. A good example of an application of this principle is the simplification of the complex dynamic load equation 8.7 in Chapter 8.

2.7 Some Examples of Methods for Regenerative Conversions—Curve Fitting

In the conduct of explicit optimum design, it is necessary to convert function expressions from available tabulated and graphical data to mathematical equation form. For either case, regeneration from discrete sets of values to a mathematical equation will be required. A desirable objective for such equations is simplicity, since complex equations might very well be too unwieldy for the conduct of explicit design. In fact, it might be highly desirable to convert an original complex equation to a simpler equation which is acceptably accurate over anticipated ranges of variation for certain parameters in order to conduct explicit optimum design. The original complex equation first would be converted to tabulated data at selected intervals on the anticipated ranges of variation for the equation parameters. Hence the problem really resolves itself into the conversion of a function expression from tabulated data to equation form, with the stipulation that the equation form should be as simple as acceptable accuracy will allow.

There are many ways of expressing tabulated data by mathematical equations. Some of the methods which the author has found very useful in *design* work are covered in this section. In addition, the reader may be able to draw from his own mathematical background in coping with unique design problems.

Function Conversion to a Polynomial Series. A very powerful but somewhat complex method for converting from tabulated data to a mathematical equation is the employment of a polynomial equation. A polynomial equation of degree n can be written as given in equation 2.9. In this equation $a_0, a_1, a_2, \ldots, a_n$ are numerical coefficients whose values will be determined by a method to be explained shortly.

$$y = a_0 + a_1 x + a_2 x^2 + a_3 x^3 + \cdots + a_n x^n \tag{2.9}$$

In this polynomial equation the dependent variable y is a function of the independent variable x. Given tabulated data will include discrete pairs of corresponding values for the variables, and the design engineer can initially choose which variable of the pair he considers to be the dependent variable and which he considers the independent variable. The conduct of explicit design might be greatly simplified by the proper decision in this choice, and this will be evident for the particular problem at hand.

Suppose that tabulated data are available which relate variables x and y in corresponding given values for $(n + 1)$ points. Generally, these corresponding known values can be expressed as

$$(x_1, y_1), (x_2, y_2), (x_3, y_3), \ldots, (x_n, y_n), (x_{n+1}, y_{n+1}).$$

Hence, if we consider passing a polynomial equation of the form given by equation 2.9 through these $(n + 1)$ points, we can write $(n + 1)$ equations, which must be satisfied simultaneously, as follows:

$$\begin{aligned}
y_1 &= a_0 + a_1 x_1 + a_2 x_1^{\,2} + a_3 x_1^{\,3} + \cdots + a_n x_1^{\,n} \\
y_2 &= a_0 + a_1 x_2 + a_2 x_2^{\,2} + a_3 x_2^{\,3} + \cdots + a_n x_2^{\,n} \\
y_3 &= a_0 + a_1 x_3 + a_2 x_3^{\,2} + a_3 x_3^{\,3} + \cdots + a_n x_3^{\,n} \\
&\;\;\vdots \\
y_{n+1} &= a_0 + a_1 x_{n+1} + a_2 x_{n+1}^2 + a_3 x_{n+1}^3 + \cdots + a_n x_{n+1}^n
\end{aligned}$$

Since the x and y values in these equations are known values from the tabulated data, we have $(n + 1)$ linear equations in unknowns a_0, a_1, a_2, $a_3, \ldots, a_n$, which can be solved simultaneously to determine their numerical values. These simultaneous linear equations can be solved in any manner with which the reader is familiar; and the solution of such equations is commonly expressed by means of determinants. Hence coefficient a_0 would be obtained as follows:

$$a_0 = \frac{\begin{vmatrix} y_1 & x_1 & x_1^2 & \cdots & x_1^n \\ y_2 & x_2 & x_2^2 & \cdots & x_2^n \\ y_3 & x_3 & x_3^2 & \cdots & x_3^n \\ \cdot & \cdot & \cdot & \cdots & \cdot \\ \cdot & \cdot & \cdot & \cdots & \cdot \\ \cdot & \cdot & \cdot & \cdots & \cdot \\ y_{n+1} & x_{n+1} & x_{n+1}^2 & \cdots & x_{n+1}^n \end{vmatrix}}{\begin{vmatrix} 1 & x_1 & x_1^2 & \cdots & x_1^n \\ 1 & x_2 & x_2^2 & \cdots & x_2^n \\ 1 & x_3 & x_3^2 & \cdots & x_3^n \\ \cdot & \cdot & \cdot & \cdots & \cdot \\ \cdot & \cdot & \cdot & \cdots & \cdot \\ \cdot & \cdot & \cdot & \cdots & \cdot \\ 1 & x_{n+1} & x_{n+1}^2 & \cdots & x_{n+1}^n \end{vmatrix}} \qquad (2.10a)$$

For the general coefficient a_i, where $i = 0, 1, 2, 3, \ldots, n$, the coefficient is expressed similarly:

$$a_i = \frac{\begin{vmatrix} 1 & x_1 & x_1^2 & \cdots & y_1 & \cdots & x_1^n \\ 1 & x_2 & x_2^2 & \cdots & y_2 & \cdots & x_2^n \\ 1 & x_3 & x_3^2 & \cdots & y_3 & \cdots & x_3^n \\ \cdot & \cdot & \cdot & \cdots & \cdot & \cdots & \cdot \\ \cdot & \cdot & \cdot & \cdots & \cdot & \cdots & \cdot \\ \cdot & \cdot & \cdot & \cdots & \cdot & \cdots & \cdot \\ 1 & x_{n+1} & x_{n+1}^2 & \cdots & y_{n+1} & \cdots & x_{n+1}^n \end{vmatrix}}{\begin{vmatrix} 1 & x_1 & x_1^2 & \cdots & x_1^i & \cdots & x_1^n \\ 1 & x_2 & x_2^2 & \cdots & x_2^i & \cdots & x_2^n \\ 1 & x_3 & x_3^2 & \cdots & x_3^i & \cdots & x_3^n \\ \cdot & \cdot & \cdot & \cdots & \cdot & \cdots & \cdot \\ \cdot & \cdot & \cdot & \cdots & \cdot & \cdots & \cdot \\ \cdot & \cdot & \cdot & \cdots & \cdot & \cdots & \cdot \\ 1 & x_{n+1} & x_{n+1}^2 & \cdots & x_{n+1}^i & \cdots & x_{n+1}^n \end{vmatrix}} \qquad (2.10b)$$

Thus we can calculate values for all coefficients $a_0, a_1, \ldots, a_n$ in equation 2.9 from the given tabulated data. Therefore a polynomial equation of degree n can be written which satisfies the tabulated data of corresponding values for $(n + 1)$ points. This method of expressing given tabulated data is very powerful because it can *always* be used for the explicit regenerative conversion to an accurate equation. However, for polynomials of high degree, the method becomes *unwieldy and impractical* for explicit design work, and generally it is better to look for a simpler approach. Before considering other simpler methods of curve fitting, let us consider a numerical example of the polynomial series method.

EXAMPLE 2.1

As a numerical example, suppose that the graph in figure 7.6 in Chapter 7 represents the available information for expressing K_{ts} as a function of (a/d). Suppose that we desire to express this function by means of a polynomial equation, with K_{ts} as the dependent variable and (a/d) as the independent variable.

From figure 7.6 the variation of the curve is not great, and the curve should be defined quite accurately by means of four points. This is especially true if we consider that the graph in that figure is not precise, since it was obtained from experimental measurements using strain gages. Hence, from the graph in figure 7.6, the accompanying tabulated data are obtained for conducting the conversion to a polynomial equation.

(a/d)	K_{ts}
0.0	2.0
0.1	1.65
0.2	1.50
0.3	1.41

Notice that the tabulated data have been obtained at reasonable intervals over the anticipated range of interest for the variables K_{ts} and (a/d). Also notice that, since four pairs of corresponding values are tabulated, the polynomial equation which is defined will be of the third degree. This is obvious because the degree of the polynomial is always n if there are $(n + 1)$ sets of corresponding values in the tabulated data.

The general form of the polynomial equation given below is similar to that of equation 2.9.

$$K_{ts} = \alpha + \beta\left(\frac{a}{d}\right) + \gamma\left(\frac{a}{d}\right)^2 + \delta\left(\frac{a}{d}\right)^3$$

Here α, β, γ, and δ are the coefficients to be determined. Obviously, if $(a/d) = 0$ in this equation, $K_{ts} = \alpha$ at that point. Hence, from the first point in the tabulated data, we have $K_{ts} = 2.0 = \alpha$. Thus the polynomial equation can now be written in the following form, with the β, γ, and δ coefficients still to be determined.

$$K_{ts} = 2.0 + \beta\left(\frac{a}{d}\right) + \gamma\left(\frac{a}{d}\right)^2 + \delta\left(\frac{a}{d}\right)^3$$

Now insert in this equation the corresponding values of the three remaining points from the tabulated data; the following three linear equations in β, γ, and δ are obtained.

$$1.65 = 2.0 + \beta(0.1) + \gamma(0.1)^2 + \delta(0.1)^3$$

$$1.50 = 2.0 + \beta(0.2) + \gamma(0.2)^2 + \delta(0.2)^3$$

$$1.41 = 2.0 + \beta(0.3) + \gamma(0.3)^2 + \delta(0.3)^3$$

Simultaneous solutions were made for these three equations using the determinant method of equation 2.10b, and the following values were found for β, γ, and δ in that manner.

$$\beta = -4.97$$

$$\gamma = +17.0$$

$$\delta = -23.3$$

Hence the polynomial equation relating K_{ts} to (a/d) is

$$K_{ts} = 2.0 - 4.97\left(\frac{a}{d}\right) + 17.0\left(\frac{a}{d}\right)^2 - 23.3\left(\frac{a}{d}\right)^3 \tag{2.11}$$

Equation 2.11 was derived for the range $0 \leq (a/d) \leq 0.3$, but undoubtedly it would be reasonably accurate for values of (a/d) somewhat greater than 0.3. Incidentally, accuracy checks of equation 2.11 for random values of (a/d) correlate very well with the graph in figure 7.6. Hence equation 2.11 is an accurate equation expressing the function of K_{ts} versus (a/d) as given originally in the graph.

Function Conversion to a Simple Exponential Equation. In converting a function from given tabulated data to a mathematical equation two significant objectives must be kept in mind. First, the mathematical equation must express the tabulated data with an acceptable degree of accuracy. Second, to be best suited for use in explicit optimum design, the mathematical equation must be as simple as possible. These two objectives are generally not compatible with each other, and therefore an acceptable compromise must often be used in practice. Hence, although the polynomial series method for function conversion can always be applied to tabulated data with preciseness for the tabulated values, the method becomes unwieldy for polynomial equations of high degree. For such cases great simplification can be obtained in many problems with an acceptable degree of accuracy over the range of interest by resorting to a different method of approximation. One of the most powerful and widely applicable methods the author has used in practice is the simple exponential equation, which we now consider.

Suppose that the variables x and y are related to each other by equation 2.12, where C and p are any constant real numbers which can be either rational or irrational. By definition, an equation of this form is a simple exponential equation.

$$y = Cx^p \tag{2.12}$$

Such equations can be classified into four basic types, depending on the value for the exponent p. The four basic *types* are shown in the graphical sketches in figures 2.4a, 2.4b, 2.4c, and 2.4d, which also show the effects of the value for the constant C. These typical sketches become obvious from a consideration of equation 2.12 and the first derivative of y with respect to x for that equation.

The types of variation depicted in figure 2.4 cover a broad range of possibilities from a general standpoint. Varying degrees of curvature can be accounted for by choice of the specific value for the exponent p. The magnitude of y with respect to x can be accounted for by choice of the specific value for the constant C. Curves concave upward can be obtained for either y increasing or y decreasing as x increases. Curves concave downward can be obtained for y increasing as x increases. The breadth of the range of possible variations which can be generally covered by a simple exponential equation is significantly increased by considering that the coordinate axes corresponding to figure 2.4 might be different from the coordinate axes to which the given information is referred.

Consider now a more specific study of equation 2.12 by taking the logarithm of both sides of that equation, which gives us

$$\log y = p \log x + \log C \tag{2.13}$$

Let us now introduce a change in variable by letting $\beta = \log y$ and $\alpha = \log x$. Hence equation 2.13 becomes

$$\beta = p\alpha + C' \tag{2.14}$$

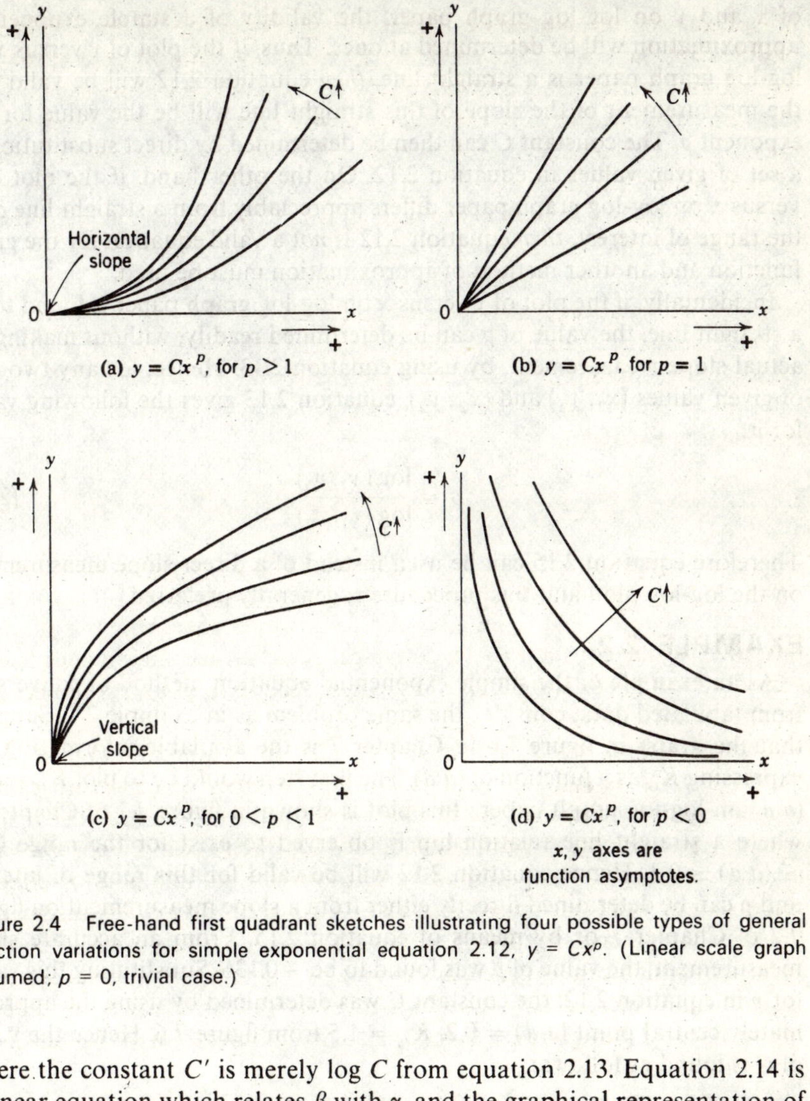

Figure 2.4 Free-hand first quadrant sketches illustrating four possible types of general function variations for simple exponential equation 2.12, $y = Cx^p$. (Linear scale graph assumed; $p = 0$, trivial case.)

where the constant C' is merely log C from equation 2.13. Equation 2.14 is a linear equation which relates β with α, and the graphical representation of β versus α would be a straight line having a slope p. Hence a plot of y versus x on a log-log graph paper *would* result in a *straight* line having a slope p, *if* equation 2.12 is valid for relating x with y.

Let us now apply the converse of this study to the problem of converting a function from given tabulated data to a mathematical equation. Assume that we have available corresponding tabular values of x and y which define the given function over the range of interest. If we now plot these given values of x and y on log-log graph paper, the validity of a simple exponential approximation will be determined at once. Thus, *if* the plot of y versus x on log-log graph paper is a straight line, *then* equation 2.12 will be valid and the measurement of the slope of this straight line will be the value for the exponent p. The constant C can then be determined by direct substitution of a set of given values in equation 2.12. On the other hand, *if* the plot of y versus x on log-log graph paper differs appreciably from a straight line over the range of interest, *then* equation 2.12 is *not* a valid equation for the given function and another method of approximation must be used.

Incidentally, if the plot of y versus x on log-log graph paper is found to be a straight line, the value of p can be determined readily, without making an actual slope measurement, by using equation 2.13. Hence, for any two sets of given values (x_1, y_1) and (x_2, y_2), equation 2.13 gives the following value for p:

$$p = \frac{\log(y_1/y_2)}{\log(x_1/x_2)} \tag{2.15}$$

Therefore equation 2.15 can be used instead of a direct slope measurement on the log-log plot, and this procedure is generally preferred.

EXAMPLE 2.2

As an example of the simple exponential equation method of conversion from tabulated data, consider the same problem as in example 2.1. Suppose that the graph in figure 7:6 in Chapter 7 is the available information for expressing K_{ts} as a function of (a/d). The first step would be to plot K_{ts} versus (a/d) on log-log graph paper; this plot is shown in figure 7.7 of Chapter 7, where a straight line relationship is observed to exist for the range $0.04 \leq (a/d) \leq 0.3$. Hence equation 2.12 will be valid for this range of interest and p can be determined directly either from a slope measurement on figure 7.7 of Chapter 7 or by means of equation 2.15. From an accurate slope measurement, the value of p was found to be -0.136. Substituting this value for p in equation 2.12, the constant C was determined by using the approximately central point $(a/d) = 0.2$, $K_{ts} = 1.5$ from figure 7.6. Hence the value of C is found as follows:

$$1.5 = C(0.2)^{-0.136}$$

$$\therefore \quad C = 1.20$$

The simple exponential equation for the function under consideration has now been uniquely determined, and it is

$$K_{ts} = \frac{1.20}{(a/d)^{0.136}} \tag{2.16}$$

The range of validity for this equation is $0.04 \leq (a/d) \leq 0.3$, and depending on the problem at hand, its accuracy might very well be acceptable outside of this range. If questions exist, random accuracy checks of equation 2.16 with respect to the given information in figure 7.6 should be made. Incidentally, the simple exponential equation 2.16 which has now been derived expresses the given function from figure 7.6 in a much simpler manner than the polynomial equation 2.11 which was derived previously from figure 7.6 also. Equation 2.16 would be much simpler to work with in conducting explicit optimum design, and hence it is the preferred equation.

Incidentally, the given graphical information for the range $0.04 \leq (a/d) \leq 0.3$, from figure 7.6, is of the exponential type depicted in figure 2.4d. This is borne out by the negative value for p (-0.136), which has been determined. Many experimentally obtained stress concentration factor graphs in the literature are of this type, and they generally can be approximated with a simple exponential equation with good accuracy over the range of significant interest.

Function Conversion to Simple Exponential Equation Referred to Shifted Coordinate Axes. Occasionally a function is originally expressed in tabulated data or graphical representation which obviously is not compatible with the possible general types of simple exponential equations depicted in figure 2.4. Such was the case for the *entire* given range of figure 7.6, where the function intersects the vertical coordinate axis $(a/d) = 0$. Hence figure 7.6 was expressed with the simple exponential equation 2.16 which is valid only over the range $0.04 \leq (a/d) \leq 0.3$, whereas the original information in figure 7.6 is given over the entire range $0.00 \leq (a/d) \leq 0.3$. As another illustration, a function which has a variation concave downward with y decreasing as x increases would not be compatible with the possible simple exponential variations shown in figure 2.4. For many of these incompatible cases, a simple exponential equation can be written with good accuracy by referring the original information to a set of coordinate axes which are *shifted* with respect to the original information axes. This method of function expression extends the simple exponential technique to a far greater range of possible function variations.

The extended method which we will now describe is based on the observation that simple exponential equations are characterized by the existence of one or two limits, except for the obvious case of a straight inclined line on a linear plot, as in figure 2.4b. This characteristic is obvious from figure 2.4 where a lower limit on y exists at $y = 0$ for figure 2.4a if $p > 1$, a lower limit exists on x at $x = 0$ for figure 2.4c if $0 < p < 1$, and lower limits exist on both x and y at $x = 0$, $y = 0$ for figure 2.4d if $p < 0$. Hence at this time let us consider the expression, by means of a simple exponential equation referred to shifted coordinate axes, of a general function whose limit is either known or anticipated.

Suppose that a given function expresses corresponding values between x and y in either tabular or graphical fashion. A finite number of corresponding values could be used for plotting the variation, at least hypothetically, on linear graph paper. They are indicated by crosses in figure 2.5 for typical variations. These sketches imply simple exponential equations referred to *new* coordinate axes u and v, from a study of the general types in figure 2.4.

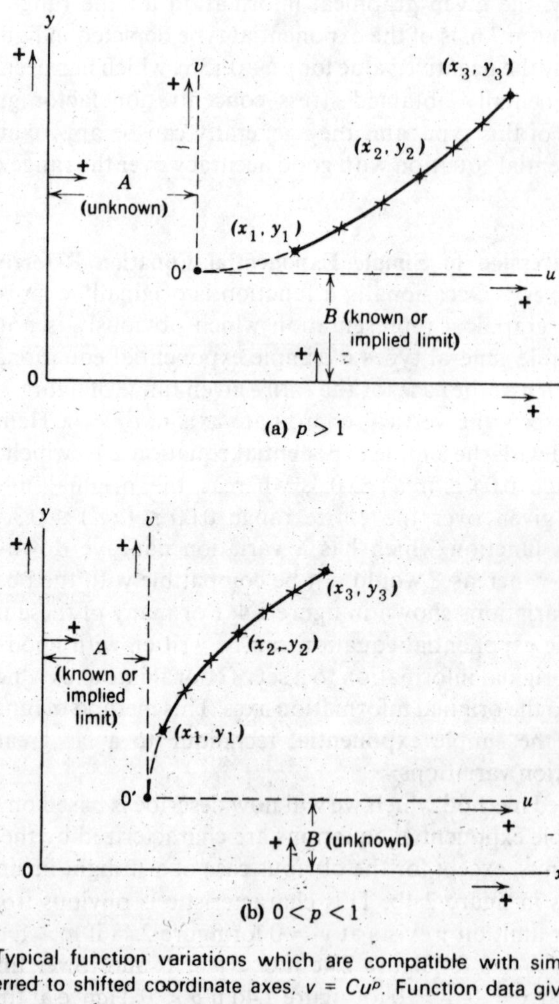

Figure 2.5 Typical function variations which are compatible with simple exponential equations referred to shifted coordinate axes, $v = Cu^p$. Function data given with respect to x, y axes.

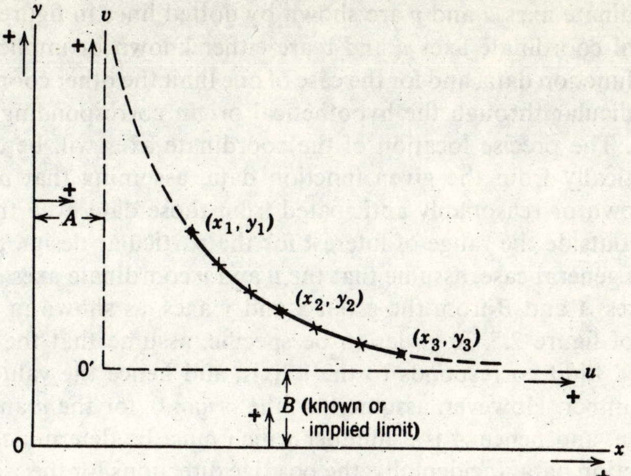

(c) $p < 0$

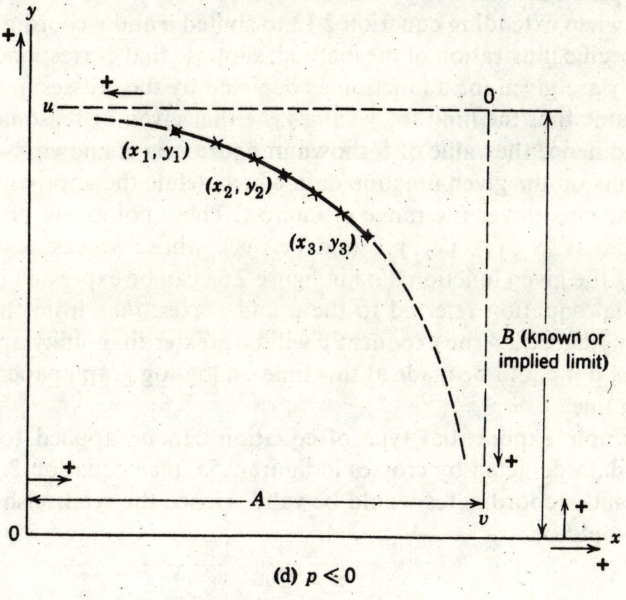

(d) $p < 0$

Figure 2.5 (*Continued*)

The coordinate axes u and v are shown by dotted lines in figure 2.5. These orthogonal coordinate axes u and v are either known or implied limits of the given function data, and for the case of one limit the other coordinate axis is perpendicular through the hypothetical origin corresponding to that of figure 2.4. The precise location of the coordinate axes will be determined mathematically from the given function data, assuming that *one* limit is either known or reasonably anticipated from those data even though they might be outside the range of interest for the particular design problem.

For the general case, assume that the u and v coordinate axes are located at distances A and B from the given x and y axes, as shown in the typical sketches of figure 2.5. In order to be specific, assume that the known or anticipated limit corresponds to the u axis, and hence the value of B is a known number. However, assume that the *origin* 0′ for the u and v axes is not known, and hence A is a number which must be determined from the given function data. Incidentally, the positive directions for the x and y axes, the relative u and v axes, and the A and B values have been chosen as shown for the typical sketches in figure 2.5. It is important to watch algebraic signs carefully when extending equation 2.12 to shifted u and v coordinate axes.

As a specific illustration of the method, suppose that corresponding values of x and y are given for a function as depicted by the crosses in figure 2.5a. Also, assume that the limit for y values is either given or reasonably anticipated, and hence the value of B shown in figure 2.5a is known. Now choose three points on the given function data which define the approximate variation of the curve over the range of interest. These points are designated in figure 2.5a as (x_1, y_1), (x_2, y_2), and (x_3, y_3), whose values obviously are known. *If* the given function data in figure 2.5a can be expressed by a simple exponential equation referred to the u and v axes, *then* from the previous study using figure 2.4 the exponent p will be greater than unity and a plot of u versus v, if it could be made at this time on log-log graph paper, would be a straight line.

If a simple exponential type of equation can be applied to the given function data depicted by crosses in figure 2.5a, *then* equation 2.12 referred to the u and v coordinates would be valid. Hence the relationship between u and v would be

$$v = Cu^p \tag{2.17}$$

where C and p are real numbers to be determined. However, from figure 2.5a, taking into account the assumed positive directions, we have the obvious relations

$$v = y - B \tag{2.18a}$$

$$u = x - A \tag{2.18b}$$

which assumed that A and B are real numbers depicting the shift of the u and v axes from the x and y axes. Hence the simple exponential equation for the case in figure 2.5a becomes

$$y = B + C(x - A)^p \tag{2.19}$$

Equation 2.19 would be exactly the same for the cases depicted in figures 2.5b and 2.5c, and it is a simple matter to extend the derivation using equation 2.17 to other cases, as that of figure 2.5d.

Referring to figure 2.5a, the known values (x_1, y_1), (x_2, y_2), and (x_3, y_3) can now be placed in equation 2.19, and three equations with three unknowns result:

$$y_1 = B + C(x_1 - A)^p$$
$$y_2 = B + C(x_2 - A)^p$$
$$y_3 = B + C(x_3 - A)^p$$

For figure 2.5a, B is assumed as a known value, and hence the three unknowns which must be determined are the values for C, A, and p. This requires the simultaneous solution of the three equations, which can be effected as follows:

$$\frac{(y_1 - B)}{(x_1 - A)^p} = C = \frac{(y_2 - B)}{(x_2 - A)^p} = \frac{(y_3 - B)}{(x_3 - A)^p}$$

$$\therefore \quad \left(\frac{y_2 - B}{y_1 - B}\right)^{1/p} (x_1 - A) = (x_2 - A)$$

$$\left(\frac{y_3 - B}{y_1 - B}\right)^{1/p} (x_1 - A) = (x_3 - A)$$

Let $U = (x_1 - A)$.

$$\therefore \quad \left(\frac{y_2 - B}{y_1 - B}\right)^{1/p} = \frac{U + (x_2 - x_1)}{U} = 1 + \frac{(x_2 - x_1)}{U}$$

$$\left(\frac{y_3 - B}{y_1 - B}\right)^{1/p} = \frac{U + (x_3 - x_1)}{U} = 1 + \frac{(x_3 - x_1)}{U}$$

$$\therefore \quad \frac{(x_3 - x_1)}{(x_2 - x_1)}\left(\frac{y_2 - B}{y_1 - B}\right)^{1/p} - \left(\frac{y_3 - B}{y_1 - B}\right)^{1/p} = \frac{(x_3 - x_1)}{(x_2 - x_1)} - 1 \tag{2.20}$$

In equation 2.20, the only unknown value is p, and this equation was derived for the case of figure 2.5a. Hence the value for p could be determined directly from equation 2.20, and an iterative approach would be required. After the value for p has been determined, the values of A and C can be found quite readily and explicitly. Thus A is determined from a linear equation such as equation 2.21, where x_1, x_3, y_1, y_3, B, and p are known values; equation 2.21 was obtained near the beginning of the derivation of equation 2.20.

$$\left(\frac{y_3 - B}{y_1 - B}\right)^{1/p}(x_1 - A) = (x_3 - A) \tag{2.21}$$

Hence, with values for both p and A now unknown, the value of C can be found directly from equation 2.19, known values for x, y such as x_2, y_2 and the known value for B being used.

Values for p, A, and C having been determined, equation 2.19 can be written as the exponential equation relating x with y. Accuracy checks with respect to the given tabular data should be made for equation 2.19 so obtained. As a quick check on accuracy, the given functional data of (x, y) values can now be plotted as corresponding (u, v) values on log-log graph paper. If the accuracy is good, the log-log plot of u versus v should fall nearly on a straight line, confirming the validity of equation 2.17. The actual equation 2.17 on that plot will be the straight line which by the derivation presented must necessarily pass through the *three* points $(u_1, v_1), (u_2, v_2), (u_3, v_3)$, which correspond to the three points $(x_1, y_1), (x_2, y_2), (x_3, y_3)$ in figure 2.5a. For a specific case such as that in figure 2.5a, the corresponding (u, v) values are obtained from the given (x, y) values using equations 2.18a and 2.18b, where both the A and B values are now known. Hence the accuracy of derived equation 2.19 can readily be studied over the range of interest from a simple log-log plot of v versus u. The slope of the straight line joining the *three* points $(u_1, v_1), (u_2, v_2), (u_3, v_3)$ on this log-log plot should agree with the value of p which was found from equation 2.20.

The method which has been outlined can be applied to a wide range of function variations. If given function data do not satisfy one of the basic types of simple variation depicted in figure 2.4, an extension to shifted coordinate axes u and v as outlined here will probably result in an equation which is *simpler* than the polynomial series type of approximation. Hence, after it has been determined that a simple exponential equation of the form $y = Cx^p$ is not valid for expressing a function, an equation of the form of equation 2.19 should be considered, as outlined. However, the student is warned not to substitute arbitrarily into equations 2.18a, 2.18b, 2.19, 2.20, or 2.21, which have been derived here for the specific case of figure 2.5a as a typical illustration of the general method. Unique problems should be attacked more fundamentally, starting with the basic equation 2.17 relating (u, v), and considering the graphical representation of the given (x, y) information as illustrated for typical cases in figure 2.5.

EXAMPLE 2.3

As a specific example of the simple exponential equation referred to shifted coordinate axes method of conversion from tabulated data, consider the same problem presented in examples 2.1 and 2.2. Suppose that the graph in figure 7.6 of Chapter 7 gives the available information for expressing K_{ts} as a function of (a/d). In example 2.1 we expressed this function in a polynomial series, the results of which are presented in equation 2.11. That equation is valid over the entire range given for (a/d), which in figure 7.6 is $0.00 \le (a/d) \le 0.30$. However, in example 2.2 we expressed the same function in a simple exponential equation, 2.16. That equation is significantly simpler than the polynomial equation 2.11, although its range of validity is curtailed to $0.04 \le (a/d) \le 0.30$. In conducting explicit optimum design for many cases, it would be far more desirable to use the simple exponential equation 2.16, provided that the curtailed range of validity is satisfactory.

In conducting explicit optimum design the twofold objective in curve fitting is acceptable accuracy over the range of interest and simplicity. We shall find from this example that the simple exponential equation referred to shifted coordinate axes method of regenerative conversion for figure 7.6 will result in an equation which is acceptably accurate over the *entire* given range $0 \le (a/d) \le 0.3$ and which is appreciably simpler than the polynomial equation 2.11 previously derived. The first step in this example will be to refer to the given graph in figure 7.6.

The given information in figure 7.6 is similar to the function variation typified in figure 2.5c, where the x and y axes correspond to the (a/d) and K_{ts} axes in figure 7.6 respectively. From the log-log plot of K_{ts} versus (a/d) in figure 7.7, as made from figure 7.6, we see that the given information is very linear on the log-log graph for $(a/d) \ge 0.04$ within the range of interest, $0 \le (a/d) \le 0.3$. This *implies* that the given information could be hypothetically projected outside the presented range of known variation so as to have a lower limit on K_{ts}, which would be K_{ts} tending to zero for very large hypothetical values of (a/d). Hence we can assume an anticipated known limit of zero for K_{ts} which hypothetically would be approached far outside the actual given range, $0 \le (a/d) \le 0.3$. Therefore, with respect to the location of the u axis corresponding to figure 2.5c, the known limit B will be assumed to be zero. The specific value chosen for B will have an insignificant effect on the accuracy of the equation being derived, provided that the limit on K_{ts} is approached far outside the range of interest for (a/d). Hence the implied choice of $B = 0$ will be very convenient for our derivation, and also it will be acceptable from the standpoint of accuracy within the range of interest, $0 \le (a/d) \le 0.3$.

The given function in figure 7.6 cuts the vertical axis $(a/d) = 0$ at a value $K_{ts} = 2.0$, with finite negative slope at that point. Considering the v axis of

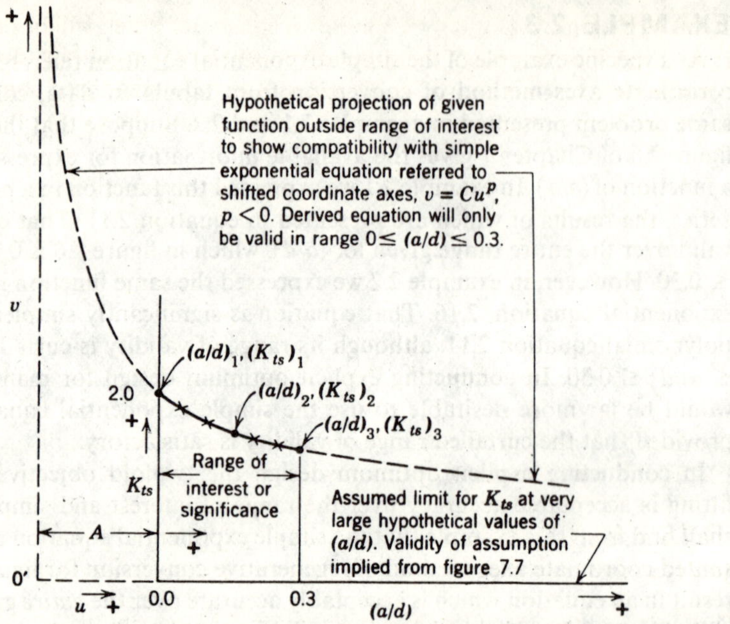

Figure 2.6 K_{ts} versus (a/d) from figure 7.6 showing shifted coordinate axes u and v to which simple exponential equation will be referred.

figure 2.5c, it would be located to the left of the K_{ts} axis which corresponds to the y axis, for this example. Hence, if the direction for the displacement A for the v axis is positive to the right of the K_{ts} axis, the value of A for this example should merely be negative. The relationship of the u and v axes to the (a/d) and K_{ts} axes for this specific example are shown in figure 2.6, which corresponds to the typical general case depicted in figure 2.5c. Therefore, at this time we will assume a function variation of the form $v = cu^p$ by equation 2.17, with respect to the u, v axes in figure 2.6.

Referring to figure 2.6, for this example the coordinate relationships for the given information between the $[(a/d), K_{ts}]$ and (u, v) coordinate axes is obviously as follows, where the specific value for A will be a negative number.

$$v = K_{ts}$$

$$u = \left(\frac{a}{d}\right) - A$$

Hence the simple exponential equation $v = Cu^p$ referred to the given coordinates would be

$$K_{ts} = C\left[\left(\frac{a}{d}\right) - A\right]^p \tag{2.22}$$

In this equation C, A, and p are unknown constants which will be determined from three specific points from the given information within the range of interest. Hence, from figure 7.6, we obtain the following three points, which define the approximate variation of the curve within the range of interest.

$$(K_{ts})_1 = 2.0, \qquad \left(\frac{a}{d}\right)_1 = 0.00$$

$$(K_{ts})_2 = 1.555, \qquad \left(\frac{a}{d}\right)_2 = 0.16$$

$$(K_{ts})_3 = 1.41, \qquad \left(\frac{a}{d}\right)_3 = 0.30$$

Conducting a derivation similar to that for equation 2.20 results in equation 2.23, where p will be the only unknown:

$$\left[\frac{(a/d)_3 - (a/d)_1}{(a/d)_2 - (a/d)_1}\right]\left[\frac{(K_{ts})_2}{(K_{ts})_1}\right]^{1/p} - \left[\frac{(K_{ts})_3}{(K_{ts})_1}\right]^{1/p} = \left[\frac{(a/d)_3 - (a/d)_1}{(a/d)_2 - (a/d)_1}\right] - 1 \qquad (2.23)$$

Thus, placing the specific values in equation 2.23, we obtain

$$\left(\frac{0.3 - 0.0}{0.16 - 0.0}\right)\left(\frac{1.555}{2.0}\right)^{1/p} - \left(\frac{1.41}{2.0}\right)^{1/p} = \left(\frac{0.3 - 0.0}{0.16 - 0.0}\right) - 1$$

$$\therefore \quad (1.875)(0.779)^{1/p} - (0.705)^{1/p} = 0.875 \qquad \text{(a)}$$

From our study of simple exponential variations in conjunction with figures 2.4 and 2.5, we realize from figure 2.6 that the case of $p < 0$ exists and the numerical value for p will be negative. At $p \to -\infty$, equation a will be satisfied, but this solution is trivial. Another solution to equation a will exist *within* the range $-\infty < p < 0$, and it can be determined quite readily with a few simple calculations which are summarized typically in the accompanying table. Hence, for equation a, $(1/p) = -4.80$ or $p = -0.208$.

$(1/p)$	$(1.875)(0.779)^{1/p} - (0.705)^{1/p}$	
0	+0.875	Trivial solution
−2	+1.08	
−4	+1.05	
−6	+0.24	
−5	+0.80	
−4.70	+0.92	
−4.80	+0.87	Desired solution

The specific value for the constant A can now be determined from equation 2.24 the derivation of which is similar to that of equation 2.21.

$$\left[\frac{(K_{ts})_3}{(K_{ts})_1}\right]^{1/p}\left[\left(\frac{a}{d}\right)_1 - A\right] = \left[\left(\frac{a}{d}\right)_3 - A\right] \tag{2.24}$$

Thus, placing the specific values in equation 2.24, we obtain

$$\left(\frac{1.41}{2.0}\right)^{-4.80}(0.0 - A) = (0.30 - A)$$

$$\therefore \quad -5.38A = 0.30 - A$$

$$\therefore \quad A = -0.0685$$

The constant C can now be determined directly from equation 2.22, using $p = -0.208$, $A = -0.0685$, $(a/d)_2 = 0.16$, and $(K_{ts})_2 = 1.555$, which are approximately central values for the variables. Therefore equation 2.22 becomes

$$1.555 = C[(0.16) - (-0.0685)]^{-0.208}$$

$$\therefore \quad C = (1.555)(0.2285)^{0.208} = 1.15$$

Thus the final equation relating the given information from figure 7.6 over the *entire* range $0.0 \leq (a/d) \leq 0.3$ is

$$K_{ts} = (1.15)\left[\left(\frac{a}{d}\right) - (-0.0685)\right]^{-0.208}$$

$$\therefore \quad K_{ts} = 1.15\bigg/\left[\left(\frac{a}{d}\right) + 0.069\right]^{0.208} \tag{2.25}$$

Accuracy checks for equation 2.25 with respect to the given information in figure 7.6 were conducted over the range $0.0 \leq (a/d) \leq 0.3$, and the error was found to be less than 1% for that entire range of interest.

Let us summarize the comparable results from examples 2.1, 2.2, and 2.3. For all of them the problem was to derive a mathematical equation expressing K_{ts} as a function of (a/d) from the graphical information given in figure 7.6. Equation 2.11 was derived from the polynomial series type of approach, and it is accurate over the entire given range $0 \leq (a/d) \leq 0.3$. A much simpler equation was derived using the simple exponential equation type of approach; it was equation 2.16. However, this equation is accurate only over the range $0.04 \leq (a/d) \leq 0.3$, which for some problems might be a significant limitation. Finally, equation 2.25 was derived from the simple exponential equation referred to shifted coordinate axes type of approach, and it is accurate over the entire given range $0 \leq (a/d) \leq 0.3$. However, it is somewhat more complex than equation 2.16, although it is simpler than the polynomial equation 2.11. These equations with their ranges of accuracy are summarized for a quick comparison as follows:

$$K_{ts} = 2.0 - 4.97\left(\frac{a}{d}\right) + 17.0\left(\frac{a}{d}\right)^2 - 23.3\left(\frac{a}{d}\right)^3 \tag{2.11}$$

valid for $0 \leq (a/d) \leq 0.3$;

$$K_{ts} = \frac{1.20}{(a/d)^{0.136}} \tag{2.16}$$

valid for $0.04 \leq (a/d) \leq 0.3$;

$$K_{ts} = \frac{1.15}{[(a/d) + 0.069]^{0.208}} \tag{2.25}$$

valid for $0.0 \leq (a/d) \leq 0.3$.

Equations 2.11, 2.16, and 2.25 all express the same functional data originally given in the graph of figure 7.6. Since only one of these equations would be used for conducting explicit optimum design, the significant question is: How to select the simplest approach in unique problems for determining which *type* of equation to use in the conversion from given graphical or tabular information? This can best be answered by considering the twofold objectives, *simplicity* and *accuracy*, for equations which will be used in explicit optimum design procedures. Hence the first step would be to look at the given graphical information to see whether or not its variation is *compatible* with one of the possible general types depicted for simple exponential variations in figure 2.4. If compatibility is a possibility over the *range of interest*, the given function should be plotted on log-log graph paper within that range as the next step. Hence, if a straight line is obtained on this plot over the range of interest, the simple exponential equation referred to the coordinate axes of the given information can easily be derived. A simple exponential equation having the form given by equation 2.12 would then be used to express the function in the design procedure. However, if the plot on log-log graph paper is *not* a reasonably straight line over the range of interest, the original information on linear graph paper should be studied to determine whether or not a simple exponential approximation referred to shifted coordinate axes is a possibility, as typified in figure 2.5. If this possibility exists, an equation should be derived having the form given by equation 2.19. If an accuracy check on that equation is satisfactory over the range of interest, the equation would be used in the explicit design procedure. However, if the accuracy check is not satisfactory, the polynomial equation approach will always work at the sacrifice of increased complexity. In that event an equation of the form given by equation 2.9 would be derived for the range of interest, and it would be used in the explicit design procedure. For such cases, we could also consider an approach using graphical iteration, which might result in a simpler equation than the polynomial equation, with acceptable accuracy over the range of interest. Therefore, as another example of curve-fitting methods, we now consider the graphical iteration method for equation derivation.

Function Conversion by Graphical Iteration. Consider a general function which relates values of x and y over a certain range of interest by means of tabulated data or graphical representation. Suppose it has been determined that it is impossible to relate x with y by using a simple exponential equation referred either to the x, y coordinate axes or to shifted u, v coordinate axes. Also, the explicit design procedure which must be followed for the specific problem at hand makes it highly desirable to obtain a mathematical equation relating y with x which is appreciably simpler than the required polynomial equation for defining this function. A desirable solution to this problem might be obtained by using a process of graphical iteration based on the simple exponential equation, and we shall now consider this method in general terms briefly.

The first step in the procedure will be to plot the given function values of x and y on log-log graph paper over the range of interest. From previous considerations, the graph so obtained will not be a straight line on the log-log plot. However, consider approximating the given information as closely as possible with a straight line on the log-log plot which either falls entirely above or entirely below the actual variation of y versus x. Denote the ordinates for this straight line by y_1, which can then be expressed by a simple exponential equation such as equation 2.12, $y_1 = Cx^p$. However, because of the variation of the actual function, the function ordinates y will differ from y_1 by an amount which should be reasonably small if possible.

The second step will be to plot the difference $(y - y_1)$ versus x on log-log graph paper over the range of interest. Most likely, the graph so obtained on the log-log plot will not be a straight line. However, consider *approximating* this second plot of $(y - y_1)$ versus x as closely as possible with a straight line which either falls entirely above or entirely below the actual variation of $(y - y_1)$ versus x. Denote the ordinates for this straight line by y_2, which can then be expressed by a simple exponential equation similar to equation 2.12, such as $y_2 = Dx^q$, where D and q are real numbers corresponding to C and p used previously. If care is exercised in selecting the straight lines on the log-log plots, the actual variation of $(y - y_1)$ will differ only a small amount from the y_2 ordinates, the differences being very small relative to magnitudes of the given y values. Hence at this point we might write a very close approximation for the given function by equating y_2 to $(y - y_1)$.

$$\therefore \quad (y - y_1) = y_2$$

$$y = y_1 + y_2$$

$$y = Cx^p + Dx^q \tag{2.26}$$

In equation 2.26, values for C and p were obtained from the straight line on the first log-log plot, whereas values for D and q were obtained from the straight line on the second log-log plot. Equation 2.26 is appreciably simpler than a polynomial equation of the fifth degree say, and it might very well be acceptably accurate over the range of interest for x and y.

For more complex functions, the graphical iteration method could be extended to have more than two terms. Hence the general expression for the graphical iteration method would be

$$y = Cx^p + Dx^q + Ex^r + \cdots + Fx^s \tag{2.27}$$

For cases which would *require* more than two or three terms with the graphical iteration method, the polynomial equation method would probably be very unwieldy. In equations 2.26 and 2.27, the values for the coefficients $C, D, \ldots,$ and the values for the exponents $p, q, \ldots$ are real numbers which can be either positive or negative, rational or irrational.

Function Conversion by Least Squares Regression Analysis. Curve-fitting techniques previously presented are based on relatively simple equations which precisely pass through very few points of a given data set. For given graphs and tabular data which vary smoothly if plotted, oftentimes desired shapes can be matched with good accuracy for the range of interest in a design problem. However, sometimes it is desired to derive a simple equation which matches an arbitrarily large number of given data points with the best possible accuracy. A statistical approach can be attempted in such a case [2.3–2.5].

Linear regression analysis with a least squares fit is appropriate for derivation of a straight line equation of the form

$$y = mx + b \tag{2.28}$$

Suppose we have n given data points (x_i, y_i) for $i = 1, 2, \ldots, n$ and a graphical plot of these data indicates that a straight line equation is appropriate. Slope coefficient m is then determined by

$$m = \frac{[\sum x_i y_i - (\sum x_i \sum y_i/n)]}{[\sum x_i^2 - (\sum x_i)^2/n]} \tag{2.29}$$

Intercept b is determined by the relation

$$b = \frac{[\sum y_i - m \sum x_i]}{n} \tag{2.30}$$

With the preceding values for m and b in equation 2.28, a measure of accuracy of fit for the given data set is the correlation coefficient r.

$$r = \frac{[\sum x_i y_i - \sum x_i \sum y_i/n]}{\{[\sum x_i^2 - (\sum x_i)^2/n][\sum y_i^2 - (\sum y_i)^2/n]\}^{1/2}} \tag{2.31}$$

The closer r is to unity the better the fit. Incidentally, for equations 2.29 through 2.31, the summations of course are all for $i = 1, 2, \ldots, n$ of the given data points.

Suppose that the straight line equation 2.28 is not appropriate for the given data set, but that the (x_i, y_i) points are nearly linear on logarithmic graph paper. For this case the simple exponential equation 2.12 would be appropriate.

$$y = Cx^p \tag{2.12}$$

Assume that x and y have only positive values in the given data set. By equation 2.13 an equivalent linear equation is obtained by logarithms, and the preceding least squares linear regression analysis can be directly applied as follows, referring to equations 2.28 through 2.31.

$$\ln y = p \ln x + \ln C$$

$$\therefore \quad p = \frac{\sum [(\ln x_i)(\ln y_i)] - [\sum (\ln x_i) \sum (\ln y_i)/n]}{\sum (\ln x_i)^2 - [(\sum \ln x_i)^2/n]} \tag{2.32}$$

and

$$C = \exp\left[\frac{\sum \ln y_i - p \sum \ln x_i}{n}\right] \tag{2.33}$$

The preceding values of p and C are calculated for use in equation 2.12, and again, if close to unity, the correlation coefficient r indicates that a good fit has been achieved.

$$r = \frac{\sum [(\ln x_i)(\ln y_i)] - [\sum (\ln x_i) \sum (\ln y_i)/n]}{\{[\sum (\ln x_i)^2 - (\sum \ln x_i)^2/n][\sum (\ln y_i)^2 - (\sum \ln y_i)^2/n]\}^{1/2}} \tag{2.34}$$

The example which follows illustrates application of the least squares regression analysis technique to curve fitting.

EXAMPLE 2.4

Consider the same problem as for examples 2.1 through 2.3 where the graph of figure 7.6 in Chapter 7 is to be converted to an equation $K_{ts} = f(a/d)$.

From the straight line log-log plot in figure 7.7 we see that the simple exponential equation 2.12 should be appropriate for the range $0.04 \leq (a/d) \leq 0.3$. Hence the following data point set is taken from figure 7.6.

i	$(a/d)_i$	$(K_{ts})_i$
1	0.05	1.78
2	0.10	1.65
3	0.15	1.57
4	0.20	1.50
5	0.25	1.45
6	0.30	1.41

A programmable calculator is used to solve the least squares regression analysis equations 2.32, 2.33, and 2.34, giving the following values.

$$p = -0.1305; \qquad C = 1.2138; \qquad r = .9929$$

Hence the equation derived in curve fitting by this approach is

$$K_{ts} = \frac{1.2138}{(a/d)^{0.1305}} \tag{2.35}$$

This equation compares quite favorably with equation 2.16 derived previously in example 2.2. Also, good accuracy is indicated by correlation coefficient r close to unity.

Function Conversion for Cases of Many Variables. In the design of real mechanical elements, we generally encounter functions of many variables; for example,

$$y = f(x, u, v, w) \tag{2.8}$$

Suppose such a differentiable function is available either in complex equation, graphical, or tabulated data form. Hence, to use this information in the conduct of explicit design studies, we convert to a simple mathematical equation which is acceptably accurate over the anticipated ranges of significance for the independent variables x, u, v, and w. Incidentally, if the choice exists, these ranges of significant interest should be selected as small as permissible, but one must be sure not to exclude independent variable values for any reasonable design possibility. Let us now choose approximately *central values* for x, u, v, and w within their anticipated ranges of significance.

Hence it is possible for us to consider the variation of y with respect to, say, x with u, v, and w held constant at their central values. Thus, using curve-fitting techniques, we could derive an equation $y = g(x)$, reasonably simple in form and acceptably accurate over the anticipated range of significance for the independent variable x. This method could be extended to a consideration of the variation of y with respect to, say, u for its range of significance, with x, v, and w held constant at their central values. In this general manner, curve-fitting techniques would give us independent equations $y = g(x)$, $y = h(u)$, $y = i(v)$, $y = j(w)$, which would be acceptably accurate over the anticipated ranges of significance for the independent variables. Finally, a relatively simple equation which follows can be written expressing y in terms of *products* of these independent functions, and this will be our desired equation for design use.

$$y = C[g(x)][h(u)][i(v)][j(w)] \tag{2.36a}$$

In this equation C is a specific constant the value of which establishes the required equality for the case where x, u, v, and w are at their central values and y is at the corresponding value determined from the given function information.

The validity of deriving a simple product equation such as 2.36a from the given function information can be proved from some of the salient results in previous parts of this chapter. In section 2.6, using the method of finite differences, we showed that any differentiable function of many variables is characterized by families of essentially parallel curves within a limited range of variation for the independent variables, such a portrayal being obtained by considering the dependent variable versus one of the independent variables with each curve of the family determined by incremental changes for any of the other independent variables. Using this principle for differentiable functions, the validity of equation 2.36a can be established.

The simple function product equation 2.36a will be valid for a limited neighborhood of the central value point, and the region of acceptable accuracy (for design use) with respect to the given information will be surprisingly large for many specific cases. If these ranges of acceptable accuracy are compatible with the ranges of significance for the independent variables, the equation is perfectly valid for use in the conduct of explicit design. On the other hand, if the ranges for acceptable accuracy are not compatible with the ranges of significance for the independent variables, two or more design equations can be derived over separate ranges of interest which will complement the ranges of significance. Hence, for such cases, the design procedure will be resolved into a consideration of two or more corresponding design equations. Fortunately, such a resolution is generally not required for most real design problems.

As a more specific example of conversions for functions of many variables, consider the case where the independent function variations can be expressed by means of simple exponential equations. For each independent variable, the coefficient of its simple exponential equation can be included in a single overall constant C, which is determined with all variables at their central values. Hence the function of several variables would be expressed in the form given by equation 2.36b, where C is the overall constant coefficient, determined as a final step using the central values for x, u, v, w, and the corresponding value for the dependent variable y as determined from the given function information.

$$y = Cx^p u^q v^r w^s \tag{2.36b}$$

In this equation, p, q, r, and s are exponents which are real numbers determined either from slope measurements on log-log plots of y versus x for p, of y versus u for q, and so on, or from equations such as 2.15 or 2.32, with the other independent variables held at their central range values for each case. Finally, the value of the coefficient C is obtained by using equation 2.36b with the determined values for p, q, r, and s, with the central values for x, u, v, and w, and with the corresponding value of y from the given information.

The author has employed simple exponential equations of the form given by equation 2.36b for innumerable real problems involving conversions for functions of many variables. An excellent example of such an application is the derivation of equation 8.29a in Chapter 8. For this example the ranges of significance for the independent variables are widespread, as shown in table 8.7, which also summarizes the chosen central values as approximate averages for the independent parameter possibilities. The log-log plots used for finding the values of the exponents are given in figures 8.5 through 8.10. Random accuracy checks for the derived equation are summarized in table 8.8, proving an acceptable degree of accuracy for its intended design study use. Incidentally, accuracy values in that equation are with respect to true values as calculated by the original complex equation 8.7, from which the tabulated data for deriving equation 8.29a were determined.

2.8 Some Examples of Methods for Degenerative Conversions—Simplification

For many real mechanical elements, the explicit method of optimum design requires the manipulation of numerous design equations. Hence the desirability of simplicity in design equations is often a very significant consideration. However, for any design equation, simplicity can be obtained only to a limited degree because of considerations of acceptable accuracy. In this section we shall consider some typical examples of equation simplification, along with the resulting errors for each case.

Neglect of Terms in Finite Series. Occasionally a mathematical equation can be significantly simplified by neglect of one or more terms, which results in an acceptable degree of accuracy over a limited range of variation for the variables. If this range is compatible with the range of interest for the problem at hand, the simplified equation can be used in the conduct of explicit design. Let us now consider a specific example in which a design equation is appreciably simplified by the neglect of an insignificant term.

EXAMPLE 2.5

Consider a circular arc having a radius of curvature R as shown in figure 2.7. A problem which has innumerable practical applications is the determination of the sag y for this circular arc with respect to a chord of length $2x$. Referring to figure 2.7, we can relate x, y, and R with equation 2.37 which follows, solving explicitly for the unknown sag y.

$$R^2 = x^2 + (R - y)^2$$

$$\therefore \quad y^2 - 2Ry + x^2 = 0$$

$$\therefore \quad y = \frac{2R \pm \sqrt{(2R)^2 - 4x^2}}{2}$$

$$\therefore \quad y = R\left[1 - \sqrt{1 - \left(\frac{x}{R}\right)^2}\right] \tag{2.37}$$

Equation 2.37 enables us to determine a *precise* value for y from given values of x and R. Having chosen the negative sign in the derivation merely limits the equation to values of y in the semicircle, $0 \leq y \leq R$ or $-90° \leq \theta \leq +90°$, referring to figure 2.7. However, let us now consider a *simplified* relationship between x, R, and y which is acceptably accurate over a more limited range of variation.

From figure 2.7 we have obtained the exact relationship $y^2 - 2Ry + x^2 = 0$. Hence we obtain the exact equation

$$y = \frac{(x^2 + y^2)}{2R} \tag{a}$$

However, consider the fact that y is small compared to x for reasonably small angles θ (refer to figure 2.7). Hence y^2 will be much smaller than x^2 for such

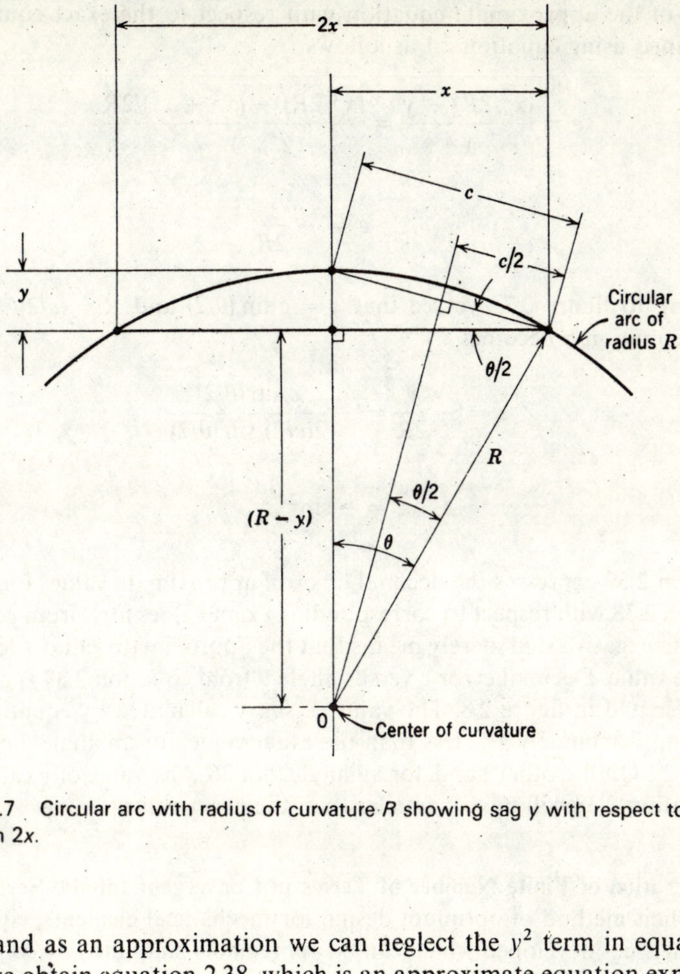

Figure 2.7 Circular arc with radius of curvature R showing sag y with respect to a chord of length $2x$.

cases, and as an approximation we can neglect the y^2 term in equation a. Thus we obtain equation 2.38, which is an approximate equation expressing y as a function of x and R in an appreciably simpler manner than exact equation 2.37.

$$y \approx \frac{x^2}{2R} \tag{2.38}$$

The problem now is to determine the accuracy of equation 2.38, which will define its range of usefulness for design work.

The true value for y as a function of x and R is given by exact equation a, whereas the approximate value is given by equation 2.38. Hence the decimal error e of the approximate equation with respect to the exact equation is determined using equation 2.1 as follows.

$$e = \frac{(x^2/2R) - y}{y} = \frac{(x^2/2R) - (x^2 + y^2)/2R}{y}$$

$$\therefore \quad e = -\frac{y}{2R} \tag{b}$$

Referring to figure 2.7, we see that $y = c \sin(\theta/2)$ and $R = (c/2)/\sin(\theta/2)$. Hence equation b becomes

$$e = -\frac{y}{2R} = -\frac{c \sin(\theta/2)}{2(c/2)/\sin(\theta/2)}$$

$$\therefore \quad e = -\sin^2\frac{\theta}{2} \tag{2.39}$$

Equation 2.39 expresses the decimal error of approximate values for y from equation 2.38 with respect to corresponding exact values for y from equation 2.37. The negative sign merely means that the approximate value is less than the true value. Decimal error e versus angle θ from equation 2.39 is graphically presented in figure 2.8. The value of sag y calculated by equation 2.38 will be approximately 1% less than the exact value, for an angle θ equal to about 12°. On the other hand, for an angle θ of 26°, the value of y calculated by equation 2.38 will be 5% smaller than the exact value.

Consideration of Finite Number of Terms in Convergent Infinite Series. In the explicit method of optimum design for mechanical elements, equations are sometimes developed which can be appreciably simplified by converting a factor to an infinite series of which only the significant terms will be used. In order to apply this method to a factor of an equation, its infinite series expression must be convergent and, preferably, its rate of convergence should be high. There are innumerable examples of this method in practice, and we shall consider some of them along with studies of the associated errors for the approximations.

EXAMPLE 2.6

Consider a factor $[1/(x + y)]$ in a mathematical equation, where x and y are independent variables. Incidentally, as a special case either x or y can be constants of known value. Suppose that $|y| < |x|$ over the entire range of interest for the variation of x and y. Hence we could expand $[1/(x + y)]$ into

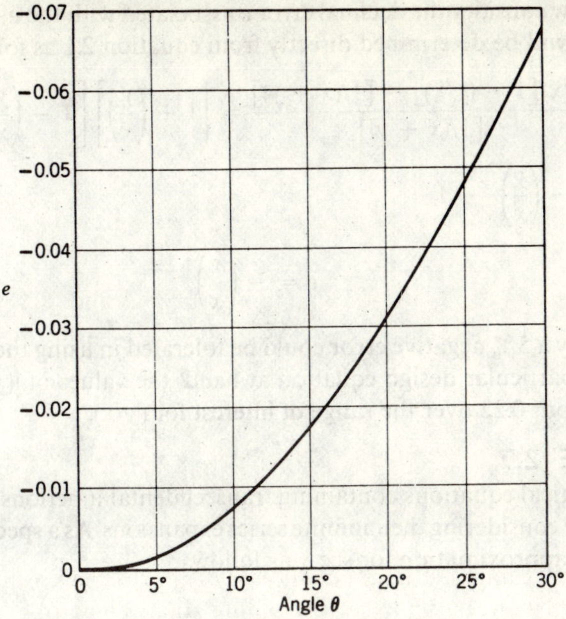

Figure 2.8 Decimal error e versus angle θ for values of sag y calculated with simple approximate equation 2.38. (See example 2.5 and figure 2.7.)

a convergent infinite series, which can be determined by simple division. Thus we obtain for $[1/(x+y)]$ the infinite series expression

$$\left(\frac{1}{x+y}\right) = \frac{1}{x}\left[1 - \left(\frac{y}{x}\right) + \left(\frac{y}{x}\right)^2 - \left(\frac{y}{x}\right)^3 + \cdots\right]$$

If we assume that $|y| < |x|$ for this problem, $y^2 \ll x^2$, $|y|^3 \lll |x|^3$, and so on, are valid relationships. Thus, for the infinite series expansion we consider only the first two terms and obtain the approximation

$$\left(\frac{1}{x+y}\right) \approx \frac{1}{x}\left[1 - \left(\frac{y}{x}\right)\right] \tag{2.40}$$

At first glance, one might wonder how equation 2.40 could result in a simplification of a design equation. It depends, of course, on the particular equation at hand, and an overall simplification of the entire design equation might result from such a transposition. This might be particularly true for the case where the entire equation, having other terms of y also, must be integrated with respect to y.

Let us now consider the decimal error e associated with the use of equation 2.40, which will be determined directly from equation 2.1 as follows:

$$e = \frac{(1/x)[1 - (y/x)] - [1/(x + y)]}{[1/(x + y)]} = \left[1 + \left(\frac{y}{x}\right)\right]\left[1 - \left(\frac{y}{x}\right) - 1\right]$$

$$= 1 - \left(\frac{y}{x}\right)^2 - 1$$

$$\therefore \quad e = -\left(\frac{y}{x}\right)^2 \tag{2.41}$$

Hence if only a 5% negative error could be tolerated in using the approximation in the particular design equation at hand, the values of $|(y/x)|$ must be less than about 0.22 over the range of interest for (y/x).

EXAMPLE 2.7

Mathematical equations containing transcendental functions can often be simplified by considering their infinite series expansions. As a specific example, consider an approximation for sec x as follows:

$$\sec x = \frac{1}{\cos x} = \frac{1}{[1 - (x^2/2!) + (x^4/4!) - (x^6/6!) + \cdots]}$$

$$= 1 + \left(\frac{x^2}{2}\right) + \left(\frac{5x^4}{24}\right) + \cdots$$

Hence, for small values of x expressed in radians, we need merely consider the first two terms in the infinite series expansion as a good approximation. Thus

$$\sec x \approx \left(1 + \frac{x^2}{2}\right) \tag{2.42}$$

Using equation 2.1, the error for the approximation in equation 2.42 is

$$e = \frac{[1 + (x^2/2)] - \sec x}{\sec x} = \left[\left(1 + \frac{x^2}{2}\right)\cos x\right] - 1 \tag{2.43}$$

In equation 2.43, e is the decimal error for using $[1 + (x^2/2)]$ instead of sec x, and x is angle expressed in radians. Thus an angle x of 25° results in an error of only −0.01.

Consider now a further simplification of equation 2.42 effected by using merely the first term in that equation. Hence, for very small values of x, we can use the approximation

$$\sec x \approx 1 \tag{2.44}$$

By equation 2.1, the error associated with equation 2.44 is

$$e = \frac{1 - \sec x}{\sec x} = (\cos x) - 1$$

$$e = (\cos x) - 1 \tag{2.45}$$

Equation 2.45 gives the decimal error for using unity instead of sec x, and typically for an angle x of about 8° we would have an error of -0.01.

The results of studies similar to this example are summarized for cos x, sin x, and tan x in figures 2.9, 2.10, and 2.11 respectively. The details of these derivations are straightforward, and they will not be repeated here.

Simplification of Complex Equations by Curve Fitting on Tabulated or Graphical Data. Often it is not possible to neglect terms in a complex unwieldy equation, because of error considerations. However, the complex equation might be unsuitable for the explicit conduct of the method of optimum design. For such cases, a simplified approximation can generally

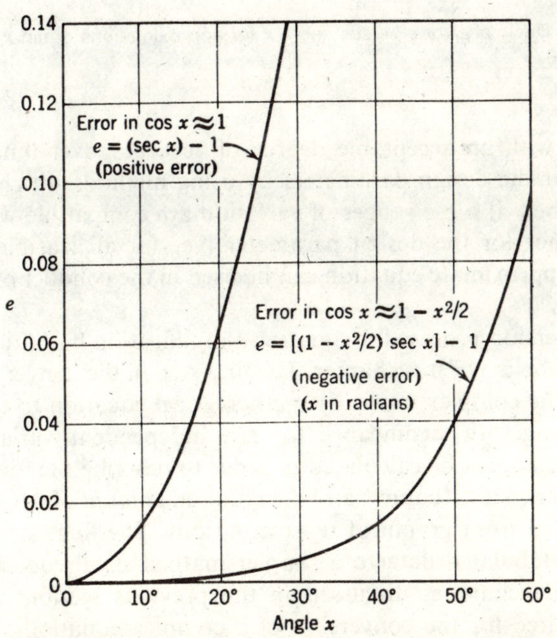

Figure 2.9 Decimal error e versus angle x for approximations of cos x: cos $x \approx 1$; cos $x \approx 1 - x^2/2$.

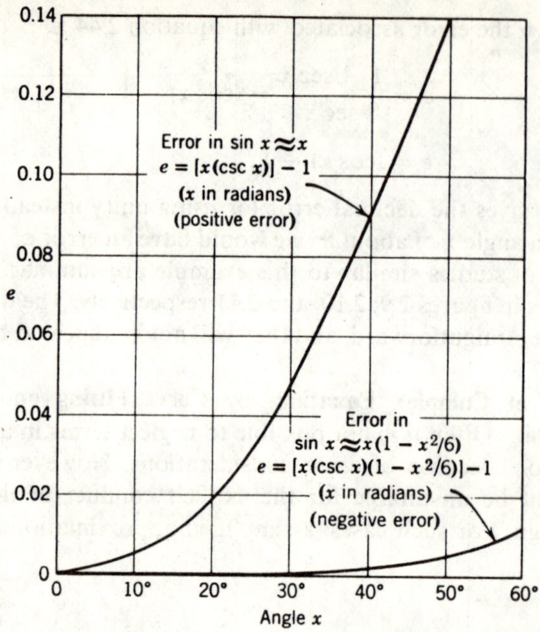

Figure 2.10 Decimal error e versus angle x for approximations of sin x: $\sin x \approx x$; $\sin x \approx x\,(1 - x^2/6)$.

be derived with an acceptable degree of accuracy over limited ranges of variation for the design parameters by using methods which have already been described. If these ranges of variation are compatible with the ranges of significance for the design parameter (i.e., for all feasible designs), the simplified approximate equation can be used in the conduct of the optimum design study.

The conversion process for equation simplification has already been outlined in section 2.3 of this chapter. The first step in the conversion process is to convert the complex unwieldy mathematical equation to tabulated data over the ranges of significance for the independent variables. Enough tabulated data must be calculated in order to describe the functional variations with adequate accuracy, and the required amount of data can generally be determined from graphical representations. The final step is merely to convert the tabulated data to a simpler mathematical equation, using the curve-fitting techniques discussed in the previous section. The pertinent details required for the conversion of a complex equation to a simplified equation with acceptable accuracy over limited ranges have already been thoroughly discussed and they will not be repeated here. An excellent example

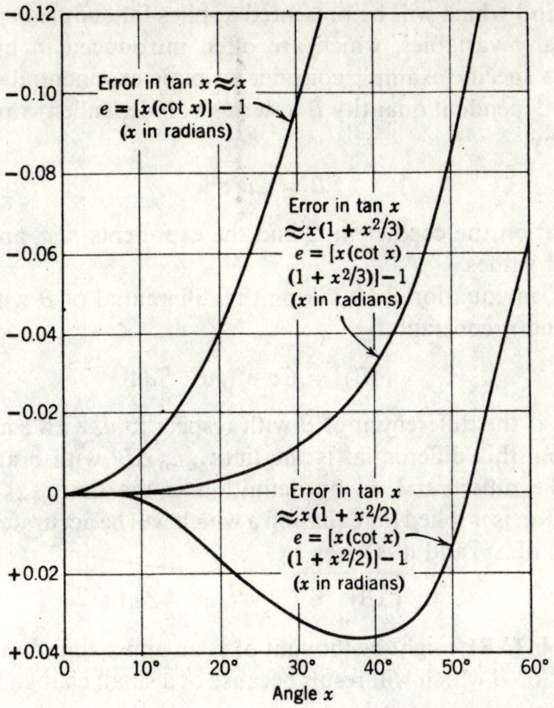

Figure 2.11 Decimal error e versus angle x for approximations of tan x: $\tan x \approx x$; $\tan x \approx x(1 + x^2/3)$; $(\tan x) \approx x(1 + x^2/2)$.

is the conversion of equation 8.7 to equation 8.29a, which is discussed in the preceding section in this chapter and in Chapter 8.

Estimating the Effects of Relative Changes in Independent Variables for Simple Exponential Equations. In the conduct of the explicit method of optimum design, a primary design equation will be developed which expresses the optimum design quantity in terms of selected independent design parameters. *Ideal* specific numerical values will be determined for these design parameters from the developed primary design equation. However, occasionally *ideal* numerical values for design parameters cannot be realized in practice because of the availability of standard sizes only—a restriction beyond the control of the design engineer. It is often necessary for the engineer to make a choice among several standard sizes which are in the *neighborhood* of an ideal optimum design parameter value. The approximation presented here is a useful technique for *rapidly* estimating the effect of small changes in independent design parameter values on the optimum design quantity.

The method which will be presented applies for simple exponential equations of many variables, which are often introduced in optimum design studies. As a specific example consider a simple exponential equation which expresses a dependent quantity B in terms of independent variables u, v, and w as given by

$$B = Cu^p v^q w^r \tag{2.46}$$

In this equation the coefficient C and the exponents p, q, and r can be any real number values.

Consider, in equation 2.46, taking the differential of B with respect to u, keeping v and w constant:

$$(dB)_u = Cv^q w^r (pu^{p-1}\, du) \tag{a}$$

where $(dB)_u$ is the differential of B with respect to u, with v and w constant. By definition, this differential is the $\text{limit}_{\Delta u \to 0}\, \Delta B$, with v and w constant. Likewise, the differential du by definition is the $\text{limit}_{\Delta u \to 0}\, \Delta u$. Hence an approximation is implied for equation a which will be accurate for reasonably small values of Δu, and it is given by

$$(\Delta B)_u \approx Cv^q w^r (pu^{p-1}\, \Delta u) \tag{b}$$

In equation b, $(\Delta B)_u$ might be thought of as an approximation for the change in the value for B which will result because of a small change Δu in the value for u. If we now divide equation b by equation 2.46, we obtain

$$\frac{(\Delta B)_u}{B} \approx \frac{Cv^q w^r (pu^{p-1}\, \Delta u)}{Cu^p v^q w^r}$$

$$\therefore \quad \frac{(\Delta B)_u}{B} \approx p\left(\frac{\Delta u}{u}\right) \tag{c}$$

By definition $(\Delta u/u)$ is the *relative change* in u, which is a unitless decimal percent number. Also, the relative change in B caused by the relative change in u is $(\Delta B)_u/B$, which is a unitless decimal percent number. Hence equation c is a simple approximation which expresses the relative change in B caused by the relative change in u.

A similar discussion for the other independent variables v and w would result in equations similar to a, b, and c. Thus the *total relative change* in B caused by relative changes in u, v, and w would be approximated by

$$\frac{\Delta B}{B} = \frac{(\Delta B)_u}{B} + \frac{(\Delta B)_v}{B} + \frac{(\Delta B)_w}{B}$$

$$\therefore \quad \frac{\Delta B}{B} \approx p\left(\frac{\Delta u}{u}\right) + q\left(\frac{\Delta v}{v}\right) + r\left(\frac{\Delta w}{w}\right) \tag{2.47}$$

This is an approximate equation for estimating rapidly the decimal percent change in B from decimal percent changes in u, v, and w, the function between variables being given by equation 2.46. If a quantity increases algebraically, the relative change is considered positive. On the other hand, if a quantity decreases algebraically, the relative change is considered negative. Expression 2.47 will estimate with good accuracy the relative change in B caused by the relative changes in u, v, and w occurring simultaneously, provided the relative changes are not unreasonably large. Approximation 2.47 is simpler to work with than equation 2.46 for that purpose.

Some Basic Search Techniques for Automated Optimal Design [2.6–2.11]

2.9 Single Variable Search for the Extreme

Suppose quantity Q is expressed as a given function of a single variable x.

$$Q = f(x) \tag{2.48}$$

A commonly encountered case is for Q to have a single minimum value Q_{min} at $x = x_{MN}$ to be determined within a range $x_{min} \leq x \leq x_{max}$, as typically illustrated in figure 2.12. The unimodal function $f(x)$ is not necessarily a continuous one. The problem is to determine a reasonably accurate estimate for the value of x_{MN} for which $Q \approx Q_{min}$, and we wish to do this by an automated sequence of calculations reasonably small in number using a digital computer.

An exhaustive search plan would find the approximate value of x_{MN}, but for a small interval of uncertainty a very large number of function evaluations would generally be necessary. For a function $f(x)$ having several minima in the range of interest, an exhaustive search plan might very well be necessary in such a case. However, for a unimodal function $f(x)$ only one minimum exists for Q and a more efficient sequential search plan could be used requiring relatively few function evaluations. Hence highly efficient search plans have been developed for such cases [2.6–2.8, 2.10]. Interval halving, golden section search, and Fibonacci search plans are typical examples of such techniques.

In addition to $f(x)$ being unimodal, suppose the range of interest for x is uncertain or not readily determined and assume that a digital computer is available for the search calculations. In this event a large step size increment $(\Delta x)_L$ can be chosen as well as a base point x_B for starting the calculations. Hence function evaluations can be made sequentially starting from the base point B at the increment $\Delta x = (\Delta x)_L$ with x either increasing or decreasing

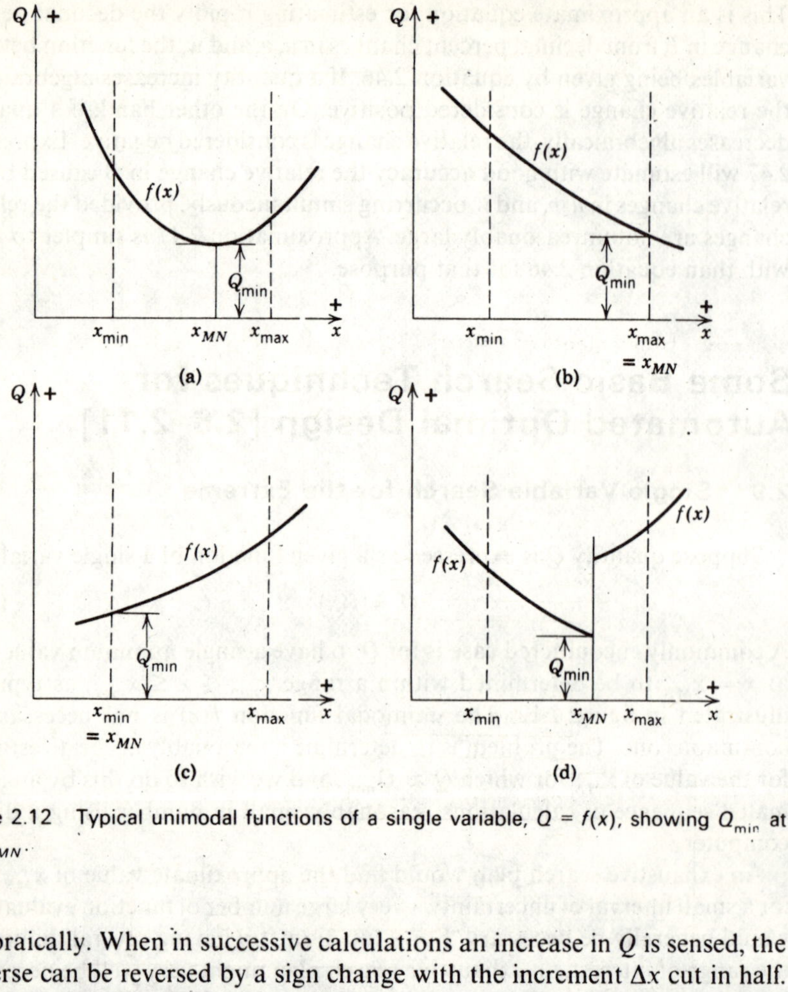

Figure 2.12 Typical unimodal functions of a single variable, $Q = f(x)$, showing Q_{min} at $x = x_{MN}$.

algebraically. When in successive calculations an increase in Q is sensed, the traverse can be reversed by a sign change with the increment Δx cut in half. This procedure is repeated until the interval Δx is reduced to a prescribed small size increment $(\Delta x)_s$ compatible with the desired degree of accuracy for the determination of x_{MN}. In this way the values of x_{MN} and its corresponding Q_{min} are approximately determined in a reasonably efficient automated sequential search plan. The general procedure is depicted in figure 2.13, which also gives a flow chart segment for the sequence of calculations. Incidentally, the type of function readily handled by this simple search plan is typically shown in figure 2.13 with Q_{min} at $dQ/dx = 0$. Specifically, this will be the case for the penalized optimization quantity in the algorithm OPTIGO presented in Chapter 6 for application to automated optimal design of mechanical elements.

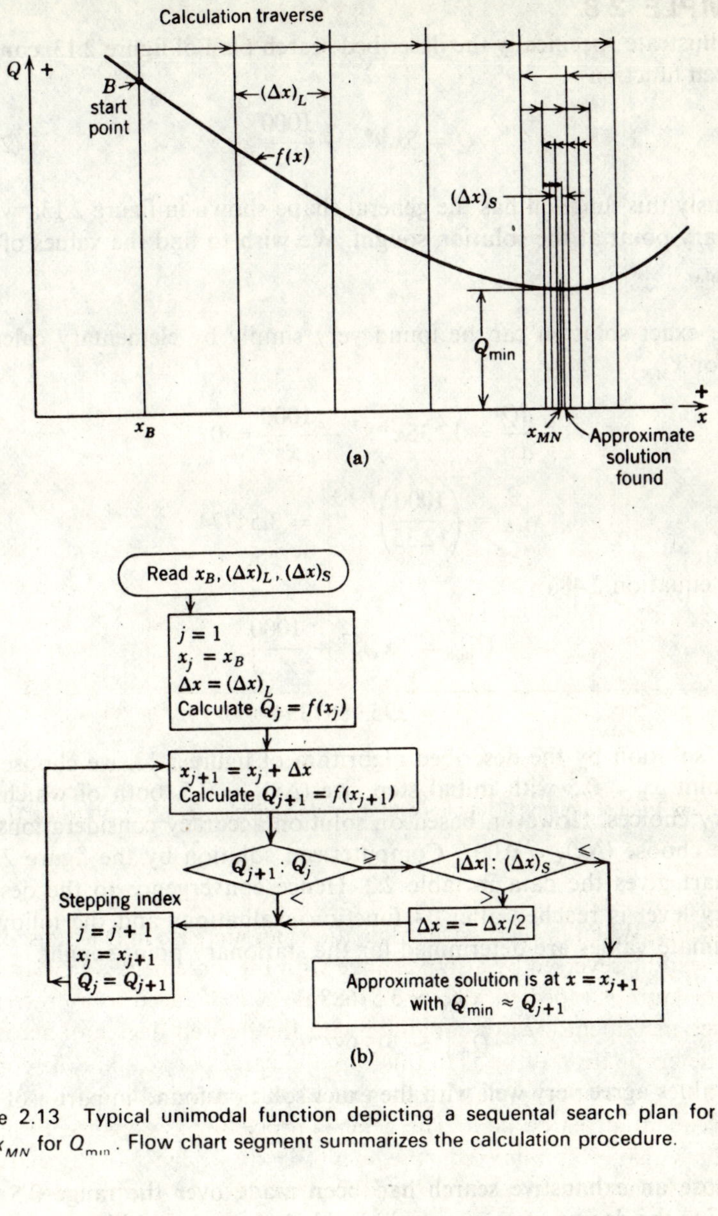

Figure 2.13 Typical unimodal function depicting a sequental search plan for finding $x \approx x_{MN}$ for Q_{min}. Flow chart segment summarizes the calculation procedure.

EXAMPLE 2.8

To illustrate specifically the described search plan of figure 2.13, consider the given function

$$Q = 5x^{2.47} + \frac{1000}{x} \tag{2.48a}$$

Obviously this function has the general shape shown in figure 2.13a with a stationary point at the solution sought. We wish to find the values of x_{MN} and $Q_{\min}$.

(a) The exact solution can be found very simply by elementary calculus. Thus, for x_{MN}

$$\frac{dQ}{dx} = 12.35x^{1.47} - \frac{1000}{x^2} = 0$$

$$x_{MN} = \left(\frac{1000}{12.35}\right)^{1/3.47} = 3.54774$$

and by equation 2.48a

$$Q_{\min} = 5x_{MN}^{2.47} + \frac{1000}{x_{MN}}$$

$$= 395.9867334$$

(b) For solution by the described algorithm of figure 2.13, we choose the start point $x_B = 0.5$ with initial step size $(\Delta x)_L = 0.5$, both of which are arbitrary choices. However, based on solution accuracy considerations for x_{MN} we choose $(\Delta x)_S = 0.001$. Computerized solution by the figure 2.13b flow chart gives the data in table 2.2. Hence convergence to the desired accuracy level is reached after 34 function evaluations and the following approximate values are determined for the stationary point sought.

$$x_{MN} \approx 3.54688$$

$$Q_{\min} \approx 395.9867624$$

These values agree very well with the exact solution found in part a of this example.

Suppose an exhaustive search had been made over the range $0.5 \le x \le 5.0$. For the degree of accuracy obtained above we would have required 4500 function evaluations by such an approach, which would be highly inefficient. On the other hand, the traditional interval halving technique over the same range for this degree of accuracy would have required only 26 func-

TABLE 2.2

Example 2.8 Solution Data from Computer Output

j	x_j	Δx (next step)	Q
1	0.5	0.5	2000.9
2	1.0	0.5	1005.0
3	1.5	0.5	680.28
4	2.0	0.5	527.70
5	2.5	0.5	448.07
6	3.0	0.5	408.75
7	3.5	0.5	396.08
8	4.0	−0.25	403.48
9	3.75	−0.25	397.53
10	3.5	−0.25	396.08
11	3.25	0.125	399.59
12	3.375	0.125	397.18
13	3.5	0.125	396.08
14	3.625	−0.0625	396.22
15	3.5625	−0.0625	395.995
16	3.5	0.03125	396.076
⋮	⋮	⋮	⋮
32	3.548828	−0.0009765	395.9868
33	3.547852	−0.0009765	395.9867
34	3.546875		395.9867624

tion evaluations. A golden section search would be still more efficient with only 19 function evaluations required for this degree of accuracy. However, the very simple described search plan of figure 2.13 is of reasonable efficiency for computer applications, and a simple determination of stationary points with no requirements for range estimates is what we need in the automated optimal design algorithm OPTIGO described in Chapter 6.

2.10 Multivariable Search for the Extreme

For most practical problems we wish to find the coordinates of the design point for which optimization quantity Q is either a maximum or a minimum, where Q is a function of several variables.

$$Q = f(x_1, x_2, \ldots, x_N) = f(x_i), \qquad i = 1, \ldots, N \tag{2.49}$$

Because of later applications in Chapter 6, we will specifically consider the problem of minimizing Q. Incidentally, maximization problems can then be handled very simply by transforming to an equivalent minimization problem by either a sign change or a reciprocation of the optimization function for Q.

Also, because of the introduction of a penalty function as explained in Chapter 6, we will be concerned with the type of problem where the solution point is at a *stationary point*. By definition, a stationary point for the function of equation 2.49 is one where the partial derivatives are all zero, stated as $\partial Q/\partial x_i = 0$ for $i = 1, \ldots, N$. Incidentally, a stationary point might be a local minimum, a local maximum, a saddle point, or an inflection point. Of course, in automated optimal design we search for the particular local minimum which is the global minimum.

A very powerful technique for multivariable search is the method of steepest descent [2.8, 2.11], which we will present in this section. This gradient type technique will be of basic importance for the algorithm OPTIGO which will be covered in Chapter 6 for automated optimal design of mechanical elements. A very general and brief description of the basic approach is given next before considering specific details.

Suppose we have a three dimensional variation expressed by equation 2.49a defining a three dimensional surface with respect to coordinate axes x_1, x_2, and Q.

$$Q = f(x_1, x_2) \tag{2.49a}$$

Hence a two dimensional search would be required to find the independent coordinate values of (x_1, x_2) which will minimize Q. Contour curves of constant Q value are sketched in figure 2.14, where P_0 is the stationary point sought for minimization of Q. An arbitrary start point called a base point B_1 could be chosen, from which we will establish the direction of steepest descent along ray $r = 1$ in the figure. A search for minimum Q could be made along this first gradient line, thereby giving a second base point B_2 as shown. From this base point a second gradient line of steepest descent is established, as depicted by ray $r = 2$ in the figure. A search for minimum Q along that ray determines the third base point B_3, from which the third gradient line or ray $r = 3$ is established. The procedure is repeated successively until we converge to the neighborhood of P_0, the point sought in figure 2.14.

In brief summary, there are two basic steps in the iterative procedure described. First, the *direction of steepest descent* must be determined from a given base point B_r, and this direction line we will call a gradient line or a ray r. Second, we must *search for the extreme along such a ray* to find the point of minimum Q on the ray. Let us now consider the details of these basic steps, using the two dimensional case only for purposes of simplifying the explanations.

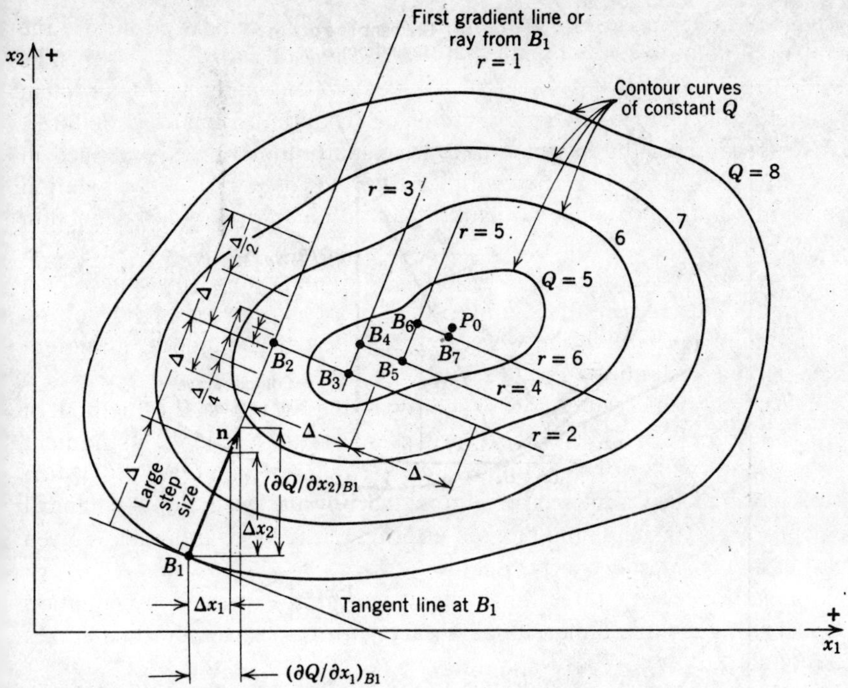

Figure 2.14 Contour curves of constant Q showing two dimensional search plan by method of steepest descent for three dimensional function $Q = f(x_1, x_2)$.

Direction of Steepest Descent. Along a two dimensional contour curve of constant Q we have

$$dQ = \frac{\partial Q}{\partial x_1} dx_1 + \frac{\partial Q}{\partial x_2} dx_2 = 0$$

which gives the special relation

$$dx_1 = -dx_2 \frac{\partial Q/\partial x_2}{\partial Q/\partial x_1}$$

Hence in terms of the unit vectors $\mathbf{i}_1$ and $\mathbf{i}_2$ we can define a differential tangential vector $\mathbf{t}$ as follows.

$$\mathbf{t} = dx_1 \mathbf{i}_1 + dx_2 \mathbf{i}_2$$

$$\mathbf{t} = \left(-\frac{\partial Q/\partial x_2}{\partial Q/\partial x_1} \mathbf{i}_1 + \mathbf{i}_2 \right) dx_2 \tag{2.50}$$

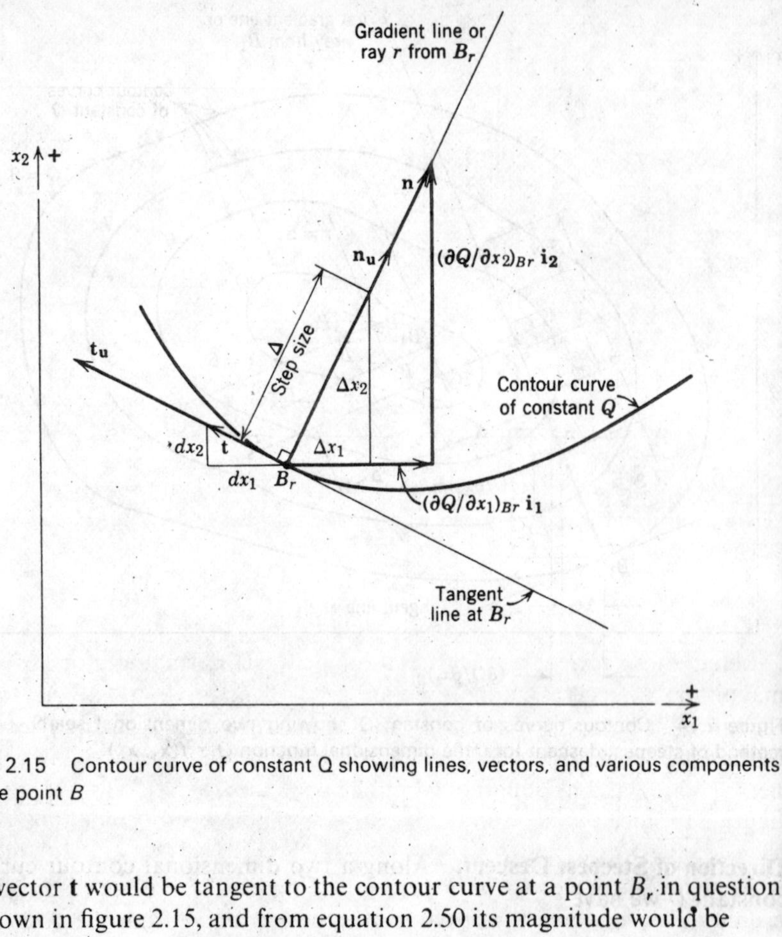

Figure 2.15 Contour curve of constant Q showing lines, vectors, and various components at base point B

The vector **t** would be tangent to the contour curve at a point B_r in question as shown in figure 2.15, and from equation 2.50 its magnitude would be

$$t = dx_2\sqrt{\left(\frac{\partial Q/\partial x_2}{\partial Q/\partial x_1}\right)^2 + 1} \tag{a}$$

Next we divide equation 2.50 by equation a and simplify, giving the tangential *unit* vector $\mathbf{t_u}$.

$$\mathbf{t_u} = \frac{-(\partial Q/\partial x_2)\mathbf{i_1} + (\partial Q/\partial x_1)\mathbf{i_2}}{\sqrt{(\partial Q/\partial x_2)^2 + (\partial Q/\partial x_1)^2}} \tag{2.51}$$

The unit vector $\mathbf{t_u}$ would also be tangent to the contour curve at point B_r as shown in figure 2.15, since it was derived directly from **t** with only a magnitude change.

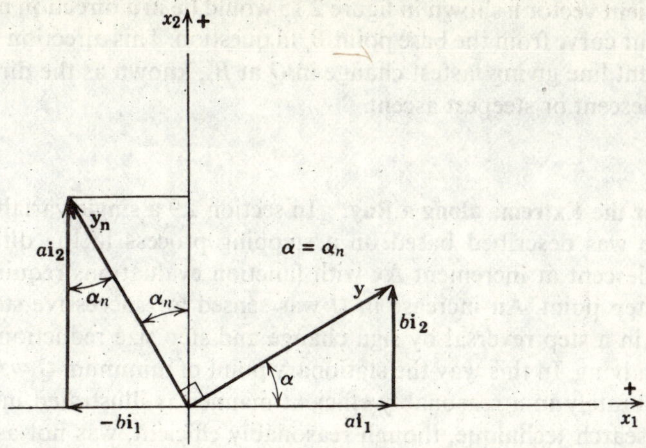

Figure 2.16 Normal vectors **y** and $\mathbf{y_n}$ of equal magnitudes showing relation between components.

Let us now review a basic relation between normal vectors of equal magnitude. In general, suppose we are given a vector $\mathbf{y} = a\mathbf{i}_1 + b\mathbf{i}_2$, having therefore the components (a, b) as shown in figure 2.16. A normal vector $\mathbf{y_n}$ of the same magnitude would be formed as shown in the figure with components $(-b, a)$, and its equation would be $\mathbf{y_n} = -b\mathbf{i}_1 + a\mathbf{i}_2$. This we seę from figure 2.16 since $\alpha = \alpha_n$, and the magnitude of both vectors would be $y = y_n = \sqrt{a^2 + b^2}$.

Applying the preceding basic principle to figure 2.15, we can readily derive a unit vector $\mathbf{n_u}$ which is normal to the contour curve from the point B_r in question. We merely exchange the components of $\mathbf{t_u}$ in equation 2.51 and change the sign of one. Hence the normal *unit* vector $\mathbf{n_u}$ is

$$\mathbf{n_u} = \frac{(\partial Q/\partial x_1)\,\mathbf{i}_1 + (\partial Q/\partial x_2)\,\mathbf{i}_2}{\sqrt{(\partial Q/\partial x_1)^2 + (\partial Q/\partial x_2)^2}} \tag{2.52}$$

A simpler vector also normal to the contour curve would be obtained by dropping the denominator in equation 2.52, which is merely a magnitude change for $\mathbf{n_u}$. Hence we obtain the normal vector **n** at point B_r expressed by equation 2.53, which could also be referred to as the gradient vector **del Q** (i.e., $\nabla \mathbf{Q}$).

$$\mathbf{n} = \left(\frac{\partial Q}{\partial x_1}\right)_{B_r} \mathbf{i}_1 + \left(\frac{\partial Q}{\partial x_2}\right)_{B_r} \mathbf{i}_2 \tag{2.53}$$

This gradient vector **n** shown in figure 2.15 would be in a direction normal to the contour curve from the base point B_r in question. This direction would be the gradient line giving fastest change in Q at B_r, known as the direction of steepest descent or steepest ascent.

Search for the Extreme along a Ray. In section 2.9 a single variable search technique was described based on a stepping process in the direction of steepest descent at increment Δx with function evaluations required for Q at each step point. An increase in Q was sensed for successive step points resulting in a step reversal by sign change and step size reduction through interval halving. In this way the stationary point of minimum Q was located quite accurately in a reasonably efficient manner as illustrated in example 2.8. The search technique, though reasonably efficient, was not as efficient as possible, but it has the advantage of being very simple and not requiring an estimate for the initial interval of uncertainty.

A similar simple search plan can be extended to the multivariable function problem by stepping from a base point B_r along a ray r of steepest descent at increment Δ. Referring again to figure 2.15, suppose we wish to take a given step size Δ along the gradient line or ray r. The step size of magnitude Δ could be resolved into components Δx_1 and Δx_2 for the two dimensional case as shown in the figure. Similarly, the magnitude of gradient vector **n** by equation 2.53 is $\sqrt{(\partial Q/\partial x_1)^2_{B_r} + (\partial Q/\partial x_2)^2_{B_r}}$ with components $(\partial Q/\partial x_1)_{B_r}$ and $(\partial Q/\partial x_2)_{B_r}$. Hence, by similar triangles in figure 2.15, step size components along a ray in the direction of steepest descent from B_r could be calculated as follows.

$$\Delta x_1 = \frac{-(\partial Q/\partial x_1)_{B_r} \cdot \Delta}{\sqrt{(\partial Q/\partial x_1)^2_{B_r} + (\partial Q/\partial x_2)^2_{B_r}}} \tag{2.54a}$$

$$\Delta x_2 = \frac{-(\partial Q/\partial x_2)_{B_r} \cdot \Delta}{\sqrt{(\partial Q/\partial x_1)^2_{B_r} + (\partial Q/\partial x_2)^2_{B_r}}} \tag{2.54b}$$

Incidentally, the minus sign is introduced in these equations to assure a direction of steepest descent instead of steepest ascent from base point B_r to start in the search.

The described procedure of searching for minimum Q along a ray of steepest descent is summarized for the two dimensional case in the flow chart segment of figure 2.17 similar to that in figure 2.13. The basic procedure is specifically illustrated in the following two simple examples, both having exact solutions readily determined by elementary calculus for comparative purposes.

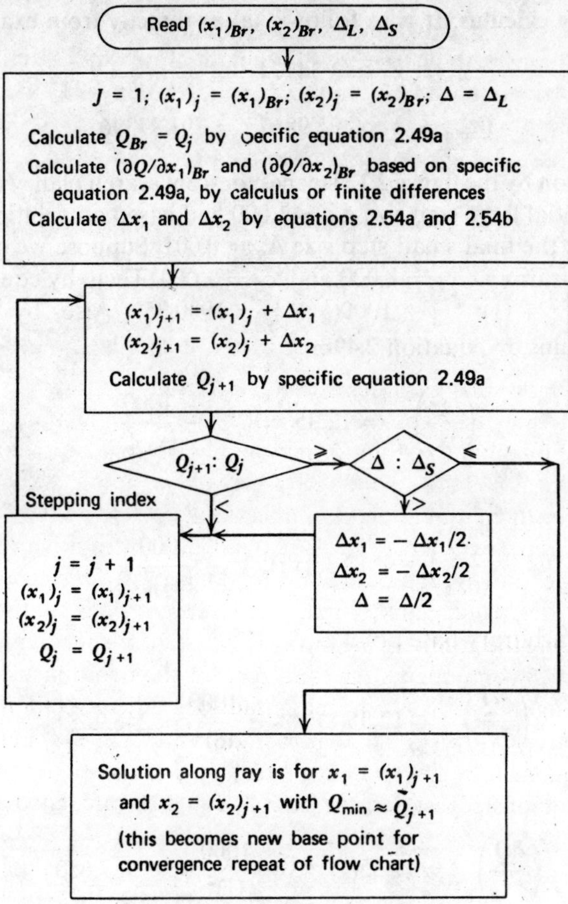

Figure 2.17 Flow chart segment to determine Q_{min} along a ray r in two dimensional search case.

EXAMPLE 2.9

We wish to find the stationary point of Q_{min} for the given function

$$Q = 5(x_1^{2.47} + x_2^{2.47}) + 1000\left(\frac{1}{x_1} + \frac{1}{x_2}\right) \tag{2.49b}$$

(a) In reference to example 2.8 part a, comparing equations 2.48a and 2.49b, and recognizing symmetry in the latter, we at once know the exact solution by elementary calculus. It is as follows, taken directly from example 2.8.

$$x_1 = x_2 = 3.54774$$

$$Q_{\min} = 2 \times 395.98673 = 791.97346$$

(b) For solution by the figure 2.17 steepest descent search plan, we arbitrarily choose the initial large step size $\Delta_L = 1.000$ and based on solution accuracy requirements, the final small step size $\Delta_S = 0.001$. Suppose we choose base point B_1 arbitrarily at $x_1 = 8.000$ and $x_2 = 1.000$. Then, by equation 2.49b $Q_{B_1} = 5[(8)^{2.47} + (1)^{2.47}] + 1000(\frac{1}{8} + \frac{1}{1}) = 1980.359$. Also, by use of elementary calculus on equation 2.49b,

$$\left(\frac{\partial Q}{\partial x_1}\right)_{B_r} = 12.35(x_1)_{B_r}^{1.47} - \frac{1000}{(x_1)_{B_r}^2}$$

and

$$\left(\frac{\partial Q}{\partial x_2}\right)_{B_r} = 12.35(x_2)_{B_r}^{1.47} - \frac{1000}{(x_2)_{B_r}^2}$$

Hence, at the arbitrary base point B_1,

$$\left(\frac{\partial Q}{\partial x_1}\right)_{B_1} = 12.35(8)^{1.47} - \frac{1000}{(8)^2} = 246.923$$

and

$$\left(\frac{\partial Q}{\partial x_2}\right)_{B_1} = 12.35(1)^{1.47} - \frac{1000}{(1)^2} = -987.650$$

By equations 2.54a and 2.54b, the component step increments for the first ray are therefore as follows.

$$\Delta x_1 = \frac{-(246.923) \times (1.000)}{\sqrt{(246.923)^2 + (-987.650)^2}} = -0.242545$$

$$\Delta x_2 = \frac{-(-987.650) \times (1.000)}{\sqrt{(246.923)^2 + (-987.650)^2}} = 0.970140$$

The preceding large step increment components Δx_1 and Δx_2 are used to start for the first ray $r = 1$, and the proportions of these components for succeeding steps will keep us on this first gradient line. Successive calculations by the figure 2.17 flow chart applied for three rays are summarized in table

2.3. Hence, if the iterations are stopped at this point, solution to the problem by the steepest descent search plan is as follows.

$$x_1 = 3.5489, \qquad x_2 = 3.5494, \qquad Q_{\min} = 791.974$$

This approximate solution agrees very well with the exact solution found independently in part a.

TABLE 2.3

Summary of Iterative Calculations for Example 2.9b

j	Δ (next step)	$(x_1)_j$	$(x_2)_j$	Q_j
1	1.000	8.000	1.000	1980.359
2	1.000	7.757	1.970	1451.269
3	1.000	7.515	2.940	1273.555
4	1.000	7.272	3.910	1210.272
5	−0.500	7.030	4.881	1215.945
6	−0.500	7.151	4.395	1205.644
7	0.250	7.272	3.910	1210.272
⋮	⋮	⋮	⋮	⋮
28	−0.001953	7.1757	4.2970	1205.329
29	0.0009766	7.1762	4.2951	1205.329
30	0.0009766	7.1760	4.2960	1205.329
31		7.1757	4.2970	1205.329
For next ray, to start $\Delta x_i = -0.970085$ and $\Delta x_2 = -0.242764$*				
1	1.000	7.1757	4.2970	1205.329
2	1.000	6.2056	4.0542	1020.573
3	1.000	5.2356	3.8114	888.002
⋮	⋮	⋮	⋮	⋮
36	0.0009766	3.5872	3.3989	792.913
37	0.0009766	3.5862	3.3987	792.913
38		3.5853	3.3985	792.913
For next ray, to start $\Delta x_1 = -0.234395$ and $\Delta x_2 = 0.972141$*				
1	1.000	3.5853	3.3985	792.913
2	−0.500	3.3509	4.3706	817.371
3	−0.500	3.4681	3.8845	796.437
⋮	⋮	⋮	⋮	⋮
29	0.0009766	3.5493	3.5475	791.974
30	0.0009766	3.5491	3.5484	791.974
31		3.5489	3.5494	791.974

Calculated as illustrated in text for first ray.

EXAMPLE 2.10

We wish to find the values of x_1 and x_2 defining the point of $Q_{\min}$ for the given function

$$Q = \frac{25}{x_1 x_2} + 3x_1 x_2 \tag{2.49c}$$

(a) Exact solution is readily determined by calculus as follows.

$$\frac{\partial Q}{\partial x_1} = -\frac{25}{(x_1{}^2 x_2)} + 3x_2 = 0$$

$$\frac{\partial Q}{\partial x_2} = -\frac{25}{(x_1 x_2{}^2)} + 3x_1 = 0$$

$$\therefore \quad (x_1 x_2) = (\tfrac{25}{3})^{1/2} = 2.886751$$

Hence *any* values of x_1 and x_2 which have this product value will give the same minimum value for $Q_{\min} = 25/(\frac{25}{3})^{1/2} + 3(\frac{25}{3})^{1/2} = 17.320508$. Thus for this problem there are an infinite number of solutions all giving the same value for $Q_{\min}$.

(b) For solution by the steepest descent search plan of figure 2.17, suppose we choose the initial large step size $\Delta_L = 0.200$ and based on solution accuracy requirements the final small step size $\Delta_S = 0.001$. As determined in part a, there are an infinite number of (x_1, x_2) combinations all giving the same $Q_{\min}$ value. Hence the particular combination of (x_1, x_2) found by the figure 2.17 search plan will depend on the initial base point arbitrarily chosen.

1. Suppose we choose $(x_1)_{B1} = 4.000$ and $(x_2)_{B1} = 1.000$. Then by equation 2.49c

$$Q_{B1} = \frac{25}{(4 \times 1)} + 3(4 \times 1) = 18.25$$

$$\left(\frac{\partial Q}{\partial x_1}\right)_{B1} = -\frac{25}{(4^2 \times 1)} + 3(1) = 1.4375$$

$$\left(\frac{\partial Q}{\partial x_2}\right)_{B1} = -\frac{25}{(4 \times 1^2)} + 3(4) = 5.7500$$

By equations 2.54a and 2.54b,

$$\Delta x_1 = -\frac{(1.4375)(0.2)}{\sqrt{(1.4375)^2 + (5.75)^2}}$$

$$= -0.0485071$$

$$\Delta x_2 = -\frac{(5.75)(0.2)}{\sqrt{(1.4375)^2 + (5.75)^2}}$$

$$= -0.1940285$$

The preceding large step increment components Δx_1 and Δx_2 are used to start for the first ray, and successive calculations by the figure 2.17 flow chart are summarized in table 2.4. Note for the solution point on the first ray, $x_1 x_2 = 3.9335 \times 0.7340 = 2.887$ and $Q_{min} = 17.3205$. These values agree very well with the exact solution found independently in part a.

2. Suppose we choose a different initial base point with coordinates $(x_1)_{B1} = 3.000$ and $(x_2)_{B1} = 5.000$. Applying the figure 2.17 flow chart as for part b.1, we obtain $Q_{B1} = 46.667$; $(\partial Q/\partial x_1)_{B1} = 14.4444$; $(\partial Q/\partial x_2)_{B1} = 8.6667$; $\Delta x_1 = -0.1714986$; and $\Delta x_2 = -0.1028992$. With these large step increment components Δx_1 and Δx_2 to start for the first ray, we obtain the iterative calculations by the figure 2.17 flow chart as summarized in table 2.5. Note for the solution point on the first ray, $x_1 x_2 = 0.7853 \times 3.6712 = 2.883$ and $Q_{min} = 17.3205$. These values agree quite well with the exact solution found independently in part a.

Multivariable Search Summary. The steepest descent search plan described above and illustrated for the two dimensional search case can be directly applied to an N-dimensional search for the stationary point of minimum Q of equation 2.49. Coordinates x_i are chosen for a start point called base point B_1. From this point steps are taken of given magnitude Δ in the N-dimensional direction of steepest descent along the first gradient line or ray $r = 1$. When successive steps indicate an increase in Q, the step size Δ of Δx_i components is cut in half and the direction of traverse is reversed by a sign change for the Δx_i components. This is continued until step size Δ has been reduced to a

TABLE 2.4

Summary of Iterative Calculations for Example 2.10, Part b.1

r	j	Δ (next step)	$(x_1)_j$	$(x_2)_j$	Q_j
1	1	0.200	4.0000	1.0000	18.2500
	2	0.200	3.9515	0.8060	17.4042
	3	−0.100	3.9030	0.6119	17.6325
	4	−0.100	3.9272	0.7090	17,3318
	5	0.050	3.9515	0.8060	17.4042
	6	0.050	3.9394	0.7575	17.3300
	7	−0.025	3.9272	0.7090	17.3318
	⋮	⋮	⋮	⋮	⋮
	20	−0.0015625	3.9333	0.7332	17.3205
	21	0.00078125	3.9337	0.7347	17.3205
	22		3.9335	0.7340	17.3205

TABLE 2.5

Summary of Iterative Calculations for Example 2.10, Part b.2

r	j	Δ (next step)	$(x_1)_j$	$(x_2)_j$	Q_j
1	1	0.200	3.0000	5.0000	46.667
	2	0.200	2.8285	4.8971	43.359
	3	0.200	2.6570	4.7942	40.177
	4	0.200	2.4855	4.6913	37.125
⋮	⋮	⋮	⋮	⋮	⋮
	14	0.200	0.7705	3.6623	17.325
	15	−0.100	0.5990	3.5594	18.121
	16	−0.100	0.6848	3.6109	17.529
⋮	⋮	⋮	⋮	⋮	⋮
	41	0.00078125	0.7866	3.6720	17.3205
	42	0.00078125	0.7859	3.6716	17.3205
	43		0.7853	3.6712	17.3205

prescribed small value Δ_S comparable with the desired degree of accuracy for the independent coordinates of the solution point. Thus the point of minimum Q is located along the first ray, and this becomes the base point B_2 for the second ray $r = 2$. The procedure is repeated until successive base points B_r and B_{r+1} are sufficiently close to assure accuracy of location for the stationary point sought.

For the general case of an N-dimensional search as described above, step size components Δx_i along a ray from a base point B_r would be calculated in terms of given Δ and the partial derivatives $(\partial Q/\partial x_i)_{B_r}$ by equation 2.55 for $i = 1, 2, \ldots, N$. The procedure would be similar to the illustrated applications of equations 2.54a and 2.54b.

$$\Delta x_i = \frac{-(\partial Q/\partial x_i)_{B_r} \cdot \Delta}{\left[\sum_{i=1}^{N} (\partial Q/\partial x_i)_{B_r}^2\right]^{1/2}} \tag{2.55}$$

In later applications for the algorithm OPTIGO of Chapter 6, the partial derivatives $(\partial Q/\partial x_i)_{B_r}$ will be calculated by the method of finite differences which was reviewed in section 2.6. Incidentally, given step size magnitude Δ is related to the Δx_i components by equation 2.56.

$$\Delta = \left[\sum_{i=1}^{N} (\Delta x_i)^2\right]^{1/2} \tag{2.56}$$

The mathematical background now covered for an N-dimensional gradient based search plan will be of basic importance for development of the algorithm OPTIGO in Chapter 6, which will be applied to automated optimal design.

REFERENCES

2.1 Walter Rudin, *Principles of Mathematical Analysis*, McGraw-Hill, New York, 1953.

2.2 R. C. Johnson, "Method of Finite Difference in Cam Design—Accuracy—Applications," *Machine Design*, Nov. 14, 1957, pp. 159–161.

2.3 W. Volk, *Applied Statistics for Engineers*, McGraw-Hill, New York, 1958, Chapter 8.

2.4 H. J. Larson, *Statistics: An Introduction*, Wiley, New York, 1975, Chapter 10.

2.5 *HP-25 Applications Programs*, Hewlett-Packard Co., Cupertino, California, 1975, pp. 87–100.

2.6 C. R. Mischke, *An Introduction to Computer-Aided Design*, Prentice-Hall, Englewood Cliffs, N.J., 1968, Chapter 4.

2.7 G. R. Walsh, *Methods of Optimization*, Wiley, New York, 1975, pp. 84–93.

2.8 W. F. Stoecker, *Design of Thermal Systems*, McGraw-Hill, New York, 1971, Chapter 8.

2.9 R. C. Johnson, *Mechanical Design Synthesis—Creative Design and Optimization*, 2nd ed., R. E. Krieger Publishing Co., Huntington, N.Y. 11743, 1978, Chapter 8 and Appendix C.

2.10 D. J. Wilde, *Globally Optimal Design*, Wiley, New York, 1978, pp. 94, 221–225.

2.11 J. N. Sidall, *Analytical Decision-Making in Engineering Design*, Prentice-Hall, Englewood Cliffs, N.J., 1972, pp. 163–173.

EXERCISES

2.1. For the given functions find the first and second derivatives of y with respect to x at the specified points in two ways as follows: (1) by elementary calculus, and (2) by the method of finite differences. For each case calculate the error of the finite difference solution.

(a) $y = 6x^3 - 4x^2 + 9x - 17$

1. at $x = 83$, using $\Delta x = 7$ for finite differences
2. at $x = -5$, using $\Delta x = 23$ for finite differences

(b) $y = 3.5 \cos x$

1. at $x = 34°$, using $\Delta x = 5°$ for finite differences
2. at $x = 79°$, using $\Delta x = 3°$ for finite differences

(Hint: For trigonometric functions with finite differences, be sure to express both x and Δx in radians.)

(c) $y = 1.75 \sin^2 x + 3x^5 - 5x^3 + 10$

1. at $x = 0.85521$ rad, using $\Delta x = 0.0349$ radian for finite differences
2. at $x = 0.5$ rad, using $\Delta x = 0.05$ radian for finite differences

2.2. For variation studies in optimum design, it is desirable to derive simple equations valid over a given range of interest for expressing given functions available in the form of complex equations, graphs, or tabulated data. Using curve-fitting techniques of this chapter, derive simple equations for the following given functions. Assume that an error $|e| \leq 0.05$ is acceptable for the intended use of the derived simple equation. For each case calculate the greatest error e_{max} for your derived simple equation.

(a) For helical springs, Wahl correction factor K_{ts} is often expressed as a function of spring index (D/d) either in the form of equation 8.34a of Chapter 8 or a given graph commonly found in the machine design literature. The range of interest is $3 \leq (D/d) \leq 20$.

Answer. As one possibility, see equation 8.34b of Chapter 8 for author's solution.

(b) The error e for approximation $\cos x \approx 1 - x^2/2$ is given by the equation and graph shown in figure 2.9. Derive a simpler approximate equation for the error e as a function of angle x in degrees, valid over the range $0 \leq x \leq 60°$.

(c) Stress concentration factor K_{ts} versus (r/d) for a torsion bar with filleted keyway is available in the form of the graph shown in figure 7.10. The range of interest is $0.01 \leq (r/d) \leq 0.12$.

(d) Stress concentration factor K_t versus (r/d) and (D/d) for a notched tensile bar is available in the graph of figure 3.3. Derive a simple equation $K_t = f[(r/d), (D/d)]$ valid for the range of interest $0.02 \leq (r/d) \leq 0.28$ and $1.01 \leq (D/d) \leq 3$.

(e) Suppose the following tabulated data are available for the function $y = f(x)$. The range of interest is $0 \leq x \leq 5$.

x	y
0	7.340
1	9.902
2	11.798
3	13.359
4	14.711
5	15.917

Answer. A very accurate solution is $y = 7.34(1 + x)^{0.432}$.

2.3. Write a computer program for the simple interval halving steepest descent search plan of figure 2.13. (If you wish, a program for a programmable calculator would be acceptable.) Solve the problem of example 2.8, choosing different start points and different step increments, and compare your solutions with those of example 2.8.

2.4. Write a computer program for the simple two dimensional search plan of figure 2.17. (If you wish, a program for a programmable calculator would be acceptable.) Solve the problems of examples 2.9 and 2.10, choosing different start points and different stepping increments. Compare your answers with those in examples 2.9 and 2.10.

2.5. For the tray design problem of Chapter 1 we have specified $V = 0.4\ \text{m}^3$ and $H = 0.2$ m. Using the figure 2.13 search plan, find the ideal optimum design based on equation 1.3b for minimizing $C_m/(cT)$.

Answer. Exact solution is $b = 1.41421m$, with $l = 1.41421m$ then obtained from equation 1.1.

2.6. Annual operating cost U is to be minimized for an electrical transmission line system as described in example 8.7 of reference 2.9, with

$$U = \frac{21.9 \times 10^7}{E^2C} + 3.9 \times 10^6 C + 10^3 E$$

In this equation, E is the line voltage in kilovolts, and C is the line conductance in mhos. Apply the steepest descent search plan of figure 2.17 with several successive rays if necessary to find the optimum values for E and C.

Answer. Exact solution, found in example 8.7 of reference 2.9, is $E = 242$ kV and $C = 0.0310$ mho.

2.7. A solid heated cylinder is to be designed for minimization of initial heat transfer rate q (Btu/hr) by convection and radiation. Heat to be stored in the cylinder is a specified amount Q_s (Btu). The problem is reduced to finding the diameter D (in feet) which minimizes $(q/C_3) = Q_s/(C_2D) + D^2/4$. [This is equation (8.50) of reference 2.9. Although with a different objective for optimization, the background for this exercise is presented in that reference.] Using the method of steepest descent (figure 2.13 suggested), find the optimum value of D for specified values of $Q_s = 2$ Btu and $C_2 = 0.4\ \text{Btu/ft}^3$.

Answer. Exact solution is by equation (8.51) of reference 2.9, giving $D = (2 \times 2/0.4)^{1/3} = 2.1544$ ft.

3

Manufacturing Errors and Product Performance

The purpose of this chapter is to present some information pertaining to several important practical problems which confront practicing design engineers. First, this concentrated review should be helpful in acquiring a recognition for places in mechanical design where the effects of manufacturing errors are significant. Hence the perplexing problem of knowing whether or not to specify high degrees of manufacturing accuracy will be better understood. Second, this review showing the possible gross effects from minute manufacturing errors should provide a foundation for a better understanding of the need for factor of safety, discussed in Chapter 5, which is often a confusing problem to student engineers. Finally, the material in this chapter will provide some basis for estimating required surface finishes and manufacturing accuracies on mechanical elements for cases where the effects of minute errors are significant.

3.1 Significant Types of Manufacturing Errors

To define a mechanical element uniquely, it is necessary to specify its geometry completely as well as its material. To manufacture a mechanical element satisfactorily, it is necessary to obtain a finished product whose geometry is acceptably close to the theoretically perfect shape. Associated with any manufacturing process is an inherently unavoidable degree of error which can either be significant or insignificant. Perfection in the absolute sense is physically unobtainable, and every mechanical element will have deviations from the theoretical geometry to some extent at least, even though these deviations might be microscopic or even submicroscopic in magnitude.

Whether or not a manufacturing error is significant depends on two basic considerations. The first consideration depends on whether the manufacturing error significantly affects product performance. The second depends on whether the manufacturing error significantly affects product life. Hence the criterion for establishing whether a manufacturing error is significant is based on whether its *effect* on product behavior is significant. Thus, for many cases, gross manufacturing errors can exist on a mechanical element with no significant effects on product performance. On the other hand, for many other cases, the existence of *minute* manufacturing errors can result in gross effects on product performance. Hence the *effect* rather than the cause is of primary importance in determining whether or not a manufacturing error is significant.

In the design of real mechanical elements, the objective should be to specify and to control manufacturing errors to an extent which will guarantee acceptable product performance and product life. In addition, product cost is of paramount importance. Hence it is important to keep in mind that manufacturing errors which are difficult to control should be limited by close tolerances *only* if their *effect* on product performance or on product life is significant.

Let us consider some general types of places in mechanical elements where minute manufacturing errors can result in gross effects on product performance or on product life. First, there are many places in mechanical assemblages where fits and relative dimensions between mechanical elements are very important. They are characterized by sliding members or geometrical relationships where the possibility of both interference and looseness must be controlled. The problems of fit and relative dimensions are very important in mechanical design work, but generally they can be solved merely by using practical knowledge and good judgment. Therefore we shall not consider such problems in this book. Second, manufacturing errors can be very significant for places where mating curved surfaces in contact transmit loads. Specific examples of this type are innumerable in mechanical design: mating gear teeth, mating surfaces of balls or rollers against their inner and outer races in bearings, and mating surfaces between cams and followers in general are a few. This type is characterized by the existence of contact stresses, which might be grossly affected by the actual presence of minute displacement errors normal to the ideal mating surfaces introduced during manufacturing. The third type of place of interest is characterized by bodies under confined motion, as in mechanisms for machinery. For such cases, minute manufacturing errors on the confining mechanical elements can result in gross accelerations of the moving bodies, which are, of course, possibly accompanied by large inertia forces and dynamic stresses. Again, examples of this type are innumerable, as in gears, cams, Geneva mechanisms, and mechanisms in general where bearings are required. The final type of place of interest is characterized by manufacturing flaws on free surfaces. Such flaws might be introduced during almost any stage of the fabrication process for the mechanical element. Typical of such flaws is the inherently unavoidable introduction of manufacturing tool marks, which might be very small scratch-type imperfections. As another example, minute flaws might exist because of imperfections in the material, which can either be at the free surfaces or within the material itself. However, minute flaws resulting from nonperfect material homogeneousness are generally more significant at surface points than at internal points. This final type of place of interest is significant because of fatigue failure possibilities in springs, shafts, tensile bars, gears, and innumerable other mechanical elements.

The manufacturing errors of interest in this book are significant because of their *effects* on important stresses and on accelerations, including *dynamic* forces and stresses. Fundamentally related to stresses and accelerations is a consideration of curvature. We shall now consider the relationship of stress and acceleration to curvature for some general examples. Then we shall show that minute uncontrollable displacement errors on surfaces can result in gross variations in curvature for those surfaces. Finally, we shall consider the effects of such gross variations in curvature on stresses and on accelerations for real mechanical elements, including such important possibly alleviating considerations as notch sensitivity for stresses and elastic deformations for accelerations, as related to *actual* product performance.

3.2 Some Relationships of Stress and Acceleration to Curvature

Let us begin by considering the radius of curvature of a given plane curved line as shown in figure 3.1, where 0 designates the point under consideration. Now draw a straight line tangent to the curve at point 0 and arbitrarily orient the x, y axes so that the x axis is parallel to this tangent line, as shown in figure 3.1. From basic calculus, the radius of curvature, at point 0 ρ_0, for the given curved line is

$$\rho_0 = \frac{[1 + (dy/dx)_0{}^2]^{3/2}}{(d^2y/dx^2)_0} \tag{a}$$

However, from our choice of the coordinate axes, at the point under consideration the slope of the given curve is zero. Therefore

$$\left(\frac{dy}{dx}\right)_0 = 0$$

Hence ρ_0 is given by equation 3.1, provided the x axis is parallel to the tangent straight line to the curve at that point.

$$\rho_0 = \frac{1}{(d^2y/dx^2)_0} \tag{3.1}$$

Equation 3.1 is a good approximation for radius of curvature even if (dy/dx) is not exactly zero at the point under consideration. This is obvious from equation a for exact radius of curvature, which shows that the error in equation 3.1 is small for reasonably small values of $(dy/dx)_0$.

Let us now define what we mean by *curvature*, as related to *radius of curvature*. We shall define curvature C at a point under consideration as the reciprocal of radius of curvature:

$$C = \frac{1}{\rho_0} = \left(\frac{d^2y}{dx^2}\right)_0 \tag{3.2}$$

Hence, if we consider points 1 and 2 on the given curved line located at equal increments Δx from the point 0 under consideration as shown in figure 3.1, we can apply directly the second derivative finite difference approximation

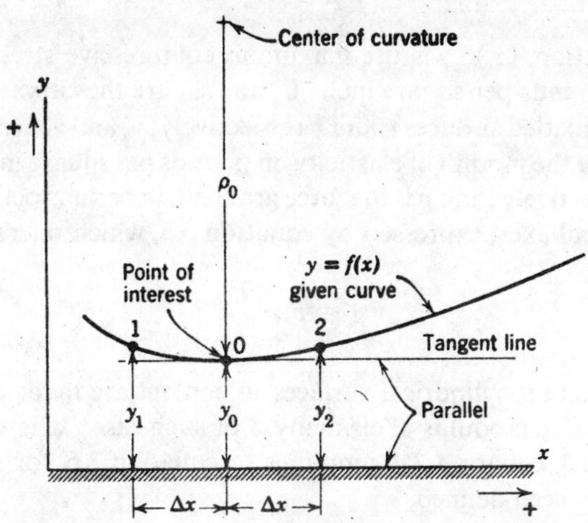

Figure 3.1 Given curved line in space showing radius of curvature ρ_0 in xy plane at point 0 under consideration.

with equation 2.5 for expressing curvature C in terms of values of displacements y_1, y_0, and y_2. Thus we obtain

$$\frac{1}{\rho_0} = C \approx \frac{y_1 + y_2 - 2y_0}{(\Delta x)^2} \tag{3.3}$$

Consider now the general case where two cylindrical surfaces with parallel axes are in contact and transmitting a radial force F, as shown in figure 3.2 [3.1, 3.2]. Consider that the radius of curvature for surface a is ρ_a and that the radius of curvature for surface b is ρ_b, at the vicinity of contact for the unloaded geometrical configuration. Because of elastic deformations at the vicinity of contact, the transmitted radial force F will be distributed over a finite area rectangular in shape as shown in figure 3.2c. A maximum compressive stress $(\sigma_c)_{\max}$ will exist between the mating surfaces in the center of that contact area, and, the existence of good lubrication being assumed, this will be a principal normal stress. From photoelastic results we know that stress levels fall off rapidly when moving away from the contact area. Hence significant contact stresses are highly localized in the immediate vicinity of the contact area, and the accompanying elastic deformations are also highly localized in such a small neighborhood. With respect to the maximum compressive stress $(\sigma_c)_{\max}$ at the contact region for parallel axes cylindrical surfaces as shown in figure 3.2, we have equation 3.4, commonly called the Hertz stress equation for cylinders.

$$(\sigma_c)_{\max} = 0.564\sqrt{\frac{p(C_a + C_b)}{[(1 - \nu_a{}^2)/E_a + (1 - \nu_b{}^2)/E_b]}} \tag{3.4}$$

In this equation, $(\sigma_c)_{\max}$ is the maximum compressive stress at the contact region, in pounds per square inch; C_a and C_b are the curvatures, in per inch units, for unloaded surfaces a and b respectively; ν_a and ν_b are Poisson's ratio; E_a and E_b are the moduli of elasticity, in pounds per square inch, for materials a and b respectively; and p is the force gradient, in pounds per inch units along the cylindrical axes, expressed by equation 3.5, which refers to figure 3.2.

$$p = \frac{F}{l} \tag{3.5}$$

Often in practice, cylindrical surfaces in contact are made of like materials having the same modulus of elasticity. For such cases, where Poisson's ratio is taken as 0.3, equation 3.4 simplifies to equation 3.6, for which the terms have already been defined.

$$(\sigma_c)_{\max} = 0.418\sqrt{pE(C_a + C_b)} \tag{3.6}$$

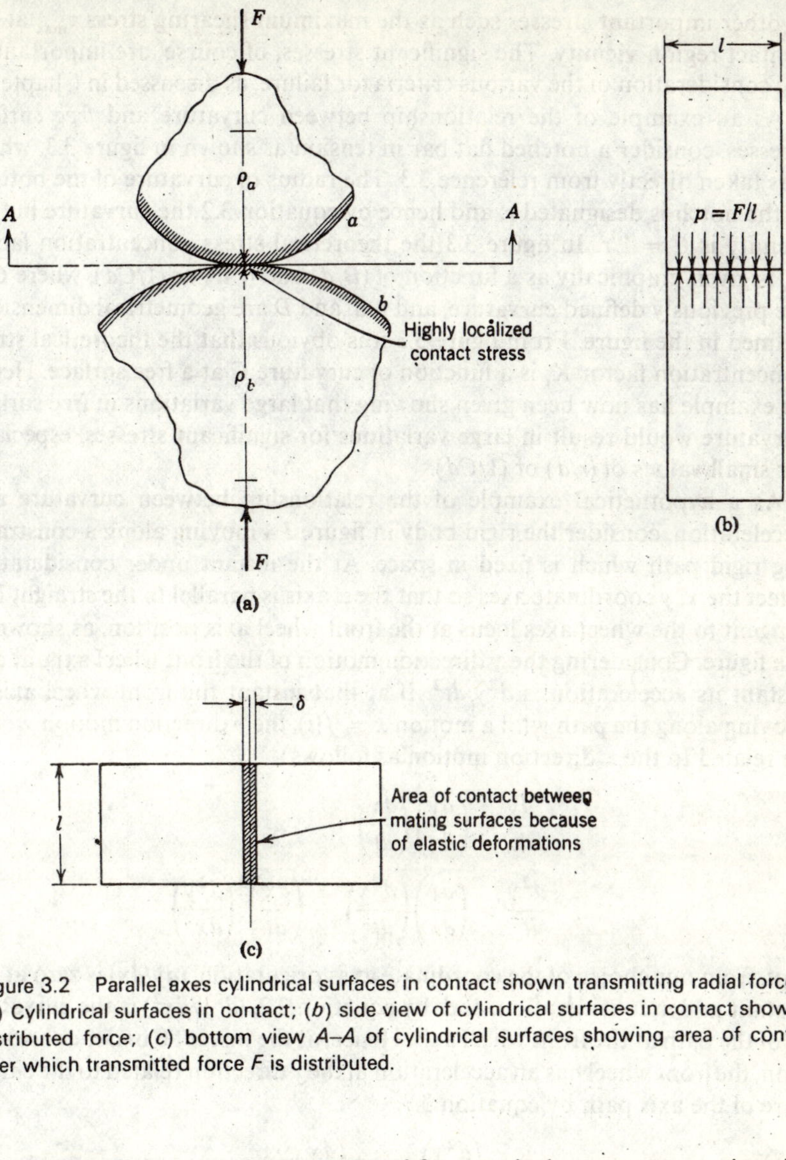

Figure 3.2 Parallel axes cylindrical surfaces in contact shown transmitting radial force F. (*a*) Cylindrical surfaces in contact; (*b*) side view of cylindrical surfaces in contact showing distributed force; (*c*) bottom view *A–A* of cylindrical surfaces showing area of contact over which transmitted force F is distributed.

An example has now been presented for a typical contact stress situation showing the relationship between maximum compressive stress $(\sigma_c)_{\max}$ and curvatures C_a and C_b. The significance of this relationship is seen from equations 3.4 and 3.6. Hence gross variations in curvatures C_a and C_b will result in significant variations in maximum compressive stress $(\sigma_c)_{\max}$. Likewise, large variations in curvatures C_a and C_b will result in significant variations in other important stresses such as the maximum shearing stress τ_{max} at the contact region vicinity. The significant stresses, of course, are important in the consideration of the various criteria for failure, as discussed in Chapter 4.

As an example of the relationship between curvature and *free* surface stresses, consider a notched flat bar in tension as shown in figure 3.3, which was taken directly from reference 3.3. The radius of curvature of the bottom of the notch is designated r, and hence by equation 3.2 the curvature in that vicinity is $C = 1/r$. In figure 3.3 the theoretical stress concentration factor K_t is given graphically as a function of (D/d) and (r/d) or $(1/Cd)$, where C is the previously defined curvature, and r, d, and D are geometrical dimensions defined in the figure. From figure 3.3 it is obvious that the theoretical stress concentration factor K_t is a function of curvature C at a free surface. Hence an example has now been given showing that large variations in free surface curvature would result in large variations for significant stresses, especially for small values of (r/d) or $(1/Cd)$.

As a hypothetical example of the relationship between curvature and acceleration, consider the rigid body in figure 3.4 moving along a constraining rigid path which is fixed in space. At the instant under consideration select the x, y coordinate axes so that the x axis is parallel to the straight line tangent to the wheel axes locus at the front wheel axis position, as shown in the figure. Considering the y direction motion of the front wheel axis, at any instant its acceleration is d^2y/dt^2. If at that instant the front wheel axis is moving along the path with a motion $x = f(t)$, the y direction motion would be related to the x direction motion as follows:

$$\frac{dy}{dt} = \left(\frac{dy}{dx}\right)\left(\frac{dx}{dt}\right)$$

$$\therefore \quad \frac{d^2y}{dt^2} = \left(\frac{dy}{dx}\right)\left(\frac{d^2x}{dt^2}\right) + \left(\frac{dx}{dt}\right)^2\left(\frac{d^2y}{dx^2}\right)$$

But, from our choice of the coordinate axes orientation, (dy/dx) is zero at the instant under consideration, and, by equation 3.2, (d^2y/dx^2) is the curvature C of the path at the front wheel axis. Hence, at the instant 0 under consideration, the front wheel has an acceleration in the y direction related to the curvature of the axis path by equation 3.7.

$$\left(\frac{d^2y}{dt^2}\right)_0 = C\left(\frac{dx}{dt}\right)_0^2 \tag{3.7}$$

In this equation, $(d^2y/dt^2)_0$ is the acceleration in the y direction, C is the path curvature, and $(dx/dt)_0$ is the velocity in the x direction, all being for the front wheel axis in figure 3.4 at the instant under consideration. Obviously, from equation 3.7, large variations in curvature C will result in large variations in

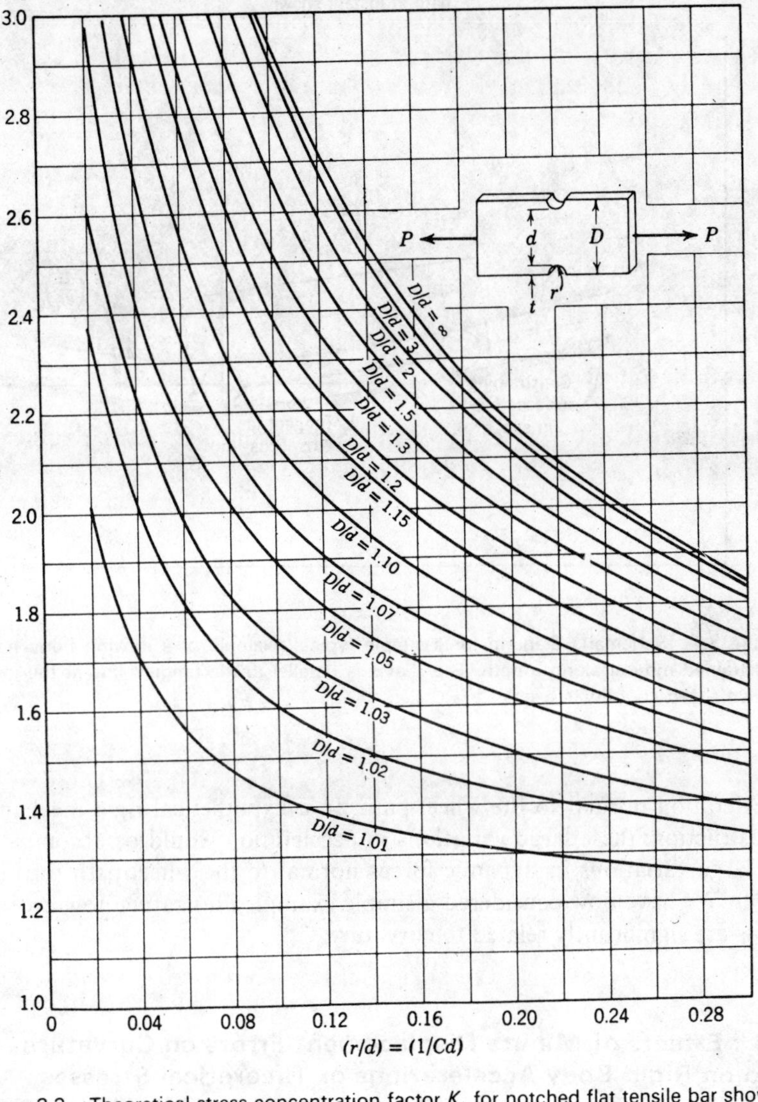

Figure 3.3 Theoretical stress concentration factor K_t for notched flat tensile bar showing K_t as a function of D/d and r/d or $1/Cd$, where r, the radius of curvature of the notch bottom, is equal to $1/C$ and C is the curvature of the notch bottom. Curves represent calculated values using Neuber theory. (Taken from reference 3.3. By permission of John Wiley and Sons, Inc.)

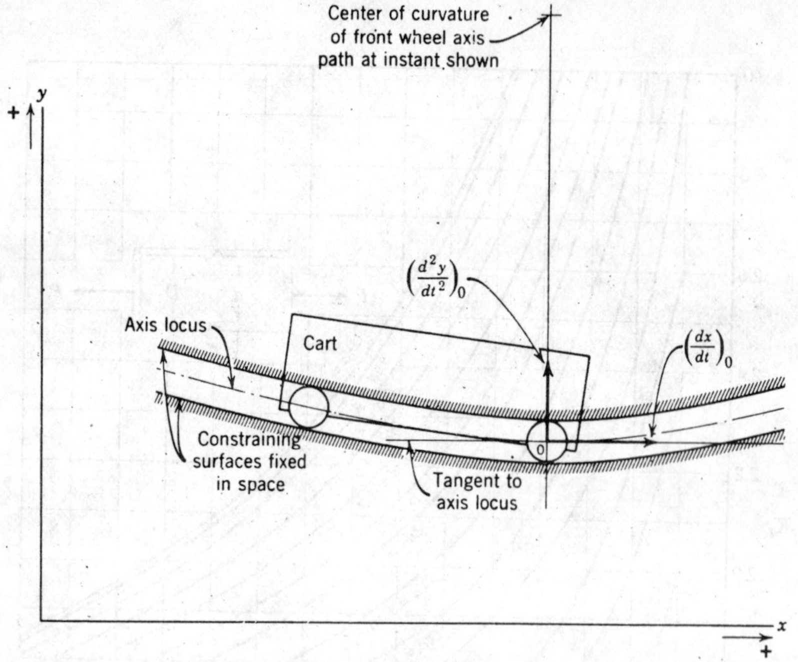

Figure 3.4 Schematic diagram for a simple typical example of a moving body having constrained motion along a path. The x axis is parallel to the tangent line at the instant shown (fixed coordinate axes).

acceleration normal to the wheel path. In a hypothetical rigid mechanical construction, these large variations in acceleration would be accompanied by large variations in dynamic forces normal to the wheel path for many cases. We have now considered a simple example illustrating how accelerations are significantly related to curvature.

3.3 Effects of Minute Displacement Errors on Curvature and on Rigid Body Accelerations or Theoretical Stresses

Assume that the given curved line in figure 3.1 represents a theoretically perfect surface for a mechanical element. Hence, if we consider points 1, 0, and 2 on that curved line located (Δx) equal increments apart, we can estimate the curvature C at point 0 under consideration using finite difference equation 3.3. Thus, if the theoretically perfect y displacements are designated $(y_1)_t$,

$(y_2)_t$, and $(y_0)_t$, the theoretically perfect curvature at point 0 would be designated C_t, and it would be closely approximated by

$$C_t \approx \frac{(y_1)_t + (y_2)_t - 2(y_0)_t}{(\Delta x)^2} \tag{3.8}$$

Assume now that it is possible for any actual displacement value y to have an error associated with it which is introduced during the manufacturing process. Hence, if the subscript t designates a theoretically perfect y displacement, let the subscript m designate a corresponding y displacement on the actual manufactured part. Therefore, if the possible error in y displacements is $\pm e_d$, we can express the manufactured part displacement y_m for any point on the curved line in terms of the theoretically perfect displacement y_t with a tolerance of $\pm e_d$, according to

$$y_m = y_t \pm e_d \tag{3.9}$$

Assume now that a mechanical element has been manufactured and that alternating errors in y displacements exist for points 1, 0, and 2 in figure 3.1. Assume that these errors are related to the displacements by

$$(y_1)_m = (y_1)_t \pm e_d$$

$$(y_2)_m = (y_2)_t \pm e_d$$

$$(y_0)_m = (y_0)_t \mp e_d$$

Assume now that a smoothly varying curve exists between all points on the manufactured surface. Hence we shall apply equation 3.3 to estimate the curvature C_m of the *manufactured* surface at point 0 under consideration. Thus

$$C_m \approx \frac{(y_1)_m + (y_2)_m - 2(y_0)_m}{(\Delta x)^2}$$

$$= \frac{(y_1)_t + (y_2)_t - 2(y_0)_t \pm 4e_d}{(\Delta x)^2}$$

Therefore, using equation 3.8 for theoretical curvature C_t, we obtain

$$C_m \approx C_t \pm \frac{4e_d}{(\Delta x)^2} \tag{3.10}$$

Hence curvature of the manufactured surface C_m will differ from the theoretically perfect curvature C_t by an error in curvature which we will designate e_c. Thus an estimate for this error in curvature can be made from equation 3.10

in terms of the increment (Δx) and the y displacement errors e_d. This is expressed by

$$e_c \approx \pm \frac{4e_d}{(\Delta x)^2} \tag{3.11}$$

Equation 3.11 estimates approximately the variation in surface curvature e_c which might result because of errors in displacements e_d measured normal to the surface, for known increments (Δx) measured along the surface. However, since exact natures of errors are uncontrollable, it is impossible to predict *accurately* what variations in curvature will result because of known displacement errors normal to a surface. However, equation 3.11 can at least be used for estimating significant orders of magnitude. Such calculations are generally far more enlightening than no calculations at all, provided it is realized that preciseness just cannot be obtained.

EXAMPLE 3.1

As a specific example, consider a curved line typified in figure 3.1 which has a theoretically perfect radius of curvature $(\rho_0)_t = 5$ in at the point under consideration. Hence the theoretically perfect curvature would be $C_t = \frac{1}{5}$ in per inch units, according to equation 3.2. Assume that very small displacement errors in the range $e_d = \pm 0.001$ in can exist on the manufactured surface for y displacements corresponding to figure 3.1, with the increment $(\Delta x) = 0.100$ in. Assume that a smooth curve exists between all points on the manufactured surface (Δx) increments apart. Thus we can estimate the error in curvature, by using equation 3.11, as follows:

$$e_c \approx \pm \frac{4e_d}{(\Delta x)^2} = \pm \frac{4(0.001)}{(0.1)^2} = \pm 0.40$$

However, the theoretically perfect curvature is $C_t = \frac{1}{5} = 0.20$. Thus, by equation 3.10, the variation for the curvature of the manufactured surface would be

$$C_m \approx C_t \pm e_c = 0.20 \pm 0.40$$

Hence the likely range for the manufactured surface curvature C_m at the point under consideration, because of extremely small displacement errors normal to the surface, would be

$$-0.20 \leq C_m \leq +0.60$$

The negative sign for the lower end of this range merely means that the curvature could be concave or convex in a direction opposite that of the theoretically perfect surface. From the results of this specific example we can conclude that very minute displacement errors normal to a surface can result in gross variations in curvature of the surface. Thus, in the manufacturing of real mechanical elements, it is extremely difficult to control curvature accurately. For cases where curvature is significant, it might be difficult to control product performance; we shall consider this point further in the next section.

Consider now the case where cylindrical surfaces in contact are transmitting a radial force F as shown in figure 3.2. Assume that the theoretically perfect curvatures at the region of contact are $(C_a)_t$ and $(C_b)_t$ for unloaded surfaces a and b respectively. Now assume that, *on the manufactured parts*, the curvatures at the contact area are $(C_a)_m$ and $(C_b)_m$ for unloaded surfaces a and b respectively, differing from the theoretically perfect curvatures because of inherently unavoidable surface displacement errors. Let us now consider the maximum compressive stress $(\sigma_c)_{max}$ in the contact area between the cylinders. $(\sigma_{c\,max})_t$ for the theoretically perfect surfaces could be calculated readily by using the theoretically perfect curvatures $(C_a)_t$ and $(C_b)_t$ in equation 3.4. Similarly, $(\sigma_{c\,max})_m$ for the manufactured surfaces could be expressed in terms of the manufactured surface curvatures $(C_a)_m$ and $(C_b)_m$ by using equation 3.4. Consideration of the ratio of maximum compressive stresses for the manufactured surfaces with respect to the theoretically perfect surfaces gives

$$\frac{(\sigma_{c\,max})_m}{(\sigma_{c\,max})_t} = \sqrt{\frac{(C_a)_m + (C_b)_m}{(C_a)_t + (C_b)_t}} \tag{3.12}$$

Equation 3.12 enables us to estimate the increase in significant stresses caused by likely variations in surface curvature resulting from minute surface displacement errors. Using equations 3.10 and 3.12, we can estimate at least an order of magnitude for the manufactured to theoretical stress ratio.

EXAMPLE 3.2

Consider two parallel axes cylindrical surfaces in contact transmitting a radial load F as depicted in figure 3.2. Assume that the theoretically perfect radius of curvature for each cylinder is 5 in, and that displacement errors $e_d = \pm 0.001$ in exist for increments $(\Delta x) = 0.100$ in at the contact region, which is numerically exactly the same as for example 3.1. From the results of example 3.1 we saw that the likely curvature for such a manufactured surface could be as great as $C_m = 0.6$, compared with the theoretically perfect curvature $C_t = 0.2$. For this numerical example we will assume that the manufactured and theoretical curvatures are as follows:

$$(C_a)_m = (C_b)_m = 0.6$$

$$(C_a)_t = (C_b)_t = 0.2$$

From equation 3.12, the significant stress ratio is

$$\frac{(\sigma_{c\,\max})_m}{(\sigma_{c\,\max})_t} = \sqrt{\frac{0.6 + 0.6}{0.2 + 0.2}} = 1.73$$

From the results of this example we can conclude that extremely small displacement errors on cylindrical surfaces in contact *can* result in gross variations in significant stresses. The alleviating effects of yielding and notch sensitivity will be considered in the next section.

Let us now consider an example which illustrates the effects of minute flaws on free surfaces with respect to significant stresses. Again we shall neglect the effects of localized yielding and fatigue notch sensitivity, both of which will be considered in section 3.4. Let us consider the theoretical stress concentration factor K_t for a typical manufacturing flaw on the surface of a theoretically flat tensile bar as shown in figure 3.5. In order to make use of the available information from figure 3.3, consider two small surface flaws of identical shape, one on each side of the tensile bar as depicted in figure 3.5. To determine the theoretical stress concentration factor K_t for the small imperfections on the tensile bar surface, it will be necessary for us to assume very small values for (r/d) or $(1/Cd)$ and values for (D/d) close to unity. Assume that the only available information which the design engineer has at his disposal is the graphical function expression given in figure 3.3, which

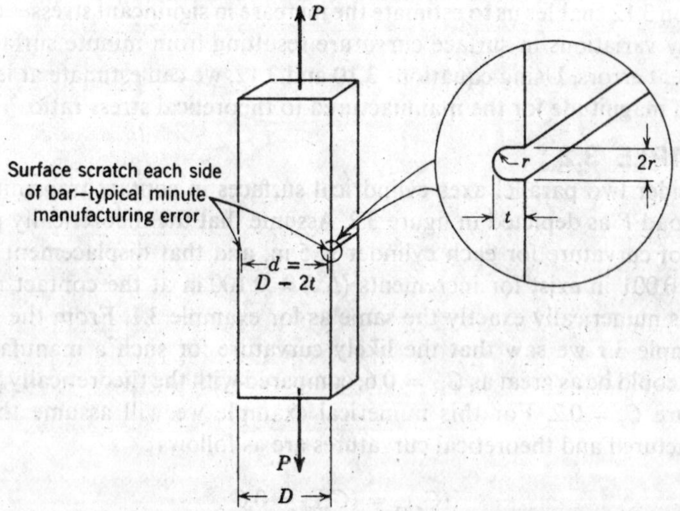

Figure 3.5 Flat tensile bar showing a typical minute manufacturing flaw on free surface.

covers only values of (r/d) in the range $0.02 \leq (r/d) \leq 0.30$. Therefore, for values of (r/d) smaller than 0.02, we shall have to project the given information outside the range of knowledge to the range of interest with an *anticipated* variation estimated from the given information. For such cases the error will be indeterminate since true values are simply not known outside of the range of knowledge for the given graphical function expression. However, projecting the given information outside of the range of knowledge to the range of interest will enable us to obtain at least an order of magnitude for K_t, and in the next section, when considering notch sensitivity, we shall see that the overall inaccuracy will be very small for the particular example. At this time we shall use curve-fitting techniques from Chapter 2 to derive an equation for K_t which will be accurate for the neighborhood at the lowest value of $(r/d) = (1/Cd) = 0.02$ and at the value closest to unity of $(D/d) = 1.01$, from the given information in figure 3.3. Therefore log-log plots based on studies in Chapter 2 were made from figure 3.3 for K_t versus $[(D/d) - 1]$ with $(1/Cd) = 0.02$ and for K_t versus $(1/Cd)$ with $(D/d) = 1.01$; the plots are shown in figure 3.6. Exponents were determined from slope measurements on these plots, and the constant overall coefficient was determined using the "central" values $(D/d) = 1.01$, $(1/Cd) = 0.02$, and $K_t = 2.0$ from figure 3.3. In this

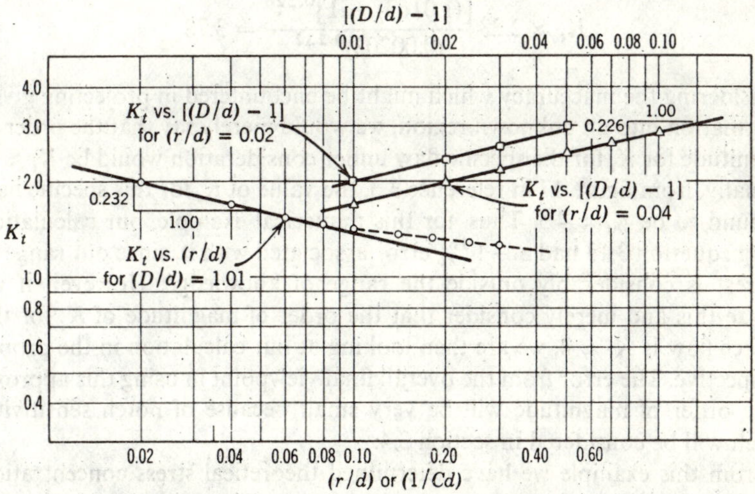

Figure 3.6 Log–log plots of K_t versus $[(D/d) - 1]$ and (r/d) from given curves in figure 3.3 for deriving exponential equation (equation 3.13). "Central" values of $(D/d) = 1.01$ and $(r/d) = 0.02$ were chosen because given information will be projected into unknown range of interest for small values of $r/d < 0.02$.

manner equation 3.13 was derived, using the principles covered in Chapter 2 on curve fitting.

$$K_t \approx 2.28 \frac{[(D/d) - 1]^{0.226}}{(1/Cd)^{0.232}} \tag{3.13}$$

This equation will enable us to project the given information from figure 3.3 to values of $(1/Cd) < 0.02$, with an anticipated variation and an unknown degree of accuracy. We have purposely chosen this specific example of function projection under the hypothetical assumption that information was not available for values of $(1/Cd)$ smaller than 0.02. However, such information in reality is available from figure 17 in reference 3.3, which will enable us to check the accuracy of our function projection technique.

EXAMPLE 3.3

Consider a tensile bar with a minute surface imperfection as shown in figure 3.5. Assume that specific values are as follows: $D = 0.250$ in, $r = 0.0005$ in, and $t = 0.002$ in. Hence $d = D - 2t = 0.250 - 2(0.002) = 0.246$ in. Thus $(D/d) = (0.250/0.246) = 1.0163$ and $(r/d) = (1/Cd) = (0.0005/0.246) = 0.00203$. Therefore, from figure 3.3, we see that the value of $(D/d) = 1.0163$ is within the range of the given information, whereas the value of $(1/Cd) = 0.00203$ is considerably outside that range. Hence, using equation 3.13, we obtain

$$K_t \approx 2.28 \frac{[(1.0163) - 1]^{0.226}}{(0.00203)^{0.232}} = 3.78$$

Considering the inaccuracy which might be encountered in projecting given information into an unknown region, we would merely say that the order of magnitude for K_t for the specific flaw under consideration would be $K_t \approx 4$. Actually, from figure 17 in reference 3.3, the value of K_t for this specific flaw is found to be $K_t \approx 4.5$. Thus, for this numerical example, our calculation using equation 3.13 had a -17% error associated with it, since our range of interest is considerably outside the range of knowledge. However, if we realize this and merely consider that the order of magnitude of K_t for the surface flaw is $K_t \approx 4$, we are then looking at our calculation in the proper perspective. The error from the overall final viewpoint in using this approximate order of magnitude will be very small because of notch sensitivity, which will be considered in section 3.4.

From this example we have determined theoretical stress concentration factor $K_t \approx 4$ for a tensile bar caused by a very small surface imperfection 0.001 in wide by 0.002 in deep. Thus we can conclude that very minute displacement errors on free surfaces can cause gross variations in the theoretical stress concentration factor. The alleviating effects of yielding and of notch sensitivity on actual product performance will be treated in the next section

Let us now consider a numerical example illustrating the effects of minute surface displacement errors on accelerations of bodies in constrained motion.

EXAMPLE 3.4

Assume that the cart and constraining path in figure 3.4 are of rigid mechanical structure. Suppose that the constraining path has minute displacement errors $e_d = \pm 0.001$ in normal to the theoretically perfect path at increments $(\Delta x) = 0.100$ in measured along the path. Thus, using equation 3.11, the associated error in curvature can be estimated as follows:

$$e_c \approx \pm \frac{4e_d}{(\Delta x)^2} = \pm \frac{4(0.001)}{(0.100)^2}$$

$$e_c \approx \pm 0.400 \text{ in}^{-1}$$

Assume that the front wheel of the cart in figure 3.4 has a modest velocity of 30 mph, which is $(dx/dt)_0$. Using equation 3.7, we can estimate the resulting error in acceleration for the y direction, e_{ay}, by inserting the error in curvature, e_c, for C in that equation. Thus

$$e_{ay} = e_c \left(\frac{dx}{dt}\right)_0^2 \tag{3.14}$$

For the specific problem,

$$\left(\frac{dx}{dt}\right)_0 = 30 \times \frac{5280 \times 12}{3600} = 528 \text{ in/sec}$$

and $e_c \approx \pm 0.400 \text{ in}^{-1}$, already calculated. Thus, by equation 3.14,

$$e_{ay} \approx \pm(0.4)(528)^2 = \pm 111{,}000 \text{ in/sec}^2$$

$$\therefore \quad e_{ay} \approx \pm \frac{111{,}000}{386} = \pm 289g$$

where $g = 386$ in/sec², the gravitational acceleration constant. Thus associated with a *mass* of 1 lb weight at the front wheel axis would be a fluctuating dynamic *force* of ± 289 lbf. This, of course, assumes a perfectly rigid mechanical structure, neglecting possibly alleviating effects of elastic deformations which will be considered in section 3.4.

From the study which has now been made in this specific example, we can conclude that very minute displacement errors on constraining rigid surfaces of rigid bodies in motion *can* result in gross variations in accelerations and gross variations in the accompanying dynamic forces. The author has considered this problem more specifically with respect to cam mechanisms in references 3.4 and 3.5, in which the effects of cam angle error as well as follower displacement error for rigid-type mechanisms are studied. Of particular interest is a correlation which was obtained between acceleration variations

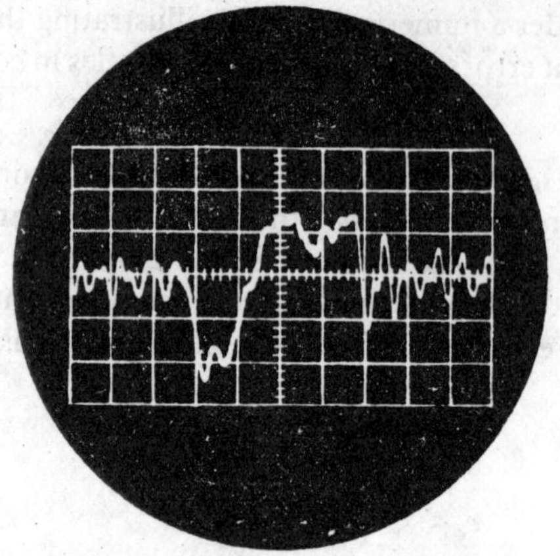

(a)

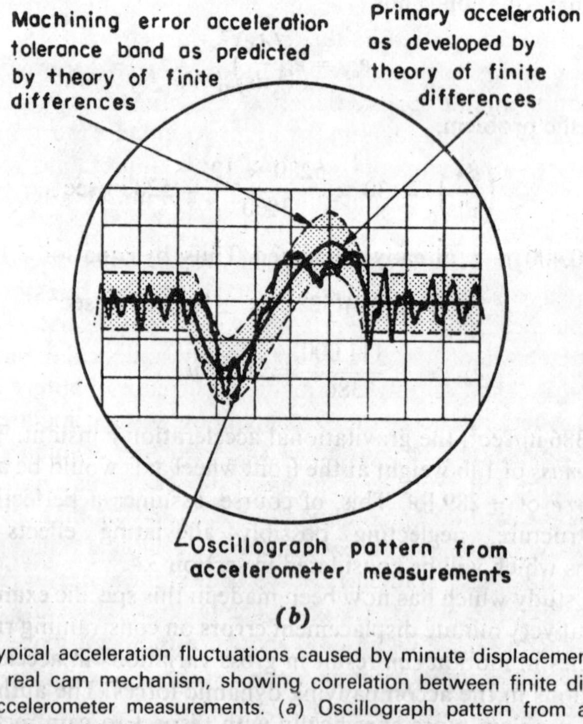

(b)

Figure 3.7 Typical acceleration fluctuations caused by minute displacement errors on a relatively rigid real cam mechanism, showing correlation between finite difference prediction and accelerometer measurements. (*a*) Oscillograph pattern from accelerometer measurement on inertial load of high speed cam mechanism. (*b*) Comparison of accelerometer measurements from *a* with theoretical accelerations as predicted by method of finite differences. (Taken from reference 3.5. By permission of Machine Design.)

predicted by the method of finite differences from known displacement errors and experimental results from accelerometer measurements on an actual mechanism. The results of this study are summarized in figure 3.7, which was taken directly from figures 6 and 7 of reference 3.5.

3.4 Effects of Minute Displacement Errors on Actual Product Performance and Product Life

Free Surface Flaws. The geometry of a minute surface flaw is extremely small, and therefore its effect on significant stresses is limited to a very small neighborhood of the imperfection. Hence we say that a minute surface imperfection is a highly localized region of stress concentration.

For the case where mechanical elements are subjected to static external loads, static stresses will exist and for most materials minute surface imperfections will have an insignificant effect. Plastic flow or yielding of the material will occur at the highly localized region of stress concentration, resulting in a minute redistribution of stress in the mechanical element.

For the case where mechanical elements are subjected to varying loads of a cyclical nature, minute surface imperfections can be very significant from the standpoint of a fatigue failure. In order to understand thoroughly the effects of minute surface imperfections on fatigue failures, one would have to understand the very complex micromechanism of initial fatigue damage for real materials [3.6, 3.12].

At the present time it is known that many factors affect the occurrence of fatigue failures, and some of them, such as residual stresses, corrosion, material impurities, and minute surface flaws, are not precisely controllable. Hence the prediction of a fatigue failure is really statistical in nature and for real mechanical elements only a probable occurrence of failure can be forecast. However, with respect to the effect of minute surface imperfections three important factors should be considered. First, whether or not a given surface flaw is significant will depend on the grain size of the material. Second, the number of grains within the volume of material having high levels of significant stresses will influence the occurrence of a fatigue failure. Third, it is believed that the stress gradient relative to the grain size is an important factor influencing the occurrence of fatigue failures. These are some of the significant general factors which will influence the occurrence of a fatigue failure initiating at a minute surface imperfection.

Theoretical stress concentration factors are obtained by assuming a perfectly homogeneous and isotropic material and neglecting the effects of grain size. However, for cases where there are relatively few numbers of grains within the high level significant stress volume, the theoretical stress concentration factor K_t is no longer significant and a fatigue notch factor K_f should be used instead. This is the case for minute surface imperfections, and we shall now consider to a greater extent this effect, which is accounted for by *notch sensitivity* [3.6].

The definition of notch sensitivity q is

$$q = \frac{K_f - 1}{K_t - 1} \tag{3.15}$$

In equation 3.15 K_f is the fatigue notch factor corresponding in usage to the theoretical stress concentration factor K_t. For most well-designed real mechanical elements, the fatigue notch factor K_f will be nearly the same as the theoretical stress concentration factor K_t, for *purposely introduced* geometrical changes such as fillets in stepped shafts. In fact, for most cases of this kind it will be slightly conservative in design work to use K_t instead of K_f. However, in studying the typical effects of *extremely small* surface imperfections we must consider notch sensitivity for a proper perspective of the real influence.

A simple equation which can be used for estimating notch sensitivity q for bending or axial loading is as follows.

$$q = \frac{1}{1 + (a/r)} \tag{3.16}$$

In equation 3.16, q is the notch sensitivity; r is the radius of the surface notch, in inches; and a is a material factor influenced by grain size. Equation 3.16 is valid for bending or axially loaded members where the depth of notch does not exceed four times the notch radius. A typical value for aluminum alloys is $a = 0.025$. With respect to steels, typical values are $a = 0.0004$ for a tensile strength of $S_t \approx 198{,}000$ psi and $a = 0.010$ for $S_t \approx 63{,}000$ psi. Thus we see that the material constant a decreases with a decrease in grain size.

EXAMPLE 3.5

Consider a tensile bar with a minute surface imperfection as shown in figure 3.5, assuming specific values of $D = 0.250$ in, $r = 0.0005$ in, and $t = 0.002$ in, which are exactly the same as in example 3.3. In that example we estimated the theoretical stress concentration factor as $K_t \approx 3.78$ by projecting the given graphical information from figure 3.3 into the unknown range of interest by use of curve-fitting techniques. Hence, under the hypothetical assumption that information other than that given in figure 3.3 was not available, we accounted for the possible inaccuracy of such a projection by merely saying that K_t has an order of magnitude of 4. However, in reality the exact value for K_t was found to be 4.5 from figure 17 of reference 3.3.

If the tensile bar is subjected to a varying axial load, a fatigue failure might very well be initiated at the surface imperfection region. However, since the flaw is very small, it will be appropriate for us to consider the fatigue notch factor K_f rather than the theoretical stress concentration factor K_t in assessing the actual effect of the imperfection on product performance.

First, consider the case where the tensile bar is made of an aluminum alloy so that the material constant is $a = 0.025$. Thus, since the notch radius $r = 0.0005$ in, the notch sensitivity q is found from equation 3.16 as follows:

$$q = \frac{1}{1 + (a/r)} = \frac{1}{1 + (0.025/0.0005)} = \frac{1}{51} = 0.0196$$

Therefore, from equation 3.15, the fatigue notch factor is determined as follows:

$$K_f = 1 + q(K_t - 1) = 1 + (0.0196)(K_t - 1)$$

Hence, for $K_t \approx 4$, we obtain

$$K_f \approx 1 + (0.0196)(4 - 1) = 1.0588 \approx 1.06$$

Therefore a fatigue notch factor of $K_f \approx 1.06$ would be in order for the specific surface flaw on the tensile bar made of aluminium. Obviously this value for K_f would change very little if K_t were changed to 3.78 or 4.5, corresponding to K_f values of 1.05 and 1.09. Thus we can conclude that the actual effect of the minute surface flaw on the aluminum alloy tensile bar would be relatively small, and it would be reasonably accurate to assume for this study a fatigue notch factor of $K_f \approx 1.06$.

Consider now the case where the tensile bar is made of annealed steel having a tensile strength close to 63,000 psi, so that the material constant is $a = 0.010$. The notch sensitivity from equation 3.16 is

$$q = \frac{1}{1 + (a/r)} = \frac{1}{1 + (0.010/0.0005)} = \frac{1}{21} = 0.0477$$

Therefore, from equation 3.15, the fatigue notch factor is determined as follows:

$$K_f = 1 + q(K_t - 1) = 1 + (0.0477)(K_t - 1)$$

Hence, for $K_t \approx 4$, $K_f \approx 1 + (0.0477)(3) = 1.14$, and for $K_t = 4.5$, $K_f \approx 1 + (0.0477)(3.5) = 1.17$. We see that a minute surface flaw of the type being considered on the tensile bar made of the annealed steel would be appreciably more significant than on the bar made of aluminum. A fatigue notch factor having an order of magnitude of $K_f = 1.17$ would be very appropriate, and using a value of $K_t = 4$ instead of 4.5 would change the calculated value for K_f only by a small amount.

Consider now the case where the tensile bar is made of quenched an tempered steel having a tensile strength close to 198,000 psi, so that th material constant is $a = 0.0004$. The notch sensitivity from equation 3.16 i

$$q = \frac{1}{1 + (a/r)} = \frac{1}{1 + (0.0004/0.0005)} = 0.555$$

Therefore, from equation 3.15, the fatigue notch factor is determined a follows:

$$K_f = 1 + q(K_t - 1) = 1 + (0.555)(K_t - 1)$$

Hence, for $K_t \approx 4$, $K_f \approx 1 + (0.555)(3) = 2.67$. On the other hand, if th exact value of $K_t = 4.5$ is used, $K_f = 1 + (0.555)(3.5) = 2.94$. We see that minute surface flaw of the type being considered on the tensile bar made o the quenched and tempered steel would be very much more significant tha on the bar made of either the annealed steel or the aluminum alloy. Thus, i would be advisable to have the surface ground after hardening.

From the results of this example we can conclude that very small surfac imperfections *can be* very significant for mechanical elements subjected t varying loads. This is especially true for hardened steel mechanical element which should be ground and possibly polished on critical surfaces. Althoug the effect of surface finish is not nearly so great for ductile materials, never theless its effect is significant in fatigue failures.

Surface Finish Effects. Minute displacement errors in the form of closel spaced surface grooves are introduced by machine tools in manufacturin processes for mechanical elements. The effect is particularly significant o the endurance limit for hardened steels, and in practice it is normally accounted for by a modifying surface factor k_a as follows [3.10].

$$S_e = k_a k_b k_c S'_e \tag{3.17}$$

In equation 3.17 S_e is the infinite life reversed bending endurance limit in psi k_a is the surface finish factor, k_b is a size factor, k_c is a miscellaneous effect factor, and S'_e is the ideal endurance limit in psi. For steels of tensile strengt $S_t \leq 200{,}000$ psi, $S'_e = 0.5S_t$ is appropriate. However, for steels of $S_t > 200{,}00$ psi a leveling effect normally occurs and $S'_e = 100{,}000$ psi should be used.

Surface factor k_a was analyzed by assuming typical surface grooves transverse to the stress trajectories, including the effects of notch sensitivity. Thi analysis resulted in figure 3.8, taken directly from reference 3.9, which shoul have value for design application in mechanical elements.

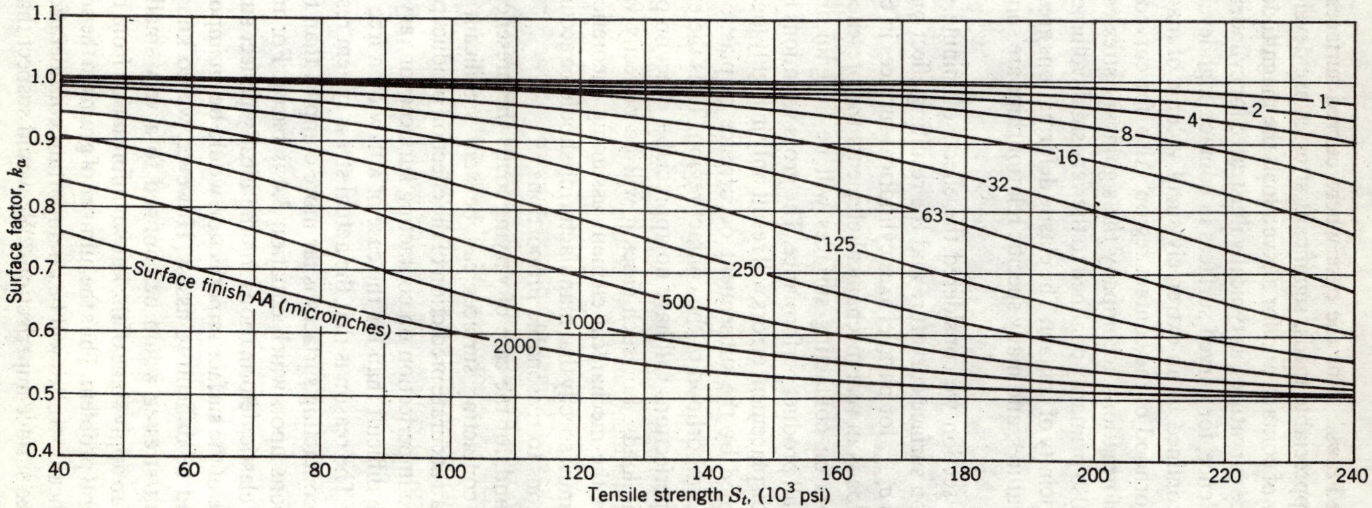

Figure 3.8 Surface finish factor k_a as a function of tensile strength S_t and surface roughness AA measurement for steel. (Taken from reference 3.9. By permission of Machine Design.)

Contacting Surface Flaws. For the case where curved surfaces in contact are transmitting appreciable loads, significant stresses are developed at the immediate vicinity of contact, and for a given load the magnitude of stress is a function of surface curvatures, as typically illustrated for cylindrical surfaces by equation 3.4. Hence, for curved surfaces in contact, high levels of significant stresses are confined to an extremely small volume of material in the immediate neighborhood of the contact region. Also, for curved surfaces in contact, elastic deformations accompany the significant stresses, and these deformations are also primarily confined to the very small volume of material in the immediate vicinity of contact. The elastic deformations are also related to the surface curvatures, and many specific relationships are summarized in reference 3.2.

In the previous section we considered the effects of minute surface displacement errors on surface curvature and the resulting effect on maximum compressive stress $\sigma_{c\,\max}$ for parallel axes cylindrical surfaces in contact. For the case of static loads in real mechanical elements the presence of minute displacement errors on contacting surfaces will be insignificant from the standpoint of overall product performance. The gross variations in curvature due to the minute displacement errors will result only in highly localized static stresses accompanied by the subsequent occurrence of plastic flow in an extremely small neighborhood of the contact region. Thus the effects of the minute surface imperfections will have no appreciable effect on product performance for static loads. For such cases it will be perfectly valid for the engineer to design the mechanical element assuming theoretically perfect curvatures, but he must specify that manufacturing should be accurate enough to limit imperfections to reasonable proportions.

On the other hand, for the case of varying loads the presence of minute imperfections on contacting surfaces can be very significant because of fatigue. However, an accurate prediction of the effect on product performance of a known minute imperfection on contacting surfaces for varying loads is considerably more difficult than for the case of a flaw on a free surface discussed previously. The reason is that the actual stress system resulting from a minute flaw on contacting surfaces is far more complex than for the ideal cylinder assumptions upon which equation 3.4 is based. For instance, the magnitude of the elastic deformations for the ideally perfect surfaces compared with the size of the surface imperfection would be an important factor whose effect would be difficult to assess. However, we do know that high levels of significant stresses would be confined to a very small volume of material, and we can apply some of the known information from basic fatigue studies to the general problem. Thus the number of grains in the small volume of high level significant stresses will be an important consideration, as it was for the study of free surface imperfections. We would suspect that the effects of minute imperfections on contacting surfaces for coarse grain materials would not be as great as they would be for fine grain materials. This should not be misconstrued as saying that coarse grain materials will have a higher fatigue life for a given load and given curvatures, since for steels the reverse is true. However, the statement does imply that we must exercise a greater amount of care with respect to surface finish for fine grain materials in order to take full advantage of the fatigue strength *possibilities*. Therefore very little will be gained by having polished surfaces for contacting elements made of annealed steel. However, a great deal of improvement in fatigue wear resistance will be realized by having an excellent finish on contacting surfaces for quenched and tempered steels. Thus hardened gear teeth should be ground and lapped if possible for many applications, and hardened rollers should be ground and polished if maximum fatigue strength is significant.

Minute Imperfections on Constraining Elements for Bodies in Motion. For a perfectly rigid mechanical structure, the effect of minute displacement errors on the constraining surface was typically illustrated in example 3.4. For such cases, minute displacement errors can have gross effects on significant accelerations and the accompanying dynamic forces. However, real mechanical structures are never perfectly rigid, and we shall briefly consider the effects of dynamic deformations at this time for structures having minute imperfections on constraining surfaces.

There are innumerable examples of real mechanical systems where moving bodies are constrained in motion by mechanical elements having minute imperfections. For instance, for a rotating motor, eccentricity in the mounting bearings of the rotor is unavoidable at least to some extent. In general, whereever bearings are required for high speed applications, minute bearing imperfections can be significant. As another example, bodies which are driven through gear trains are influenced to some extent by minute imperfections on the gear teeth. In general, cam mechanisms are always characterized by the existence of minute imperfections on constraining surfaces.

As a typical general example of a simple real mechanical system, consider the mass M in figure 3.9 which is being driven with constrained motion through an elastic region of force gradient k. Assume that displacements of the constrained motion deviate from their theoretically perfect values by a minute uncontrollable amount of error. Hence, associated with the commanding motion, we have minute displacement errors, and the problem is to determine the effects on the responding motion of the mass M. If the errors on the driving motion are transmitted to an appreciable extent to the responding motion of the driven mass, significant dynamic force variations *can* result.

Minute displacement errors on constraining elements can be thought of as a component part of the total forcing function acting on the moving system.

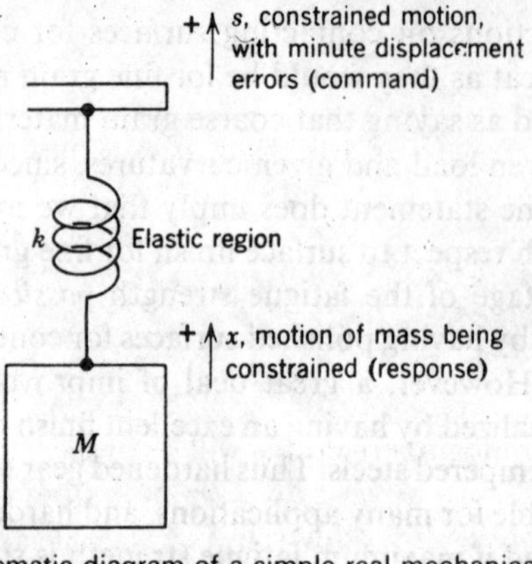

Figure 3.9 Typical schematic diagram of a simple real mechanical system where a moving body is under constrained motion.

The effect of such errors on the responding motion of the driven mass depends on many significant factors. For one, the exact nature of the error on the commanding motion can be important. Hence the magnitude of the error, its rate of change, and the frequency of occurrence, if repetitive, are important factors. The other important considerations are related to the exact nature of the mechanical system being driven. Of paramount importance in this respect are the natural frequencies of the mechanical structure, since natural frequency can be considered an index to dynamic response capability. Space does not permit a review of vibrational analysis in this book, but it is assumed that the reader has at least a basic knowledge in that field [3.11].

Whether or not a minute displacement error on a constraining motion is transmitted to the driven mass depends on the relationship between the exact nature of the error and the characteristics of the mechanical structure. Three types of relationship have particular significance when the errors occur repetitiously, as they often do in practice. The first is that for a "soft" spring mechanical system, where (in figure 3.9) the natural frequency is much lower than the frequency of error occurrence. For such cases very little of the error is transmitted to the responding mass motion, and dynamic forces *due to the error* are insignificant. The second type is that for a "hard" spring mechanical system, where in (figure 3.9) the natural frequency is much higher than the frequency of error occurrence. For these cases the majority of the error is transmitted to the responding mass motion, and dynamic forces due to the error can be very significant. This was the case for example 3.4, where a minute displacement error resulted in a gross effect on acceleration. The third type is that for the condition near resonance, where (in figure 3.9) the natural frequency of the mechanical system is nearly the same as the frequency of error occurrence. For such cases the original error can be appreciably amplified on the responding mass motion, and the dynamic forces which result can be very large and troublesome. Obviously such conditions of resonance should be avoided if at all possible.

For most real mechanical systems with bodies in constrained motion, *extremely* small imperfections occurring at frequent intervals will be insignificant because they will fall in the "soft" spring category. Hence the very small imperfections will not be transmitted to the responding mass motion, and its acceleration will be virtually unaffected. However, there are many types of problems where minute imperfections will be of the "hard" spring type, and very significant acceleration variations on the responding mass motion will result. Experimental proof is shown in the accelerometer measurements in figure 3.7, which were taken for an actual cam mechanism where the master cam had been step-milled and thus a repetitious type of error occurrence had been introduced.

The method of finite differences can be employed for studying the effects of typical errors on real mechanical systems in constrained motion. The bases for this application are covered in references 3.7 and 3.8 for mechanical systems having one or more degrees of freedom. The reader is referred to those articles because the method of finite differences is particularly well suited for analyzing the effects of typical manufacturing displacement errors on dynamic response.

From the study which has now been made on constrained motion of real mechanical systems, we can conclude that minute manufacturing errors can either be significant or insignificant with respect to dynamic forces. This depends on whether or not the error is transmitted to the responding mass motion, which is determined by the nature of the error and the characteristics of the mechanical structure. In addition to accelerations and dynamic forces, the complex effects of dynamic stresses are very important for real bodies in constrained motion, and they should be considered in the light of possible fatigue failure, including the effects of minute imperfections as previously discussed.

3.5 Conclusions

The effects of manufacturing errors which have been studied in this chapter are commonly encountered in innumerable real mechanical elements. From this basic study we can state the following three general conclusions:

1. For real mechanical bodies which are in constrained motion or which are subjected to varying loads, minute manufacturing errors *can* have gross effects on actual product performance.

2. For many real mechanical elements, significant manufacturing errors are inherently unavoidable, primarily because of the relationship between curvature, stresses, and accelerations, and the extreme difficulty in controlling curvature, which is grossly affected by extremely minute displacement errors.

3. Because of the statistical nature of fatigue failures and the limited amount of information now available, the exact effects even of controlled manufacturing errors on product performance cannot be predicted accurately. At best, the fatigue life of an original mechanical element can be only approximately predicted. In order to do this, significant factors have to be controlled to specified limits and an appropriate criterion for failure such as that described in Chapter 4 must be used.

REFERENCES

3.1 S. Timoshenko and J. N. Goodier, *Theory of Elasticity*, 3rd ed., McGraw-Hill, New York, 1970, pp. 409–420.

3.2 R. J. Roark and W. C. Young, *Formulas for Stress and Strain*, 5th ed., McGraw-Hill, New York, 1975, Chapter 13.

3.3 R. E. Peterson, *Stress Concentration Design Factors*, Wiley, 1953, Figure 16. (Condensed version given in figure 17 of 1974 edition.)

3.4 R. C. Johnson, "How Profile Errors Affect Cam Dynamics," *Machine Design*, Feb. 7, 1957, pp. 105–108.

3.5 R. C. Johnson, "Development of a High Speed Indexing Mechanism," *Transactions of the Fourth Conference on Mechanisms*, Penton, Cleveland, Ohio, 1958, pp. 39–43; *Machine Design*, Sept. 4, 1958, pp. 134–138.

3.6 George Sines and J. L. Waisman, *Metal Fatigue*, McGraw-Hill, New York, 1959, 415 pp.

3.7 R. C. Johnson, "Dynamic Analysis of Cam Mechanisms," *Transactions of the Fifth Conference on Mechanisms*, Penton, Cleveland, Ohio, 1958, pp. 21–35.

3.8 R. C. Johnson, "The Dynamic Analysis and Design of Relatively Flexible Cam Mechanisms Having More than One Degree of Freedom," ASME Paper 58-A-192, *Journal of Engineering for Industry*, Vol. 81, Series B, No. 4, Nov. 1959, pp. 323–331.

3.9 R. C. Johnson, "Specifying a Surface Finish that Won't Fail in Fatigue," *Machine Design*, May 3, 1973, pp. 106–107.

3.10 J. E. Shigley, *Mechanical Engineering Design*, 2nd ed., McGraw-Hill, New York, 1972, pp. 504–505.

3.11 J. P. Den Hartog, *Mechanical Vibrations*, 4th ed., McGraw-Hill, New York, 1956, Chapter 2.

3.12 A. F. Madayag, *Metal Fatigue: Theory and Design*, Wiley, New York, 1969, 419 pp.

EXERCISES

3.1. Two slow speed steel cams in rolling contact transmit force F as shown in figure 3.2. For the theoretical design we have specified $\rho_a = 50$ mm, $\rho_b = 70$ mm, $l = 10$ mm, and $F = 2200$ N.

(a) Calculate contact stress$(\sigma_c)_{max}$ in (Pa), and width δ of contact area (figure 3.2).

(b) An alternating step milling manufacturing error $e_d = \pm 0.012$ mm normal to the cam surfaces is possible for both cams, at an angular increment of $\Delta\theta = 2°$ on the cam. Calculate the effect on cam curvature for both and calculate the manufactured to theoretical stress ratio by equation 3.12.

3.2. Rigid mass M is to travel along a theoretically straight path at velocity V. However, because of manufacturing errors, small bumps may be introduced on the constraining path. Suppose $M = 1000$ lb-sec^2/in and $V = 100$ in/sec.

(a) Estimate the order of magnitude of dynamic force resulting from a smooth bump of height $y = 0.010$ in with its base 0.100 in. (*Hint*: Apply equation 2.38 of Chapter 2 and equation 3.11.)

(b) A periodically varying bump is approximated by the equation $y = 0.005 \sin(\pi x/0.05)$ in.

1. Estimate maximum dynamic force on mass M for the rigid system case.
2. Estimate maximum dynamic force on mass M for the case of M being separated from the constraining path by an elastic region of force gradient $k = 4000$ lbf/in with no damping.

3.3. A simple beam of span $L = 10$ in has a central fully reversed load of $F = \pm 1000$ lbf. The beam cross section is a solid square 1.00 in on the side. The material is steel of $S_t = 100{,}000$ psi tensile strength.

(a) Calculate the theoretical bending stress σ_b and compare with the endurance limit S_e of the material. A surface finish of 125 μ in AA is anticipated with machining grooves normal to the beam span length direction.

(b) Calculate the stress concentration effect of a single groove normal to the span length direction of dimensions 0.002 in wide and 0.003 in deep as follows.

1. Theoretical stress concentration factor K_t.
2. Fatigue notch factor K_f, based on an estimated value for notch sensitivity q taken from the literature.
3. What would be the realistic effect on bending stress σ_b from this type of groove imperfection?

4

Material Properties and Failure Theories

The purpose of this chapter is to review very briefly some of the material properties as related to failure theories, which is very important background for optimal design of mechanical elements. The first part of the chapter will be devoted to certain significant material properties. The second part will review some significant considerations of combined stress systems, including the general triaxial state. The final part will review some significant theories of failure and their correlation with experimental results. The chapter is *not* intended to be a study of the problem of *material selection*, which will be thoroughly covered in later chapters in conjunction with optimization techniques applied to specific mechanical elements.

4.1 Stress and Strength

Before discussing some specific material properties, it is appropriate to introduce some very important notation which will be adhered to rigorously throughout the book. This notation is for the terms *stress* and *strength*, which must be uniquely designated by *different* symbols during conduct of the method of optimum design. First, by definition, *stress* is the intensity of force at a point in question with respect to a reference plane. Hence stress is a vector quantity whose magnitude and direction must be expressed for unique definition. In this book we shall denote *normal stresses* by the Greek letter sigma, σ, and *shearing stresses* by the Greek letter tau, τ. Thus, from elementary theory of elasticity [4.3], at a point in question the *state of stress* will be uniquely defined by three normal components, σ_x, σ_y, σ_z, and three shearing components, τ_{xy}, τ_{xz}, and τ_{yz}. At a point in question for any mechanical element, the values of the stress components are dependent upon the geometry of the element, the external loading to which it is subjected, and some material properties such as Poisson's ratio and occasionally the modulus of elasticity. Let us now consider the term *strength*, which is a *material property* always determined by using a standard test specimen in a standard test procedure, by the occurrence of a particular type of predefined failure phenomenon. Thus, from a simple tensile test, the yield *strength* of a ductile material is the value of tensile stress which exists when the curve of stress versus unit elongation flattens out; indicating appreciable yielding as the failure phenomenon. Or, from a simple tensile test, the ultimate *strength* of a brittle material is the value of tensile stress which exists when the test specimen breaks, and for this case material separation is the failure phenomenon. In this book we shall *always* denote material *strengths* by the letter S. Therefore S_y is the yield strength of a material from a simple tensile test, and S_t is the ultimate *strength* of a material from a simple tensile test. Since material *strengths* are determined with standard test specimens in standard test procedures, strengths are properties depending on material characteristics alone. In the conduct of explicit optimum design it is very important to differentiate between material *strengths* determined by standard test procedures and the state of *stress* existing at a point in question in a mechanical element. For explicit optimum design work the two terms must not be confused with each other, and it should be emphasized that published material *strengths* can be considered dependent upon materials only. The state of *stress* at a point in question for a mechanical element is unaffected by the material *strengths*, although the state of *stress* might be *limited* by certain material *strengths*; the limit equation can be written by selecting an appropriate criterion for failure as reviewed in the last part of this chapter. Hence in this book *stresses* will be designated by σ and τ, and material *strengths* will be designated by S.

Material Properties

4.2 Some Material Properties Related to Strength

The mechanical elements of interest for this book are generally characterized by design equations which are expressed in part by terms known as *material properties*. Such material parameters as modulus of elasticity, tensile strength, and yield strength are typical properties having great importance in design work. Values of various material parameters are published in several types of available sources of information such as textbooks, handbooks, manufacturers' brochures, as well as literature which is circulated only within a particular company [4.5–4.9]. In practice, a design engineer will be able to compile a list of *available* materials with tabulated values for significant mechanical properties; a *typical* condensed version of such a list is given in table 4.1. Such a table will summarize the significant information pertaining to materials which are available at a particular *time* in a particular *locality*, and it will be a very important piece of information for the design engineer.

Fatigue Strength of Steel. The endurance limit of steel can be estimated by equation 3.17, repeated as 4.1 below.

$$S_e = k_a k_b k_c S'_e \tag{4.1}$$

The terms have already been defined following equation 3.17 in Chapter 3. Incidentally, the miscellaneous effects factor k_c accounts for any significant factors affecting S_e not otherwise accounted for, such as effects from unusual temperatures, corrosion, and plating.

It is well known that the infinite life endurance limit of steel is reached after approximately 10^6 cycles of stress variation, as shown in figure 6-23 of reference 4.11. At 10^3 cycles a virtually static stress situation can be assumed, with fatigue strength then being approximately 0.9 S_t. A linear relationship exists on a log-log plot of fatigue strength S_f versus number of cycles n_c between these extremes. Hence, assuming that surface finish factor k_a is fully effective on endurance limit but not significant for static stress cases, finite life fatigue strength S_f of steel can be estimated by the following equation simply derived by curve fitting from the described plot.

$$S_f = k_b k_c \frac{0.9S_t}{(n_c/1000)^{\log\{[0.9S_t/(k_a S'_e)]^{1/3}\}}} \tag{4.2}$$

Equation 4.2 is valid for the finite life range of $10^3 \leq n_c \leq 10^6$, where n_c is the number of cycles, S_f is the fully reversed finite life fatigue strength of the steel (psi or Pa), and k_a, k_b, k_c, S_t, and S'_e are as defined following equation 3.17 in Chapter 3.

For cases of finite life with steel, merely use S_f in the failure criterion instead of endurance limit S_e. However, for $n_c \leq 1000$ cycles assume static loading. On the other hand, for $n_c \geq 10^6$ cycles the value of endurance limit S_e from equation 4.1 should be used in the failure criterion.

4.3 Some Material Properties Other than Strengths

At this time let us very briefly consider some material properties other than those which are determined from standard load types of test procedures. Because these properties are generally not so significant in design work as some of the characteristics determined from load tests, we shall treat them in a cursory manner.

TABLE 4.1

Some Mechanical Properties Related to Strength for Representative Engineering Materials

Material	Yield Strength S_y		Tensile Strength S_t		Elongation, %	Endurance Limit* (500×10^6 cycles)		Modulus of Elasticity E	
	kpsi	MPa	kpsi	MPa		kpsi	MPa	Mpsi	GPa
AISI 1020 HR steel	35	241	60	414	24			29	200
AISI 1020 CR steel	60	414	80	552	13			29	200
AISI 2340 steel						For steels see equations 4.1 and 4.2			
Oil Q & D 400°F	232	1600	260	1793	10			29	200
Oil Q & D 800°F	167	1151	185	1276	14			29	200
AISI 4130 steel									
Water Q & D 400°F	197	1358	232	1600	10			29	200
Water Q & D 800°F	168	1158	190	1310	14			29	200
304 stainless steel	33	228	75	517	55			29	200
Titanium	70	483	80	552	20	38	262	16	110
Titanium Alloy	130	896	150	1034	18	85	586	16	110
2024-T4 Aluminum	47	324	68	469	20	20	138	10.6	73
7075-T6 Aluminum	73	503	83	572	11	23	159	10.4	72
AZ61A Magnesium	33	228	45	310	16	17	117	6.5	45
ZK20A Magnesium	28	193	38	262	4	10.5	72	6.5	45
Plastics									
Acetal homopolymer			10	69	45	3.8	26	0.4	2.8
Type 6/6 nylon			10	69	100	2.4	16	0.3	2.1
Phenolic			8	55	1.8			0.45	3.1
Polyester with chopped roving			27	186	3			1.4	9.6

* For completely reversed tensile and compressive stress.

TABLE 4.2

Approximate Values of Some Material Properties Other than Strengths

Class of Material	Weight Density w		Unit Material Cost c		Temp. Coeff. for Therm. Expans. α		Thermal Conductivity k		Specific Heat c		Poisson's Ratio ν
	(lbm/in^3)	(kg/m^3)	(\$/in^3)	(\$/m^3)	$\mu(°F)^{-1}$	$\mu(°K)^{-1}$	$\frac{Btu}{hr\text{-}ft\text{-}°F}$	$\frac{W}{m\text{-}°K}$	$\frac{Btu}{lbm\text{-}°F}$	$\frac{J}{kg\text{-}K}$	
Steels											
(C HR, C CR)	0.283	7833	0.04–0.06	2441–3661	6.5	11.7	27	47	0.10–0.11	419–461	0.27
Titanium and its alloys	0.163	4512	24–34	1.5×10^6–2.1×10^6	5.3	9.5	7.0	12	0.12–0.13	503–544	0.33
Aluminum alloys (wrought)	0.100	2768	0.08–0.11	4882–6712	13	23	67–135	116–234	0.22–0.23	921–963	0.33
Magnesium alloys (wrought)	0.066	1827	0.14	8543	14.5	26	24–80	42–138	0.24	1005	0.35
Plastics											
Acetal	0.051	1412	0.048	2929	45	81	0.13	0.22–0.24	0.35	1466	0.35
Nylon	0.041	1135	0.048	2929	50	90	0.14	0.24	0.4	1675	0.40

In table 4.2 values of weight density w are given in the first column for the typical available materials. These values are determined from measurements of weight and volume, which can be performed quite accurately. Hence, assuming isotropic materials of uniform density, the available values of w can be considered very accurate for design usage.

In the second column of table 4.2, values of unit material cost c are given for some of the typically available materials. We shall assume that this is the cost of the material to the manufacturer of the mechanical element as purchased from the vendor in average standard size shapes. It does not include machining costs or handling costs within the plant manufacturing the element; these should be accounted for differently as typically illustrated in section 7.6 of Chapter 7. Unit material cost c is seldom published in textbooks, reference handbooks, or manufacturers' brochures because its value for any material is significantly influenced by *time* and *locality*. Hence, unless information is available at a particular time in a particular locality, published values of unit material cost might be highly inaccurate. However, for mechanical elements produced in mass quantity, unit material cost c can be a very significant factor in the method of optimum design, and the material choice might very well be influenced by relative costs which vary with time and locality.

Temperature coefficient for linear thermal expansion, α, is given in the third column of table 4.2. Published values for this parameter are generally average values over a specified temperature range. Hence the values given in table 4.2 can be considered inaccurate for precision work in many cases, except for a narrow range of temperature variation. For good accuracy of the temperature coefficient α, it should be expressed as a function of temperature for most materials if large temperature variations are expected. When such considerations are significant, generally the design engineer will have to seek advice from a metallurgist or a materials specialist to obtain information which will define this function. The fourth and fifth columns of table 4.2 are for thermal conductivity k and specific heat c, also required for some thermodynamically related calculations for mechanical elements.

The final column in table 4.2 is for Poisson's ratio, which is required for some stress calculations, such as those with equation 3.4.

Review of Combined Stress Systems [4.3]

4.4 Unique Definition and Distribution of Stress Systems

Stress is a tensor quantity which is uniquely defined at a point in question by the specification of six components. Thus, with respect to rectangular coordinates, stress at a point is uniquely defined by the specification of three normal stress components, σ_x, σ_y, and σ_z, and three shearing stress components, τ_{xy}, τ_{yz}, and τ_{xz}. For any mechanical element, the state of stress at a point is completely considered only if all six stress components are known and included in the problem of design; generally, several of the stress components are zero.

The distribution of stress for all points in a mechanical element is generally a very complex problem which can be solved by the theory of elasticity or by experimental methods. This distribution of stress is affected by material properties, boundary conditions, considerations of continuity for internal displacements, and considerations of internal and external equilibrium. The problem is far too extensive for complete coverage in this book. Hence we shall assume that the general stress distribution at significant regions of interest for mechanical elements being designed is either known or determinable from available sources in the literature; this will be illustrated in the specific examples in Chapters 7 and 8 [4.4, 4.11, 4.12].

4.5 Principal Stresses

If the state of stress is known at a point in question in a mechanical element, it can be completely defined by the magnitudes *and* orientations of the three principal normal stresses, which we shall designate σ_1, σ_2, and σ_3 throughout this book. These stresses act on orthogonal planes which have zero shearing stresses on them; these principal normal stresses are of extreme importance in design work since they are directly involved in all the significant criteria for failure. We shall begin by considering plane stress, known as the *biaxial state of stress*, which is merely a special case of the general triaxial state.

We shall say that a state of stress is biaxial at a point in question if it is possible to select a differential cube so oriented at the point that the normal *and* tangential stresses acting on a pair of opposite parallel faces are *both* zero. Thus the state of stress at the point in question is typically defined by σ_x, σ_y, and τ_{xy} as shown in figure 4.1a, if it is assumed that σ_z, τ_{xz}, and τ_{yz} are all zero by the definition of the biaxial state. Throughout this book we shall assume that normal stresses are positive if tensile and negative if compressive. Also, with respect to notation, we shall assume that σ_x is algebraically larger than σ_y.

From a consideration of force equilibrium, it could be shown that it is *always* possible to orient a differential cube at a point so that *zero* shearing stresses act on its faces. Hence, if we consider a differential cube which is oriented at a certain angle θ to the arbitrarily chosen x axis as shown in figure 4.1a, we obtain a differential cube having only the principal *normal*

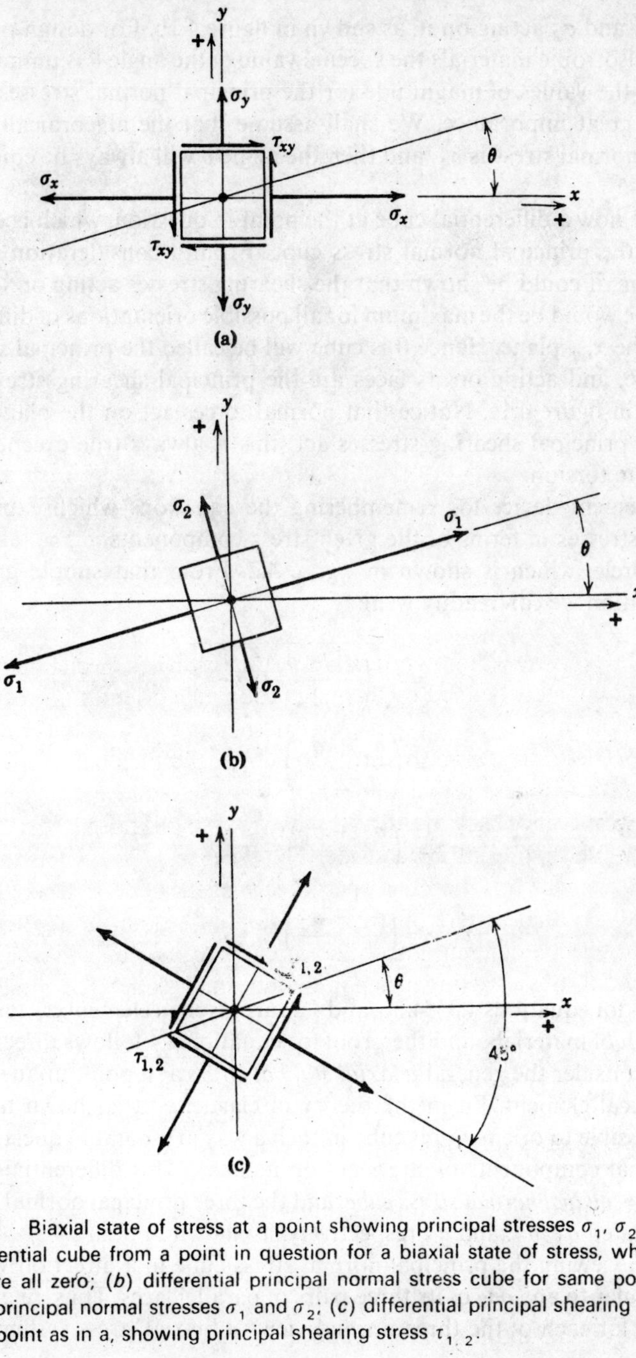

Figure 4.1 Biaxial state of stress at a point showing principal stresses σ_1, σ_2, and $\tau_{1,2}$. (*a*) Differential cube from a point in question for a biaxial state of stress, where σ_z, τ_{xz}, and τ_{yz} are all zero; (*b*) differential principal normal stress cube for same point as in *a*, showing principal normal stresses σ_1 and σ_2; (*c*) differential principal shearing stress cube for same point as in a, showing principal shearing stress $\tau_{1,2}$.

stresses σ_1 and σ_2 acting on it, as shown in figure 4.1b. For design problems involving isotropic materials the specific value of the angle θ is unimportant. However, the values of magnitude for the principal normal stresses σ_1 and σ_2 are of great importance. We shall assume that the algebraically larger principal normal stress is σ_1, and then the angle θ will always be equal to or less than 45 .

Imagine now a differential cube at the point in question, which is oriented at 45° to the principal normal stress cube. From a consideration of force equilibrium, it could be shown that the shearing stresses acting on the faces of this cube would be the maximum for all possible orientations of differential cubes in the x, y plane. Hence this cube will be called the principal shearing stress cube, and acting on its faces are the principal shearing stresses $\tau_{1,2}$ as shown in figure 4.1c. Notice that normal forces act on the planes upon which the principal shearing stresses act; this is always true except for the case of pure torsion.

A convenient device for remembering the equations which express the principal stresses in terms of the given stress components σ_x, σ_y, and τ_{xy} is Mohr's circle, which is shown in figure 4.2. From that simple graphical representation we can readily write

$$\tau_{1,2} = \sqrt{\left(\frac{\sigma_x - \sigma_y}{2}\right)^2 + \tau_{xy}^2} \tag{4.3a}$$

$$\sigma_1 = \left(\frac{\sigma_x + \sigma_y}{2}\right) + \tau_{1,2} \tag{4.3b}$$

$$\sigma_2 = \left(\frac{\sigma_x + \sigma_y}{2}\right) - \tau_{1,2} \tag{4.3c}$$

$$\tau_{1,2} = \left(\frac{\sigma_1 - \sigma_2}{2}\right) \tag{4.4}$$

The proofs for equations 4.3a, 4.3b, and 4.3c are given in elementary textbooks on strength of materials, and the proof for equation 4.4 follows directly.

Let us consider the general *triaxial state of stress* at a point in question in a mechanical element. From the theory of elasticity, it is shown that it is always possible to orient a unit cube in such a way at a point in question that only normal components of stress act on its faces. This differential cube is called the *principal normal stress cube*, and the three principal normal stresses are designated σ_1, σ_2, and σ_3 respectively, as shown in figure 4.3.

Imagine viewing the principal normal stress cube in a direction which is perpendicular to any *one* of its three pairs of parallel faces. Thus, referring to figure 4.3, for each of the three normal views a biaxial stress system would

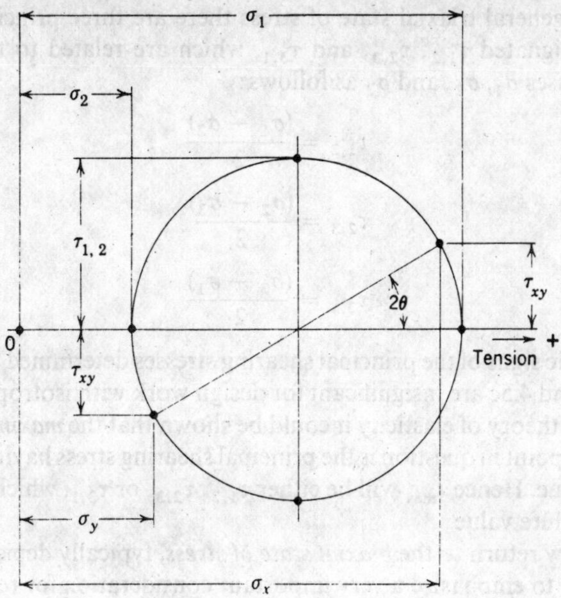

Figure 4.2 Mohr's circle for biaxial state of stress. (See figure 4.1)

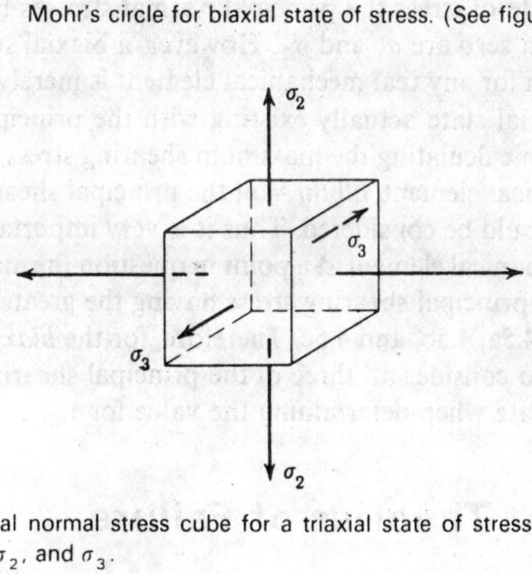

Figure 4.3 Principal normal stress cube for a triaxial state of stress, showing principal normal stresses σ_1, σ_2, and σ_3.

appear. The pairs of principal normal stresses associated with each of the three normal views would be (σ_1, σ_2), (σ_2, σ_3), and (σ_3, σ_1), respectively. From an extension of the biaxial case, associated with each of the three normal views of the principal normal stress cube there would be a *principal shearing stress* acting on planes which are oriented at 45° to the principal normal stresses, as in figures 4.1b and 4.1c. Therefore, from an extension of equation

4.4, for the general triaxial state of stress there are three principal shearing stresses, designated $\tau_{1,2}$, $\tau_{2,3}$, and $\tau_{3,1}$, which are related to the principal normal stresses σ_1, σ_2, and σ_3 as follows:

$$\tau_{1,2} = \frac{(\sigma_1 - \sigma_2)}{2} \tag{4.5a}$$

$$\tau_{2,3} = \frac{(\sigma_2 - \sigma_3)}{2} \tag{4.5b}$$

$$\tau_{3,1} = \frac{(\sigma_3 - \sigma_1)}{2} \tag{4.5c}$$

The algebraic signs of the principal shearing stresses determined by equations 4.5a, 4.5b, and 4.5c are insignificant for design work with isotropic materials.

From the theory of elasticity it could be shown that the *maximum* shearing stress at the point in question is the principal shearing stress having the largest absolute value. Hence τ_{max} will be either $\tau_{1,2}$, $\tau_{2,3}$, or $\tau_{3,1}$, whichever has the largest absolute value.

Let us now return to the *biaxial state of stress*, typically depicted in figure 4.1, in order to emphasize a very important consideration for real problems. For a biaxial state of stress the principal normal stresses having numerical values other than zero are σ_1 and σ_2. However, a biaxial state of stress at a point in question for any real mechanical element is merely a *special case* of the general triaxial state actually existing with the principal normal stress $\sigma_3 = 0$. Hence, in calculating the maximum shearing stress τ_{max} at a point in any real mechanical element, *all three* of the principal shearing stresses $\tau_{1,2}$, $\tau_{2,3}$, and $\tau_{3,1}$ should be considered. Thus it is very important to realize that for any real mechanical element at a point in question the maximum shearing stress τ_{max} is the principal shearing stress having the greatest absolute value from equations 4.5a, 4.5b, and 4.5c. Therefore, for the *biaxial state of stress*, it is important to consider all three of the principal shearing stresses of the actual triaxial state when determining the value for τ_{max}.

Significant Theories of Failure

4.6 Purpose and Choice of the Failure Criterion for Design

In the explicit method of optimum design for unique mechanical elements the engineer must be able to express mathematically the relationship between significant *stresses* and significant material characteristic *strengths*, for the prediction of the occurrence of a predefined failure phenomenon. In the determination of significant *stresses*, there are several approaches which the engineer might follow. If it is appropriate and feasible, the many available sources of stress concentration factors such as reference 4.4, should be used. On the other hand, for unique situations it might be necessary to employ the theory of elasticity or to conduct experimental studies with, for example, photoelasticity. Often a combination of these approaches will be necessary in order to relate significant stresses to external loads on unique types of mechanical elements. However, for most *design* situations the conduct of general experimental studies will be infeasible, and the design engineer will have to rely on other sources of information available at the time of design [4.3, 4.11, 4.12]. With respect to the significant material characteristic *strengths*, parameter values such as those summarized in table 4.1 will be accessible. Thus, for most design situations, the available material characteristics will be limited to those which have been published.

In the literature many criteria for failure have been advanced for the purpose of predicting the occurrence of a failure phenomenon in unique mechanical elements. Some of the particularly significant criteria which the author has found useful in design work will be summarized briefly in the remainder of the chapter. For a more complete coverage of the subject the reader should consult other sources, such as references 4.1, 4.2, 4.8, 4.10, and 4.11.

Choosing the appropriate criterion of failure for a unique design problem is based on a consideration of four fundamental factors. First, the question of what phenomenon constitutes the occurrence of failure for the specific mechanical element being designed should be considered. For the condition of few cycles of load application in a very accurate mechanism, a static theory of failure based on the *yielding* of the material probably will be appropriate. On the other hand, if the mechanism is to withstand millions of cycles of load application, a criterion for *fatigue* failure should be used. Second, the correlation of the theory of failure with known experimental results for the type of material being considered should be taken into account. Some of these results are summarized in references 4.1, 4.2, and 4.13, and they are generally depicted in figure 4.4 for static stresses based on yielding. The chosen criterion for failure to be used in design should correlate reasonably well with past experimental evidence, and, if an appreciable discrepancy does exist for certain materials, it should be on the conservative side. Third, the availability and the accuracies of the material parameters which are involved in the expression of the criterion for failure should be considered. Material parameter values for the selected criterion for failure must be either available or obtainable with acceptable accuracy for the significant available materials if the design engineer is to arrive at a final optimum design. Finally, the availability or the determinability of the significant *stresses* involved in the expression of the criterion for failure must be taken into account. Significant stresses at critical regions must be either available or determinable as related to external loads on the element being designed in order to apply properly the criterion for failure. On the basis of these four factors, an appropriate criterion for failure should be chosen for the specific design problem under consideration.

4.7 Static Stress Criteria for Failure

Maximum Normal Stress Theory. The maximum normal stress theory of failure is based on the assumption that a predefined failure phenomenon occurs when the largest principal normal stress reaches a certain critical value. For reasons of uniqueness in notation, assume that the three principal normal stresses have the algebraic order $\sigma_1 > \sigma_2 > \sigma_3$. Hence, according to the maximum normal stress theory of failure, undesirable yielding of the material in a mechanical element will not occur if equations 4.6 and 4.7 are satisfied at all significant points. In those equations N_y is a factor of safety based on the occurrence of yielding, and the selection of N_y is based on the factors discussed in Chapter 5.

$$\sigma_1 \leq \frac{S_y}{N_y} \tag{4.6}$$

$$|\sigma_3| \leq \frac{(S_c)_y}{N_y} \tag{4.7}$$

In equation 4.6, S_y is the yield strength of the material determined by a simple tension test; and for most materials the compressive yield strength $(S_c)_y$ can be taken as the same numerical value, for equation 4.7.

Similarly, according to the maximum normal stress theory of failure the material would separate in a mechanical element at a critical point if $\sigma_1 = S_t$. Hence, to avoid the undesirable occurrence of such a failure phenomenon, equation 4.8 should be satisfied at all points in the mechanical element, where N is the factor of safety based on the ultimate strength S_t.

$$\sigma_1 \leq \frac{S_t}{N} \tag{4.8}$$

Similarly, if a certain failure phenomenon occurs in a material in a standard simple compressive test at a certain compressive normal stress of value S_c, we would say that equation 4.9 should be satisfied at all points in order to avoid the undesirable occurrence of that phenomenon in the mechanical element being designed. For this case N_c is a factor of safety based on the occurrence of the failure phenomenon, which, for example, might be crumbling of a brittle material.

$$|\sigma_3| \leq \frac{|S_c|}{N_c} \tag{4.9}$$

Maximum Shearing Stress Theory. The maximum shearing stress theory of failure is based on the assumption that a predefined failure phenomenon occurs at a point when the largest principal shearing stress τ_{max} reaches a certain critical value. Algebraic sign is unimportant for isotropic materials, and therefore we shall consider the absolute value, $|\tau_{max}|$. Hence, according to the maximum shearing stress theory, yielding of the material in a mechanical element will not occur if equation 4.10 is satisfied at every point, where again, N_y is a factor of safety based on the occurrence of yielding and its selection is based on the discussion in Chapter 5. In equation 4.10 the published yield strength from a simple tension test S_y is used as the material parameter instead of the generally less accurate published value of shear yield strength $(S_s)_y$.

$$|\tau_{max}| \leq \frac{S_y}{2N_y} \tag{4.10}$$

Similar equations could be written for the occurrences of other types of failure phenomena besides yielding by using material parameters such as S_t.

Distortion Energy Theory. The distortion energy theory of failure is based on the assumption that a predefined failure phenomenon will occur at a point when the distortion energy per unit volume reaches a certain level. For the triaxial state of stress the distortion energy theory states that yielding of the material will not occur if equation 4.11 is satisfied at every point.

$$\sigma_1{}^2 + \sigma_2{}^2 + \sigma_3{}^2 - \sigma_1\sigma_2 - \sigma_2\sigma_3 - \sigma_3\sigma_1 \leq \left(\frac{S_y}{N_y}\right)^2 \tag{4.11}$$

In this equation, σ_1, σ_2, and σ_3 are the principal normal stresses, S_y is the yield strength of the material from a simple tensile test, and N_y is the previously mentioned factor of safety based on the occurrence of yielding.

Comparison of Significant Theories with Experimental Evidence. The great amount of experimental evidence obtained for many materials from test specimens subjected to biaxial states of static stress can be compared with the various significant theories of failure. However, a very limited amount of experimental evidence for triaxial states of stress exists. For the biaxial state one of the three principal normal stresses, σ_1, σ_2, or σ_3, is zero, and as before we shall assume that the algebraic order for the principal normal stresses is $\sigma_1 > \sigma_2 > \sigma_3$. Therefore we can graphically depict the significant theories of failure based on yielding for the biaxial state as shown in figure 4.4. In this figure regions from typical experimental results summarized in reference 4.1 are generally depicted for some significant types of materials. In the figure the vertical axes are for the algebraically smaller of the two existing principal normal *stresses*, divided by S_y from a simple tensile test in order to obtain a unitless number. The horizontal axes are for the algebraically larger of the two existing principal normal *stresses* divided by S_y, as a unitless number. With the chosen notation for the principal normal stresses which defines their algebraic order, it is necessary to consider only points to the right of the 45° line in figure 4.4.

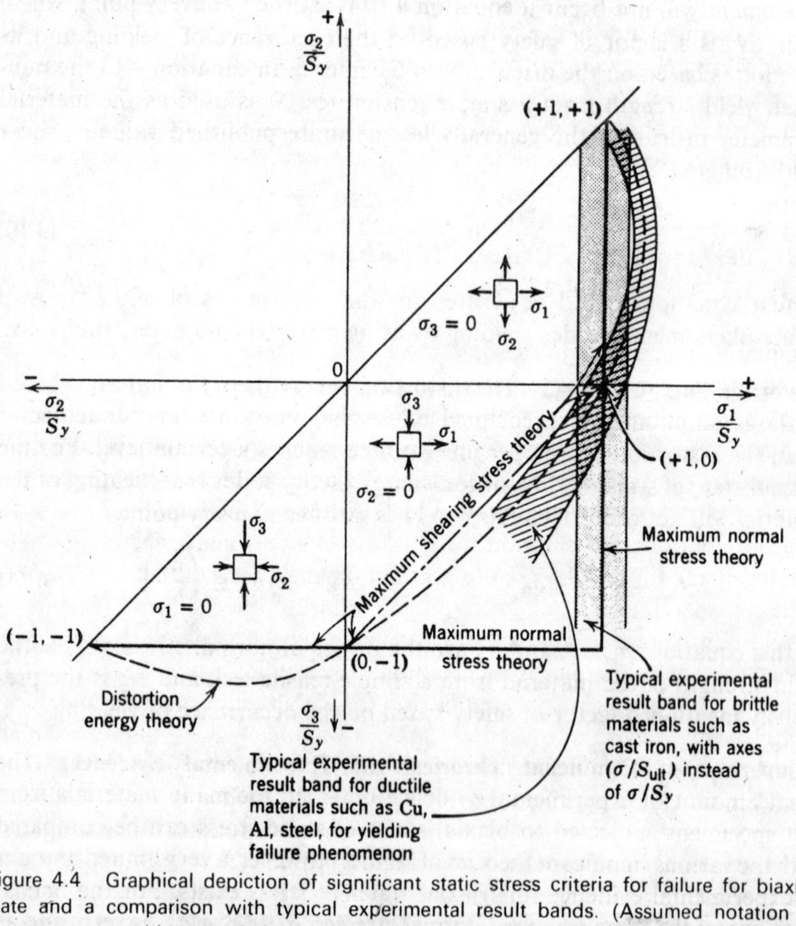

Figure 4.4 Graphical depiction of significant static stress criteria for failure for biaxial state and a comparison with typical experimental result bands. (Assumed notation is $\sigma_1 > \sigma_2 > \sigma_3$, with one of these principal normal stresses equal to zero.) Note: See reference 4.1 for more specific experimental results.

From the figure we see that the distortion energy criterion for failure falls most closely to the experimentally obtained evidence. However, for many of the experimental results it has been found to be nonconservative, especially in the first quadrant where both of the existing principal normal stresses are tensile. Hence, as a reasonable and more conservative criterion for design usage, we shall apply the maximum shearing stress theory of failure for static stresses for yielding of ductile materials as the failure phenomenon.

Incidentally, for materials which are very brittle, such as cast iron, experimental evidence has correlated very well with the maximum normal stress theory of failure. Hence, for such materials, the criterion for proper design will be expressed by equation 4.8 or 4.9. If there is a question whether a material is brittle or ductile, we shall adhere to the more conservative maximum shearing stress theory of failure for design usage. It is well known that most real materials are neither purely ductile nor purely brittle. Hence the reasonably conservative criterion for proper design expressed by equation 4.10, based on the maximum shearing stress theory of failure, will be very important for static stress design problems in the chapters to follow.

4.8 Varying Stress Criteria for Failure [4.1, 4.2, 4.8, 4.10, 4.11, 4.13]

Often it is necessary to design a mechanical element which must be subjected to many cycles of repeated loading, and failures due to fatigue must be avoided. Unfortunately, because of factors which are out of the design engineer's control, it is impossible to predict precisely when a fatigue failure will occur, especially in original design problems. Hence, designing mechanical elements subjected to varying stresses will be based on the most appropriate available criterion for failure and a reasonable factor of safety greater than unity will be used to account for the uncertainties.

In considering the problem of fatigue failure in mechanical elements, it should be *emphasized* that stresses vary with *location* as well as with *time*. Hence the equations which follow should be applied to *critical regions* of mechanical elements where significant stresses are concentrated and have relatively high levels of amplitude. Also, in applying the equations which follow at critical regions, it should be realized that the occurrence of fatigue damage is primarily influenced by the maximum and the mean values of significant stress variation and by the number of cycles of repeated load. Thus fatigue damage is virtually unaffected by either the exact shape of the stress variation or the amount of time between stress cycles.

Let us briefly consider some of the difficulties connected with the accurate prediction of fatigue failures for unique design problems. First, there is a lack of conclusive experimental verification for the various criteria, and that which is available has been obtained only under ideal test conditions. Second, some of the ideal test conditions are not truly representative of some of the actual operating conditions to which real mechanical elements are subjected. For instance, the criteria for failure are based on the assumption that the principal normal stress axes are fixed in direction with respect to the mechanical element. Fortunately, for most mechanical elements in practice this is a valid assumption. Third, there are many uncontrollable elusive factors which can significantly affect fatigue failures. For instance, it is well known that size effect is very important, although its exact effect for many cases is not predictable in *design* work. Also, such factors as residual stresses, surface finish, and corrosion can be very influential, although their exact effects on fatigue damage for most cases are not predictable. Hence, in original design work, the design engineer is limited in the accuracy of predicting *when* a fatigue failure will occur, although this will *not* limit the conduct of optimum design studies, as we shall see in later chapters.

Maximum Shearing Stress Theory of Fatigue Failure. The most significant criterion which the author has found for application to original mechanical element design problems where varying stress is important is the maximum shearing stress theory of fatigue failure. Assume that the principal normal stresses σ_1, σ_2, and σ_3 vary cyclically at a critical point in the mechanical element, as depicted in figure 4.5 for σ_1. Hence, at the critical region, the maximum values for the principal normal stresses from a time variation standpoint will be denoted by $\sigma_{1\,\max}$, $\sigma_{2\,\max}$, and $\sigma_{3\,\max}$. Also, the mean values for the principal normal stresses from a time variation standpoint will be designated σ_{1M}, σ_{2M}, and σ_{3M}. Thus the maximum shearing stress theory of fatigue failure states that a satisfactory design exists only if equations 4.12, 4.13, and 4.14 are *all* satisfied at *all* points in the mechanical element.

$$|(\sigma_{1\,\max} - \sigma_{2\,\max})| - (1 - p)\,|(\sigma_{1M} - \sigma_{2M})| \leq \frac{S_e}{N} \tag{4.12}$$

$$|(\sigma_{2\,\max} - \sigma_{3\,\max})| - (1 - p)\,|(\sigma_{2M} - \sigma_{3M})| \leq \frac{S_e}{N} \tag{4.13}$$

$$|(\sigma_{3\,\max} - \sigma_{1\,\max})| - (1 - p)\,|(\sigma_{3M} - \sigma_{1M})| \leq \frac{S_e}{N} \tag{4.14}$$

In equations 4.12, 4.13, and 4.14 S_e is the fatigue strength of the material from a simple standard reversed bending type of fatigue test for the desired number of load cycles. Also, N is the factor of safety greater than unity whose value is chosen for a specific problem from the considerations of factors covered in Chapter 5. In equations 4.12, 4.13, and 4.14, p is defined as

$$p = \frac{S_e}{S_y} \tag{4.15a}$$

where S_e is the fatigue strength and S_y is the yield strength of the material. If we incorporate different factors of safety N_e and N_y on S_e and S_y respectively, the definition for p becomes appropriately as follows.

$$p = \frac{S_e/N_e}{S_y/N_y} \tag{4.15b}$$

In this event, we use $N = N_e$ for the right sides of equations 4.12, 4.13, and 4.14.

Suppose we now divide both sides of equations 4.12, 4.13, and 4.14 by 2 and refer to equations 4.5a, 4.5b, and 4.5c. In this way the criterion is expressed in terms of the three principal shearing stresses, and the governing one we designate as τ_G. Thus the maximum shearing stress criterion can be expressed compactly by the single equation 4.16.

$$(\tau_G)_{\max} - (1 - p)(\tau_G)_M \leq \frac{S_e}{2N_e} \tag{4.16}$$

In this equation, p is determined by equation 4.15b, and the subscripts max and M refer to maximum and mean values respectively from a time variation standpoint similar to that seen in figure 4.5.

The maximum shearing stress criterion for fatigue failure is particularly suitable for original design problems for several important reasons. First, the correlation with available experimental evidence is reasonably good and it is generally on the conservative side. Second, the material parameters S_e and

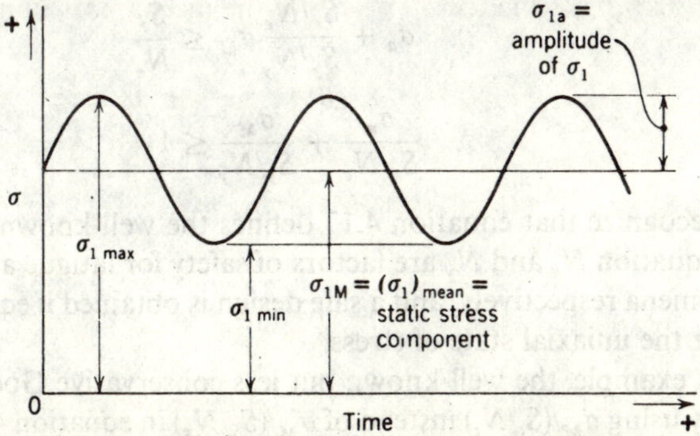

Figure 4.5. Typical variation of a principal normal stress with respect to time at a point of significance.

S_y which are involved are commonly available as published values for most significant engineering materials. Also, S_y generally obtained from a standard tensile test should be an *accurate* published material parameter, as previously discussed. In addition, the value of S_e is generally the *only* published material parameter which is readily available for most materials from a simple standard fatigue test. Third, the maximum shearing stress criterion for fatigue failure requires the determination of *differences* in principal normal stresses which are directly related to the principal shearing stresses by equations 4.5a, 4.5b, and 4.5c. Thus, from most of the published stress concentration factors, it is possible to relate τ_G to the external loading on the mechanical element. Hence for these reasons in this book we shall generally apply the maximum shearing stress theory of fatigue failure for the specific original *design* examples having varying stresses.

Incidentally, many fatigue failure criteria are available in the literature for special stress cases, such as the uniaxial state [4.1, 4.8, 4.10, 4.11]. In general, these criteria are but special case applications of the basic triaxial stress criterion given by equations 4.12, 4.13, and 4.14. For example, the well-known Soderberg line for the uniaxial varying stress case can be directly derived from these equations. More specifically, for the uniaxial case $\sigma_1 = \sigma$; $\sigma_2 = \sigma_3 = 0$, and equations 4.12 through 4.14 give the following.

$$\sigma_{\max} - (1 - p)\sigma_M \leq \frac{S_e}{N_e}$$

However, from figure 4.5 we see that $\sigma_{\max} = \sigma_M + \sigma_a$ and we apply equation 4.15b for p. Thus the preceding equation becomes equation 4.17 as follows.

$$\sigma_M + \sigma_a - \sigma_M + \frac{S_e/N_e}{S_y/N_y}\sigma_M \leq \frac{S_e}{N_e}$$

$$\therefore \quad \sigma_a + \frac{S_e/N_e}{S_y/N_y}\sigma_M \leq \frac{S_e}{N_e}$$

$$\therefore \quad \frac{\sigma_a}{S_e/N_e} + \frac{\sigma_M}{S_y/N_y} \leq 1 \tag{4.17}$$

We readily recognize that equation 4.17 defines the well-known Soderberg line. In this equation N_e and N_y are factors of safety for fatigue and yielding failure phenomena respectively, and a safe design is obtained if equation 4.17 is satisfied for the uniaxial state of stress.

As another example, the well-known but less conservative Goodman line is obtained by using $\sigma_M/(S_t/N_t)$ instead of $\sigma_M/(S_y/N_y)$ in equation 4.17. In this substitution, S_t is the tensile strength of the material and N_t is a factor of safety on it.

As a final example, for static load cases $\sigma_a = 0$ with $\sigma_M = \sigma$. Thus equation 4.17 becomes $\sigma \leq S_y/N_y$, which is merely the criterion of equation 4.6 previously presented.

Von Mises-Hencky Criterion for Fatigue Failure [4.11]. For a biaxial state of varying stress (typical variations for σ_1 and σ_2 as in figure 4.5), we define mean and alternating stress components σ'_M and σ'_a in terms of the principal stresses as follows.

$$\sigma'_M = [(\sigma_1)_M{}^2 - (\sigma_1)_M(\sigma_2)_M + (\sigma_2)_M{}^2]^{1/2} \tag{4.18a}$$

$$\sigma'_a = [(\sigma_1)_a{}^2 - (\sigma_1)_a(\sigma_2)_a + (\sigma_2)_a{}^2]^{1/2} \tag{4.18b}$$

These von Mises stresses can then be used in equation 4.17 to obtain the following criterion.

$$\frac{\sigma'_a}{(S_e/N_e)} + \frac{\sigma'_M}{(S_y/N_y)} \leq 1 \tag{4.19}$$

By this criterion, satisfaction of equation 4.19 will avoid failure for the biaxial state.

Octahedral Shear Stress Criterion for Triaxial State Fatigue Failure. Experimental results indicate that an almost linear relationship exists between the *permissible* amplitude of alternating octahedral shear stress and the invariant sum of the orthogonal normal mean stresses. This criterion is presented in reference 4.2, and it is given by equation 4.20, which should be satisfied at all points in the mechanical element to avoid fatigue damage.

$$\tfrac{1}{3}[(\sigma_{1a} - \sigma_{2a})^2 + (\sigma_{2a} - \sigma_{3a})^2 + (\sigma_{3a} - \sigma_{1a})^2]^{1/2} \leq A - \alpha(\sigma_{1M} + \sigma_{2M} + \sigma_{3M}) \tag{4.20}$$

where σ_{1a}, σ_{2a}, and σ_{3a} are the amplitudes of the principal normal stresses, as typified for σ_1 in figure 4.5; $(\sigma_{1M} + \sigma_{2M} + \sigma_{3M})$ is the invariant sum of the mean normal stresses, which from the theory of elasticity is the same for any orientation of rectangular coordinate axes at the point in question. Hence, in general, $(\sigma_{1M} + \sigma_{2M} + \sigma_{3M}) = (\sigma_{xM} + \sigma_{yM} + \sigma_{zM})$. Incidentally, the mean value for the principal normal stress σ_1 is typified as σ_{1M} in figure 4.5. In equation 4.20 the values for A and α are material parameters which can be determined from special tests as described in reference 4.2.

The application of equation 4.20 to original optimum *design* studies at this time is seriously handicapped because of the material parameters A and α in that equation. Numerical values for these parameters are simply not available at this stage for most significant engineering materials, and therefore the application of equation 4.20 to most *design* problems is presently infeasible for most design engineers. As a final point of consideration, complete stress systems are either not known or not reasonably determinable at stress concentration regions for many cases, as would be required for the use of equation 4.20 in original design work. However, for design usage of the maximum shearing stress criterion, only the time variation of that particular stress component needs to be known.

EXAMPLE 4.1

At a free surface of a mechanical element, principal normal stresses $\sigma_1 = \sigma_2 = \sigma$ vary periodically in phase between 0 and σ_{max}. The material is 2024-T4 aluminum alloy. Factors of safety are to be $N_e = 1.5$ and $N_y = 1.2$, based on studies such as those in Chapter 5. Calculate the allowable value for σ_{max} based on 500×10^6 cycles of life.

Solution. Any real mechanical element problem of stress analysis should be considered in light of the triaxial state. Hence, because of the free surface stipulation we have $\sigma_3 = 0$, and the three principal shearing stresses by equations 4.5a, 4.5b, and 4.5c are as follows.

$$\tau_{1,2} = \frac{(\sigma_1 - \sigma_2)}{2} = 0$$

$$\tau_{2,3} = \frac{(\sigma_2 - \sigma_3)}{2} = \frac{\sigma}{2}$$

$$\tau_{3,1} = \frac{(\sigma_3 - \sigma_1)}{2} = -\frac{\sigma}{2}$$

Hence for equation 4.16 we use $\tau_G = \sigma/2$ as governing, from the principal shearing stresses. Thus from a time variation standpoint we would have $(\tau_G)_{max} = \sigma_{max}/2$ and $(\tau_G)_M = (\tau_G)_{max}/2 = \sigma_{max}/4$. Hence the maximum shearing stress criterion of equation 4.16 gives the relation

$$\frac{\sigma_{max}}{2} - (1-p)\left(\frac{\sigma_{max}}{4}\right) \leq \frac{S_e}{2N_e}$$

Thus we require $\sigma_{max} \leq 2S_e/(1+p)N_e$, where from table 4.1 $S_e = 20{,}000$ psi, which also gives $p = (20{,}000/1.5)/(47{,}000/1.2) = 0.34$ using equation 4.15b. Therefore, for a safe design we should have

$$\sigma_{max} \leq \frac{2 \times 20{,}000}{(1.34 \times 1.5)} = 19{,}900 \text{ psi}$$

4.9 Surface Wear Criteria for Failure [4.14, 4.15, 4.16]

In many machinery design applications, connected mechanical elements transmit load with relative rolling and/or sliding between the contacting surfaces. In such cases surface endurance and wear are of primary concern to the design engineer. Three main categories of wear will be reviewed briefly here, since they are most significant to the engineer in original design situations.

Abrasive Wear. This category of wear is generally to be avoided in good design practice, since a large amount of surface material is lost in a relatively short period of time. Hence gross plowing of this type generally makes the machine inoperative after an unacceptably short life span. Generally it is characterized by convex surfaces in contact transmitting force with high levels of contact stress and large amounts of relative sliding. The contact stress $(\sigma_c)_{max}$ is generally not far from or even greater than the yield strength of the material. To avoid the failure of abrasive wear, the design is usually modified to obtain one of the remaining two categories of wear controlled by design to an acceptable life span for the machine.

Adhesive Wear. Two surfaces in contact develop attractive forces between atoms, which when accompanied by sliding cause removal of wear fragments in the form of loose particles. In applications of good design, this category of adhesive wear is generally characterized by relatively low levels of contact stress and large amounts of surface sliding. Contact stress is reduced to an acceptable level by designing for close conformity of the contacting surfaces, such as by having a convex surface on one part mating with a concave surface on another. In this way the transmitted load is distributed over a relatively large surface contact area. A good example in machine design is the semi-lubricated sleeve bearing, commonly used in low cost applications. However, it has the disadvantage of relatively large frictional power loss at the sliding surfaces compared with rolling contact designs.

For general application, estimates of wear can be made by applying the fundamental law of adhesive wear stated by equation 4.21.

$$V \approx \frac{kFx}{3p} \tag{4.21}$$

In this equation V is the volume of material lost (cubic inches or cubic meters), k is a dimensionless wear coefficient, F is the normal contact force (lbf or newtons), x is the distance slid (inches or meters), and p is the hardness or flow pressure for the surface being worn (psi or pascals).

Flow pressure p is a material parameter which may be roughly estimated by equation 4.22 [4.17].

$$p \approx 3S_y \tag{4.22}$$

In this equation S_y is the material yield strength (psi or pascals).

TABLE 4.3

Typical Values* of Wear Coefficient k

Surface Lubrication	Metal on Metal: Like	Metal on Metal: Unlike	Nonmetal on Metal
Clean	5×10^{-3}	2×10^{-4}	5×10^{-6}
Poor	2×10^{-4}	2×10^{-4}	5×10^{-6}
Average	2×10^{-5}	2×10^{-5}	5×10^{-6}
Excellent	2×10^{-6} to 10^{-7}	2×10^{-6} to 10^{-7}	2×10^{-6}

* From Table 6.6 of reference 4.14. (By permission of John Wiley and Sons, Inc.)

Wear coefficient k is a material combination parameter affected by surface conditions. Some approximate values which may be used in design are given in table 4.3 taken from reference 4.14. Other values of k for more specific cases are also found in that source.

The adhesive wear failure criterion for acceptable design is based on equation 4.21. For the configuration of the design being considered, an acceptable amount of wear is related to the volume V of material worn, and the life for the particular application is related to the distance slid x. Hence, with estimates for material parameters p and k as previously described, allowable contact force F is determined by equation 4.21. A specific illustration of application to a design problem is example 8.3 covered on pages 233 through 241 of reference 4.18. Let us next consider a simple example.

EXAMPLE 4.2

A rectangular block of OILITE* Bronze is to reciprocate at 30 cpm with an amplitude of 0.5 in on steel while transmitting a normal load of 20 lbf. The condition of lubrication should be good with periodic oiling anticipated during the desired life span of 1000 hr. Estimate the required contact area on the rectangular block for 0.002 in wear normal to the sliding surface.

Solution. Adhesive wear is assumed appropriate and equation 4.21 will be applied. Volume of material worn will be $V = 0.002A$, where A is the area of contact on the block. For good lubrication with unlike metal on metal, we interpolate in table 4.3 and use wear coefficient $k = (2 \times 10^{-5} + 2 \times 10^{-6})/2 = 1.1 \times 10^{-5}$. As given $F = 20$ lbf, and we calculate distance slid to be $x = (2 \times 0.5 \text{ in})(30 \text{ cpm} \times 60 \text{ min/hr} \times 1000 \text{ hr}) = 1.8 \times 10^{6}$ in. Flow

* OILITE is a registered trademark of the Amplex Division of Chrysler Corporation.

pressure p is estimated by equation 4.22, where we use compressive yield strength $S_y = 11{,}000$ psi obtained from the literature. Hence $p \approx 3 \times 11{,}000 = 33{,}000$ psi. Now applying equation 4.21,

$$V = 0.002\text{A} \approx \frac{kFx}{3p}$$

$$\approx \frac{(1.1 \times 10^{-5})(20)(1.8 \times 10^{6})}{(3 \times 33{,}000)}$$

$$\therefore \quad A \approx 2.00 \text{ in}^2 \text{ is the required contact area.}$$

Surface Fatigue. High levels of combined stress are developed slightly beneath the surface of contact, resulting in initial fatigue damage at that location. Subsequently a crack propagates to the surface and progresses to a form of wear known as surface fatigue. In this category of wear, surface contact stress $(\sigma_c)_{max}$ may be reasonably large, but the elements are in rolling contact combined with perhaps a small amount of relative sliding. If stresses in the contact region are of sufficiently low level, fatigue damage does not develop and the life span can be very long. Good examples in machine design are rolling element bearings of many types and meshing gear teeth. Many experiments have been conducted resulting in publication of load-stress factors K_1 for various material combinations of design interest. Typical values of K_1 so obtained are given in table 4.4 for a virtually infinite life. The published load-stress factors K_1 have been experimentally determined, and they are the basis for wear factors K used for gear design such as in table 8.2 of Chapter 8. However, more general application of the load-stress factors K_1 can be made to other machinery design cases, such as for rolling followers in contact with cams.

The basis for the K_1 load-stress factor value is the Hertz stress equation 3.4 for cylinders, as presented in Chapter 3. Hence, with contact stress $(\sigma_c)_{max}$ at its allowable value S_{ca}, for a given material combination we see from equation 3.4 that $p(C_a + C_b)$ would be a constant, known as the load-stress factor K_1. Thus referring to figure 3.2 and equation 3.5 of Chapter 3, using $C_a = 1/\rho_a$ and $C_b = 1/\rho_b$, we derive equation 4.23.

$$F_a = \frac{K_1 l}{(1/\rho_a) + (1/\rho_b)} \tag{4.23}$$

In this equation F_a is the allowable load (lbf or newtons), K_1 is the experimental load-stress factor (psi or pascals) as given in table 4.4, l is the length of contact (inches or meters), and ρ_a and ρ_b are the radii of curvature of the contacting cylinders (inches or meters) as in figure 3.2. Hence, for avoidance of

TABLE 4.4

Typical Values* of Load-Stress Factors K_1 for 10^8 Stress Cycles with Pure Rolling

1.05 Carbon Tool Steel, Hardened through to 60–62 R_C and:	Load-Stress Factor K_1	
	psi	MPa
Gray iron casting, ASTM A 339-51 T, Grade 20, 140–160 BHN	1000	6.9
Gray iron casting, ASTM A 48-48, Class 30, 200–220 BHN	1600	11.0
Nodular iron casting, ASTM A 339-51 T, Grade 80-60-03, 207–241 BHN	2000	13.8
Gray iron casting, ASTM A 48-48, Class 35, 225–255 BHN	2300	15.9
AISI 1340 steel, induction-hardened to 45–55 R_C	10,000	69
AISI 1020 steel, carburized to 0.045 in depth of case, 50–58 R_C	13,000	90
AISI 4340 steel, induction-hardened to 50–55 R_C	13,000	90
Random fiber cotton base phenolic	1000	6.9

* From Chapter 4 of reference 4.15; see that source for much additional data of value. (By permission of The University of Michigan Press.)

the surface fatigue type of failure, the criterion is transmitted force $F \leq F_a$ by equation 4.23.

The allowable contact stress S_{ca} can be calculated based on experimental load-stress factors K_1 by means of equation 4.24, derived from equation 3.4 as follows.

$$K_1 = p(C_a + C_b)$$

$$(\sigma_c)_{\max} = S_{ca}$$

$$\therefore \quad S_{ca} = 0.564\sqrt{\frac{K_1}{(1 - \nu_a^2)/E_a + (1 - \nu_b^2)/E_b}} \tag{4.24}$$

In this equation K_1 and S_{ca} in (psi or pascals) are as defined above, whereas the other terms are defined following equation 3.4 in Chapter 3.

By equation 4.24 we can calculate the value of allowable compressive stress S_{ca} from published load-stress factors K_1 as in table 4.4 for use in contact cases other than parallel axis cylinders. Compressive stress equations for contacting bodies in general can be found in the literature, such as in references

4.3 and 4.12. Hence allowable loads based on surface fatigue can be estimated for mechanical elements having special curved surface contact conditions.

EXAMPLE 4.3

A cam made of grade 80-60-03 (207–241 BHN) nodular iron has a self-aligning rolling element follower with an outer shell of hardened steel (R_C 60) for contact with the cam. The follower shell outer diameter is 90 mm and its width is 23 mm. For a critical contact position, cam radius of curvature is 30 mm. Calculate the allowable load for this position based on surface fatigue using a factor of safety 1.5 on K_1.

Solution. Assuming a long life of 10^8 stress cycles and a condition of pure rolling between cam and follower, we obtain the value of $K_1 = 13.8 \times 10^6$ Pa from table 4.4. This we reduce by 1.5 for factor of safety and apply equation 4.23 to calculate allowable load F_a as follows.

$$F_a = \frac{(13.8 \times 10^6/1.5)(23 \times 10^{-3})}{1/[(90/2) \times 10^{-3}] + 1/(30 \times 10^{-3})}$$

$$\therefore \quad F_a = 3809 \text{ N allowable load}$$

(In English units this becomes $F_a = 856$ lbf.)

REFERENCES

4.1 Joseph Marin, *Engineering Materials*, Prentice-Hall, Englewood Cliffs, N.J., 1952, 491 pp.

4.2 George Sines and J. L. Waisman, *Metal Fatigue*, McGraw-Hill, New York, 1959, 415 pp.

4.3 S. Timoshenko and J. N. Goodier, *Theory of Elasticity*, 3rd ed., McGraw-Hill, New York, 1970, 567 pp.

4.4 R. E. Peterson, *Stress Concentration Factors*, Wiley-Interscience, New York, 1974, 317 pp.

4.5 "Materials Selector Issue," *Materials Engineering*, Reinhold, New York, published yearly.

4.6 E. A. Parker, *Materials Data Book*, McGraw-Hill, New York, 1967, 398 pp.

4.7 R. L. Peters, *Materials Data Nomographs*, Reinhold, New York, 1965, 224 pp.

4.8 J. Marin, *Mechanical Behavior of Engineering Materials*, Prentice-Hall, Englewood Cliffs, 1962, 502 pp.

4.9 J. Datsko, *Material Properties and Manufacturing Processes*, Wiley, New York, 1966, 543 pp.

4.10 R. C. Johnson, "Predicting Part Failures," Part 1, *Machine Design*, Jan. 7, 1965, pp. 137–142.

4.11 J. E. Shigley, *Mechanical Engineering Design*, 2nd ed., McGraw-Hill, New York, 1972, 753 pp.

4.12 R. J. Roark and W. C. Young, *Formulas for Stress and Strain*, 5th ed., McGraw-Hill, New York, 1975, 624 pp.

4.13 R. B. Heywood, *Designing Against Fatigue of Metals*, Reinhold, New York, 1962, 436 pp.

4.14 E. Rabinowicz, *Friction and Wear of Materials*, Wiley, New York, 1965, 244 pp.

4.15 C. Lipson and L. V. Colwell, *Handbook of Mechanical Wear*, The University of Michigan Press, Ann Arbor, 1961, 469 pp.

4.16 C. Lipson, *Wear Considerations in Design*, Prentice-Hall, Englewood Cliffs, N.J., 1967, 145 pp.

4.17 E. Rabinowicz, "Surface Energy Approach to Friction and Wear," *Product Engineering*, McGraw-Hill, New York, March 15, 1965, pp. 95–99.

4.18 R. C. Johnson, *Mechanical Design Synthesis—Creative Design and Optimization*, 2nd ed., Krieger, Huntington, N.Y., 1978, 448 pp.

EXERCISES

4.1. As reviewed in section 4.2, fatigue strength S_f for steel varies linearly with number of cycles n_c on a log-log plot, with the extremes $S_f = 0.9S_t$ at $n_c = 1000$ cycles and $S_f = k_a S'_e$ at $n_c = 10^6$ cycles. Factors k_b and k_c have not been included in these end point values. Hence the simple exponential type of equation 2.12 of Chapter 2 should be appropriate for curve fitting. Based on the relation now described, derive equation 4.2 for fatigue strength S_f of steel versus number of cycles n_c.

4.2. Endurance limit for steel is calculated by equation 4.1, and figure 3.8 of Chapter 3 may be used to estimate surface finish factor k_a. Assume a uniaxial state of stress, with size factor $k_b = 0.75$ and miscellaneous effects factor $k_c = 1.0$ for a particular application. Assume that we are interested in the possibility of finite life as well as various degrees of fluctuating stress. The latter will be accounted for by an η factor defined as

$$\eta = \frac{\sigma_{\min}}{\sigma_{\max}}$$

referring to figure 4.5 for the minimum–maximum time variation notation. Choose an appropriate failure criterion and derive an equation for allowable $\sigma_{\max}$ stress S_a as a function of $n_c (1000 \le n_c \le 10^6)$, η, S_t, S_y, k_a, k_b, and k_c. As a refinement for your analysis include a factor of safety N_y on S_y and a factor of safety N_f on fatigue strength S_f of equation 4.2. Also, assuming a ductile steel denote the nominal stress by σ and apply a stress concentration factor K_f to the alternating component with a static plastic stress concentration factor $K_p = 1.00$ applied to the mean component. Write a computer program (programmable calculator type is fine) for determining allowable $\sigma_{\max}$ denoted as S_a in terms of input data for S_t, S_y, η, k_a, k_b, k_c, N_y, N_f, n_c, and K_f. Apply your program for various combinations of input data values of interest.

4.3. In mechanical elements critical stress points often occur on free surfaces. For such a case assume that the state of stress is defined by known values of σ_x, σ_y, and τ_{xy}, where the xy plane is tangent to the free surface at the point in question. Write a computer program (programmable calculator type is fine) for determining the three principal normal stresses, the three principal shearing stresses, and the governing shearing stress τ_G. Apply your program to various combinations of input data values for σ_x, σ_y, and τ_{xy}.

4.4 Fluctuating stress is often critical at a free surface of a mechanical element, and from a time variation standpoint we have minimum and maximum values for loading F and for nominal stress components σ_x, σ_y, and τ_{xy} similar to those in figure 4.5. The fluctuating load F is such that we can define $\eta = F_{min}/F_{max}$, giving for the mechanical element the same stress proportions $\eta = (\sigma_x)_{min}/(\sigma_x)_{max} = (\sigma_y)_{min}/(\sigma_y)_{max} = (\tau_{xy})_{min}/(\tau_{xy})_{max}$. Assume a ductile steel with stress concentration factor K_f applied only to the alternating component for nominal value of governing principal shearing stress τ_G. Choose an appropriate failure criterion and derive an equation for allowable number of stress cycles n_c in terms of independent variables $(\sigma_x)_{max}$, $(\sigma_y)_{max}$, $(\tau_{xy})_{max}$, K_f, η, k_a, k_b, k_c, S_t, S_y, and factors of safety N_f and N_y (see exercise 4.2 for N_f definition). Write a computer program (programmable calculator type is fine) for calculating n_c in terms of the independent variables as input data. Apply your program to various combinations of input data values. (Incidentally, if equation 4.2 is used in your derivation, a value for $n_c \geq 10^6$ means merely infinite life, whereas if we obtain $n_c \leq 1000$ the design is not safe for even static loads.)

4.5. This exercise will serve as a check for the programs which you derived in exercises 4.2 and 4.4. For a numerical application of your program in exercise 4.2 you assumed an n_c value and determined the allowable uniaxial stress S_a for a ductile steel. This stress you may consider as $(\sigma_x)_{max}$ in a special triaxial state case where $\sigma_y = \tau_{xy} = 0$. Hence, with this as part of the input data for the program which you derived in exercise 4.4, all other common input data values being the same as for exercise 4.2, determine the number of allowable stress cycles n_c. The value of n_c so obtained from the exercise 4.4 program should agree with the input value of n_c used for the exercise 4.2 problem.

4.6. A simple beam of span L has a fluctuating central load F applied, and various beam cross sections of known section modulus (I/c) value are of interest. Suppose from a time variation standpoint that F varies periodically between F_{min} and F_{max} and we define $\eta = F_{min}/F_{max}$. Choose an appropriate failure criterion and derive an equation for allowable load $(F_{max})_a$ in terms of fatigue strength S_f; factor of safety N_f; yield strength S_y; factor of safety N_y; η, L, (I/c), and stress concentration factor K_f applied only to the alternating component of stress. Assume that the material will be ductile. Derive a computer program for calculating $(F_{max})_a$ and apply to various combinations of input data values.

4.7. In the design of semilubricated sleeve bearings a (PV) limit concept is often used as the failure criterion, where a value such as $(PV) = 20{,}000$ psi–fpm might be given for a bearing material of interest. In this criterion P is the projected area pressure calculated by $P = R/(DL)$ where R is the radial bearing load (lbf), D is the bearing diameter (inches) and L is the bearing length (inches). Also, V is the sliding velocity (fpm) calculated by $V = \pi DN/12$, where N is journal speed in rpm and D is the same as before. Assume that a specified amount of radial wear Δ is acceptable over a specified period of time T (hours). Prove that the (PV) limit concept failure criterion is compatible with the fundamental law of adhesive wear.

4.8. A class 30 gray cast iron block is to slide with a reciprocating motion on hardened steel ways with average lubrication while transmitting a normal force of 50 lbf. Contact area is $A = 3\ \text{in}^2$, and amplitude of motion is 1.00 in at a rate of 60 cycles/min. Estimate the amount of wear Δ on the block normal to the contact area after 500 h of operation.

4.9. A R_C 60 hardened steel roller of 3-in diameter is to roll with no sliding on class 30 gray cast iron plane ways while transmitting a radial load of $F = 300$ lbf. Factor of safety $N_{K_1} = 1.8$ is to be used on load-stress factor K_1 with 10^8 contact stress cycles required. Calculate the required width b of the roller and the maximum contact stress $(\sigma_c)_{max}$.

5

Factor of Safety in Design

The concept of *factor of safety* has been employed in mechanical design work for many years, and, although the necessity of its usage is well realized by practicing engineers, the basis for its selection is often nebulous, especially to engineering students. For example, factor of safety is many times erroneously interpreted as a "factor of carelessness." Hence this chapter has been devoted to this vital subject for the four basic reasons stated in the next paragraph.

First, this study should result in a better understanding, from a scientific standpoint, of the *meaning* of factor of safety as an index for product reliability. Second, the material in this chapter presents a clearer scientific picture explaining the *need* for factor of safety in real design work, especially when considered in light of the discussion of the possible gross effects resulting from minute uncontrollable manufacturing errors in Chapter 3. Third, the background of this analytical study should make possible the selection of a more realistic numerical value *estimate* for factor of safety in original design work, based on reasonable and conservative estimates of decimal percentage tolerance bands for applied load and load capability as discussed in section 5.7. Fourth, for critical design problems where the required additional effort is deemed profitable, the general analytical method of this chapter should make it possible to select an *accurate* minimum value for factor of safety, if adequate statistical information from past experience is available and if accurate information is available regarding performance requirements of the part to be designed.

At this stage of development the fourth item in the previous paragraph is often highly impractical for original design work, because of the lack of adequate available information. However, the selection of a numerical value for factor of safety based on *reasonable* and *conservative* estimates for anticipated decimal percentage tolerance bands seems to be the best which we can, in general, do for original design work at this time. Thus, the third item mentioned in the previous paragraph is feasible from a practical standpoint, especially when based on much knowledge gained from experience and analytical studies such as illustrated in Chapter 3. Certainly, in view of the present state of knowledge, the selection of a numerical value for factor of safety based on such a quasi-scientific approach is better than a completely arbitrary choice based on tradition; unfortunately, *design* engineers must work as well as possible *within* the confines of the present scientific knowledge!

5.1 Actual Load and Load Capability for a Mechanical Element

After a mechanical element has been manufactured, it is assembled in a machine or structural setting and during its useful life it will be subjected to certain *actual loads*. Hence, as shown in figure 5.1, a tensile bar might be subjected to an actual force P, a helical spring might be deflected to transmit an actual force F, a gearset might be subjected to an actual twisting moment $(M_t)_p$ at an actual angular velocity of ω_p for the pinion shaft, and a roller or ball bearing might be subjected to an actual radial load F_r. In general, the *actual load* to which a mechanical element will be subjected in usage is a function of factors *external* to the element determined by functional requirements of the entire machine or structure.

On the other hand, mechanical elements are designed for certain *load capabilities* based on anticipated loads causing the occurrence of specific types of failure phenomena. For instance, on the basis of the occurrence of yielding, the *load capability* for the tensile bar in figure 5.1a would be $(S_t)_y A$, where $(S_t)_y$ is the yield strength of the material from a simple tensile test and A is the cross sectional area of the bar. As another example, on the basis of the occurrence of unsmooth operation resulting from surface fatigue damage, the *load capability* for the ball or roller bearing in figure 5.1d would be its radial load rating R_r, which is a function of its material and geometrical characteristics. In general, the load capability of a manufactured mechanical element is a function of factors inherent in the element itself broadly classified as material and geometrical characteristics. These factors are affected by the design and by the manufacturing processes used for the mechanical element.

Generally speaking, the *actual load* to which a mechanical element will be subjected in usage is independent of the corresponding *load capability* of the element, because actual load and load capability are generally affected by factors which are independent of each other. However, for a design to be

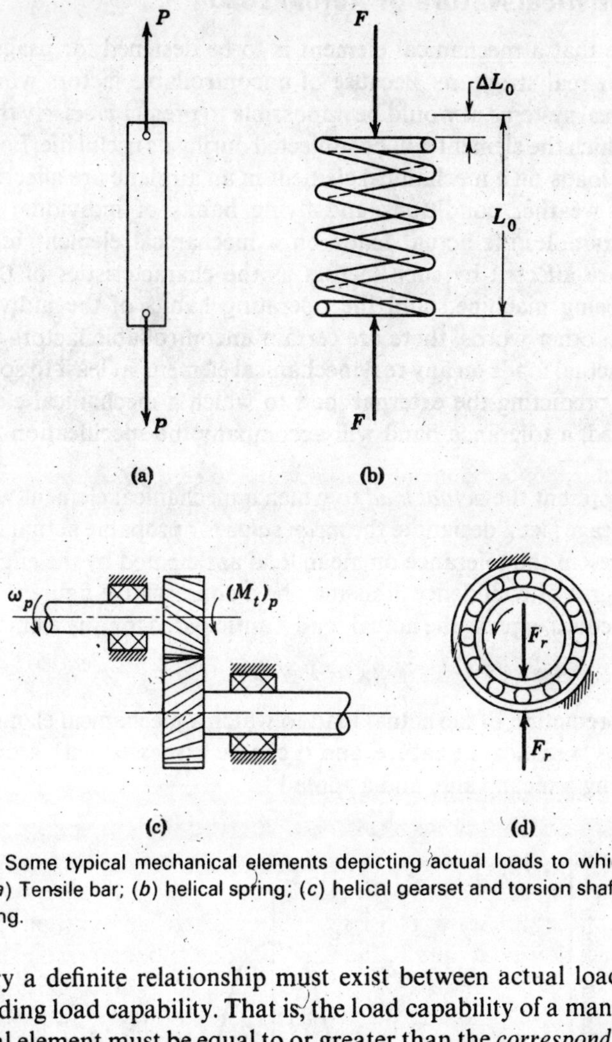

Figure 5.1 Some typical mechanical elements depicting actual loads to which they are subjected. (*a*) Tensile bar; (*b*) helical spring; (*c*) helical gearset and torsion shaft; (*d*) roller or ball bearing.

satisfactory a definite relationship must exist between actual load and the corresponding load capability. That is, the load capability of a manufactured mechanical element must be equal to or greater than the *corresponding* actual load to which the element will be subjected, in order to avoid the occurrence of the significant failure phenomenon. The specific relationship between actual load and load capability for a specific problem can be expressed by employing an appropriate criterion for failure, using stress analysis for relating external loads to significant stresses, as reviewed in the previous chapter. However, to make the discussion in this chapter general, we shall consider the relationship between actual load and load capability for a general mechanical element, only illustrating the general theory with some numerical examples.

5.2 Statistical Nature of Actual Load

Suppose that a mechanical element is to be designed for usage in a real machine or real structure. Because of uncontrollable factors which are inherent in real systems, it would be impossible to predict *precisely* the external loads to which the element will be subjected during its useful life. For instance, the actual loads on a mechanical element in an airplane are affected by such factors as weather conditions and flying habits of individual pilots. As another example, the actual loads on a mechanical element in a milling machine are affected by such factors as the characteristics of the specific material being machined and the operating habits of the individual machinists. In other words, there are certain uncontrollable factors which will affect the actual loads on any real mechanical element, at least to some extent. Hence, in predicting the external load to which a mechanical element will be subjected, a tolerance band will accompany the specification of a mean load value.

Let λ represent the *actual load* to which a mechanical element will be subjected in usage; let $\bar{\lambda}$ designate the *mean value* for probable actual loads; and let $\Delta\lambda$ represent the tolerance on mean load anticipated by the engineer with a good degree of confidence; this may be a conservative figure if necessary. Hence we could express the actual load λ anticipated for any element by

$$\lambda = \bar{\lambda} \pm \Delta\lambda \tag{5.1}$$

Thus the prediction of the actual load to which a mechanical element will be subjected is statistical in nature, and it can be expressed with a distribution curve having a mean value and a spread.

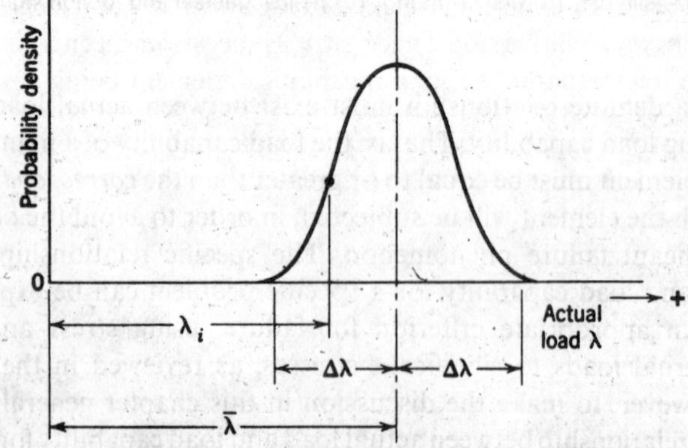

Figure 5.2 Typical statistics distribution curve for actual load λ which might be expected on a mechanical element in usage.

A typical statistical distribution curve for actual load λ is depicted in figure 5.2. The specific shape of the distribution curve depends on the possible environment to which the element will be subjected for the particular problem being considered, and it will not necessarily be symmetrical about a vertical axis as it is in figure 5.2 for reasons of simplicity. The amount of spread $\Delta\lambda$ for an actual load distribution curve will also depend on the particular problem at hand.

5.3 Statistical Nature of Load Capability

Corresponding to the actual load on a mechanical element is its *load capability*, which is based on the occurrence of a certain significant failure phenomenon. The load capability of a mechanical element is affected by geometrical and material characteristics, both of which are susceptible to minute uncontrollable manufacturing flaws which nevertheless can be significant, as discussed in Chapter 3. Hence the load capability of a real mechanical element in all probability will not be precisely the same as that predicted from ideal geometry assuming an ideal material. Thus the mean value for load capability will be accompanied by a tolerance band when the load capability of a manufactured mechanical element is being predicted.

Let $\bar{L}$ represent the *mean value* for the anticipated *load capability* of a manufactured mechanical element, which can generally be thought of as the theoretical design load capability. Let ΔL represent the tolerance on load capability, which exists because of inherently unavoidable but significant manufacturing errors. The magnitude of ΔL depends on the particular problem being considered, and it is the anticipated variation in strength estimated by the engineer from experimental knowledge with a good degree of confidence, on the conservative side if necessary. Hence, from a general viewpoint, the predicted load capability L of a mechanical element being designed can be expressed by

$$L = \bar{L} \pm \Delta L \tag{5.2}$$

Thus, associated with the prediction of load capability for a mechanical element is a statistical distribution curve, characterized by a mean value and a spread.

A typical statistical distribution curve for the predicted load capability of a manufactured element is depicted in figure 5.3. The specific shape of the distribution curve depends on the anticipated strength for the particular element being designed, and it will not necessarily be symmetrical about a vertical axis, as it is shown in figure 5.3 for simplicity. The amount of spread,

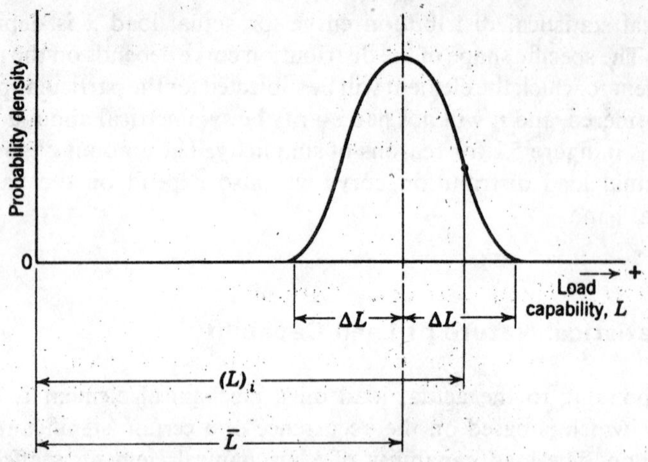

Figure 5.3 Typical statistics distribution curve for load capability L of a manufactured mechanical element.

ΔL, for a load capability distribution curve will also depend on the particular mechanical element being designed and the effects of the entire manufacturing process.

5.4 Basic Relationship between Statistical Distributions for Actual Load and Load Capability of a General Mechanical Element [5.1]

The relationship between the distribution curves for actual load and load capability of a manufactured mechanical element is very important in design work. The specific relationship which *should* exist for a *satisfactory design* depends on the particular mechanical element under consideration and the significance attached to the occurrence of the particular failure phenomenon. The importance of the occurrence of a particular failure phenomenon depends on the specific application of the machine or structure and the consequence of the occurrence of the failure phenomenon with respect to human life and economics.

Where the occurrence of the failure phenomenon would be disastrous, the relationship between the distribution curves for the actual load and the corresponding load capability should be as shown in figure 5.4. The mean value for the actual load $\bar{\lambda}$ should be smaller than the mean value of the corresponding load capability $\bar{L}$, to an extent which will guarantee no realistic overlapping of the distribution curves. The amount of separation

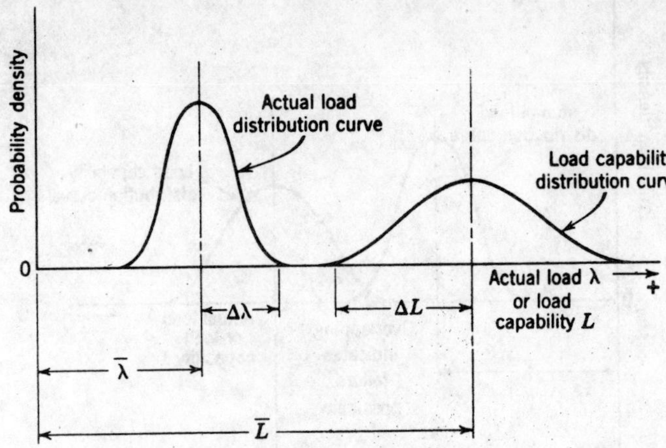

Figure 5.4 Proper relationship of typical distribution curves for actual load λ and load capability L for a mechanical element for the case where the occurrence of the significant failure phenomenon would be disastrous.

which must exist between $\bar{\lambda}$ and $\bar{L}$ for such cases depends on the spread associated with both distribution curves.

On the other hand, there are many real design cases where the occurrence of a particular failure phenomenon could be tolerated occasionally with no significant consequences, although it would be highly desirable to avoid such an occurrence. Such might be the case if a mechanical element will wear after much usage because of fatigue at contacting surfaces. For example, in designing a gearset for a particular machine, a life of 100 million cycles might be desirable. However, if *occasionally* a gearset wears significantly after, say, 90 million cycles, the consequences might not be grave and then the situation could be tolerated. For such cases the relationship between distribution curves for actual load and load capability could be such that overlapping exists, as depicted in figure 5.5. In fact, for such cases where the failure phenomenon can be tolerated occasionally, the mean values for actual load and load capability might possibly coincide with no grave consequences resulting, depending on the particular problem at hand.

At this time it will be appropriate to make a general definition of what we mean by the term *factor of safety*. Factor of safety can be defined as the ratio of the mean load capability to the corresponding mean actual load for a mechanical element, as expressed by

$$N = \frac{\bar{L}}{\bar{\lambda}} \tag{5.3}$$

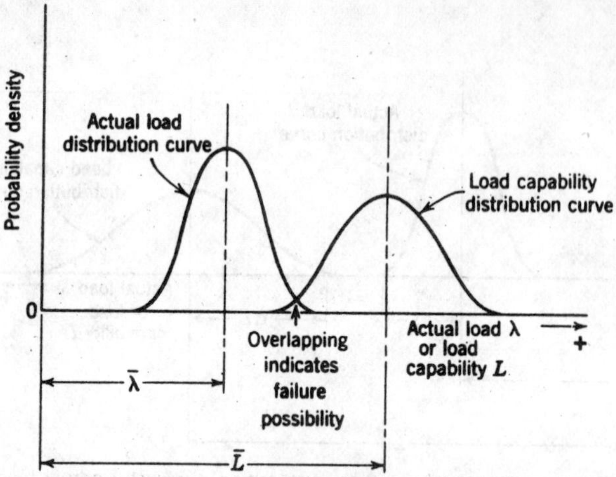

Figure 5.5 Relationship of typical distribution curves for actual load λ and load capability L for a mechanical element for the case where the occurrence of the failure phenomenon can be tolerated occasionally.

where N is the factor of safety, $\bar{L}$ is the mean value for load capability of the manufactured mechanical element, and $\bar{\lambda}$ is the mean value for the corresponding actual load to which the element will be subjected in usage. Hence factor of safety N is related to the occurrence of a particular failure phenomenon, since load capability is always referred to such an occurrence. Therefore factor of safety referred to yielding of the material might be designated N_y, whereas factor of safety referred to an ultimate strength with the occurrence of breakage or material separation might be designated N_{ult}. The selection of factor of safety for design use will be discussed at the end of this chapter. Choosing an appropriate numerical value depends on the spread and shape of each distribution curve, the mean value for the actual load, and the significance attached to the occurrence of the failure phenomenon.

For a mechanical element in a machine or structure being used, the particular failure phenomenon will occur if the load capability based on that phenomenon is less than the corresponding actual load. The probability density or frequency of failure occurrence will be of great interest in design work. Hence the distribution curve for the *difference* between load capability and actual load will be very significant, since the probability of failure occurrence can be estimated from it. The distribution curve for $(L - \lambda)$ is typically depicted in figure 5.6; negative values correspond to the occurrence of the failure phenomenon, whereas positive values correspond to no failure occurrence. Therefore, for cases where the occurrence of the failure phenomenon would be disastrous, the distribution curve for $(L - \lambda)$ should be so

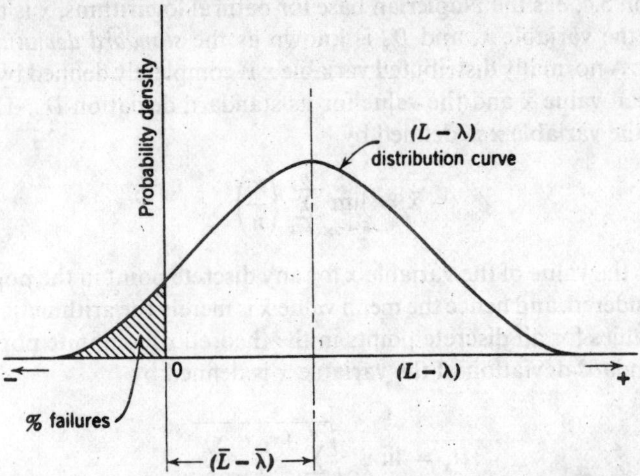

Figure 5.6 Typical distribution curve for $(L - \lambda)$ (where L = load capability of a manufactured element with respect to the occurrence of a particular failure phenomenon and λ = actual load on element in usage) for a mechanical element, showing the possible occurrence of the significant failure phenomenon.

located that all *realistic* values for $(L - \lambda)$ are positive. On the other hand, for cases where the occurrence of the failure phenomenon could be occasionally tolerated, realistic values for $(L - \lambda)$ can be negative, and the area under the $(L - \lambda)$ distribution curve for this negative portion would be the decimal percentage of failures which should be expected.

5.5 Normal or Gaussian Distribution of a Variable [5.3, 5.5, 5.6]

A variable x is said to be normally distributed if it is a linear function of several other random variables, all of approximately equal significance. According to the central-limit theorem in statistics, a variable x is often normally distributed even if the random variables on which it is dependent are not themselves normally distributed. Hence it is reasonable to assume a normal or Gaussian distribution for many real cases.

If a variable x is assumed to be normally distributed, the frequency of occurrence, y, can be expressed by equation 5.4, which is commonly found in books on statistics.

$$y = \frac{1}{D_x\sqrt{2\pi}} e^{-(x-\bar{x})^2 / 2D_x^2} \tag{5.4}$$

In equation 5.4, e is the Napierian base for natural logarithms, $\bar{x}$ is the mean value for the variable x, and D_x is known as the *standard deviation* of the variable x. A normally distributed variable x is completely defined by specifying its mean value $\bar{x}$ and the value for its standard deviation D_x. The mean value for the variable x is defined by

$$\bar{x} = \lim_{n \to \infty} \sum_{i=1}^{n} \left(\frac{x_i}{n}\right) \tag{5.5}$$

where x_i is the value of the variable x for any discrete point in the population being considered, and hence the mean value $\bar{x}$ is merely the arithmetic average of the x values for all discrete points in the theoretically infinite population.

The standard deviation of the variable x is defined by

$$D_x = \lim_{n \to \infty} \sqrt{\sum_{i=1}^{n} \frac{(x_i - \bar{x})^2}{n}} \tag{5.6}$$

In this equation x_i is the value of the variable x for any discrete point in the infinite population being considered and $\bar{x}$ is the mean value for the variable x as defined by equation 5.5. Hence, from equation 5.6, we see that the standard deviation D_x is merely the rms average of the error values $(x_i - \bar{x})$ for all discrete points in the theoretically infinite population being considered. In statistics textbooks the standard deviation is generally designated σ, but we shall use D_x for standard deviation in this book in order to avoid confusion with our *normal stress* σ designation introduced in Chapter 4.

The normal or Gaussian distribution curve as expressed by equation 5.4 is typically depicted in figure 5.7, and it is symmetrical with respect to its mean value $\bar{x}$. The area under the total distribution curve for $-\infty < x < +\infty$ represents 100% of the total population for all x values. The area under the distribution curve for $(\bar{x} - D_x) < x < (\bar{x} + D_x)$ represents 68.3% of the total population for all x values. The area under the distribution curve for $(\bar{x} - 3D_x) < x < (\bar{x} + 3D_x)$ represents 99.73% of the total population for all x values. In quality control work, a tolerance specified for a dimension is generally assumed to be three times the standard deviation, and we shall abide by this assumption for general application in this book. Therefore, if a variable x is expressed with good confidence as a mean value $\bar{x}$ plus or minus a tolerance Δx, we shall consider Δx to be $3D_x$, where D_x is the standard deviation with respect to x. In equation form, this expression is

$$x = \bar{x} \pm \Delta x = \bar{x} \pm 3D_x \tag{5.7}$$

In reality, if a variable x is expressed as a mean value $\bar{x}$ plus or minus an increment Δx, the engineer has good confidence that the variable x will fall within the tolerance limits. If the increment Δx is three times the standard

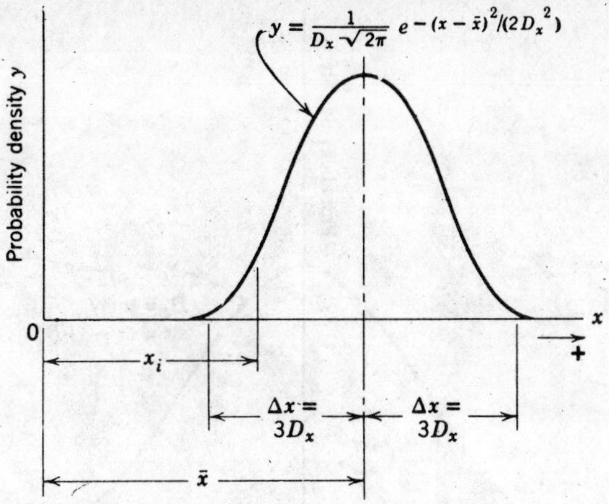

Figure 5.7 Typical normal or Gaussian distribution curve for a variable x, showing assumed relationship of commonly used tolerance band to standard deviation D_x.

deviation of the normal distribution (i.e., $3D_x$), then the variable x will fall within the tolerance range for 99.73% of the time.

In order to take advantage of available statistics tables, a new unitless variable t will be defined now for expressing the original variable x. The linear relationship between t and x is expressed by the following equation, which introduces the unitless variable t.

$$t = \frac{(x - \bar{x})}{D_x} \tag{5.8}$$

In equation 5.8, $\bar{x}$ is the mean value for the given distribution from equation 5.5, and D_x is the standard deviation for the given distribution from equation 5.6.

If the given variable x is normally distributed, the new variable t will also be normally distributed, and its standard deviation D_t will correspond to $(x - \bar{x}) = D_x$ in equation 5.8. Hence the standard deviation D_t for the variable t is merely unity. Also, the mean value for the variable t is merely zero. Therefore the distribution curve for the variable t is symmetrical about $t = 0$, as shown in figure 5.8. With the transformation of variables according to equation 5.8, the equation for the distribution curve in figure 5.8 is, from equation 5.4,

$$y_t = \frac{1}{\sqrt{2\pi}} e^{-t^2/2} \tag{5.9}$$

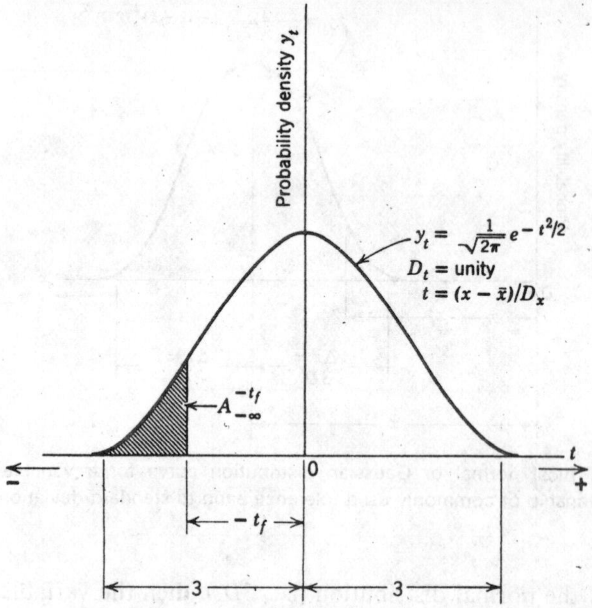

Figure 5.8 Normal or Gaussian distribution curve for unitless variable t, which is related to given variable x of figure 5.7 by equation 5.8.

For the distribution curve associated with the variable t, the area under the entire curve for $-\infty < t < +\infty$ is merely unity. Statistics tables are generally given for relating significant areas under the distribution curve to the variable t. Of particular interest for later studies in this chapter will be the area under the distribution curve for $-\infty < t < -t_f$, where t_f is a specific positive number of significance. This area of interest is expressed by

$$A_{-\infty}^{-t_f} = \int_{-\infty}^{-t_f} y_t \, dt = \int_{-\infty}^{-t_f} \frac{e^{-t^2/2}}{\sqrt{2\pi}} \, dt \tag{5.10}$$

The area expressed by equation 5.10 is the shaded portion in figure 5.8; it equals the decimal percentage of all points having values of t algebraically less than $-t_f$. Statistics tables are readily available for determining the area $A_{-\infty}^{-t_f}$ as a function of t_f, and later in this chapter it will be desirable to express this function by a mathematical equation valid over a certain range of interest. Hence we can convert the tabulated data from statistics tables to a mathematical equation using curve-fitting techniques from Chapter 2. In this way equation 5.11, which is valid for $0.001 \leq A_{-\infty}^{-t_f} \leq 0.015$ with less than 2 % error, was derived. Later in this chapter we shall show that this is com-

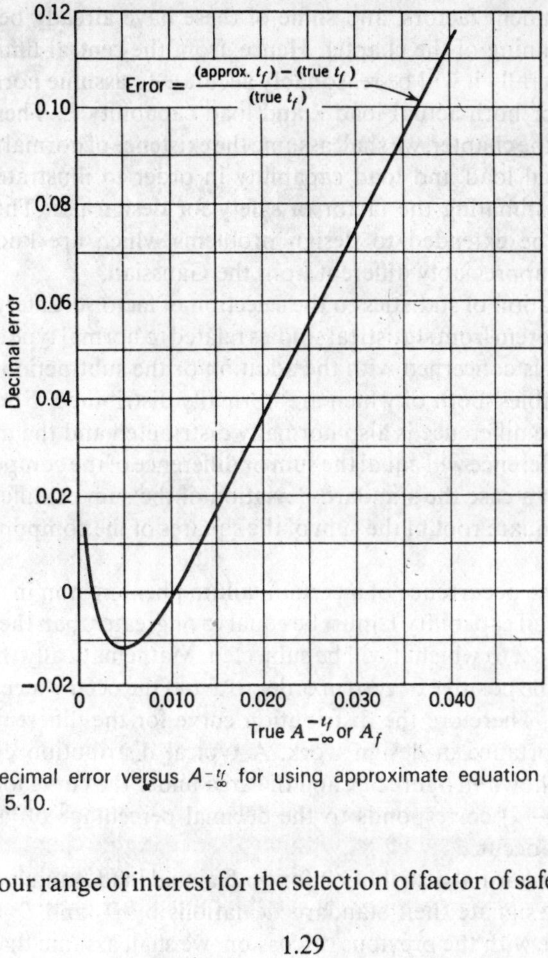

Figure 5.9 Decimal error versus $A_{-\infty}^{-t_f}$ for using approximate equation 5.11 instead of exact equation 5.10.

patible with our range of interest for the selection of factor of safety for critical problems.

$$t_f = \frac{1.29}{(A_{-\infty}^{-t_f})^{0.128}} \tag{5.11}$$

The decimal percentage error which is associated with using approximate equation 5.11 instead of exact equation 5.10 was calculated for true values of $A_{-\infty}^{-t_f}$, and the results are graphically given in figure 5.9.

5.6 Relationship of Actual Load λ to Load Capability *L*, Assuming Normal Distributions

The actual load to which a mechanical element will be subjected and the corresponding load capability for the occurrence of a significant failure phenomenon in a manufactured mechanical element are both dependent on many independent factors, and some of these have already been discussed near the beginning of the chapter. Hence, from the central-limit theorem in statistics, generally it will be reasonably accurate to assume normal distribution curves for both actual load λ and load capability L. Therefore, in the remainder of the chapter, we shall assume the existence of normal distributions for both actual load and load capability in order to illustrate the general method for estimating the factor of safety for design use. The same basic method can be extended to design problems which are known to have distributions appreciably different from the Gaussian.

The application of statistics to the selection of factor of safety will depend on a basic theorem from statistical studies related to normal type distributions. This theorem is concerned with the addition or the subtraction of two independent variables, both of which are normally distributed. For such a case, the sum or the difference is also normally distributed and the mean value of the sum or difference will equal the sum or difference of the component means. Also, for such a case the standard deviation of the sum or difference will be equal to the square root of the sum of the squares of the component standard deviations.

To avoid the occurrence of a certain failure phenomenon in a mechanical element, its load capability L must be equal to or greater than the corresponding actual load λ to which it will be subjected. Mathematically this states that $(L - \lambda)$ must be positive or zero in order to avoid the occurrence of the failure phenomenon. Therefore the distribution curve for the difference $(L - \lambda)$ is of great importance in design work. A typical distribution curve for this difference is shown in figure 5.6, and the area under the curve for all negative values of $(L - \lambda)$ corresponds to the decimal percentage of failures which will probably occur.

Assuming a normal distribution for both actual load λ and load capability L, we shall designate their standard deviations by D_λ and D_L respectively. In accordance with the previous discussion, we shall assume that a tolerance of $\pm\Delta\lambda$ on actual load λ with respect to the mean load $\bar{\lambda}$ will correspond to $\pm 3D_\lambda$, as in quality control. In the same way, we shall assume that a tolerance of $\pm\Delta L$ on load capability L with respect to mean value $\bar{L}$ will correspond to $\pm 3D_L$. Considering now the distribution curve for the difference $(L - \lambda)$, its standard deviation $D_{(L-\lambda)}$ could be calculated, from the basic theorem stated near the beginning of this section, according to the equation

$$D_{(L-\lambda)} = \sqrt{(D_L)^2 + (D_\lambda)^2} \tag{5.12}$$

Equation 5.12 expresses the standard deviation of the very important distribution for the *difference* $(L - \lambda)$ as a function of the standard deviations of the load capability distribution and the actual load distribution.

In order to take advantage of available statistics information, we shall introduce a new unitless variable t defined by equation 5.8, where x is considered to be merely $(L - \lambda)$. From figure 5.6, the point of interest is $(L - \lambda) = 0$, since the failure phenomenon will occur for all negative values of $(L - \lambda)$. Thus, considering x to be $(L - \lambda)$ in equation 5.8, we have

$$t = \frac{(L - \lambda) - (\bar{L} - \bar{\lambda})}{D_{(L-\lambda)}}$$

Therefore the point of interest on the unitless t axis is $-t_f$, corresponding to $(L - \lambda) = 0$, where

$$t = -t_f = \frac{0 - (\bar{L} - \bar{\lambda})}{D_{(L-\lambda)}}$$

Thus the value for t below which failure will occur is $-t_f$, and, using equation 5.12 in the preceding equation, we obtain

$$t_f = \frac{(\bar{L} - \bar{\lambda})}{\sqrt{D_L{}^2 + D_\lambda{}^2}} \tag{5.13}$$

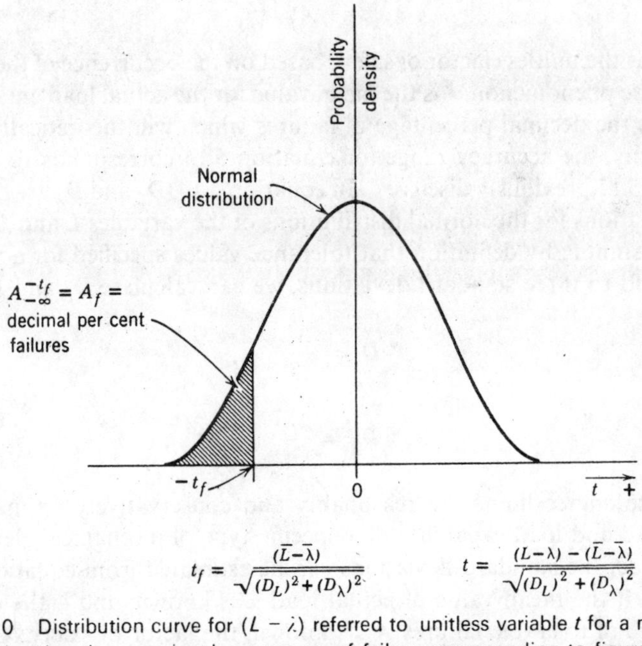

Figure 5.10 Distribution curve for $(L - \lambda)$ referred to unitless variable t for a mechanical element, showing the occasional occurrence of failure corresponding to figure 5.6. (L = load capability of manufactured element for occurrence of a particular failure phenomenon; λ = actual load anticipated on element in usage; D_L = standard deviation of variable L; D_λ = standard deviation of variable λ.)

The distribution curve for the difference $(L - \lambda)$ referred to the unitless t axis is shown in figure 5.10. The area under this curve for $-\infty < t < -t_f$ gives the decimal percent of mechanical elements for which the failure phenomenon will theoretically occur, expressed as $A_{-\infty}^{-t_f}$. To simplify notation we shall designate this significant area A_f, which is the decimal percentage of failures that will theoretically occur. Using this notation, we combine equations 5.11 and 5.13:

$$t_f = \frac{1.29}{A_f^{0.128}} = \frac{(\bar{L} - \bar{\lambda})}{\sqrt{D_L^2 + D_\lambda^2}}$$

$$\therefore \quad \frac{\bar{L}}{\bar{\lambda}} = 1 + \frac{1.29\sqrt{D_L^2 + D_\lambda^2}}{A_f^{0.128}\,\bar{\lambda}}$$

However, from the definition expressed by equation 5.3, $\bar{L}/\bar{\lambda}$ is merely the factor of safety N. Therefore we have the following equation, which can be used for calculating the factor of safety.

$$N = 1 + \frac{1.29\sqrt{D_L^2 + D_\lambda^2}}{A_f^{0.128}\,\bar{\lambda}} \tag{5.14}$$

where N is the unitless factor of safety based on the occurrence of the significant failure phenomenon, $\bar{\lambda}$ is the mean value for the actual load anticipated, and A_f is the decimal percentage of failures which will theoretically occur. Incidentally, the accuracy range for equation 5.14 corresponds to that for equation 5.11 previously discussed. In equation 5.14, D_L and D_λ are the standard deviations for the normal distributions of the variables L and λ respectively. Assuming by definition that tolerance values specified for a variable correspond to three standard deviations, we can calculate D_L and D_λ:

$$D_L = \frac{\Delta L}{3} \tag{5.15}$$

$$D_\lambda = \frac{\Delta \lambda}{3} \tag{5.16}$$

Thus, if tolerance limits are reasonably and conservatively estimated for actual load and load capability of a specific type of mechanical element to be designed, the standard deviations can be estimated from equations 5.15 and 5.16. If the mean value of actual load $\bar{\lambda}$ is known, and if the decimal percentage of tolerable failures A_f is known, the factor of safety N can be calculated reasonably and conservatively from equation 5.14. Then the required mean load capability $\bar{L}$ can be calculated from equation 5.3, and this will be the basis for design of the mechanical element.

For the remainder of the chapter, we shall be concerned with the problem of applying the statistics theory which has been summarized to the selection of a reasonable and conservative factor of safety for subsequent *design* usage. Hence, in the examples presented, we are merely illustrating how the previously summarized theory might be used for selecting a reasonable and conservative appropriate numerical value for the factor of safety *before* starting the final design of the element. Thus the examples in this chapter are *not* illustrations of *mechanical design*, which will be emphasized in the remaining chapters of this book. Also, the examples in this chapter do *not* illustrate the application of the previous statistics theory to the *analysis* of parts which have already been designed, although this should be a straightforward procedure after this chapter has been studied.

EXAMPLE 5.1

A tensile bar is to be designed for transmitting an axial force P as shown in figure 5.11. Because of factors beyond the control of the design engineer, the precise value for the *actual* force P cannot be predicted. However, considering the particular structure for which the tensile bar will be used, the engineer can calculate the ideal force $\bar{P}$ with a conservative estimate for a tolerance band $\pm \Delta P$ within which the actual force P will fall. Hence the actual load P to which the bar will be subjected is expressed as $P = \bar{P} \pm \Delta P$.

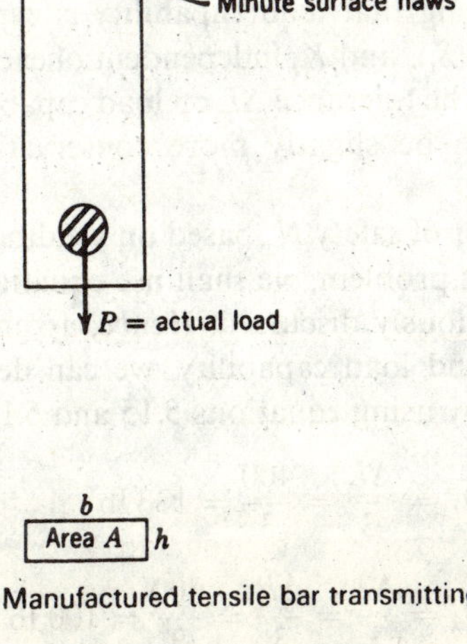

Figure 5.11 Manufactured tensile bar transmitting actual load P.

Suppose that the numerical value for the ideal load is $\bar{P} = 1000$ lb and that a reasonable but conservative tolerance for this mean load is $\pm\Delta P = \pm 300$ lb. Therefore the actual value for P which will be encountered will fall within the range expressed as $P = 1000 \pm 300$ lb. Assume that the actual force P to which the bar will be subjected will be virtually a static load within this range.

Assume that the particular structure wherein this tensile bar will be used is such that yielding of the tensile bar is tolerable but undesirable. Therefore suppose that a 1% probability for yielding of the tensile bar, which by definition of A_f corresponds to a value $A_f = 0.010$, is considered acceptable.

The ideal *load capability* for the tensile bar will be $(S_t)_y A/K_t$, where K_t is the theoretical stress concentration factor associated with the pin connections in figure 5.11, A is the cross sectional area for the tensile bar to be determined by design, and $(S_t)_y$ is the published yield strength of the material as determined by a simple tensile test. From the given load conditions, we know that the load capability for the designed tensile bar will be of the order of magnitude of approximately 1300 lb. The desired theoretical cross sectional area A for the bar can be obtained with virtually no error in manufacturing the bar, and, as shown in Chapter 3, minute surface flaws will be insignificant for static loads with real materials. However, because of uncontrollable factors in the manufacturing processes of materials, we shall assign a conservative error of $\pm 5\%$ to the published value for $(S_t)_y$. Also, the theoretical stress concentration factor K_t for a pin connection region cannot be predicted precisely. Now suppose that K_t can be determined within $\pm 25\%$ as a conservative estimate for the pin connection region. Therefore the load capability for the manufactured tensile bar can be predicted conservatively within $\pm(5 + 25) = \pm 30\%$. This estimate is based on the study of relative change for simple exponential equations in section 2.8 of Chapter 2, considering that load capability is expressed as $[(S_t)_y A/K_t]$ for the tensile bar with $(S_t)_y$ and K_t independent of each other. Therefore the order of magnitude for the tolerance ΔL on load capability will be $\approx \pm(0.30)(1300) = \pm 390$, and to be slightly more conservative we shall estimate $\Delta L = \pm 400$ lb.

To estimate the factor of safety N_y based on yielding which is appropriate to use in design for this problem, we shall use equation 5.14, which is valid for $A_f = 0.010$ as previously discussed. Hence, assuming normal distributions for actual load and load capability, we can determine the standard deviations D_λ and D_L by using equations 5.15 and 5.16 as follows.

$$D_L = \frac{\Delta L}{3} = \frac{400}{3} = 133 \text{ lb}$$

$$D_\lambda = \frac{\Delta\lambda}{3} = \frac{\Delta P}{3} = \frac{300}{3} = 100 \text{ lb}$$

Since for this problem the mean actual load $\bar{\lambda}$ is $\bar{P} = 1000$ lb, and since the tolerable decimal percentage failures resulting from yielding is $A_f = 0.010$, equation 5.14 becomes

$$N_y = 1 + \frac{1.29\sqrt{(133)^2 + (100)^2}}{(0.010)^{0.128}(1000)} = 1 + \frac{1.29 \times 167}{(0.554)(1000)}$$

$$\therefore \quad N_y = 1.39 \approx 1.4$$

For this particular problem it would be appropriate to use a factor of safety of approximately 1.4 for N_y based on yielding. Thus, from equation 5.3, the mean load capability $\bar{L}$ based on yielding should *actually* be $\bar{L} = N\bar{\lambda} = N_y\bar{P} = (1.4)(1000) = 1400$ lb for the manufactured bar. Consequently we would conduct the design of the tensile bar on the basis of this theoretical load capability.

EXAMPLE 5.2

We consider now an example of determining an appropriate factor of safety for the design of a tensile bar subjected to a varying axial load. Suppose that the tensile bar in figure 5.11 is subjected to an axial load which varies cyclically from 0 to P_{max} and back to 0, as illustrated in figure 5.12. Assume that it is impossible to predict precisely the value P_{max} will have in the actual machine being designed; then we can express only the *anticipated* value as a mean value $\bar{P}_{max}$ with a conservative tolerance band $\pm\Delta P_{max}$ as follows:

$$P_{max} = \bar{P}_{max} \pm \Delta P_{max}$$

Suppose for this problem that the mean value for the actual maximum load P_{max} is $\bar{P}_{max} = 1000$ lb and that the tolerance band is conservatively estimated as $\pm\Delta P_{max} = \pm 500$ lb.

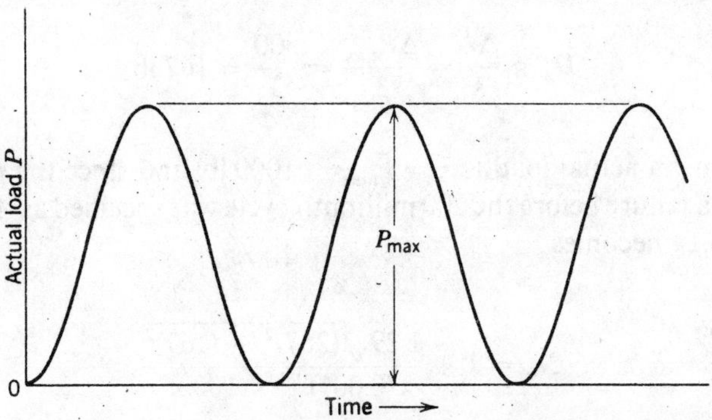

Figure 5.12 Typical time variation of axial load P for manufactured tensile bar (figure 5.11) in usage.

Suppose that the significant failure phenomenon is the occurrence of breakage because of fatigue at the 200 millionth cycle of load application. Suppose that 99.9% probability is desired for a life of 200 million cycles or longer, but that a failure occurring slightly before 200 million cycles could be tolerated with a probability of 0.1% for such an occurrence. Hence, for this problem, $A_f = 0.001$.

Because of the gross effects of minute manufacturing errors on product performance for varying stress cases, as discussed in Chapter 3, the load capability for a manufactured bar based on a fatigue failure at the 200 millionth cycle cannot be predicted precisely. The prediction of the theoretical stress concentration factor K_t at the pin connections, the effects of residual stresses in the bar, the effects of minute manufacturing and material flaws, and the accuracy of the available fatigue strength S_f for a material are all reasons for the difficulty in accurately predicting the load capability of the manufactured bar. Hence, on the basis of typical calculations of error effects similar to those in Chapter 3, of studies of published results from fatigue experiments indicating probable degrees of scatter, and of practical fatigue failure experience, suppose that the engineer conservatively estimates that the load capability for a manufactured tensile bar can be predicted within $\pm\Delta L = \pm 800$ lb with respect to the occurrence of a fatigue failure at exactly 200 million cycles.

An appropriate factor of safety N_e based on the occurrence of a fatigue failure not before 200 million cycles with 99.9% confidence in the prediction will now be estimated from equation 5.14. Assuming normal distributions for both the actual maximum load P_{max} and the load capability of a manufactured tensile bar based on a fatigue failure at the 200 millionth cycle, the standard deviation D_L and D_λ can be estimated from equations 5.15 and 5.16 as follows.

$$D_L = \frac{\Delta L}{3} = \frac{800}{3} = 267 \text{ lb}$$

$$D_\lambda = \frac{\Delta\lambda}{3} = \frac{\Delta P_{max}}{3} = \frac{500}{3} = 167 \text{ lb}$$

Since the mean actual load is $\bar{\lambda} = \bar{P}_{max} = 1000$ lb, and since the probability for having a failure before the 200 millionth cycle was specified as $A_f = 0.001$, equation 5.14 becomes

$$N_e = 1 + \frac{1.29\sqrt{(267)^2 + (167)^2}}{(0.001)^{0.128}(1000)}$$

$$\therefore N_e = 1.984$$

Hence, for this particular problem for subsequent design of the tensile bar, it would be appropriate to select a factor of safety of approximately $N_e \approx 2$, based on a fatigue failure occurring not before 200 million cycles of load application at least 99.9% of the time.

5.7 Selection of Factor of Safety Based on Percentage Estimates for Tolerances on Actual Load and Load Capability

In the previous section we determined a reasonable value for the factor of safety N based on equation 5.14, which has been thoroughly discussed and illustrated. However, for the original design of some mechanical elements, for selecting an appropriate factor of safety equation 5.14 has a serious limitation since it requires estimating the tolerance ΔL on mean load capability $\bar{L}$ which is not precisely known *before* the element has been designed. We circumvented this difficulty in example 5.1 by estimating a reasonable order of magnitude for $\bar{L}$, and a more careful study of equation 5.14 would show us that large errors in estimating $\bar{L}$ will result in small errors for the determined reasonable value of N.

An equation having more general application for selecting an appropriate factor of safety for original design usage can be derived on the basis of percentage estimates for tolerances on actual load and on load capability. Hence, by using equations 5.3, 5.14, 5.15, and 5.16, we obtain the following derivation.

$$N = 1 + \frac{1.29\sqrt{(\Delta L/3)^2 + (\Delta\lambda/3)^2}}{A_f^{0.128}\bar{\lambda}}$$

$$\therefore \quad N = 1 + \frac{0.43\sqrt{(\Delta L/\bar{L})^2 N^2 + (\Delta\lambda/\bar{\lambda})^2}}{A_f^{0.128}}$$

$$\therefore \quad [5.4A_f^{0.256} - (\Delta L/\bar{L})^2]N^2 - [10.8A_f^{0.256}]N + [5.4A_f^{0.256} - (\Delta\lambda/\bar{\lambda})^2] = 0 \quad (5.17)$$

In equation 5.17, N is the unitless factor of safety which is appropriate for design based on the occurrence of the significant failure phenomenon, A_f is the decimal percentage of tolerable failures permitted for the particular problem at hand, $(\Delta L/\bar{L})$ is the decimal percent tolerance band anticipated on load capability for the manufactured element, and $(\Delta\lambda/\bar{\lambda})$ is the corresponding decimal percent tolerance band on the anticipated actual load for the element in usage. Hence, if reasonable and conservative estimates are made for the decimal percent tolerance band $(\Delta L/\bar{L})$ on anticipated *load capability* and for the decimal percent tolerance band $(\Delta\lambda/\bar{\lambda})$ on anticipated *actual load*, equation 5.17 can be solved for the appropriate and reasonable factor of safety N by using the quadratic equation. Thus, in using equation 5.17 it is unnecessary to estimate an order of magnitude for the anticipated load capability $\bar{L}$ of the element to be designed, as we did in example 5.1. Instead, *decimal percent* estimates for $(\Delta L/\bar{L})$ can be extended from past studies and past experiences to general case types in original design problems. Therefore equation 5.17 can be applied more easily to original design problems than equation 5.14.

For best accuracy of equation 5.17, A_f should be within the range $0.001 \leq A_f \leq 0.015$, as previously discussed. For cases where failure occurrences can be tolerated to a limited extent with decimal percentage failures $A_f > 0.015$, precise accuracy of A_f is generally not a significant requirement and hence equation 5.17 will be acceptable for estimating a reasonable value of N for most real problems. On the other hand, for the case where *no* failure occurrences could ever be tolerated, A_f is equal to zero, and a different approach must be followed for determining a reasonable value of N; this will be given in section 5.8.

EXAMPLE 5.3

Consider the problem of selecting a reasonable factor of safety N_y for use in the design of the tensile bar as described in example 5.1. Using numerical values from that example, we have

$$A_f = 0.010$$

$$\frac{\Delta\lambda}{\bar{\lambda}} = \frac{\Delta P}{\bar{P}} = \frac{300}{1000} = 0.300$$

$$\frac{\Delta L}{\bar{L}} = 0.05 + 0.25 = 0.300$$

Thus, for this specific example, we make the reasonable and conservative estimates that the actual load P to which the bar will be subjected in usage can be predicted within $\pm 30\%$ and that the load capability of the manufactured bar can be predicted within $\pm 30\%$, as previously estimated for example 5.1. Also, as in example 5.1 we assume that a 1% probability for yielding of the tensile bar is considered acceptable. Therefore, from equation 5.17, we determine the following reasonable value for factor of safety N_y.

$$[5.4(0.010)^{0.256} - (0.3)^2]N_y^2 - [10.8(0.010)^{0.256}]N_y + [5.4(0.010)^{0.256} - (0.3)^2] = 0$$

$$\therefore \quad N_y^2 - (2.12)N_y + 1 = 0$$

Therefore, by the quadratic equation,

$$N_y = \frac{+2.12 + \sqrt{(2.12)^2 - 4}}{2} = 1.06 + 0.35$$

$$\therefore \quad N_y = 1.41$$

An appropriate factor of safety for this problem as determined by equation 5.17 would be $N_y = 1.41$, which compares very favorably with the results of example 5.1 as determined by equation 5.14 for which a reasonable value for $\bar{L}$ had to be estimated.

EXAMPLE 5.4

Consider the problem of selecting a reasonable factor of safety N_e based on fatigue loading of the tensile bar as described in example 5.2. However, instead of estimating that the load capability for a manufactured tensile bar can be predicted within ±800 lb with respect to the occurrence of a fatigue failure at exactly 200 million cycles, suppose that a reasonable and conservative estimate from past studies is that the load capability for a manufactured tensile bar can be predicted within ±40% with respect to the occurrence of a fatigue failure at exactly 200 million cycles of loading. Thus, referring to example 5.2, we have the following numerical values to use in equation 5.17 for this problem.

$$\frac{\Delta L}{\bar{L}} = 0.40$$

$$\frac{\Delta \lambda}{\bar{\lambda}} = \frac{\Delta P_{max}}{\bar{P}_{max}} = \frac{500}{1000} = 0.50$$

$$A_f = 0.001$$

Therefore, from equation 5.17, we can determine the following reasonable value for factor of safety N_e based on the occurrence of a fatigue failure not before 200 million cycles of loading, with 99.9% confidence in the prediction.

$$[5.4(0.001)^{0.256} - (0.40)^2]N_e^2 - [10.8(0.001)^{0.256}]N_e + [5.4(0.001)^{0.256} - (0.50)^2] = 0$$

$$\therefore \quad N_e^2 - 2.42N_e + 0.883 = 0$$

Therefore, by the quadratic equation,

$$N_e = \frac{2.42 + \sqrt{(2.42)^2 - 4(0.883)}}{2} = 1.21 + 0.76$$

$$N_e = 1.97$$

5.8 Selection of a Factor of Safety for the Case Where the Occurrence of the Failure Phenomenon Would Be Disastrous

For cases where the occurrence of the failure phenomenon would be disastrous, being on the conservative side is more important than for the cases discussed in the previous sections. If *absolutely no* failure occurrence can be tolerated, load capability L for *all* possible manufactured elements *must always* be greater than or equal to *all* possible corresponding actual loads λ to which the element might be subjected in usage. Thus, considering the extreme limits in equations 5.1 and 5.2 and referring to figure 5.4, the following equation must be satisfied in order to assure that there will be absolutely no possible occurrence of the failure phenomenon; this equation assumes that values for the limits $\Delta\lambda$ and ΔL are absolutely reliable.

$$\bar{\lambda} + \Delta\lambda \leq \bar{L} - \Delta L$$

$$\therefore \quad \bar{L} \geq \bar{\lambda} + \Delta\lambda + \Delta L$$

Thus by equation 5.3,

$$N = \frac{\bar{L}}{\bar{\lambda}} \geq 1 + \frac{\Delta\lambda + \Delta L}{\bar{\lambda}}$$

$$= 1 + \frac{\Delta\lambda}{\bar{\lambda}} + \frac{\Delta L}{\bar{\lambda}}$$

$$= 1 + \frac{\Delta\lambda}{\bar{\lambda}} + \frac{\Delta L}{\bar{L}} N$$

$$\therefore \quad N \geq \frac{1 + \Delta\lambda/\bar{\lambda}}{1 - \Delta L/\bar{L}} \tag{5.18}$$

In equation 5.18, N is an appropriate factor of safety which will assure us of having absolutely no possible failure occurrence, $(\Delta\lambda/\bar{\lambda})$ is the decimal percent absolute tolerance band which exists on the anticipated mean actual load to which the mechanical element will be subjected in usage, and $(\Delta L/\bar{L})$ is the corresponding decimal percent absolute tolerance band which is anticipated on load capability for the *manufactured* element. The decimal percent tolerance bands for equation 5.18 must be estimated from past studies and past experience, and they must be reasonable and conservative if questions exist.

EXAMPLE 5.5

Consider the problem of selecting a reasonable factor of safety N_y for the tensile bar described in examples 5.1 and 5.3, assuming, however, that absolutely no yielding could be tolerated in a manufactured bar in usage.

Thus, from examples 5.1 and 5.3, we shall use the following numerical values, assuming that the tolerance bands are absolutely reliable rather than that they correspond to three standard deviations.

$$\frac{\Delta\lambda}{\bar{\lambda}} = \frac{\Delta P}{\bar{P}} = \frac{300}{1000} = 0.300$$

$$\frac{\Delta L}{\bar{L}} = 0.05 + 0.25 = 0.300$$

Therefore, by equation 5.18,

$$N_y \geq \frac{1 + 0.3}{1 - 0.3} = \frac{1.3}{0.7} = 1.86$$

$$\therefore \quad N_y = 1.86$$

Hence, for this problem, if absolutely no yielding in a manufactured tensile bar in usage could be tolerated, an appropriate factor of safety would be $N_y = 1.86$. This is considerably greater than the value of $N_y \approx 1.4$ determined from examples 5.1 and 5.3, where a 1% probability of yielding in the bar was considered acceptable.

EXAMPLE 5.6

Consider the problem of selecting a reasonable factor of safety N_e based on fatigue loading of the tensile bar as described in examples 5.2 and 5.4, assuming, however, that absolutely no failure could be tolerated before 200 million cycles of loading for a manufactured bar in usage. Thus, from examples 5.2 and 5.4, we shall use the following numerical values, assuming that the tolerance bands are absolutely reliable rather than corresponding to three standard deviations.

$$\frac{\Delta L}{\bar{L}} = 0.40$$

$$\frac{\Delta\lambda}{\bar{\lambda}} = \frac{\Delta P_{\max}}{\bar{P}_{\max}} = 0.50$$

Therefore, by equation 5.18,

$$N_e \geq \frac{1 + 0.500}{1 - 0.400} = \frac{1.5}{0.6} = 2.50$$

$$\therefore \quad N_e = 2.50$$

Hence, for this specific problem, if absolutely no fatigue failure occurrence could be tolerated before 200 million cycles of loading, an appropriate factor of safety would be $N_e = 2.50$. This value is considerably greater than the value of $N_e \approx 1.97$ which was determined as appropriate for examples 5.2 and 5.4, where a 0.1% probability of a fatigue failure occurring before 200 million cycles was considered acceptable.

5.9 Resolution of $(\Delta\lambda/\bar{\lambda})$ and $(\Delta L/\bar{L})$ into Components

For application in practice, it is appropriate to resolve the dimensionless terms $(\Delta\lambda/\bar{\lambda})$ and $(L/\bar{L})$ into components which can more readily be estimated in original design work. Thus, based on the extended study in reference 5.4, assuming that the components are reasonably small compared with unity, equation 5.18 can be written as follows to facilitate choosing factor of safety, N.

$$N = \frac{[1 + (\Delta\lambda/\bar{\lambda})_1][1 + (\Delta\lambda/\bar{\lambda})_2][1 + (\Delta\lambda/\bar{\lambda})_3]}{[1 - (\Delta L/\bar{L})_1][1 - (\Delta L/\bar{L})_2][1 - (\Delta L/\bar{L})_3]} \tag{5.18a}$$

In equation 5.18a the percent increases in actual load terms are defined as follows, and guidelines for estimating numerical values are presented in reference 5.4.

$(\Delta\lambda/\bar{\lambda})_1$ = percent increases anticipated in actual load because of uncertainties in predicting forces or moments to which the element will be subjected in the field.

$(\Delta\lambda/\bar{\lambda})_2$ = percent increase anticipated in actual stress distribution in the mechanical element because of nonideal loadings in realistic situations. Thus, as an example, a nonuniform stress distribution may exist in the tensile bar of figure 5.11 because of a small eccentricity in loading caused by small manufacturing errors.

$(\Delta\lambda/\bar{\lambda})_3$ = percent increase in actual stresses of the mechanical element because of uncertainty in theoretical stress calculations. For example, in some practical cases it may be difficult to estimate precise values for stress concentration factors, such as for the pin loaded connection in the tensile bar of figure 5.11.

In equation 5.18a the percent decreases in load capability terms are defined as follows, and guidelines for estimating numerical values are presented herein and in reference 5.4.

$(\Delta L/\bar{L})_1$ = percent decrease in part strength anticipated because of uncertainty in material strength properties. Thus an appropriate percent tolerance on yield strength S_y might be 0.05. On the other hand, much greater scatter is usually associated with endurance limit, S_e [5.7]. In such cases, table 5.1, taken from reference 5.4, may be used as a guideline for selection of the appropriate value of $(\Delta L/\bar{L})_1$.

TABLE 5.1

Guide for Selecting $(\Delta L/\bar{L})_1$ on Mean Value of S_e [5.4]

Material	S_t (1000 psi)	S_e (1000 psi)	Standard Deviation (1000 psi)	Standard Deviation %	% Survival				
					99.99	99.9	99.0	95.0	90.0
4340 Steel (Cr–Ni–Mo)	140	71	3.5	4.9	0.181	0.152	0.113	0.079	0.064
	190	85	6.7	7.8	0.289	0.242	0.180	0.125	0.101
	230	90	5.3	5.9	0.218	0.183	0.136	0.095	0.077
	260	97	6.3	6.5	0.241	0.202	0.150	0.104	0.085
4350 Steel (Cr–Ni–Mo)	300	100	4.4	4.4	0.163	0.137	0.101	0.071	0.057
Titanium alloy (6A1–4V)	145	84	5.4	6.4	0.237	0.198	0.147	0.103	0.083
Aluminium alloy (7076-T61)	76	27	1.6	6.0	0.222	0.186	0.138	0.096	0.078
Al–Ni–Bronze (5Ni–10Al)	117	48	4.5	9.4	0.348	0.291	0.216	0.151	0.122
Beryllium copper (2Be)	175	36	2.7	7.5	0.278	0.232	0.172	0.120	0.098
Material with unknown scatter data†	—	—	—	8.0	0.296	0.248	0.184	0.128	0.104

If 50% survival rate is satisfactory, $(\Delta L/\bar{L})_1 = 0.00$ may be used.
† Assuming 8% standard deviation on scatter of fatigue-test data.

TABLE 5.2
Guide for Selecting Value of $(\Delta L/\bar{L})_2$ [5.4]

Type of Loading	Type of Cross Section	Strength Geometry Factor*	$(\Delta L/\bar{L})_2$	Less Conservative Choice† for $(\Delta L/\bar{L})_2$
Axial	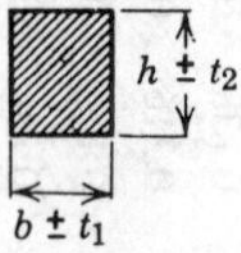	bh	$t_1/b + t_2/h$	$[(t_1/b)^2 + (t_2/h)^2]^{1/2}$
		d^2	$2(t/d)$	
Bending about the central axis shown	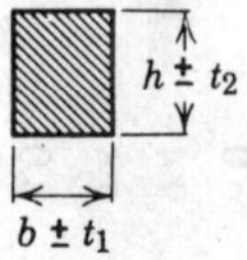	bh^2	$(t_1/b) + 2(t_2/h)$	$\{(t_1/b)^2 + [2(t_2/h)]^2\}^{1/2}$
		d^3	$3(t/d)$	
Torsion		d^3	$3(t/d)$	
Bending combined with torsion		d^3	$3(t/d)$	
Helical spring	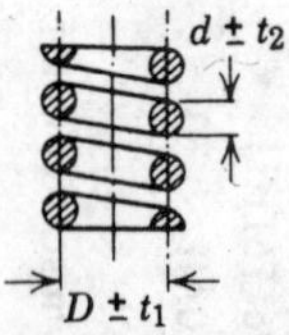	d^3/D	$(t_1/D) + 3(t_2/d)$	$\{(t_1/D)^2 + [3(t_2/d)]^2\}^{1/2}$

* All other factors being equal, strength of the cross section from a force or moment standpoint is proportional to the strength-geometry factor.

† Application of variance theory from statistics would allow use of terms in this column, although they are less conservative than those in the preceding column. These terms are valid if dimensions involved vary independently with normal distribution about the nominal dimension. Percent tolerance should correspond to at least three standard deviations.

$(\Delta L/\bar{L})_2$ = percent decrease in part strength anticipated because of tolerances on dimensions of the manufactured part. Hence table 5.2, taken from reference 5.4 may be used as a guideline for selecting the appropriate value of $(\Delta L/\bar{L})_2$.

$(\Delta L/\bar{L})_3$ = percent decrease in part strength because of other influencing factors which may exist in the manufactured part and decrease its strength. For example, if poorly controlled manufacturing processes are possible, the effect on part strength reduction should be accounted for by this term.

Application of equation 5.18a would be the same as use of equation 5.18 for selection of factor of safety N in original design work.

EXAMPLE 5.7 [5.4]

A helical compression spring is to be designed for infinite life with high reliability for survival. The mechanism design is such that spring force will be adjusted in assembly to within $\pm 5\%$ of the nominal design value. Select appropriate values to be used in design for factors of safety N_e and N_y on endurance limit and yield strength, respectively.

Solution. Equation 5.18a will be used, and from the problem statement we will use $(\Delta\lambda/\bar{\lambda})_1 = 0.05$ to account for anticipated variation in actual spring force. Next, we will consider the effect of nonideal stress distribution and for background we review the derivation of the shearing stress equation for a helical spring [5.8]. Hence spring wire is subjected to a twisting moment $T = FD/2$, where F is spring force and D is mean coil diameter. In the derivation $D/2$ is the eccentricity of F with respect to the wire center. Thus suppose our mounting is such that we estimate a variation in eccentricity of $\pm 10\%$ because of noncentral spring force possibilities. This in effect will result in a nonideal stress distribution in the spring wire, and we will use $(\Delta\lambda/\bar{\lambda})_2 = 0.10$. Finally, theoretical stress calculations for helical springs should be quite accurate, but to be conservative we will use a small factor $(\Delta\lambda/\bar{\lambda})_3 = 0.05$ to account for any possible terms overlooked or approximated in the derivations.

Let us now consider the anticipated decrease in strength possibilities for the terms in the denominator of equation 5.18a. To start, the $(\Delta L/\bar{L})_1$ value will differ for N_y and N_e. Yield strength of the material should have much less scatter than endurance limit. Hence, as previously suggested, we will use $(\Delta L/\bar{L})_{1y} = 0.05$ for the N_y calculation, and for high reliability (say 99.99% infinite life survival) we will use $(\Delta L/\bar{L})_{1e} = 0.296$ from table 5.1 for the N_e calculation. Next, we refer to table 5.2 to account for uncertainties in part strength because of dimensional tolerances. Thus from a review of the literature [5.8, 5.9, 5.10], referring to table 5.2 herein for the equation, we calculate

$$\left(\frac{\Delta L}{\bar{L}}\right)_2 = \left(\frac{t_1}{D}\right) + 3\left(\frac{t_2}{d}\right) = 0.02 + 3(0.01) = 0.05$$

which assumes that mean coil diameter D will be controlled to $\pm 2\%$ and standard wire diameter d will be accurate to $\pm 1\%$. Finally, we assume that no other significant strength reduction effects are anticipated, so $(\Delta L/\bar{L})_3 = 0$ is appropriate.

Applying equation 5.18 we calculate

$$N_y = \frac{(1.05)(1.10)(1.05)}{(1 - 0.05)(1 - 0.05)(1 - 0.00)} = 1.344$$

$$N_e = \frac{(1.05)(1.10)(1.05)}{(1 - 0.296)(1 - 0.05)(1 - 0.00)} = 1.813$$

Use of the preceding values for N_y and N_e in the design of the helical spring should be reasonable, conservative, and appropriate in this case.

5.10 Conclusions

In this chapter a general approach has been presented for the selection of factor of safety for use in design problems. For any particular original design problem a precise and unique value for factor of safety cannot be determined, because it has been clearly shown that any failure phenomenon is statistical in nature for manufactured mechanical elements in usage. The prediction of the occurrence of a certain failure phenomenon is significantly affected by many uncontrollable and indeterminable factors in real problems. Hence it can be concluded that the engineer must employ *reasonable* and *conservative* estimates for determining approximate distribution curve characteristics when selecting an appropriate factor of safety for original design usage. As we shall see in later chapters, this inaccuracy in the final selection of the numerical *value* will not affect the conduct of specific design studies using the basic *method of optimum design* for most cases.

From the study presented in this chapter, the following important general observation should be realized. That is, for many real mechanical elements the significant failure phenomenon is either yielding of the material or the occurrence of fatigue damage after a certain number of cycles of repeated loading, and for such cases equations 5.14, and 5.17 based on the statistics approach will be more appropriate for estimating a reasonable factor of safety than equation 5.18. The statistics approach will result in the selection of a *lower* factor of safety, which is generally compatible with the desires of optimum design, and this will be evident from many of the examples in later chapters. Thus, for original design problems, the appropriate value for factor of safety should be selected in view of the *significance* attached to the occurrence of the particular failure phenomenon. It is unrealistic and unreasonably conservative, for many real cases, always to assume that the occurrence of the failure phenomenon absolutely cannot be tolerated.

REFERENCES

5.1 G. D. Cohen, "Predicting Performance Failures," *Machine Design*, Oct. 3, 1957, pp. 106–111.

5.2 M. F. Spotts, "An Application of Statistics to the Dimensioning of Machine Parts," *Transactions of the ASME, Journal of Engineering for Industry*, Vol. 81, No. 4, Nov. 1959, pp. 317–322.

5.3 W. Volk, *Applied Statistics for Engineers*, McGraw-Hill, New York, 1958.

5.4 R. C. Johnson, "Predicting Part Failures," Part 2, *Machine Design*, Jan 21, 1965, pp. 157–162.

5.5 H. J. Larson, *Statistics: An Introduction*, Wiley, New York, 1975, 418 pp.

5.6 G. A. Korn and T. M. Korn, *Mathematical Handbook for Scientists and Engineers*, McGraw-Hill, New York, 1968, Chapters 18 and 19.

5.7 F. B. Stulen, H. N. Cummings, and W. C. Schulte, "Preventing Fatigue Failures," Part 5, *Machine Design*, June 22, 1961, pp. 161.

5.8 J. E. Shigley, *Mechanical Engineering Design*, 2nd ed., McGraw-Hill, New York, 1972, pp. 352–363.

5.9 N. P. Chironis, *Spring Design and Application*, McGraw-Hill, New York, 1961, p. 11.

5.10 A. M. Wahl, *Mechanical Springs*, 2nd ed., McGraw-Hill, New York, 1963, p. 95.

EXERCISES

5.1. A mechanical element is to be designed in a situation where some statistical information can be conservatively estimated from past experience in similar situations.

(a) Based on three standard deviations we estimate that increase in actual load might be $(\Delta\lambda/\bar{\lambda}) = 0.30$ and decrease in load capability might be $(\Delta L/\bar{L}) = 0.40$. Calculate an appropriate factor of safety N for new design use assuming that a failure rate of $A_f = 0.005$ could be tolerated.

(b) Calculate an appropriate factor of safety N for new design use assuming that no failure occurrences could be tolerated. Assume that total population estimates have been made on increase in actual load and decrease in part strength possibilities, giving $(\Delta\lambda/\bar{\lambda}) = 0.40$ and $(\Delta L/\bar{L}) = 0.50$. Compare the factor of safety so obtained with your answer to part a.

5.2. For the tensile bar of figure 5.11 stress σ can be expressed as $\sigma = K_t P/(bh)$. This we can equate to S/N, where S is a material strength and N is the factor of safety on S. Hence we have the expression

$$N = S\frac{bh}{K_t P} \tag{5.19}$$

which when compared with equation 5.3 has a load capability term such as $L = S(bh)$ and an actual load term which then would be $\lambda = K_t P$. Thus applying equation 2.47 of Chapter 2, the various uncertainties in equation 5.18a can be expressed in terms of the tensile bar design parameters as follows.

$$\left(\frac{\Delta\lambda}{\bar{\lambda}}\right)_1 = \frac{\Delta P}{P} \qquad \text{for uncertainty in actual load}$$

$$\left(\frac{\Delta\lambda}{\bar{\lambda}}\right)_3 = \frac{\Delta K_f}{K_f} \qquad \text{for uncertainty in stress analysis}$$

$$\left(\frac{\Delta L}{\bar{L}}\right)_1 = \frac{\Delta S}{S} \qquad \text{for uncertainty in material strength}$$

For the effects of nonuniform stress distribution, we might refer to exercise 5.3 and derive the expression for $(\Delta\lambda/\bar{\lambda})_2$ in terms of anticipated percent eccentricities in loading, (e_b/b) and (e_h/h). Finally, to account for the effect of tolerances on dimensions we refer to equation 5.19 and recognize that (bh) is what we might call a strength geometry factor. Hence we obtain

$$\left(\frac{\Delta L}{\bar{L}}\right)_2 = \frac{\Delta(bh)}{(bh)} = \frac{\Delta b}{b} + \frac{\Delta h}{h} = \frac{t_1}{b} + \frac{t_2}{h}$$

where t_1 and t_2 are the tolerances on dimensions b and h respectively. This latter expression is given in the next to last column of table 5.2 for the tensile bar. However, from a less conservative standpoint we look at it statistically and obtain $(\Delta L/\bar{L})_2 = [(t_1/b)^2 + (t_2/h)^2]^{1/2}$, which is given in the last column of table 5.2 for the tensile bar. Derive the other conservative and less conservative expressions for $(\Delta L/\bar{L})_2$ given in table 5.2, for tensile bars, beams, torsion bars, shafts in combined loading, and helical springs.

5.3. For the tensile bar of figure 5.11 it is anticipated that force P might be eccentric to the geometric axis of the bar by amounts e_b and e_h for the directions of b and h respectively. Such eccentricities will introduce bending effects which cause nonuniform stress distribution in the bar. Suppose we can estimate values for percent eccentricity terms defined by the dimensionless ratios (e_b/b) and (e_h/h). Derive an expression for $(\Delta\lambda/\bar{\lambda})_2$ of equation 5.18a in terms of (e_b/b) and (e_h/h) to account for nonuniform stress distribution in the tensile bar because of the small eccentricities which are possible as described. (*Hint*: derive an equation for N similar to equation 5.19 of exercise 5.2 which will have an additional factor containing the eccentricity terms. Assume that $e_b \ll b$ and $e_h \ll h$.)

5.4. Repeat calculations of N_y and N_e for the helical spring problem of example 5.7 using the less conservative estimates for $(\Delta L/\bar{L})_2$ in the last column of table 5.2. Compare your answers with those from example 5.7.

5.5. A simple beam of span L_s with ideal central load F is to be designed. The cross section is to be rectangular with b and h dimensions to be determined by design, applying an appropriate failure criterion from Chapter 4. The beam will bend about the neutral axis parallel to the b side of the cross section. Force F will vary periodically between $F = 0$ and $F = F_{max}$, and a design value has been estimated for load F_{max}. For various uncertainties we estimate total population tolerance bands as follows for use in equation 5.18a.

$$\left(\frac{\Delta\lambda}{\bar{\lambda}}\right)_1 = \frac{\Delta F_{max}}{F_{max}} = 0.20; \qquad \left(\frac{\Delta\lambda}{\bar{\lambda}}\right)_3 = 0.05; \qquad \left(\frac{\Delta L}{\bar{L}}\right)_3 = 0.00$$

Suppose that nonideal stress distribution will be primarily caused by a noncentral loading condition, with the possibility of load F being offset from its true central position by an eccentricity e. Hence we estimate the possibility of $(e/L_s) = 0.10$ for the specfic application, and based on this we can calculate the corresponding value for $(\Delta\lambda/\bar{\lambda})_2$ applying beam theory. Two factors of safety are to be calculated for design use: N_y on yield strength S_y, and N_e on endurance limit S_e. Hence for the N_y calculation we will use $(\Delta L/\bar{L})_{1y} = 0.05$, and for the N_e calculation we will choose the appropriate value for $(\Delta L/\bar{L})_{1e}$ from table 5.1 assuming a high degree of reliability desired for infinite life survival. Finally, to account for tolerances on cross sectional dimensions $(\Delta L/\bar{L})_2$ will be calculated from table 5.2 assuming percent tolerances $(t_1/b) = 0.02$ and $(t_2/h) = 0.01$. Using equation 5.18a calculate appropriate values of N_y and N_e to be used on S_y and S_e respectively when designing the beam.

6

Some Optimization Techniques for Mechanical Elements

In recent years many optimization techniques have either been newly developed or applied from earlier publications. Linear programming, geometric programming, dynamic programming, Lagrangian multipliers, dual variables, the method of optimum design, and many algorithms for automated optimal design are but a few of the many techniques which have been called to our attention in the broad and general field of optimization. To review and summarize all of this important work would be a mammoth task which we shall not attempt, but the interested reader may wish to see references 6.1 through 6.15 to obtain a broader background.

In this book attention is focused on *mechanical elements*, and two very powerful optimization techniques well suited for application to mechanical elements will be covered in this chapter. They are (1) the method of optimum design and (2) automated optimal design with specific consideration for a recently developed algorithm called OPTIGO. However, before describing these two optimization techniques in detail we will cover some basic background pertinent to the optimal design of mechanical elements.

Some Basic Background for Mechanical Elements

6.1 Structures and Machines

In the following paragraphs we shall develop the background which will logically lead to the principles of optimum design for mechanical elements. We shall begin by basically defining what we mean by the terms *mechanical structure* and *machine*. The part played by a mechanical element within such a structure or machine will then be quite evident.

We shall define a *mechanical structure* as a physical assemblage of mechanical elements constructed for the purpose of transmitting force, offering protection, or performing any other physical function. A special type of mechanical structure is a machine. We shall define a *machine* as a mechanical structure which is characterized by mechanical elements having relative motion and generally by the ability to transmit or to dissipate significant amounts of energy.

6.2 General Characteristics of Mechanical Elements

A general study of many mechanical elements shows that the following three characteristics are common to all.

1. *All* mechanical elements must satisfy certain *functional requirements* of the structure or machine wherein they are used. By definition, a functional requirement is a condition which must be satisfied by the mechanical element if the structure or the machine is to function properly. In the strict sense, functional requirements can either be positive and generally *specified*, or negative and generally *implied*. For instance, in the design of a mechanical extension spring the functional requirements of the machine from the spring might be a certain maximum force F_{max} and a certain force gradient k. The values F_{max} and k would be positive functional requirements, since they must be directly obtained for proper functioning of the machine and they are conditions which we want. On the other hand, we would not want the spring to break during operation of the machine. Hence, for proper functioning of the machine, we might specify that the maximum shearing *stress* in the spring wire should be equal to or less than an allowable shearing *strength* of the material, using an acceptable criterion for failure due to fatigue. However, this specification is really for a negative functional requirement, since it states a condition which we do *not* want. Negative functional requirements are

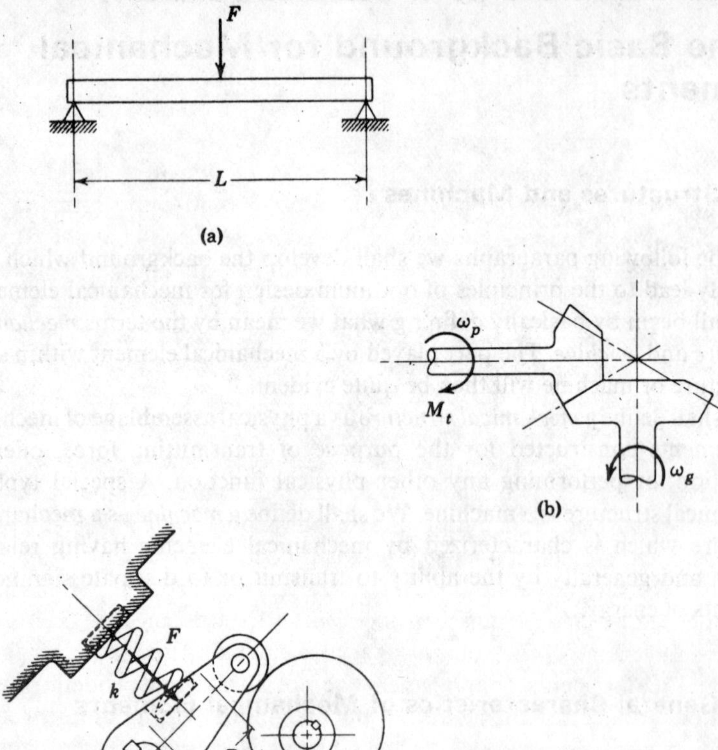

Figure 6.1 Simple sketches of some mechanical elements, giving some typical functiona requirements. (*a*) Simple beam (functional requirements F_{max} and L); (*b*) bevel gearse (functional requirements M_t, ω_p, and ω_p/ω_g); (*c*) cam load spring (functional require ments F_{max}, k, and D_{max}); (*d*) tensile bar (functional requirements T, L, and d_{max}).

really *undesirable effects,* whereas positive functional requirements are really *desirable effects.* In this book, when we speak of *functional requirements,* we shall mean the positive type or desirable effects. On the other hand, *undesirable effects* will be termed as such in our discussions.

Some examples of functional requirements are shown in figure 6.1. In figure 6.1a, in order for the mechanical structure to function properly, the beam must transmit a maximum force F_{max} and it must have a specified length L. In figure 6.1b, the machine under consideration will function properly only if the pinion shown transmits the specified twisting moment M_t at the specified angular velocity ω_p. Also, for proper functioning of the machine, the gearset must have a specified velocity ratio, ω_p/ω_g. In figure 6.1c, the cam mechanism will function properly only if the load spring can deliver a specified maximum force F_{max} with a specified force gradient k. In figure 6.1d a tensile bar is shown in a hypothetical mechanism setting transmitting a specified axial force T. Also, for proper functioning of the machine, the tensile bar must have a specified length L and its lateral dimension d cannot exceed a dimension d_{max} which is imposed by existing space restrictions.

For all the examples in figure 6.1, notice that the functional requirements are influenced by factors which are *external* to the element being designed, and this is generally true for mechanical elements. Also, notice that the functional requirements of an element are not always independent of each other. For instance, for the gearset, if the maximum twisting moment M_t is changed, proper functioning of the machine might require an accompanying change in angular velocity ω_p. As another example, for the cam mechanism to function properly, a change in spring force gradient k might require an accompanying change in maximum spring force F_{max}. In other words, because of considerations of the entire structure or machine external to the element being designed, it is often not possible to change arbitrarily one functional requirement without considering the effect of such a change on the others. Hence functional requirements, *if* they can be changed at all from their specified values, are often dependent on each other as the result of conditions external to the element. This is an important point to remember in later chapters.

Another important point to be aware of is that a functional requirement is not always limited to a specific single value to obtain a satisfactory mechanical design. There might be a satisfactory *feasible range* over which a functional requirement could have a satisfactory value. For instance, in figure 6.1a, for proper functioning of the structure, the beam length L might have any value within the feasible range $L_{min} \leq L \leq L_{max}$, where L_{min} and L_{max} are limits imposed by functional requirements of the structure and L is a geometrical parameter. Or, as another example, if a mechanical element must be *capable* of absorbing a certain specified amount of energy, $(P.E.)_{min}$, a satisfactory design would certainly be obtained if the actual maximum energy storage *capability* $(P.E.)_c$ of the designed element were greater than $(P.E.)_{min}$. Hence the satisfactory range for feasible maximum energy storage capability of *all* satisfactory designs would be defined by the limit equation $(P.E.)_c \geq (P.E.)_{min}$. Thus associated with functional requirements we often have *limit equations*, which define a satisfactory *range* for values of the functional requirement.

2. At least to some extent, *all* mechanical elements have certain *undesirable effects*. For instance, associated with all mechanical elements is the very important undesirable effect of cost. Other common undesirable effects of importance might be deflections, stresses, weight, space occupancy, vibration, force, and the like, and the degree of *significance* for any such undesirable effect depends on the particular problem at hand.

To have a satisfactory design of a mechanical element, feasible values for all undesirable effects must be limited to *tolerable* ranges. For instance, for obtainment of a satisfactory design, cost C of a mechanical element might be limited to a maximum amount which might be stated mathematically as $C \leq C_{max}$. For many cases it might not be possible to formulate mathematically the cost C of an element in the design procedure, but the engineer must keep in mind the limit equation for cost, and this consideration might very well temper his final choice in a design. Another undesirable effect, whose feasible values must always be limited to a tolerable range for obtainment of a satisfactory design, is that of stress. As an illustration, we might say that the maximum shearing *stress* τ_{max} at any point in the mechanical element must be equal to or less than an *allowable* shearing stress related to a material *strength*, using the maximum shearing stress theory of failure from Chapter 4. Stated mathematically, this limit equation might be $\tau_{max} \leq S_y/2N_y$, where S_y is the yield strength of the material from a simple tension test and N_y is the factor of safety based on the occurrence of yielding in the material. Thus similar limit equations exist for all undesirable effects, and they must be adhered to in the design procedure for obtainment of a satisfactory design.

3. *All* mechanical elements are uniquely defined by a complete specification of the material and the geometry of the element. Such a final specification might be given after completion of design in the form of the detail drawing, typically depicted in figure 6.2.

Many of the important properties of a material are given by the material parameters found in the literature, such as the modulus of elasticity E, Poisson's ratio ν, weight density w, and unit weight cost c. The usefulness of these published material parameters has been discussed in Chapter 4. For instance, there it was pointed out that certain material *strengths* determined by simple standard tests can be used as material parameters in establishing the upper limit for permissible values of *stress* in the mechanical element being designed, by means of an appropriate theory of failure. On the other hand, some important material properties cannot be defined very well

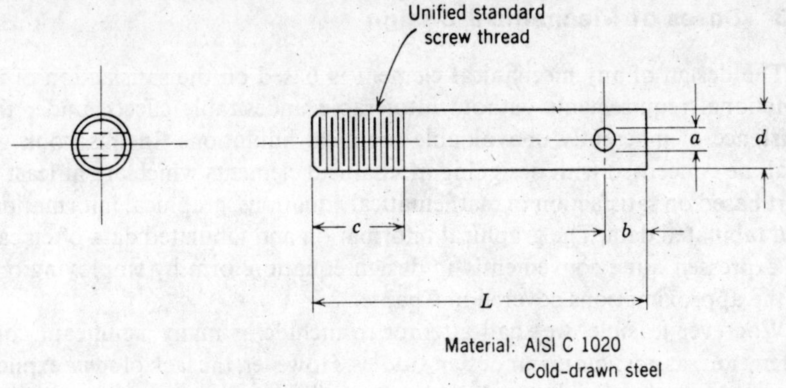

Figure 6.2 Independent parameters which uniquely define a typical mechanical element.

mathematically. For instance, various machining properties, the extent of availability, and the restriction to standard sizes are important considerations of this type. Often they are defined only vaguely, and the engineer must rely on experience and good judgment in considering such properties. Generally, in the conduct of a design study the selection of a material for a mechanical element will be based primarily on the material parameters which are explicitly defined. However, in making a final specific optimum choice the engineer might find it necessary to consider some of these hidden properties. Or, because of such practical properties, the engineer might initially eliminate certain classes of materials from consideration in the design procedure for the element. Therefore, regarding the material for a mechanical element, we might conclude that it is uniquely defined by explicit material parameters *and* by certain other properties which are not readily adaptable to mathematical definition.

The geometry of a mechanical element can always be uniquely defined by specifying certain *independent geometrical* parameters. Thus in figure 6.2 the parameters L, d, c, a, b, and specification of the standard thread system uniquely define the geometry of the element. For many mechanical elements the definition of the geometry is unique only if certain manufacturing error limits are also specified. For instance, in figure 6.2 the class of thread might be an important consideration, as might be the tolerances on the diameters d and a, depending on the particular application. In cam mechanisms, springs, gears, bearings, and many other mechanical elements the effects of manufacturing errors can be very important, as studied in Chapter 3. Hence for many cases the significant geometry is explicitly defined *only* by specifying theoretical values for geometrical parameters *and* limits on manufacturing errors, and these tolerances must be adhered to in the actual manufacturing process for the element.

6.3 Bases of Mechanical Design

The design of any mechanical element is based on the satisfaction of its functional requirements without intolerable undesirable effects, under the existence of inherently unavoidable practical limitations. In this book we shall be concerned with designing mechanical elements which are at least in part based on satisfaction of mathematical equations, graphical information, and tabulated data. The graphical information and tabulated data often can be expressed quite conveniently in design equation form by employing one of the approximations covered in Chapter 2.

Wherever feasible, we shall attempt to include as many significant considerations as *possible* in our design studies. However, the lack of *some* explicit knowledge for reasons beyond our control will not prevent us from reaching satisfactory and logical conclusions, which by necessity might be somewhat *conservative*. Mechanical design in this book will be based on the explicit inclusion of as much available information as *possible*, but the unavailability of some information will *not* cause us to abandon all theoretical rigor in the procedure of *design*.

6.4 A General Viewpoint on Typical Design Equations

From a study of many mechanical elements, it is found that the design equations generally express either functional requirements or undesirable effects in terms of parameters which can be classified according to three basic groups. These parametric groups have been discussed previously in section 6.2 and are as follows:

1. *Functional requirement* parameters (or undesirable effect parameters) other than the one explicitly expressed by the design equation.
2. *Material* parameters.
3. *Geometrical* parameters.

The functional requirement parameters in a design equation are generally specified values from an analysis of the entire mechanical structure or machine conducted before the actual design of the mechanical element. However, since they are determined by factors external to the element, we can think of the functional requirement parameter *group* in a design equation as independent of the element being designed. Hence, in a typical design equation, we can say that the specified functional requirement parameter group is independent of both the material parameter group and the geometrical parameter group.

The material parameters in a design equation are generally *not* independent of each other. For instance, weight density w, unit weight cost c, modulus of elasticity E, Poisson's ratio ν, and yield strength S_y are material parameters whose values are generally unique for any specific material. Hence in the procedure of *design* we generally cannot change the modulus of elasticity E without changing the specific weight w. An exception to this dependent relationship for material parameters might be a material strength, which can sometimes be changed independently for a specific material by heat treatment or cold working. However, for most cases material parameters *cannot* be changed individually in an arbitrary manner. The design engineer has to rely on the chemist or the metallurgist to supply him with a finite list of *available* materials whose individual characteristics are uniquely determined for any specific material. Hence, as a *group*, the characteristic parameters of any specific material can be thought of as independent of both the functional requirement parameters and the geometrical parameters in a typical mechanical element design equation.

It is also interesting to notice that in a design equation the value for a material parameter *group* cannot be changed arbitrarily or continuously. It can generally be altered only by changing materials, which are available only in relatively small numbers. Hence the value for a material parameter *group* can be changed or selected only in *discrete* steps during conduct of the design procedure. This will be clearly illustrated in examples which follow.

Regarding geometrical parameters, for the conduct of design studies it is always *possible* and *desirable* to select *independent geometrical* parameters which uniquely define the geometry of the mechanical element. Redundancy in defining the geometry of the element generally should be avoided. Hence, choosing *independent geometrical* parameters will allow the *independent* variation of each in the design procedure.

Generally speaking, in the conduct of design studies geometrical parameters can be varied far more continuously than material parameters. However, where standard sizes are involved, geometrical parameters often can be varied only in *discrete* steps. In selecting the particular *independent geometrical* parameters for use in the design study of a mechanical element, if the choice exists it is generally wise to choose parameters which are restricted to standard sizes or parameters whose limits are either known or specified by functional requirements of the entire mechanical structure or machine. Hence, for the conduct of design studies, *independent geometrical* parameters should be chosen from those which are *limited* rather than from the geometrical parameters which are *not* restricted, if the choice exists.

Although geometrical parameters which are independent of each other can always be selected, the geometrical parameter *group* in a typical design equation is often dependent on the specified functional requirement parameter group and the material parameter group. This is in contrast to the independence of the latter two *groups* in design equations.

Occasionally, a specific term in a design equation cannot be classified within any particular *one* of the three basic parameter groups. When this occurs, a relating equation could be written to express the term as a function of parameters in the three groups, but such a relating equation may not be necessary. As an example, in the next paragraph we shall briefly consider the term N for factor of safety, keeping in mind the pertinent discussions from previous chapters.

For a desired degree of reliability, typically expressed as a specified value for A_f in equation 5.17 of Chapter 5, the required factor of safety N is, precisely speaking, dependent upon functional requirements, material parameters, *and* geometrical parameters. This is obvious from a consideration of equation 5.17 and its derivation, since the percent tolerance band $(\Delta\lambda/\bar{\lambda})$ on anticipated *actual load* is dependent on functional requirements external to the mechanical element, and the percent tolerance band $(\Delta L/\bar{L})$ on anticipated *load capability* is dependent on the specific material and on the geometrical parameters. However, for most of the design problems in this book we shall consider that factor of safety N is primarily a *functional requirement* parameter, and hence for any particular design problem we are merely considering that the percentage term $(\Delta L/\bar{L})$ for load capability in equation 5.17 is essentially the same value for *all feasible* design solutions.

The dependence and independence of the various parameters and parametric groups in design equations will be extremely important in the study of the *method of optimum design* and its examples. The results of this section can be summarized simply in the accompanying typical skeleton design equation, which figuratively expresses the quantity B.

$$B = f\left[\begin{pmatrix}\text{Specified}\\ \text{functional}\\ \text{requirement}\\ \text{parameters}\end{pmatrix}, \begin{pmatrix}\text{Material}\\ \text{charac-}\\ \text{teristic}\\ \text{parameters}\end{pmatrix}, \begin{pmatrix}\text{Geometrical}\\ \text{parameters}\end{pmatrix}\right]$$

	B	Specified functional requirement parameters	Material characteristic parameters	Geometrical parameters
Above (↑)		Dependent on considerations external to element	Dependent on chemical and physical structures in available materials	Dependent on remainder of design equation if value for B is specified
Below (↓)	Either a functional requirement *or* a limited significant undesirable effect	Individual parameters are either independent *or* dependent on each other	Individual parameters are mostly dependent on each other	Individual parameters can be varied independent of each other

6.5 Adequate Design

By definition, *adequate design* is the selection of the material and the values for the independent geometrical parameters for a mechanical element so that the element *satisfies* its functional requirements and so that undesirable effects are kept to *tolerable* magnitudes. The design is accomplished by using the available engineering information in the form of equations, graphs, tabulated data, and practical experience. Adequate design is often characterized by cut-and-try methods, because of the existence of a rather loosely defined overall design objective which results in an infinite number of *possible* design solutions, as discussed in Chapter 1.

For many mechanical elements, an adequate design is really the optimum design. This is true for cases where the undesirable effects are *not* close to their tolerable limits and where the *design cost* is minimized by the application of cut-and-try methods using practical experience and good judgment with some simple *analysis* if necessary. As an example, no one will deny that it would be nonsensical to spend days designing a bolt for clamping a light weight decorative cover on a sewing machine. However, if the bolt is to be used in mass quantity in an airplane where weight is an important factor, it might pay to spend some effort in designing the bolt on the basis of minimum weight.

In many cases the adequate design of a mechanical element has resulted in a *satisfactory* performance of the mechanical structure or machine for many years. However, in many such cases an optimum design study of the mechanical element will result in invaluable savings or in an appreciable improvement in product performance or quality, either of which was hitherto an *unknown* possibility. Hence satisfactory adequate design should not be the sole criterion upon which the possibility of an optimum design study is discarded. Also, there are many situations where adequate design techniques would not readily reveal a satisfactory solution to a design problem because of certain practical limits, and the erroneous conclusion of an *impossibility* might be drawn. For such cases the *method of optimum design* might more readily reveal a *possible* solution to the problem if it exists.

6.6 Optimum Design

By definition, the *optimum design* of a mechanical element is the selection of the material and the values for the independent geometrical parameters with the explicit objective of either minimizing an undesirable effect or of maximizing a functional requirement, making certain in the design procedure that the element satisfies other functional requirements and that other undesirable effects are kept within *tolerable* limits. The optimum design for a mechanical element should be conducted on the basis of the *most significant* quantity to be minimized or to be maximized, and the *particular* basis of optimum design depends on the *particular* mechanical element under consideration and on its *particular* application in its *particular* structural or machine setting. Hence the procedure of optimum design might be conducted on the basis of minimizing any *one* of such quantities as cost, weight, or deflection; similarly, the objective in the design procedure might be to maximize any *one* of such quantities as power transmission *capability*, energy storage *capability*, force transmission *capability*, or speed *capability*.

On occasion a multiobjective situation exists where more than one optimization quantity is appropriate for a mechanical element in a particular application. In such cases one of two strategies may be followed. In the first strategy a composite criterion function is derived containing weighted components corresponding to each of the optimization quantities. In the second strategy the engineer exercises his judgment and chooses one of the optimization quantities for a new single objective problem with the necessary equations and limits established for the other quantities. For practical application it has been found that this latter strategy is easier to handle, since limit values can be estimated more readily than relative weighting values in most realistic situations of design.

The problem of optimum design for real mechanical elements requires a study of relationships among functions which generally consist of many variables. To determine an explicit solution for an optimum design problem, it is necessary to consider not only individual function *variations* but, more important, the *ranges* of *compatible variation* among several limited functions of many variables. From such a study the domains of variation with respect to all existing feasible designs will be evident, and the optimum design can be selected in an explicit manner. Hence the problem of optimum design can be extremely complex, and its resolution can be greatly facilitated by following an organized *method* of solution. In this chapter we shall present such a *method of optimum design* which is particularly suited to mechanical elements.

In sections 6.14 and 6.15 the suggested *method of optimum design* will be outlined in the form of general procedural steps which if followed will greatly facilitate the explicit solution of most *design* problems for mechanical elements. However, it should be emphasized that this is *not* meant to be a "cookbook" type of presentation which must be rigidly followed by rote for all specific design problems. It is hoped not only that in studying these fundamentally important sections the reader will learn the suggested general method, but also that he will *think* in terms of the basic underlying theoretical concepts which *must* be rigidly adhered to. If this approach to study is emphasized, the engineer will acquire the highest possible degree of facility for application of the optimum design approach to specific original problems.

In the remaining chapters of this book many examples illustrating the principles of optimum design will be specifically covered. Obviously it would be impossible to cover *all* mechanical elements for *all* bases of optimum design for *all* possible functional requirements and for *all* limitations on functional requirements, undesirable effects, geometry, and materials. The various combinations of these considerations would be astronomical in number; hence we shall discuss only a relatively few of the possibilities. The purpose of the specific examples will be to illustrate basic techniques *of optimum design* applied to mechanical elements.

From the specific examples to be covered, it is interesting to notice that optimum design based on *intuition* is often risky. Results from rigorous optimum design studies are frequently obvious; however, for many other cases such results are spectacular and occasionally even paradoxical. Thus, in order to avoid dangerous pitfalls, we must adhere to a rigorous *method* in conduct of the optimum design procedure.

6.7 Choosing the Basic Geometrical Shape

Before starting on the optimum design of a mechanical element, the engineer should consider carefully whether or not a change in the basic geometry might not be beneficial. Such a change might be *implied* by the results of the optimum design study as conducted on the original mechanical element geometry. However, before making any change in basic design, the engineer must be sure to weigh carefully all hidden practical considerations as well as the implied theoretical change considerations. Some basic changes in mechanical elements might be obviated by practical considerations, which are often unwritten design limitations. As an example, the weight of a long torsion shaft might be reduced by having a central hole along its axis. However, the cost of obtaining such a hole in the long shaft might be prohibitive, and for that reason we might have to eliminate the hollow shaft shape possibility from consideration in the design study for a particular problem. It has been proved in the past that almost anything can be made which is uniquely defined geometrically; however, it is not always *practical* to accomplish such tasks, especially in the light of economic limitations.

Often the optimum design study on a simple geometrical shape of a mechanical element will infer by induction a desirable basic design change to a more complex geometrical shape. The new more complex design might not be as susceptible to the application of an accurate and explicit optimum design procedure. For such cases the design of the complex element will be inferred from the rigorous optimum design of the simple element, and a rigorous analysis of the final complex design might be entirely feasible as a final check.

Conversely, we might encounter in practice a complex mechanical element which does not lend itself to a rigorous and explicit optimum design procedure. Then we might conduct a rigorous and explicit optimum design study on a simplified version of the complex element. The results obtained for selecting optimum values of the design parameters would then be extended from the simplified version to the actual complex design. As a final step, the complex design so determined would be subjected to a rigorous analysis as a final check. Techniques of this type are characteristic of some problems, and such procedures really constitute good practice in original design work. It would be far better to follow such a procedure than to abandon all rigor by resorting to pure guesswork. The latter choice could be disastrous for many mechanical elements.

The problem of determining the best geometrical shape can occasionally be solved in an explicit manner for extremely simple mechanical elements by using the *calculus of variations*. However, even for very simple problems the method is generally very complex, and for many complicated practical design problems it is far too unwieldy. Hence the calculus of variations would enable us to prove that the *circular* tray is the best possible shape for the ideal problem in Chapter 1, but it would have little value for solving the complicated real gear design problems in Chapter 8. Thus, because of the kind of real design problem confronting us in this book, we shall not devote space to the field of mathematics known as the calculus of variations. However, this does not obviate its value for the study of idealized design problems by advanced students.

6.8 A General Mathematical Viewpoint on Optimum Design

Optimum design for real problems requires a complex study of functional variations *and* functional relationships, generally for several limited functions of many variables. Let us now briefly consider the *general* problem of optimum design from a mathematical viewpoint.

In what follows we shall let Q designate the quantity upon which the optimum design is based. Also, we shall let u, v, w, x, y, and z designate the design variables. Assume that the basic problem is to determine the values for the design variables which will result in the *greatest possible value* for the optimum design quantity Q.

First, consider an extremely simple example, which we can readily depict graphically. Suppose that the following general equations mathematically define this problem.

$$Q = f_1(x)$$

$$A = f_2(x)$$

$$A \leq A_{max}$$

$$x \leq x_{max}$$

In these equations f_1 and f_2 designate specific functions which define the relationships between numerical values of Q, A, and x. Also, A_{max} and x_{max} are limit values which are used for defining the ranges of feasible or allowable values for A and x respectively. The functional relationships for this problem are typically depicted in figure 6.3, which is self-explanatory.

Second, consider another simple example, which we can readily depict

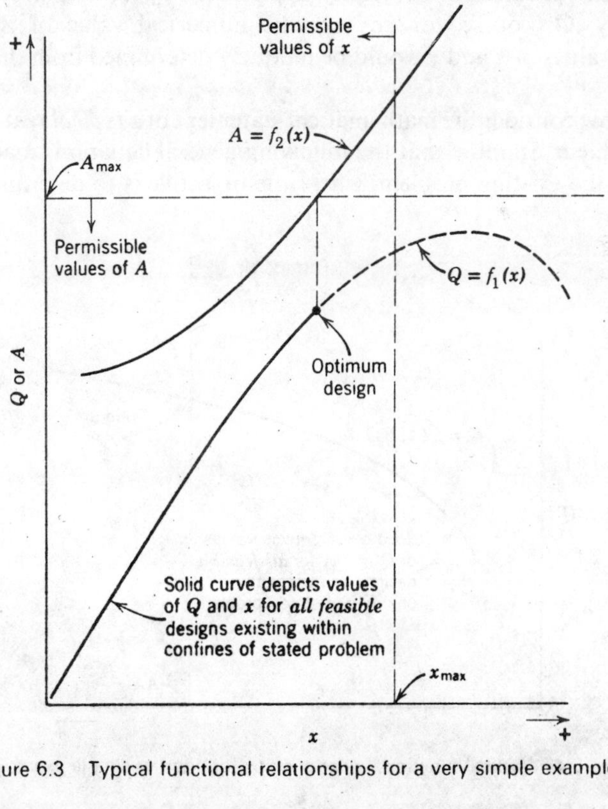

Figure 6.3 Typical functional relationships for a very simple example.

graphically. Suppose that the following general equations mathematically define this problem.

$$Q = f_1(x, y)$$
$$A = f_2(x, y)$$
$$B_1 = f_3(x, y)$$
$$A \le A_{max}$$

In these equations f_1 and f_2 designate the specific functions which define the relationships between numerical values of Q, A, x, and y. B_1 is assumed to be a specified value, and A_{max} is a limit value used for defining the range of permissible values for A. Thus we could combine the first three of these equations by eliminating x and y, thereby expressing the quantity Q as

$$Q = f_4(A, B_1)$$

where f_4 is the function which defines the relationship between numerical values of Q and A, since B_1 is a specified value. Hence the functional relationships for this problem are typically depicted in figure 6.4, which is self-explanatory. Of course, for any specific numerical value of A, specific numerical values of x and y would be uniquely determined from the original equations.

Let us now consider the mathematical statement of a *typical* real optimum design problem. Suppose that the following general equations mathematically define the existing problem, where the objective is to determine values

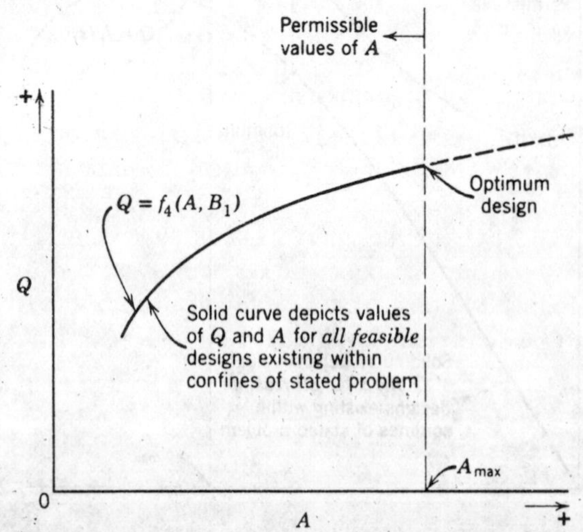

Figure 6.4 Typical functional relationships for another simple example.

for the design variables which will result in the *greatest possible value* for the optimum design quantity Q.

$$Q = f_1(u, v, w, x, y, z)$$

$$A = f_2(u, v, w, x, y, z)$$

$$B_1 = f_3(u, v, w, x, y, z)$$

$$u = f_4(M_i)$$

$$v = f_5(M_i)$$

$$A_{max} = f_6(M_i)$$

where $i = 1, 2, 3, \ldots, n$;

$$A \le A_{max}$$

$$w \le w_{max}$$

$$x \ge x_{min}$$

$$y_{min} \le y \le y_{max}$$

$$z = z_j$$

where $j = 1, 2, 3, \ldots, m$. In these equations the symbols f_1 through f_6 designate specific functions, B_1 designates a specified value, and the letters with subscript max or min designate limit values used for defining ranges of feasible or allowable values for the corresponding letters. Letter M designates *material*, where the subscript i indicates a specific material from a finite discrete list of n *available* materials. Also, the last equation $z = z_j$, represents the requirement of using only standard size values for z, which is a typical limitation for real design problems.

Because of the complexity of the typical real design problem mathematically stated in the previous paragraph, no attempt will be made to depict the functional relationships graphically. However, this problem is typical of what is encountered in real design work, and also of the design problems presented and solved in this book. The general *method of optimum design* outlined in sections 6.13 through 6.17 is predicated on the overall objective of explicitly solving problems of this type in the simplest possible manner.

6.9 Primary Design Equation

In the optimum design of a mechanical element, the most important design equation is the one which expresses the optimization quantity upon which the particular optimum design is based, since essentially it directs the procedure of design. For any particular case this criterion function will be known as the *primary design equation*, which will be designated by (P.D.E.) whenever the notation is advantageous.

For any particular mechanical element the particular primary design equation will be determined by the most significant functional requirement *or* by the most significant undesirable effect, as discussed in section 6.6. For instance, in a stone-crushing machine the primary design equation for a torsion shaft might be the one which expresses the torque transmission *capability*, which is a functional requirement to be maximized in the design procedure. On the other hand, in an airplane the primary design equation for a torsion shaft might be the one which expresses the shaft weight, which is an undesirable effect to be minimized in the design procedure, and the required torque transmission *capability* would be given as a specified functional requirement.

6.10 Subsidiary Design Equations

In the optimum design of a mechanical element, the design equations other than the primary design equation will be called the *subsidiary design equations*. Hence the subsidiary design equations generally express either functional requirements or significant undesirable effects, whether they are directly specified or indirectly implied. It is very important not to neglect a subsidiary design equation even though it is merely implied; for instance, stress equations are often very important subsidiary design equations which are merely implied. Subsidiary design equations will be designated by (S.D.E.) whenever this notation is advantageous.

6.11 Constraints

In previous sections we have explained the existence of satisfactory ranges for values of certain parameters, which are expressed mathematically by simple limit equations. Limitations on *stress* in a mechanical element are imposed by certain material properties called *strengths*, in conjunction with an appropriate theory of failure. Limitations on geometry are imposed by certain functional requirements of the mechanical structure or machine as by space restrictions, by practical manufacturing limitations in conjunction with material characteristics, and by availability of standard sizes. Limitations on functional requirements are imposed by factors peculiar to the entire mechanical structure or machine.

Constraints may be classified into two basic types: *regional constraints*, which express certain ranges of acceptability for a variable by means of limit equations, and *discrete value constraints*, which express a list of acceptable values for a variable. They are introduced to satisfy standard size limitations encountered in practice, such as a standard wire size limitation for a helical spring as an example. Also, they are introduced to satisfy other specific value limitations, such as the whole number constraint obviously necessary for gear teeth. Discrete value constraints may be infinite or finite in number. A special case is a specified single value for a functional requirement, which is really nothing more than a singular discrete value constraint.

It is important not to neglect any constraint, even though it might be only implied. Occasionally a limit equation is on two geometrical parameters, based on practical experience making use of permissible proportions. This is clearly illustrated in some of the examples which follow. Following the pattern of notation which has been established, limit equations will be designated by (L.E.) whenever the notation is advantageous.

6.12 Summary of Design Equations in Optimum Design

The three types of equations encountered in the optimum design of a mechanical element are figuratively depicted in the accompanying typical schematic representation (page 196). In these equations, f, g, and h represent functions which are, fortunately, relatively simple for many mechanical elements. It is important to notice the dependent or independent relationship between the various parameters and parameter groups in these equations.

The equation system derived and compiled for uniquely defining an optimization problem will be called the *initial formulation.* It includes the criterion function for expressing the optimization quantity, the subsidiary design equations for expressing the significant functional requirements and undesirable effects, and the various types of significant constraints as previously described. Hence the initial formulation is the first equation system summarized which mathematically ties together the entire optimization problem to be solved. At this stage, however, no attempt has been made either to organize or to simplify the equation system, and generally no conclusions of optimum design have been drawn.

From the description of the initial formulation in the preceding paragraph, we can readily see that judgment of the design engineer plays an important part. Choice of optimization quantity and decisions as to which items are significant for functional requirements, undesirable effects, and constraints all require judgment to some extent. Hence compilation of the initial formulation is not a cookbook type of problem, and such a situation is very appealing to the progressive engineer with a flair for challenge!

Quantity to be maximized or to be minimized by optimum design

Three *independent* groups

Design parameters (uniquely define element)

$$(\text{P.D.E.}) = f\left[\begin{pmatrix}\text{Functional}\\ \text{requirement}\\ \text{parameters}\end{pmatrix}, \begin{pmatrix}\text{Material}\\ \text{parameters}\end{pmatrix}, \begin{pmatrix}\text{Geometrical}\\ \text{parameters}\end{pmatrix}\right]$$

Individually specified or limited parameters

Individual parameters dependent on each other because of physics and chemistry of the material

Individually independent parameters

$$(\text{S.D.E.}) = g\left[\begin{pmatrix}\text{Functional}\\ \text{requirement}\\ \text{parameters}\end{pmatrix}, \begin{pmatrix}\text{Material}\\ \text{parameters}\end{pmatrix}, \begin{pmatrix}\text{Geometrical}\\ \text{parameters}\end{pmatrix}\right]$$

Specified or limited quantity

Independent group

Groups are dependent on specified or limited subsidiary design quantities *and* on optimum design of element

$$h_1 \leq (P) \leq h_2$$

h_1	(P)	h_2
Lower limit, depending on factors external to element	Limited design parameter or limited (S.D.E.) quantity	Upper limit, depending on factors external to element

In light of the preceding discussion, some basic strategy should be given for the compilation of the initial formulation in an optimization problem. In general, to start make the initial formulation as simple as you dare. Include only the items which you believe are of definite significance in this initial formulation. This will facilitate the steps to follow which lead to specification of the optimum design. The optimum design so determined would then be thoroughly analyzed for all the marginal items which were ignored in the initial formulation. If no unacceptable quantities are detected in this end analysis, the optimum design has been found! If violations of acceptability have been uncovered in this end analysis, the appropriate equations and constraints are added to the initial formulation as the start for another attempt at solution for the optimum design. However, in general, such an iterative approach will be unnecessary after some experience has been gained by the design engineer.

Method of Optimum Design (MOD)

Development of an optimization technique called the *method of optimum design* was started in 1954 and the first major publication on the subject was the first edition of this book in 1961. Other publications on the technique and applications are found in references 6.16 through 6.28. A general description of this technique will be presented in this section, followed by more specific details pertaining to mechanical element applications and examples in the next several sections. In studying this MOD material the reader should refer to the appropriate nomenclature summarized in Appendix A.

6.13 General Description of MOD

After sketching a model of the item to be designed, boundary conditions are reviewed, including all functional requirements and undesirable effects. Thus specifications and constraints are summarized, and equations are derived which tie the design variables together mathematically. Appropriateness of introducing an optimization objective is then considered, and if optimization of design is relevant this item becomes the optimization quantity for which the primary design equation (P.D.E.) is derived. Hence the initial formulation for an optimization problem is derived and compiled.

Some Guidelines for Problem Formulation. The general procedure for deriving the initial formulation for an optimization problem of a mechanical element can be summarized in the following steps.

1. Decide on the basic configuration for the mechanical element in the form of a free-hand sketch showing geometric parameters and other items of interest, such as loads, deflections, and critical stress regions.

2. Summarize the significant constraints for the design problem, listing specified items, discrete value constraints, and regional constraints in the form of limit equations (L.E.)'s.

3. Review the summary from step 2, and express the appropriate constrained parameters in terms of design variables. Summarize the equation system so derived or compiled as part of the initial formulation.

4. Decide on the objective for design optimization, designated in general as Q. If Q exists in one of the equations from step 3, use that equation as the (P.D.E.) with the remaining equations as the (S.D.E.)'s. If Q does not exist in the equation system of step 3, derive the equation for Q in terms of the design variables. For such a case this will be the (P.D.E.), with the equations from step 3 as the (S.D.E.)'s.

5. Review the summarized equation system from steps 2, 3, and 4, also referring to the free-hand sketch from step 1. Be sure that all variables are tied together mathematically in terms of the design parameters. If the equation system is not complete, derive or compile the necessary relations to add to the (S.D.E.) list, such as equations for geometrically dependent parameters.

The steps outlined above are the bases for derivation of the initial formulations in the examples of Chapters 6 through 8. Execution of the steps requires some skill and judgment in many cases, but this will become relatively easy after some experience is gained through practice.

Basic Procedural Steps for MOD [6.18]. Successive steps of a systematic plan lead to the specification of the optimum design, summarized as follows.

1. *Initial Formulation* (*I.F.*). This is a summary of the initial system of equations previously discussed and illustrated, including the (P.D.E.), (S.D.E.)'s, and (L.E.)'s. The (I.F.) may not be organized to the desired format and it may not be simplified as much as possible. Hence, for reaching the decisions of optimum design in an efficient and systematic manner, it may be necessary to transform the (I.F.) equation system to a more suitable form, which is the next step.

2. *Final Formulation* (*F.F.*). In this step the (I.F.) equation system is transformed if appropriate to the desired (F.F.) format suitable for use in the variation study step. The transformation to the (F.F.) equation system involves two possible steps, a and b, which follow, either one or both of which may be appropriate.

(a) *Equation Reduction Step.* Free variables of the (I.F.) are eliminated in an equation combination procedure described in sections 6.14 and 6.15. Thus the number of equations in the (I.F.) is reduced by an amount equal to the number of free variables, for the (F.F.).

(b) *Equation Organization Step.* The reduced equation system from a may not be organized as desired for the variation study which follows. Hence further mathematical manipulation may be necessary to organize the scrambled variables into the format desired for the (F.F.) equation system.

3. *Variation Study* (*V.S.*). In this step the (F.F.) equation system is considered simultaneously along with the constraints for general determination of the points potential for optimum design. This step is generally facilitated

by the sketching of a *variation diagram*, which may be either two dimensional (2-D) or three dimensional (3-D) in character. For the free-hand sketching of the variation diagram, it is necessary to consider the constraints from the (I.F.) along with the (F.F.) equations in the sketching of the pertinent curves involved. The effects of boundary constraints, the points potential for optimum design, and the possibility of incompatible specifications are all clearly revealed in this systematic (V.S.) step. The determination of the optimum design is explicitly summarized in the systematic format of the variation diagram.

4. *Execution of Results.* The logic for explicit determination of the optimum design has been established and compactly summarized in the variation diagram of the preceding step. Hence the determined decisions of optimal design are known to the user for specific execution in the various ways given below.

(a) *Flow Charts for Computation.* The flow chart (F.C.) is logically derived from an understanding of the variation diagram from step 3. The flow chart is the basis for writing of highly efficient computer programs for automated optimal design. Some general guidelines for the synthesis of (F.C.)'s are summarized in section 7.7 of Chapter 7.

(b) *Goals of Design for Layouts.* Results from the (V.S.) step indicate directions in which to head, particularly for geometric parameters, and this can be a valuable guide in the making of a design layout.

(c) *Guides for Creative Change.* Goals of design for geometric parameters in b can imply a change in configuration for creative design. Chapter 4 of reference 6.18 illustrates this with some industrial examples.

5. *Evaluation of the Optimum Design.* At the conclusion of step 4 in a numerical application, the physical design of the mechanical element has been determined and hopefully this is the optimum design for all practical purposes. Hence, to confirm our acceptance of this design, an evaluation phase is generally required. The design is analyzed to determine what has been achieved numerically for the optimization quantity. Also, we generally wish to analyze for many other phenomena of interest associated with acceptability of design, such as any marginal items ignored in the (I.F.). If the overall performance of the design is acceptable in all respects, we have been successful in finding and specifying the optimum design. If not, we must expand the (I.F.) to include the constraints violated and/or modify the pertinent boundary values for improvement of design. Such changes, of course, are followed by a repetition of the MOD procedural steps.

Types of Variables in (I.F.). Incidentally, upon review of the (I.F.) for an optimization problem, certain types of variables will be recognized. For one, there is optimization quantity Q expressed by the (P.D.E.), which is to be minimized or maximized depending on the problem. Second, there are what we will call *constrained parameters*, n_c in number. They are defined as the ones having either regional constraints or discrete value constraints directly imposed on them by (L.E.)'s in the (I.F.). Next, there are what we will call *free variables*, n_f in number. They are defined as the ones with no constraints directly imposed on them through (L.E.)'s in the (I.F.), not counting optimization quantity. This does not mean that the free variables are totally unconstrained in the optimization problem, since constraint effects will be imposed on them indirectly through the entire equation system of the (I.F.). Finally, the total number of variables in the (I.F.) will be considered as the sum of constrained parameters and free variables, expressed as $n_v = n_c + n_f$.

Types of Problems in MOD [6.18]. There are basically three classes of problems encountered in application of the MOD. They are briefly summarized below.

1. *Case of Normal Specifications* (*N.S.*). This is the simplest type of optimization problem, where the (F.F.) is merely a single developed (P.D.E.). Generally, optimization conclusions are drawn with respect to the independent variables of this developed (P.D.E.) from a simple (V.S.). In mechanical elements, often an independent material selection factor (MSF) is recognized in this developed (P.D.E.). If so, it can be used to select the optimum material from a feasible list of material properties. The test for this (N.S.) type of problem is $n_f \geq N_s$, where n_f is the number of free variables (i.e., no constraints directly imposed) and N_s is the number of (S.D.E.)'s in the (I.F.). This test is made in the (I.F.) stage prior to derivation of the (F.F.).

2. *Case of Redundant Specifications* (*R.S.*). For this type of problem, it is found that the (F.F.) must be an equation *system* 2 or more in number. For such cases, in the MOD an *ideal problem* of (N.S.) is hypothetically created by temporarily ignoring some of the constraints on selected parameters, which we call *eliminated parameters*, from the (I.F.). The developed (P.D.E.) of the ideal problem then derived is designated as equation I in the (F.F.) equation system. However, to complete the total picture, the eliminated parameters must be expressed by what we call *relating equations* and designate as equations II, III, IV, and so on in the (F.F.) equation system. These equations relate the eliminated parameters to the *related parameters*, and they are constrained parameters generally common to all the (F.F.) equations. The (V.S.) must consider the total picture with all constraints from the (I.F.) imposed, and this (V.S.) is made from a simultaneous consideration of the entire (F.F.) equation system. The number of dimensions required for a (V.S.) variation diagram is one more than the number of related parameters. The test for the (R.S.) type of problem is $n_f < N_s$, and this test is made in the (I.F.) stage prior to derivation of the (F.F.).

3. *Case of Incompatible Specifications* (*I.S.*). This is really nothing more than a special category of (R.S.), which is revealed as being the situation for certain problems in numerical application. With the particular boundary values numerically specified, there is merely no design solution which satisfies all constraints and specifications. Hence the MOD clarifies what would otherwise be a frustrating situation, also indicating which boundary values could be changed to alleviate the situation. If the pertinent boundary values can be changed, the domain of feasible designs is opened and the optimum design can then be determined by the MOD.

General Planning, (I.F.) to (F.F.) in MOD. For those learning the MOD, one of the most difficult details of execution is in the transformation of the (I.F.) to the (F.F.). Three general items helpful in this respect will be outlined briefly below for the (R.S.) type of problem.

1. *Exploratory Calculations.* Some very simple equations have been derived in reference 6.18 for rapidly predicting certain characteristics of interest for the MOD technique in specific problems. One characteristic of interest is the number of dimensions designated as D_{vs} required for a (V.S.), and this we can calculate after inspection of the (I.F.) by equation 6.1.

$$D_{vs} = n_v - N_s + 1 \tag{6.1}$$

In this equation n_v is the number of variables (not counting optimization quantity) in the (I.F.). Incidentally, the value of D_{vs} will actually be reduced by one for each truly independent parameter in the (F.F.). A *truly independent parameter* is a variable appearing only in equation I of the (F.F.).

As a second item of initial interest, in transformation from (I.F.) to (F.F.) there are A_t number of different approaches which theoretically might be taken. Generally most of these are *determinate approaches* compatible with the required mathematical manipulation of the transformation process. However, occasionally an *indeterminate approach* is uncovered and discarded as a possible approach for solving the optimization problem. Of course, the simplest determinate approach is what we would like to follow. A_t is rapidly calculated after inspection of the (I.F.) by equation 6.2.

$$A_t = \frac{n_c!}{[n_e!(n_c - n_e)!]} \tag{6.2}$$

In this equation n_c is the number of constrained variables (i.e., ones having constraints directly imposed) in the (I.F.), and n_e is the number of eliminated parameters calculated by equation 6.3.

$$n_e = N_s - n_f \tag{6.3}$$

Incidentally, from equation 6.2 we note that A_t is the number of combinations of n_c things taken n_e at a time disregarding the order. The specific groups of eliminated parameters represent the different approaches possible for different (F.F.)'s, and for each of the A_t approaches the related parameters are the complementary group from the list of constrained parameters in the (I.F.). Hence the number of related parameters n_r can be calculated from equation 6.4.

$$n_r = n_c - n_e \tag{6.4}$$

Incidentally, use of a circle diagram illustrated in later chapters may facilitate determination of the possible approaches for derivation of the (F.F.).

2. *Choosing the Approach.* As now described, the various groups of eliminated parameters possible and their corresponding related parameter groups for each of the A_t possible (F.F.)'s can be determined rapidly in the (I.F.) stage. Choice of the particular approach to take for derivation of the specific (F.F.) should be made after some thought, and extrapolation through the variation diagram stage of the (V.S.) should be visualized if possible. In choosing the best one of the A_t approaches, one should look for two features of simplicity. First, which approach will be simplest from the standpoint of equation manipulation when deriving the (F.F.)? Oftentimes the (I.F.) is already in the proper format of the (F.F.) for one of the approaches. At least, some of the possible approaches are usually simpler than the others from this equation manipulation standpoint. Second, which approach will result in the simplest variation diagram in the (V.S.) stage? Generally this is difficult to predict in the (I.F.) stage, and the first consideration becomes the primary one in choosing the particular (F.F.) approach to specifically carry through in a complicated problem.

3. *General Format for (F.F.).* When considering the various approaches possible for (F.F.)'s, summarizing the general equation system formats for each can be very helpful. Such a general summary of (F.F.) possibilities when compared with the specific (I.F.) equation system provides a clear picture of what is necessary for the transformation process in each case, and this can be helpful in selecting the particular approach to carry through. Also, after the choice has been made, the (F.F.) general format summary can serve as a valuable guide for specific execution of the actual equation manipulation process in the transformation from the particular (I.F.).

As a very simple illustration, let us consider the first (I.F.) given in section 6.8 of this chapter. The two possible approaches would be summarized by general (F.F.) formats as follows.

Approach 1:

$$Q = f_1(x) \tag{I}$$

$$A = f_2(x) \tag{II}$$

Approach 2:

$$Q = f_3(A) \tag{I}$$

$$x = f_4(A) \tag{II}$$

Obviously approach 1 would be simpler to take, since the (I.F.) of section 6.8 is already expressed in that format, with no equation manipulation then being necessary. Incidentally, for this problem A and x are constrained parameters. Also, for approach 1 the eliminated parameter is A, with x being the related parameter; whereas for approach 2 the eliminated parameter is x, with A being the related parameter.

The (F.F.) equation system general format summary step technique now described is quick and easy to use, and of course it has greatest value in complicated problems where the necessary equation manipulation procedure is not so clear.

Some Features of the MOD. The MOD has many features, some of which will be summarized very briefly below.

1. In application the MOD will determine the global optimum design point (i.e., the best possible solution point in the feasible design space).
2. The (V.S.) variation diagram in the MOD presents a vivid picture of the functional relationships and the boundary constraint effects for the total optimization problem, clearly revealing the variation of optimization quantity in feasible design spaces.
3. The (V.S.) variation diagram in the MOD explicitly reveals all possible points which are potential for optimum design. The particular one which governs for a given numerical problem depends on the specific feasible design space revealed for the case.
4. The (V.S.) variation diagram in the MOD reveals special effects of boundary conditions which may be used to advantage in certain cases. For example, occasionally the domain of optimum design is infinite in scope, such as any point on the segment of a boundary curve along the feasible design space. Such a situation would be clearly revealed by the MOD, and then introduction of a secondary objective for optimization would improve the overall design characteristics achieved.
5. The (V.S.) variation diagram in the MOD reveals the appropriate strategy for improvement of optimum design if this is desired, indicating which boundary values are most pertinent in this respect. On occasion the value of optimization quantity achieved is not good enough, but modifying the appropriate boundary values can appreciably improve the results to an acceptable level.

6. The (V.S.) variation diagram in the MOD clearly reveals the possibilitie if any, of incompatible specifications (I.S.), as well as which boundary valu to change for alleviating the situation if encountered in a numerical probler

7. The MOD results in derivation of general flow charts for certain bas types of optimization problems, resulting in highly efficient comput programs which can be used in limited applications of automated optim design. Hence highly constrained optimization problems are solved in gener by the MOD resulting in computer programs which can be used with fal limits if appropriate for solving numerical problems of a fairly broad rang The technique is particularly suitable for mechanical elements encountere often in practice, such as helical springs, sleeve bearings, and gears.

8. The MOD minimizes the number of variables involved in the optimiza tion study, since free variables have been eliminated in the (F.F.) stage. Henc if an automated optimal design search plan is combined with the MOD solution to the optimization problem is simplified and computation time reduced.

Some Limitations of the MOD. The MOD has limitations, and we woul be remiss if we did not summarize them briefly.

1. Execution of the procedural steps in the MOD is time consuming, an such expenditure must be justified by the possibility of valuable result achieved. This limitation is of minimal concern for general problem types c broad application where a computer program will be derived with much us anticipated.

2. The MOD is a complex systematic approach to optimization of desig problems, and the technique must be learned as a new subject for most desig engineers. Deficiency in understanding or even lack of facility with the tech nique can prevent its appropriate application. Worse yet, inappropriat application of the MOD can result in erroneous design conclusions with fa from optimum results, and in many cases the error is difficult to detect.

3. Equations of the (I.F.) may be too complicated for the derivation of th (F.F.) by the MOD, and/or for the making of the (V.S.), especially with in separable variables. Often this situation can be alleviated by application o appropriate curve-fitting techniques as described in Chapter 2 and as illus trated in the MOD example of Chapter 8 for gears.

In the next three sections of the chapter we will consider in more detai procedural steps for the three classes of problems encountered in the MOD previously designated as (N.S.), (R.S.), and (I.S.). The discussions will b directed towards application to mechanical elements, with a simple tensio bar optimization problem carried through in the examples of the thre sections.

6.14 Case of Normal Specifications for Mechanical Elements

For the case of *normal specifications* it will be possible to develop the primary design equation to the point where the effects of *all* subsidiary design equations are included and where the unique optimum design can be directly extracted, *all* significant specifications and limits having been taken into account. For such cases it is possible to determine uniquely the optimum material and optimum values for the geometry with a minimum amount of effort. Many real problems in optimum design belong to this category, and it will be illustrated by a simple example. As presented in section 6.13, the test for this type of problem is an adequate number of free variables relative to the number of subsidiary design equations, recognized by the relation $n_f \geq N_s$. The test is made at the end of the (I.F.) stage, which is after step 4 below.

For the case of *normal specifications*, the *method of optimum design* can be outlined in the following detailed procedural steps.

1. Draw a free-hand sketch of the mechanical element showing the significant basic geometry. Select the *independent geometrical* parameters which will be used for uniquely defining the geometry of the element. If the choice exists, select geometrical parameters whose values are either specified as functional requirements or limited to permissible ranges.

2. After having decided on the significant explicit objective for conduct of design, which will be the basis for optimum design as previously discussed, derive the *primary design equation* which expresses the *optimum design quantity*. If possible, write this equation in terms of *groups* of functional requirement parameters, material parameters, and geometrical parameters.

3. Derive *all subsidiary design equations*, which express functional requirements and significant undesirable effects, other than the primary design equation. It is important to include *all* the subsidiary design equations, whether they are directly specified or merely implied. Also, write an *entirely geometrical* subsidiary design equation for *each* geometrical parameter *other* than the previously selected *independent geometrical* parameters, *if* such other parameters exist in *any* of the design equations.

4. Write *all limit equations*, for functional requirement parameters, for significant undesirable effect parameters, for material parameters, and for geometrical parameters.

5. *Combine all* subsidiary design equations with the primary design equation by eliminating an *unlimited and unspecified* common parameter (i.e., free variable) from the primary design equation for *each* subsidiary design equation. In doing this, the primary design equation will be so *developed* that it will consist only of specified values, *independent* parameters, and *independent* parameter *groups*. In this procedure of equation combination geometrical parameters are often eliminated, since they can be *varied independently* of

each other and generally they can be *varied continuously*; these are important desired characteristics for eliminated parameters because of step 9, in view of the fact that steps 7 and 8 necessarily will be made first. Of course, equation combination by elimination of other unspecified and unlimited parameters whose values can be varied independently and continuously is equally satisfactory. On the other hand, combination by eliminating material parameters in this step generally should be avoided, since they are dependent on each other and even as a group they can be varied only in relatively few discrete steps. Do not attempt to combine by eliminating a functional requirement parameter which is a specified value. Satisfaction of a subsidiary design equation expressing the specified functional requirement would then be lost in the optimum design procedure which follows.

If it is impossible to execute step 5 because of the number of specified limits or specified values, the general case of *normal specifications* does not exist. The case of redundant specifications then exists. A more complicated procedure for optimum design will be required, and this will be explained and illustrated in section 6.15 of this chapter.

6. Using the *developed primary design equation* from step 5, roughly determine the *variation* of the optimum design quantity with respect to each *independent parameter* or *independent* parameter *group* in the primary design equation. If it is helpful, draw a free-hand sketch of each variation. If it simplifies the mathematics, determine the variation of the inverse optimum design quantity instead of the optimum design quantity itself. For example, minimizing cost C is exactly the same as maximizing the inverse cost C^{-1} for any mechanical element.

7. Considering the variations from step 6 together with the previously written limit equations, determine the optimum value for each independent parameter or independent parameter group in the *developed primary design equation*; because of the explicit objective initially chosen for conduct of *design*, these optimum values will be uniquely determined. It is *important* to remember that *limit values* can be substituted in the procedure of *design* only for parameters or parameter groups which are *independent* of the other terms in the developed primary design equation. Also, be sure to realize that the optimum value for an independent parameter or independent parameter group is *not* always a limit value; this depends on the particular variation in the developed primary design equation. At the conclusion of step 7 the primary design equation will be expressed in terms of functional requirement parameters, material parameters, limits, and, occasionally, specific optimum values other than limits; this will be called the *final primary design equation*.

8. From the *final primary design equation* at the end of step 7, select the optimum material on the basis of the material parameter group in that equation. This group will be called the *material selection factor*, and its value

will determine the optimum feasible material. Occasionally it will be impossible to separate the material parameters from the geometrical parameters in the developed primary design equation. Then an explicit determination of the optimum material will be more difficult, since the MSF will include geometrical parameters. However, the problem is not formidable, as will be illustrated in some of the specific examples in later chapters.

9. From the *final primary design equation* at the end of step 7, optimum values for many of the geometrical parameters of the mechanical element will already have been determined. However, to determine the optimum values for the geometrical parameters which were eliminated from the primary design equation in step 5, merely solve for the parameter values using the original subsidiary design equations at this time. This step will complete the optimum design of the mechanical element.

Let us now compare the preceding procedural steps with the more general steps outlined in section 6.13 of this chapter. As you will recognize, preceding steps 2 through 4 constitute the (I.F.), step 5 gives the (F.F.), and the (V.S.) is made in step 6. This is followed by execution of results in steps 7 through 9. At the conclusion of step 9 the optimum design has been uniquely determined, and of course this would be followed by an evaluation phase to establish acceptability from all standpoints.

EXAMPLE 6.1

As an academic and qualitative illustration of the method of optimum design for the case of *normal specifications*, let us consider the problem of designing a simple tensile bar which must have a specified length L and which must transmit a specified constant magnitude force P, as depicted in figure 6.5. To simplify the illustration, assume that the bar will be manufactured in large quantities; thus a logical objective for optimum design would be *minimization of cost*. Hence the problem in design is to select the available material and the value for the required cross sectional area A which will result in the smallest possible cost for manufacturing the bar.

Based on the previous assumptions, the major *variable* component of manufacturing cost when considering all feasible designs is the cost of the material to be used in the bar. To begin, write the (P.D.E.) expressing the optimum design quantity, which is material cost C_m for the bar.

$$C_m = cV = c(AL)$$

Hence the (P.D.E.) is

$$C_m = cLA \tag{6.5}$$

where C_m is the cost of the bar material, in dollars; c is the unit volume material cost, in dollars per cubic inch; L is the specified bar length, in inches; and A is the cross sectional area, in square inches.

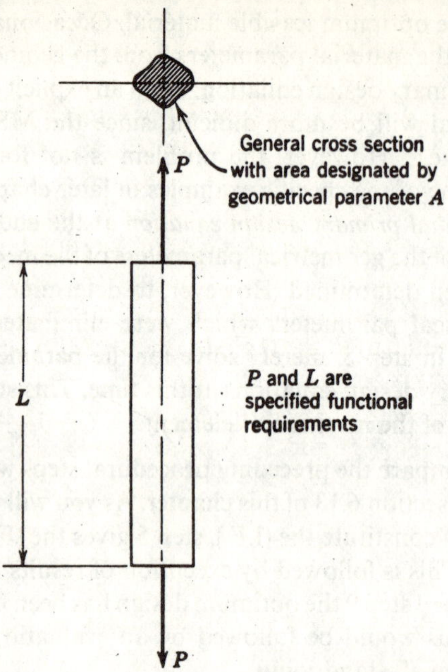

Figure 6.5 Simple tensile bar of specified length L transmitting uniformly distributed specified axial force P.

No subsidiary design equations have been directly specified for this problem. However, assume that the particular application of the tensile bar in its structural setting is such that yielding cannot be tolerated. Hence *stress* is a significant implied undesirable effect which will introduce a subsidiary design equation. Thus, we shall write the (S.D.E.) as follows from elementary strength of materials.

$$\sigma = \frac{P}{A} \tag{6.6}$$

where σ is the normal stress in the bar and P is the specified axial force.

No limit equations have been directly specified for this problem. However, the limitation on permissible values of stress σ is implied by the theory of failure. Hence, if we assume a factor of safety of N_y based on the yield strength S_y for *any* material, we have the limit equation

$$\sigma \le \frac{S_y}{N_y} \tag{6.7}$$

Summarizing to this point, equation 6.5 is the (P.D.E.), equation 6.6 is the (S.D.E.), and 6.7 is a limit equation. Also, P, L, and N_y are functional requirement parameters, c and S_y are material parameters, and A is a geometrical parameter. σ and C_m are undesirable effects, and the basis of optimum design is the minimizing of material cost C_m. Notice that the geometry has been uniquely defined by *independent* parameters A and L, and that for this problem L is really a functional requirement since it is specified as a value initially.

Hence we have finished steps 1 through 4 in the preceding outline of optimum design. We note with one free variable A and one subsidiary design equation (equation 6.6) in the (I.F.) that $n_f = N_s$. Therefore, this example is the (N.S.) type of problem.

Next, following step 5 of the outline, combine the (S.D.E.) with the (P.D.E.), eliminating the *unlimited* and *unspecified* geometrical parameter A. Thus

$$C_m = cL\left(\frac{P}{\sigma}\right) \tag{6.8}$$

From this *developed* (P.D.E.) note that C_m varies directly with each of the parameters c, P, and L. However, in this equation we see that C_m varies inversely with the stress σ, which is now really an independent parameter, having replaced the independent geometrical parameter A. Because the variation of cost C_m with respect to each of the terms in equation 6.8 is a very simple one in this problem, we need not spend much time on step 6 of the outlined method.

The next step is to consider the effect of the limit equation on the *developed* (P.D.E.) equation 6.8. Hence, when we consider the variation of C_m with respect to the *independent* parameter σ in equation 6.8, together with the effect of limit equation 6.7 which defines the range for *all* permissible values of σ, it is obvious that for minimizing cost C_m we should place σ equal to its *upper* limit S_y/N_y. This is clearly depicted in figure 6.6, which normally would not be drawn for a simple problem like this one. Thus the final (P.D.E.) at the end of step 7 of the outlined method is given by the following equation, which is obtained from equation 6.8:

$$C_m = cL\left(\frac{P}{S_y/N_y}\right)$$

$$C_m = (PLN_y)\left(\frac{c}{S_y}\right) \tag{6.9}$$

In this, the final (P.D.E.), P, L, and N_y are specified functional requirements which we probably cannot change. However, if the possibility of changing any of the specifications exists, the objective, of course, should be to minimize the independent group $[PLN_y]$. Incidentally, a reasonable and conservative

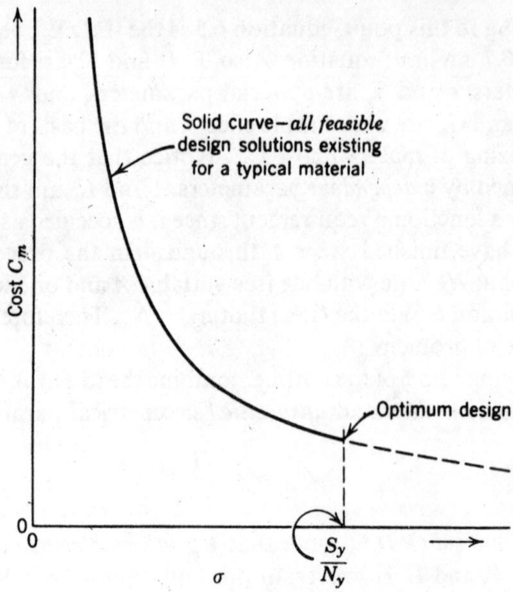

Figure 6.6 Typical variation of material cost versus stress for geometrically feasible designs of a simple tensile bar. Functional requirements P, L, and N_y are specified values.

minimum value for the factor of safety N_y could be selected on the basis of the discussion in Chapter 5, considering the significance in the structural setting attached to the occurrence of the failure phenomenon of yielding for the tensile bar in this particular problem.

In addition, the final (P.D.E.), equation, also indicates how the optimum material should be selected. From that equation, the MSF is the independent group (c/S_y). From a list of *available feasible materials*, we would calculate the value of (c/S_y) for each material, and tabulate for future use. In all practical likelihood, we would have to make this simple calculation for only a relatively few available feasible materials. Obviously the optimum material would be the one having the lowest value for the independent material selection factor (c/S_y), and the selection of this material would result in the lowest cost for the tensile bar regardless of the specific values for P, L, and N_y.

Finally, for determining the optimum value for cross sectional area A we would use the original (S.D.E.), equation 6.6, with $\sigma = S_y/N_y$. Hence equation 6.6. would become

$$A = \frac{PN_y}{S_y}$$

and thus we could calculate very simply the optimum value for the area A

The value for S_y which we would use in this equation would be that for the optimum material, which has already been selected. Thus the design would be complete at this point.

6.15 Case of Redundant Specifications (R.S.) for Mechanical Elements

For the case of normal specifications, by properly combining equations it was possible to include the effects of *all* subsidiary design equations in the primary design equation, with *all* the limit equations and specifications applying only to independent parameters still existing in the developed primary design equation. An explicit optimum design could be extracted at once for such cases. For the *case of redundant specifications* it is *not* possible by proper equation combination to include the effects of *all* subsidiary design equations and *all* limit equations in a *single* development of the primary design equation. This case is generally characterized by the necessity for eliminating a *limited* parameter in order to combine a subsidiary design equation with the primary design equation. Hence there is generally an *excess* number of limit equations for such cases, and a special optimum design procedure must be followed. Also, this type of case is characterized by an excess number of subsidiary design equations, which cannot be combined properly with the primary design equation at one time. As stated in other words in section 6.13, the test for the (R.S.) type of problem is an inadequate number of free variables compared with the number of subsidiary design equations, recognized by the relation $n_f < N_s$. The test is made at the end of the (I.F.) stage, which is after step 1 of the procedure given below.

The optimum design of mechanical elements for the case of redundant specifications *can* be extremely difficult, since it generally requires a study of variations and relationships between several functions of many variables. Fortunately, for most mechanical elements the problem is not formidable, especially when the simple exponential type of equations is involved. A general *method* for facilitating solutions to optimum design problems having redundant specifications has been formulated on the basis of many specific case studies, and the salient results outlining the general procedure will be presented in this section.

Because of the complexity of most problems, optimum design for the *case of redundant specifications* generally is more clearly understood *after* some specific examples have been studied. It is suggested that this section be reread after some examples, such as the ones following and particularly those in Chapter 7 have been carefully studied. Then the reader will be better prepared to cope with *original* optimum design problems for the case of redundant specifications.

The general *method of optimum design* for the case of redundant specifications can be outlined in the following detailed procedural steps. However, as is evident from some of the examples in this book, the reader might find it expedient to incorporate slight modifications or special procedures for particular problems based on the overall objectives and underlying theoretical concepts of this general plan.

1. Repeat procedural steps 1 through 4 in section 6.14.

2. *Temporarily ignore* one or more of the limitations or specifications, so that a *hypothetical* case of normal specifications exists. This hypothetical case will be referred to as the *ideal problem*, and any design study pertaining to it will be for what will be called an *ideal* mechanical element. Discretion based on the following steps should be exercised in deciding which particular limitations or specifications to ignore temporarily, since frequently much effort can be subsequently saved by making the proper choice.

3. For the *ideal problem* now defined, develop the (P.D.E.) by properly combining equations according to procedural step 5 in section 6.14. Thus, in performing this combination, it will be necessary to eliminate or omit from the (P.D.E.) the parameters whose limits or specified values were temporarily ignored in step 2 above. To simplify future discussions, these particular parameters will be referred to as the *eliminated parameters* from now on. However, the parameters or parameter groups *remaining in* the developed (P.D.E.) at the end of this step will be *independent* of each other *within* the confines of the *ideal problem.*

4. Determine which parameters now existing *in* the developed (P.D.E.) of the ideal problem are *related* by subsidiary design equations to the *eliminated parameters.* These particular parameters *in* the developed (P.D.E.) of the ideal problem will be designated as the *related parameters* in what follows, in order to simplify subsequent discussions. For the same reason, the particular subsidiary design equations which relate the now defined *related parameters* to the *eliminated parameters* will be designated as the *relating equations* in what follows. (Incidentally, as an exception to the rule, in the design procedure being explained, do *not* consider *material parameters* to be included in the terminology of *related parameters*, even though they might very well satisfy the definition. This exception can be made because materials will be accounted for in the design procedure by a stepping process corresponding to the discrete list of *available feasible* materials for the particular design problem at hand, and this will be illustrated in many of the examples which follow. The reason for doing this is that material parameters *cannot* be varied either independently of each other or continuously in the design procedure as previously explained, and for most problems there will be only a relatively few *available* and *feasible* significant materials to consider. As a word of caution, the reader should *not* consider *stress*, like σ as a material

parameter. As emphasized in Chapter 4, *stress* is the intensity of force with respect to a reference plane; therefore, in the conduct of design studies, *stress* should be treated as a parameter which can be varied continuously and independently, although permissible values for it would be *limited* by material parameters. Therefore, treat *stress* as a *related parameter*, if the definition for the terminology is satisfied.)

5. From the developed (P.D.E.) of the ideal problem, determine the variation of the optimum design quantity with respect to the independent parameters *in* that equation *other* than the ones which are also in the previously defined *relating equations.* Hence, as in steps 6, 7, and possibly 8 of section 6.14, final optimum values can be determined for these *truly independent* parameters or groups in the developed (P.D.E.) of the ideal problem. Thus, by considering the existing limits, we can select the optimum values for these parameters at this time and they will be the final choice.

6. From the developed (P.D.E.) of the ideal problem, determine the variation of the optimum design quantity with respect to the *related parameters*, previously defined. Determine also the effects on the optimum design quantity of the limits existing on permissible values of these *related parameters.* This step will determine the variation of the optimum design quantity with respect to the *related parameters* for all feasible designs existing within the confines of the ideal problem.

7. Using the *relating equations* previously defined, determine the variations of the *eliminated parameters* with respect to the *related parameters*, both previously defined. On these variations be sure to impose the limits or specifications existing on permissible values of the *eliminated parameters*; these limits were previously temporarily ignored.

8. Superimpose the variations from step 7 on the variation from step 6, all with respect to the *related parameters* as the "independent variable coordinates," including the effects of the existing limits. The resulting ranges of *compatible* variation for these limited functions in this step will determine the domain of variation for *all* feasible existing design solutions to the overall problem. Thus optimum values for the *related parameters* and for the *eliminated parameters* will be obvious. Therefore the optimum design of the mechanical element for the case of redundant specifications will be complete at this time.

Let us now compare the preceding procedural steps with the more general steps outlined in section 6.13 of this chapter. As you will recognize, preceding step 1 yields the (I.F.), steps 2 through 4 give the (F.F.), and steps 5 through 8 correspond to the (V.S.). Execution of results will logically follow from an understanding of the (V.S.), resulting in unique specification of the optimum design. Finally, acceptability of this design would have to be established by means of a thorough analysis.

The beauty of the general method outlined is that it requires the study c function relationships with respect to a *minimum* number of variables. Thu with the suggested method, it is necessary to consider function variatio relationships with respect to the *related parameters* only, instead of with respe to *all* parameters in the original problem. This is a tremendous simplificatio for most real design problems, even for cases where functional relationshir must be considered with respect to several *related parameters.*

An example of a specific problem having four *related parameters* is foun in Chapter 8. In that chapter the problem is to design a helical gearset on th basis of maximum power transmission *capability*. It is found that a case c redundant specifications exists, and the *related parameters* are pinion pitc diameter D_p, pitch helix angle ψ, and the two normal plane variables C_{nF} an N_{np}. The *eliminated parameters* in obtaining the developed (P.D.E.) for th *ideal problem* are diametral pitch P_{nd} referred to the normal plane, numbe of pinion teeth N_p, and face width b. The primary design equation express power transmission *capability*, as governed by tooth wear designated (HP) Therefore it is required to consider functional relationships for $(\text{HP})_w$, N b, and P_{nd}, with respect to *related parameters* D_p, ψ, C_{nF}, and N_{np}. Th design problem is solved in Chapter 8, and as can be expected it is qui complex. However, the solution is relatively simple by the MOD as outline in this section, compared with any attempt at solution by considerin functional relationships with respect to all parameters.

The purpose of the examples in this section is strictly to illustrate th application of the general *method of optimum design* which has been outline For example 6.2 it is well realized that the optimum design solution coul be determined in other ways; for instance, on a single graph we could loo at the ranges of compatible variation between the limited functions C_m versu A and σ versus A for each material, although such an approach would b somewhat more cumbersome. However, for more complex design problem as presented in later chapters, following the suggested general method highly recommended because it will *greatly* facilitate the unique determina tion of the optimum design. Incidentally, it should be emphasized that single material selection factor generally does *not* exist for the case of redur dant specifications as it generally does for the case of normal specification this point will be obvious from the examples which follow.

EXAMPLE 6.2

Let us now consider the design of the simple tensile bar illustrated i example 6.1 with another limit equation introduced. Let us say that the valu for the cross sectional area A in figure 6.5 for any satisfactory design cann be less than a specified minimum value $A_{\min}$. This might be a practical limita tion introduced by manufacturing considerations. Hence we have an add tional limit equation:

$$A \geq A_{\min} \tag{6.10}$$

Thus the (I.F.) is summarized as follows, with specified values assumed for $P, L, N_y, A_{\min}$, and for c and S_y for several materials of interest. As before, the objective is to minimize material cost C_m.

$$C_m = cLA \qquad \text{(P.D.E.) (6.5)}$$

$$\sigma = \frac{P}{A} \qquad \text{(S.D.E.) (6.6)}$$

$$\sigma \leq \frac{S_y}{N_y} \qquad \text{(L.E.) (6.7)}$$

$$A \geq A_{\min} \qquad \text{(L.E.) (6.10)}$$

It would be impossible to combine the (S.D.E.) with the (P.D.E.), by eliminating the *independent geometrical* parameter A, and still be able to account for *both* limit equations in the developed (P.D.E.). Hence with $n_f = 0$ and $N_s = 1$ we have the case of *redundant specifications*, and it will be necessary for us to ignore *temporarily* either the (S.D.E.) and the limit equation on stress *or* the limit equation on A. We could choose either of these two possible approaches, and either would lead to exactly the same final optimum design solution.

Although normally not done for problems this simple, the exploratory calculations would be made at the conclusion of this (I.F.) stage. From a review of it we recognize that $n_f = 0$, $N_s = 1$, and $n_v = n_c = 2$ (for A and σ). Thus by equations 6.3 and 6.4,

$$n_e = 1 - 0 = 1$$

$$n_r = 2 - 1 = 1$$

Also, by equations 6.1 and 6.2,

$$D_{vs} = 2 - 1 + 1 = 2$$

$$A_t = \frac{2!}{1!(2-1)!} = 2$$

Hence, in confirmation of the preceding observations, there are two approaches possible for (F.F.)'s, and our (V.S.) will be two dimensional in character. Also, with $n_f = 0$ we conclude that there is no possibility for simplification by equation reduction when transforming to the (F.F.).

In review of the (I.F.) equation summary, the simplest approach would be with σ as the eliminated parameter and A as the related parameter. For this approach the (I.F.) and the (F.F.) are the same and no equation organization is necessary. Hence for this approach the (F.F.) would be as follows.

$$C_m = cLA \tag{I}$$

$$\sigma = \frac{P}{A} \tag{II}$$

A typical or representative 2-D variation diagram for a material of interest would be as shown in figure 6.7a, and we readily see that there are two points potential főr optimum design designated as 1 and 2 in the figure. Choice of the proper one could be determined logically from a comparison of A values

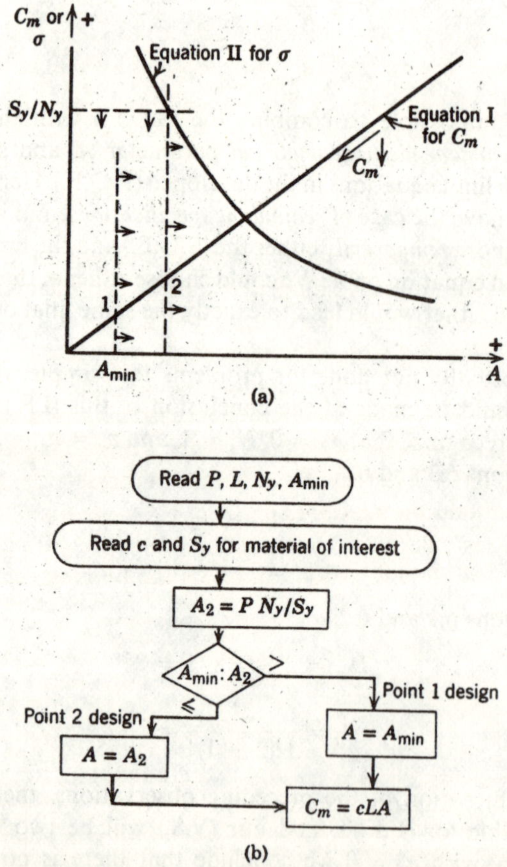

Figure 6.7 Variation diagram and flow chart for example 6.2. (*a*) Typical 2-D variation diagram for a material of interest. (*b*) Calculation flow chart (repeat for each material of interest).

t points 1 and 2; the larger A value obviously governs. Thus the calculation ow chart of figure 6.7b is derived very simply, and it summarizes compactly he procedure for finding the overall optimum design in a numerical applica- on.

As a worthwhile exercise, the reader should repeat solution of this optimiza- ion problem by the other approach. Hence choose A as the eliminated para- eter, with σ then being the related parameter. Derive and summarize the F.F.), sketch the typical 2-D variation diagram, again revealing two points otential for optimum design (corresponding to 1 and 2 of figure 6.7a). inally, derive the calculation flow chart similar to that in figure 6.7b.

XAMPLE 6.3

The solution procedure illustrated in example 6.2 is most common and owerful for the MOD, with the end result being a flow chart for computa- onal purposes. As you note the variation diagram of figure 6.7a is only for single material of interest, with the total problem involving many materials eing handled by repeated application of the figure 6.7b flow chart.

In this example we will consider the same optimization problem as for cample 6.2. However, we will illustrate a graphical staging process suitable or 2-D problems involving simple exponential type equations, where the otal picture for the solution, including all materials, is presented on the final raph. The basic procedure outlined in the detailed steps is followed. Hence, ippose we start from the (I.F.) summarized in example 6.2 by ignoring mporarily the limit equation on A, which would then be the eliminated arameter. Therefore we obtain a hypothetical case of normal specifications hich is for the *ideal problem* as defined earlier. Hence, having completed rocedural steps 1 and 2 in the suggested outline of this section, we shall now ollow step 3 and obtain the *developed* (*P.D.E.*) of the *ideal problem*, which is quation 6.8 previously derived. Thus A is the *eliminated parameter*, σ is the *lated parameter*, and (S.D.E.), equation 6.6, is the *relating equation*, in the rminology defined earlier.

$$C_m = cL\frac{P}{\sigma} \tag{6.8}$$

eferring to procedural step 4, we shall employ a discrete curve graphical ocedure in considering variations, since the *relating equation* is affected by material parameter which is the limit on stress σ. Hence material parameters ill be accounted for by the graphical consideration of discrete variations orresponding to the discrete list of available significant materials for this articular problem. Therefore, since values for both P and L were initially ecified, σ is the only design variable which must be considered mathematic- ly in the developed (P.D.E.) of the ideal problem, equation 6.8. Since σ is also the *related parameter* as previously defined, we can omit procedural step 5 and proceed directly to step 6.

We are now interested in the variation of C_m with respect to the *related parameter* σ *and* in the feasible range of C_m as governed by the limitation on σ expressed by equation 6.7. To make the situation more realistic and to bring out some important points as we proceed, we shall assume that there are three feasible materials, which we will designate material 1, material 2, and material 3 in order to be general. We shall assume that these three materials have different values for their material parameters.

The variation of C_m with respect to the *related parameter* σ could be sketched for the three materials as in figure 6.6. However, for simple exponential equations like 6.8, it is often easier to draw a log-log variation, since simple parallel straight lines are then involved. Hence we shall consider a log-log variation, and for reasons which will be obvious we shall group the right side of equation 6.8 into functional requirements, material parameters, and the *related parameter* σ. We obtain

$$\log C_m = \log(PL) + \log(c) - \log\sigma \tag{6.11}$$

where (PL) is a specified constant, and c would depend on the material. Hence for any specific material, equation 6.11 as C_m versus σ on a log-log scale graph would be of the form $y = b + mx$, the equation of a straight line on a linear scale graph of y versus x. Therefore, for various discrete materials the feasible variation of C_m versus σ would be a series of parallel straight lines having minus unity slope on a log-log scale graph, as is evident from equation 6.11. Thus the range of all *feasible design* values of C_m existing for the *ideal problem*, as governed by the limit on σ given by equation 6.7 and the restriction to three available significant materials, could be drawn simply as in figure 6.8. If it should be desired to sketch quickly the actual variation of C_m versus σ for all available significant materials, the parallel straight line log-log plots, of course, would be a great simplification.

For 2-D problems we have only one *related parameter*, and procedural steps 7 and 8 of this section can be satisfied with a simple and direct approach which will be illustrated. This simple procedure will be explained in the remaining paragraphs of this example.

In figure 6.8, the straight parallel *solid* lines depict the domain of variation of cost C_m for *all* feasible designs existing within the confines of the *ideal* problem. Hence for each material the range of feasible values for C_m can be determined readily for the ideal problem, establishing a lower limit value on C_m for each material as governed by permissible values of stress. Thus we could determine the lower limit on C_m for each material; these values will be retained for use in the next variation sketch.

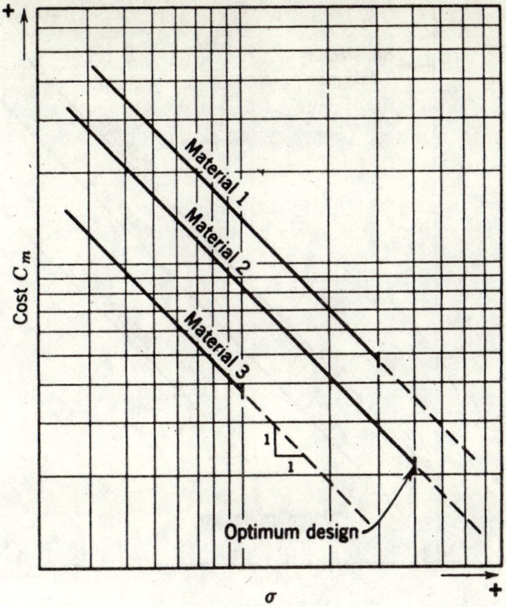

Figure 6.8 Sketch on log–log scale of cost versus stress showing all feasible designs of simple tensile bar temporarily ignoring limit on A. Lower limits on C_m for three available materials are established by upper limits on σ, S_y/N_y. Functional requirements P, L, and N_y are specified. (Refer to equation 6.11.)

The next step will be to replace the *related parameter* σ in the developed (P.D.E.) by the *eliminated parameter* A, using the *relating equation* 6.6. In doing this we shall be merely reverting the developed (P.D.E.), equation 6.8, to the original (P.D.E.), equation 6.5. The variation of the optimum design quantity C_m with respect to the previously eliminated parameter A will be sketched, the lower limits on C_m governed by stress limits as shown in figure 6.8 for each material being retained. Hence, in order to take advantage of simple parallel straight lines, we shall consider a sketch of C_m versus A from equation 6.5 on a log-log scale graph. Thus, from equation 6.5, we have

$$\log C_m = \log L + \log c + \log A \tag{6.12}$$

where L is a specified constant and c would depend on the material. Hence in a log-log scale sketch of C_m versus A equation 6.12 would give a series of parallel straight lines of slope plus unity, and the order of materials in the vertical ordinate direction would be the same as in figure 6.8 since the separation of straight lines is determined by the same term, $\log c$ in both equation 6.11 and equation 6.12. The log-log scale sketch of C_m versus A would be as

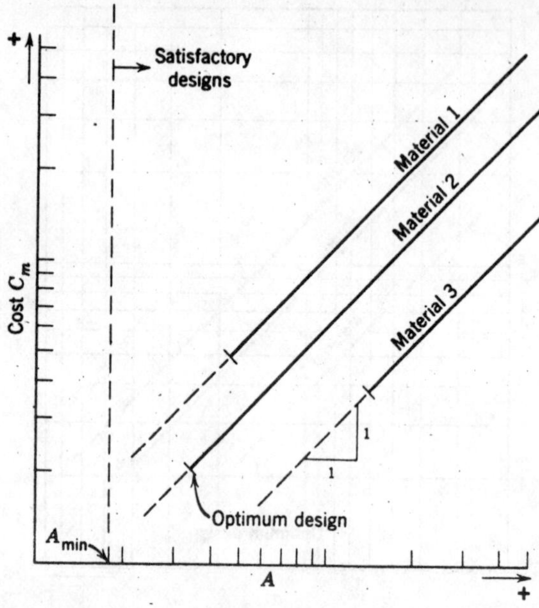

Figure 6.9 Sketch on log–log scale of cost C_m versus cross sectional area A showing all feasible designs of simple tensile bar for available materials. Lower limits on C_m as governed by stress limitation equation 6.7 are carried over from figure 6.8. Typical limit A_{min} on area A as given by equation 6.10 is depicted. Functional requirements P, L, and N_y are specified. (Refer to equation 6.12.)

typically shown in figure 6.9, the lower limits on C_m as governed by the limits on σ from figure 6.8 having been retained.

From figure 6.9 it is obvious that the optimum selection of the material will depend on the relative location of the limit A_{min} with respect to the *feasible design* solid lines of the ideal problem shown. Hence, if A_{min} falls as shown in figure 6.9, the limit equation for A could rightfully be completely ignored because equation 6.10 states that $A \geq A_{min}$ is satisfactory. Hence minimum cost would be determined solely by the stress limitation, and obviously from figure 6.9 material 2 would be the optimum material. This would be the material having the lowest value for the material selection factor $[c/S_y]$, exactly as from equation 6.9 The optimum value for A in such a situation would be determined from the (S.D.E.), equation 6.6, using $\sigma = S_y/N_y$, and obviously it would differ from the value A_{min}.

On the other hand, if A_{min} should be as shown in figure 6.10, material 3 would be the optimum material and again A would not be equal to A_{min}. In that case, A would be determined by stress, using equation 6.6 with $\sigma = S_y/N_y$, where S_y would be for material 3. If A_{min} should be appreciably to the right

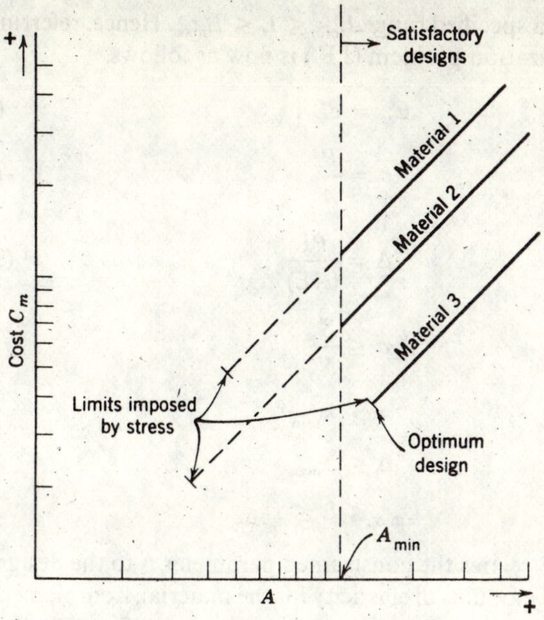

Figure 6.10 Repeat of figure 6.9 showing relocation of value for limit A_{min} resulting in different optimum design.

of the position shown in figure 6.10, material 3 would still be the optimum material, but the optimum value for A would be that new value of A_{min}, and σ would *not* be at its upper limit for the material.

All feasible design solutions to the problem in its entirety are clearly depicted in the last one of the progressive graphical steps, such as in figure 6.10, and often much can be learned from it. From this simple illustrative example we can see that *redundant specifications* can introduce some very interesting variations in optimum design. It might not be necessary to follow in detail the suggested outline of solution for such cases, which has been explained and illustrated, for the optimum design solution of this simple illustrative problem. However, redundant specifications are oftentimes unavoidably encountered in practice for more complex problems. Their effects should be anticipated and taken into account in optimum design, and the presented MOD is an orderly explicit practical way to do this.

EXAMPLE 6.4

Let us now extend the figure 6.5 tensile bar design optimization problem still further by introducing the requirement that elongation Δ must be no greater than a specified limit value Δ_{max}, as well as assuming that length L

can be within a specified range $L_{min} \leq L \leq L_{max}$. Hence, referring to exampl 6.2 our optimization problem (I.F.) is now as follows.

$$C_m = cLA \qquad \text{(P.D.E.) (6.5}$$

$$\sigma = \frac{P}{A} \qquad \text{(S.D.E.) (6.6}$$

$$\Delta = \frac{PL}{(AE)} \qquad \text{(S.D.E.) (6.13}$$

$$\sigma \leq \frac{S_y}{N_y} \qquad \text{(L.E.) (6.7}$$

$$A \geq A_{min} \qquad \text{(L.E.) (6.10}$$

$$\Delta \leq \Delta_{max} \qquad \text{(L.E.) (6.14}$$

$$L_{min} \leq L \leq L_{max} \qquad \text{(L.E.) (6.15}$$

Equation 6.13 relates the constrained parameter Δ to the design parameter where E is the modulus of elasticity for the material, from elementary strengt of materials. As before, the objective is to minimize C_m, and now the specifie values are for $P, N_y, A_{min}, \Delta_{max}, L_{min}, L_{max}$, and for c, E, and S_y for each c several materials of interest.

First, from the preceding (I.F.) we recognize that we have $n_f = 0 < N_s = 2$ so the problem type is (R.S.). Also, applying the exploratory calculations c section 6.13 with no free variables $n_v = n_c = 4$ (for L, A, σ, and Δ). Hence b equation 6.1 we have

$$D_{vs} = 4 - 2 + 1 = 3$$

which foretells that a three dimensional (3-D) variation study will be neces sary. By equation 6.3 the number of eliminated parameters for the (F.F.) wi be $n_e = 2 - 0 = 2$, and by equation 6.4 the number of related parameter will be $n_r = 4 - 2 = 2$. Also, by equation 6.2 the number of approache theoretically possible for (F.F.)'s are

$$A_t = \frac{4!}{[2!(4-2)!]} = 6$$

The circle diagram of figure 6.11 readily reveals the six possibilities; termina points of the six lines in the figure would represent the possible combination of eliminated parameters, with the remaining two for each case being th corresponding related parameter combination.

With no free variables simplification of the (I.F.) by equation reduction i not possible. Also, upon inspection of the preceding (I.F.) we readily recogniz

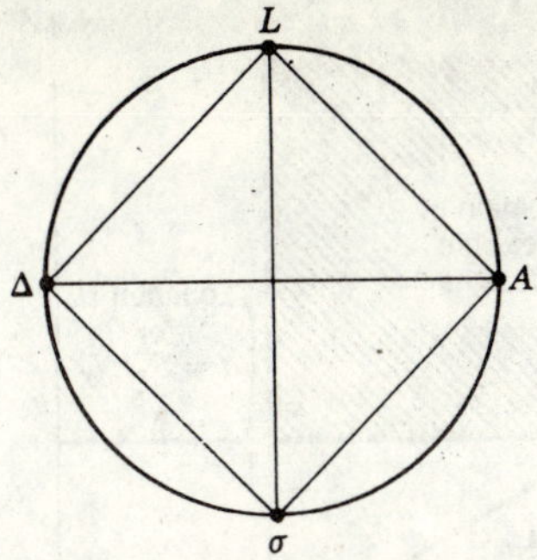

Figure 6.11 Circle diagram for example 6.4.

that no equation organization is necessary if we leave the equations as they stand for the (F.F.). Hence the simplest approach for this problem is to choose σ and Δ as the eliminated parameters with L and A then being the related parameters giving the (F.F.) as follows.

$$C_m = cLA \tag{I}$$

$$\sigma = \frac{P}{A} \tag{II}$$

$$\Delta = \frac{PL}{(AE)} \tag{III}$$

Thus with this simplest approach, the equation combination procedure of step 3 is unnecessary, and we also note that step 4 is complete at this point. Also, the related parameters L and A of equation I are both in at least one of the relating equations II and III, so there are no truly independent parameters and procedural step 5 should be skipped in this case.

However, for procedural steps 6 through 8 we require a three dimensional variation diagram for the (V.S.), and a typical or representative one for a material of interest is shown in figure 6.12. We start by sketching the coordinate axes L and A for the plan view, and make the two sectional cuts shown. For procedural step 6, referring to equation I we readily sketch the straight line side cuts shown in figure 6.12 and impose the constraints L_{min}, L_{max}, and A_{min} as shown on the plan view. Hence, for the ideal problem, point i shown in the figure is recognized as the point of optimum design. Next, we follow procedural step 7 and impose in the plan view of the figure the boundary lines for equation II with $\sigma = S_y/N_y$ and for equation III with $\Delta = \Delta_{max}$ as shown. Incidentally, the three dimensional surfaces for equations I, II, and III have C_m, σ, and Δ along a common axis normal to the plan view which is

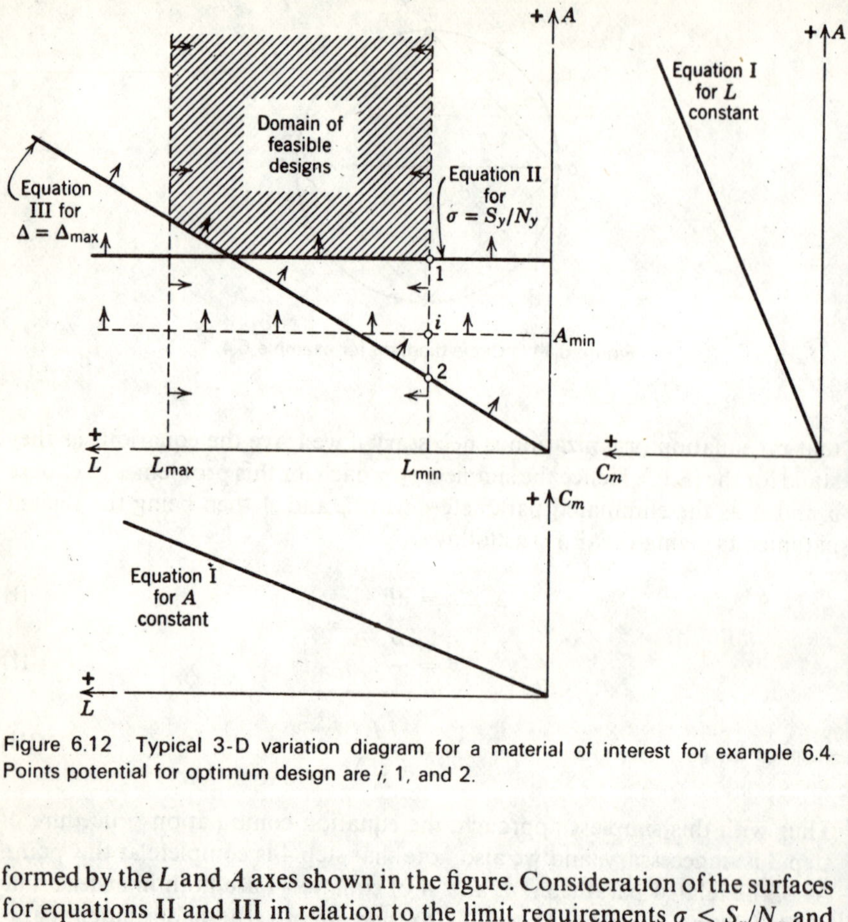

Figure 6.12 Typical 3-D variation diagram for a material of interest for example 6.4. Points potential for optimum design are *i*, 1, and 2.

formed by the L and A axes shown in the figure. Consideration of the surfaces for equations II and III in relation to the limit requirements $\sigma \leq S_y/N_y$ and $\Delta \leq \Delta_{max}$ readily reveals which side of the boundary lines are acceptable as indicated by the arrows in figure 6.12.

If the particular numbers for the problem were such that the boundaries fell relative to each other as shown in figure 6.12, ideal point i would not be accessible and instead we would choose point 1 for the optimum design. On the other hand, if both of the boundaries parallel to the L axis were below point 2 in the figure, that would be the point of optimum design. Finally, if point i were above both points 1 and 2, it would be the point of optimum design. Hence we conclude for this problem that there are three points potential for optimum design, namely, i, 1, and 2 of figure 6.12. A simple test for which one governs in a numerical application would be based on an A comparison; the one of largest A value (A_i, A_1, of A_2) would govern, and this would uniquely determine the point of optimum design.

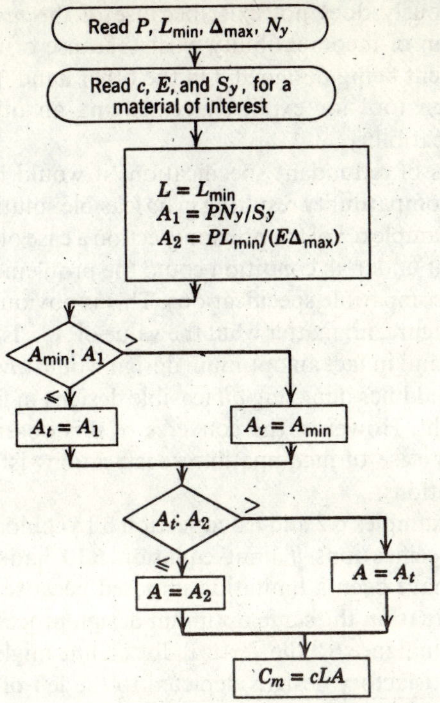

Figure 6.13 Flow chart derived from (F.F.) and variation diagram of figure 6.12, for example 6.4. Repeat flow chart for each material of interest. Overall optimum design is for one of lowest C_m value.

An understanding of the figure 6.12 (3-D) variation diagram as now explained logically leads to the calculation flow chart shown in figure 6.13, which would be applied successively for each material of interest. The overall optimum design would be recognized in the end as the one of smallest C_m value uncovered in that manner.

6.16 Case of Incompatible Specifications (I.S.) for Mechanical Elements

The case of *incompatible specifications* is really a special case of redundant specifications. For the case of incompatible specifications the subsidiary design equations and the limit equations are such that it is impossible to satisfy them *all* with *any one* physically *feasible* design of the mechanical element. As for the case of redundant specifications, the case of incompatible specifications obviously does not exist because of the *method of optimum design*; the situation of incompatibility exists because of unreasonable demands on the element being designed. On the other hand, the MOD can be a very useful design tool for explicitly remedying an otherwise hopeless situation of incompatibility.

For certain cases of redundant specifications it would be physically impossible to have incompatibility resulting in *no* feasible solution. For instance, in the illustrative example 6.3 of the previous section a case of redundant specifications existed, but under *no* condition could the problem as stated be considered a case of incompatible specifications. This is obvious from a study of figure 6.10. In that figure, no matter what the value of A_{min} is, a feasible design would *always* exist and in fact an optimum design would *always* exist. This is true because the solid lines depicting *all* feasible designs in figure 6.10 extend *infinitely* to the right. However, the converse of this observation does not hold. That is, every case of incompatible specifications is *always* a case of redundant specifications.

The illustrative examples 6.2 and 6.3 in section 6.15 could have been a case of incompatible specifications *if* limit equation 6.10 had been $A \leq A_{max}$, where A_{max} might have been a limitation imposed because of space restrictions. For such a situation the same optimum design procedure would have been followed, and in figure 6.9 the vertical dotted line might have been that for A_{max} with all satisfactory designs depicted to the left of the dotted line; the region to the right would then be unsatisfactory. *If* that were the case, there would be *no* feasible existing design solution and we would have a case of incompatible specifications. Then it would be necessary either to increase the value of A_{max} or to conduct a search to obtain a new material, or possibly to employ a combination of both approaches.

We shall illustrate the case of incompatible specifications by reverting to the example 6.3 in section 6.15 and considering an additional specification.

EXAMPLE 6.5

Let us consider exactly the same design problem as in examples 6.2 and 6.3 of 6.15, except that the force gradient or axial rigidity k of the tensile bar will be specified as a required value k_1. Therefore another (S.D.E.) will be added to the (I.F.) summarized at the start of example 6.2. Hence the additional (S.D.E.) would be

$$E = \frac{P/A}{\Delta/L} = \frac{PL}{A\Delta}$$

$$k = \frac{P}{\Delta} = \frac{AE}{L}$$

$$k = \frac{AE}{L} = k_1 \tag{6.16}$$

Thus for the (I.F.) of this example we have $n_f = 0 < N_s = 2$, and the problem type is obviously (R.S.). We could choose the specified value k_1 as a temporarily ignorable one as well as the limit A_{min} in the optimum design procedure. Hence under this assumption we could conduct the optimum design study similarly to the illustrative example 6.3 with a small extension. Thus the *developed* (P.D.E.) of the *ideal* problem would be equation 6.8 as before, where σ is the *related parameter* as defined in section 6.15.

$$C_m = cL\frac{P}{\sigma} \tag{I}$$

An *eliminated parameter*, as before, would be A with the (S.D.E.), equation 6.6, as a *relating equation*.

$$A = \frac{P}{\sigma} \tag{II}$$

Similarly, by combining the (S.D.E.)'s 6.6 and 6.16 we obtain

$$k = \frac{PE}{(L\sigma)} \tag{III}$$

Therefore, as defined in section 6.15, we could consider this another *relating equation* with k as another eliminated parameter having a singular descrete value constraint k_1.

In summary, since P and L are specified values and since available significant materials will be accounted for by a simple graphical depiction of discrete variations, for this particular problem we have only *one related parameter*, two *eliminated parameters*, and two *relating equations*. Thus, as discussed in example 6.3, for problems having only *one related parameter* in the developed (P.D.E.) of the ideal problem, it is possible to conduct the variation study based on simple two dimensional graphs by progressively looking at C_m versus σ, C_m versus A, and finally C_m versus k and retaining the determined limit effects for each successive step. This graphical staging process would be the same as that presented in example 6.3 except for an additional step, which will now be explained.

Let us say that a condition such as is depicted in figure 6.10 has been found to exist by the procedure described in example 6.3. Hence the lower limits on *feasible design* values of cost C_m are terminated for materials 1 and 2 by the existing limit A_{min} on feasible values of A, and the lower limit on *feasible design* values of cost C_m is terminated for material 3 by the existing limit $[S_y/N_y]$ on permissible values of σ carried over from figure 6.8. Therefore, at this time, we have determined the range of variation of cost C_m for *all* feasible designs existing within the confines of all restrictions *except* the specified force gradient k_1. Thus the additional step will be to consider the variation of

C_m with respect to force gradient k, including the effects of the limits shown in figure 6.10. Hence, using the second *relating equation* derived near the beginning of this section, the developed (P.D.E.), equation 6.8, becomes as follows. In order to continue working with *parallel straight lines*, we shall look at the variation of C_m with respect to k on a log-log scale sketch.

$$C_m = cL\left(\frac{P}{\sigma}\right) = cL\left[\frac{P}{(PE/kL)}\right]$$

$$\therefore \quad C_m = L^2\left(\frac{c}{E}\right)k$$

$$\therefore \quad \log C_m = 2\log L + \log\left(\frac{c}{E}\right) + \log k \qquad (6.17)$$

where L is a specified constant and the value for (c/E) would depend on the material. Hence equation 6.17 in a log-log scale graph of C_m versus k would be a series of parallel straight lines of plus unity slope. Notice from equation 6.17 that the order of materials in the vertical ordinate direction on a C_m versus k plot might be different from that existing in figures 6.8 and 6.10, since their order was determined by a log (c) term in equations 6.11 and 6.12, whereas the variations from equation 6.17 would be ordered according to the term log (c/E) for the various materials. The typical sketch of equation 6.17, carrying over the previously determined *feasible design* ranges for C_m from figure 6.10, is presented in figure 6.14. In this figure we have assumed the relative vertical order of straight lines for the various materials to be as shown, as determined by the log (c/E) term in equation 6.17, and this is a reasonable assumption for this illustration.

In figure 6.14 the specified value of k_1 is shown by the vertical dotted line. This obviously shows that *no* feasible design exists, and we have a case of incompatible specifications. However, the sketches in figure 6.14 indicate what we might do in order to obtain a feasible design solution. For instance if the value of k_1 could be increased to the limit imposed by stress for material 3 in the figure, a really good optimum design would result. On the other hand looking at figure 6.14 tells us that it might be possible to decrease $A_{\min}$ slightly (desirability of decreasing $A_{\min}$ is seen from figure 6.10) and increase k_1 slightly until a feasible design is obtained with material 2 as the optimum and only material.

As in example 6.3, for a problem of this simplicity it is generally not required to draw the actual variations, even though they would be very easy to obtain with straight and parallel lines. A few simple calculations will generally make it possible to locate the relative positions of the various lines, and free-hand sketches will be adequate. The importance of drawing free-hand sketches cannot be overemphasized, however.

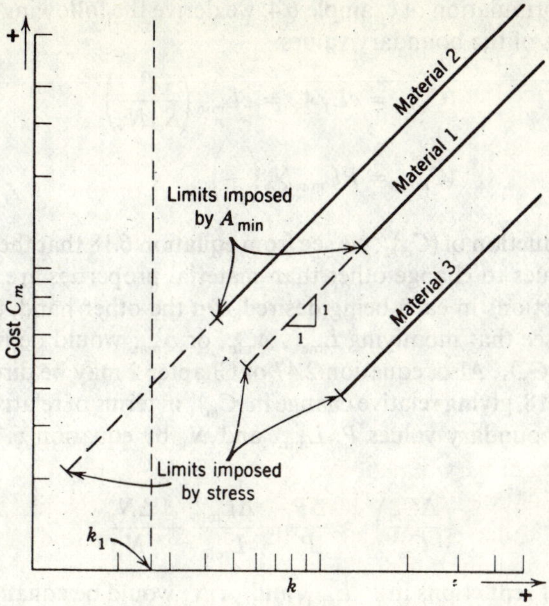

Figure 6.14 Sketch on log–log scale of cost C_m versus force gradient k for simple tensile bar for available materials. Limits on C_m imposed by stress limitation equation 6.7 and cross sectional area limitation equation 6.10 are carried over from figure 6.10. Value k_1 as depicted would result in a case of incompatible specifications with no feasible design. Functional requirements P, L, and N_y are specified. (Refer to equation 6.17.)

6.17 Modification of Boundary Values

We will define the *boundary values* for an optimization problem as the set of specified values, including limit values, of the initial formulation. In problems of optimum design there are two types of situations where the modification of boundary values might be of benefit. In both cases the MOD reveals clearly what should be done, based on an understanding of the variation diagram and the associated equations.

The first type of situation is recognized when the optimization quantity achieved is not good enough. In such cases the modification of appropriately selected boundary values can usually improve the optimization quantity appreciably. For illustrative purposes, consider example 6.4 and the specific variation diagram shown in figure 6.12. The point of optimum design would then be 1, and suppose the value of optimization quantity $(C_m)_1$ is not small enough to satisfy management. Referring to figure 6.12 and equations I and II for the final formulation of example 6.4, we derive the following equation for $(C_m)_1$ in terms of the boundary values.

$$(C_m)_1 = cL_1A_1 = cL_{\text{min}}\left(\frac{P}{S_y/N_y}\right)$$

$$\therefore \quad (C_m)_1 = PL_{\text{min}}N_y\left(\frac{c}{S_y}\right) \tag{6.18}$$

Hence, for reduction of $(C_m)_1$, we see from equation 6.18 that the appropriate boundary values to change other than material properties are P, L_{min}, and N_y with reductions in each being desired. On the other hand, based on the equation we see that modifying L_{max}, A_{min}, or Δ_{max} would be ineffective for reduction of $(C_m)_1$. Also, equation 2.47 of Chapter 2 may be directly applied to equation 6.18, giving relative change in $(C_m)_1$ in terms of relative changes in the effective boundary values P, L_{min}, and N_y by equation 6.19, assuming c/S_y constant.

$$\left(\frac{\Delta C_m}{C_m}\right)_1 \approx \frac{\Delta P}{P} + \frac{\Delta L_{\text{min}}}{L_{\text{min}}} + \frac{\Delta N_y}{N_y} \tag{6.19}$$

Hence percent reductions in P, L_{min}, and/or N_y would be equally as effective for reducing $(C_m)_1$.

The second type of situation where modification of boundary values can be of benefit is in alleviation of incompatible specifications. For illustrative purposes, suppose in the (I.F.) of example 6.4 we had the requirement $A \leq A_{\text{max}}$ instead of $A \geq A_{\text{min}}$. Thus if the variation diagram were as specifically shown in figure 6.12, except replacing A_{min} with A_{max}, we obviously would have (I.S.). To alleviate the situation we refer to the variation diagram and recognize the need to satisfy the relation

$$A_{\text{max}} \geq A_1 = \frac{PN_y}{S_y} \tag{6.20}$$

The right side of equation 6.20 is merely equation II of the (F.F.) from example 6.4 with $\sigma = S_y/N_y$, as seen from the revised figure 6.12 variation diagram. Thus we see from equation 6.20 that the boundary values to change for alleviation of (I.S.) are A_{max}, P, N_y, and S_y. On the other hand, changing L_{min}, L_{max}, Δ_{max}, E, or c would be ineffective for alleviating the (I.S.). The amount of change required for A_{max}, P, N_y, and/or S_y would be determined by satisfaction of equation 6.20.

6.18 Problems with More than One Primary Design Equation

Occasionally we encounter a problem in the optimum design of a mechanical element characterized by the existence of more than one primary design quation. This is generally the case when stress is significant in more than one lace in the mechanical element. A specific example illustrating this situation ; given in sections 7.8 and 7.9 of Chapter 7. In the strict sense, many of the ituations characterized by more than one primary design equation are really omplex cases of redundant specifications.

For mechanical elements having more than one primary design equation, he method of optimum design can be conducted in *separate stages*, each rimary design equation generally being developed in the same manner as or cases having only one primary design equation. The results from the esign study of each stage will define a material and values for independent eometrical parameters which uniquely define the optimum design of the nechanical element as governed by the considerations of that particular stage. he design results from each stage might very well be entirely compatible, nd hence the final design of the mechanical element will be determined imply and explicitly as merely the optimum design which is common to all tages. On the other hand, the optimum designs for each stage might differ ppreciably with respect to the selection of materials or certain values for ndependent geometrical parameters. Hence, with respect to any incompatible esults from the optimum design studies in the various stages, a *compromise* night be required in order to obtain a final overall optimum design of the lement. This can best be illustrated in some specific examples which will be overed in the following chapters.

Another way of handling a problem with more than one primary design quation is to treat it as a multiobjective situation having more than one ptimization quantity, as described earlier in section 6.6. However, as dis-ussed therein the creation of a composite criterion function is difficult in ractical applications because of the relative weighting terms. Also, the lternative choice of one or more of the components to be limited as discussed n section 6.6 may be difficult in problems of this type. However, the approach lescribed in the previous paragraph provides a practical solution for most ases.

Automated Optimal Design

The widespread availability of digital computers to design engineers in ecent years has given great value to techniques of automated optimal design. Through mathematical programming many algorithms have been developed within the past several years, and some background for studying the subject s given in references 6.29 through 6.40. In addition, Appendix C of reference 5.18 is on "Computers in Automated Optimal Design," and a well-known computer code named PATSH, developed by D. E. Whitney at M.I.T., is presented therein. It is based on the direct pattern search technique developed by R. Hooke and T. A. Jeeves in reference 6.38. It has been found to solve optimization problems with a high level of efficiency, it has better than average solution speed, and it is small compared with many other codes. Thus it is attractive for small computers and repeated use.

Very recently a powerful algorithm was developed by G. E. Johnson of the University of Virginia and M. A. Townsend of Vanderbilt University. It is based on a general conjugate direction recursion formula and an acceptable point univariate strategy. It has been found to be efficient, reliable, and relatively insensitive to tuning parameters as far as success or failure is concerned. This algorithm is presented in Appendix C of this book along with its application to a mechanical element optimal design problem, and results are compared with the MOD solution to the same problem.

An AOD algorithm named OPTIGO was recently developed by the author of this book, and it will be covered in some detail in the remaining part of the chapter. It is a gradient based search technique incorporating finite differences, an interval-halving plan for ray search, self-tuning, parabolic penalty functions, and a shotgun search start point strategy. Application of the computer code OPTIGO to a very complicated practical mechanical element design problem will be presented in Chapter 8 for hydrostatic journal bearings. Some basic background for the algorithm OPTIGO is presented in sections 2.9 and 2.10 of Chapter 2, and further explanations of its features and construction will be presented later in the present chapter. The listing of the Fortran IV computer program for OPTIGO is given in Appendix B. A nomenclature of some terminology for AOD and OPTIGO is presented in Appendix A, which should be referred to when necessary during the remainder of this chapter. Before describing the algorithm OPTIGO in any further detail, let us consider some basic material applicable to AOD in general.

6.19 Basic Equation System Format for AOD

In brief review, the MOD (I.F.) equation system basic format could be expressed as follows, for a typical example.

$$Q\downarrow = Q(x_i) \quad \text{or} \quad U\uparrow = U(x_i) \qquad \text{for} \quad i = 1, \ldots, N \qquad \text{(P.D.E.)}$$

$$S_n = S_n(x_i) \qquad \text{for} \quad n = 1, \ldots, N_s \qquad \text{(S.D.E.)'s}$$

$$(T_l)_{\min} \leq T_l \leq (T_l)_{\max} \qquad \text{for} \quad l = 1, \ldots, N_L \qquad \text{(L.E.)'s}$$

In this equation system, Q is optimization quantity to be minimized (or U to be maximized) expressed as a function of variables x_i of which there are N;

S_n is an (S.D.E.) expressing a functional requirement or undesirable effect, of which there are N_s; and T_l is a limited quantity which may be an S_n or a x_i, of which there are N_L limited items. The optimization problem is mathematically tied together by this equation system, and the solution is found explicitly by the MOD procedure based on derivation of a (F.F.) and (V.S.) as explained in sections 6.13 through 6.15. The intelligence of the engineer is integrated with the MOD procedure in the derivation of the solution for specific applications.

In automated optimal design the optimization problem equation system has to be more rigidly controlled to a specific format, since a digital computer lacks basic intelligence and it is capable only of following instructions which are clearly presented to it by the programmer. Hence the standard AOD equation system format used here is as described below.

Basic Format for AOD Equation Systems.

$$Q\downarrow = Q(x_i) \qquad \text{for} \quad i = 1, \ldots, N \qquad \text{Criterion function}$$

$$E_j = E_j(x_i) = 0 \qquad \text{for} \quad j = 1, \ldots, N_E \qquad \text{Equality constraints}$$

$$R_k = R_k(x_i) \geq 0 \qquad \text{for} \quad k = 1, \ldots, N_R \qquad \text{Regional constraints}$$

In the preceding equation system, Q is optimization quantity *always to be minimized* expressed as a function of independent variables x_i, of which there are N; E_j is an equality constraint function, of which there are N_E, expressed as a function of the x_i's; and R_k is a regional constraint function, of which there are N_R, expressed as a function of the x_i's.

We might initially express an optimization problem through a MOD (I.F.) before deciding that an AOD solution would be desired. Hence it will be necessary to transpose from the MOD (I.F.) to an AOD basic format as summarized herein. This is usually straightforward, requiring some simple mathematical manipulation. However, before transposing to the AOD basic format it is recommended that the following steps of preparation be made if appropriate on the MOD (I.F.). This will simplify solution by AOD.

1. Eliminate optimization quantity from all (S.D.E.)'s by equation combination, if the situation exists.
2. Eliminate all free variables from the (I.F.) if possible, by equation combination.
3. Avoid equality constraints if possible. Eliminating free variables is a step in this direction.

Let us now consider a simple example illustrating the transposition from MOD (I.F.) to AOD basic format.

EXAMPLE 6.6

Suppose a fenced yard of rectangular shape is to be designed for maximizing enclosed area A for a specified length L of fence. Dimensions x and y of figure 6.15 are constrained because of space limitations x_{max} and y_{max} on x and y respectively. The (I.F.) for solution by the MOD is readily derived and summarized as follows:

$$A\uparrow = xy \qquad \text{(P.D.E.)} \quad (6.21)$$

$$L = 2(x + y) \qquad \text{(S.D.E.)} \quad (6.22)$$

$$x \le x_{max} \qquad \text{(L.E.)} \quad (6.23)$$

$$y \le y_{max} \qquad \text{(L.E.)} \quad (6.24)$$

Specified: L, x_{max}, and y_{max}.
Find: values of x and y which will maximize enclosed area A.

Suppose that we wish to transform the preceding MOD (I.F.) to the more rigidly specified format of AOD.

To start we will use the notation $x_1 = x$ and $x_2 = y$. Also, we must convert the given maximization problem to an equivalent minimization problem. As will be explained later, this can be done by defining optimization quantity Q as the reciprocal of A. Also, negative values for x and y were assumed to be unacceptable in the (I.F.) as presented, but this must be rigorously expressed for understanding by the computer in AOD. Hence, incorporating what has now been discussed, the equation system satisfying the AOD basic format is readily derived as follows for the fenced yard problem.

$$Q\downarrow = \frac{1}{(x_1 x_2)} \quad (6.25)$$

$$E_1 = L - 2(x_1 + x_2) = 0 \quad (6.26)$$

$$R_1 = x_{max} - x_1 \ge 0 \quad (6.27)$$

$$R_2 = y_{max} - x_2 \ge 0 \quad (6.28)$$

$$R_3 = x_1 \ge 0 \quad (6.29)$$

$$R_4 = x_2 \ge 0 \quad (6.30)$$

Note that the required basic format for AOD previously summarized has been rigidly adhered to in the specific equation system. Incidentally, the last two regional constraints have been introduced for avoidance of negative values for x and y as previously discussed.

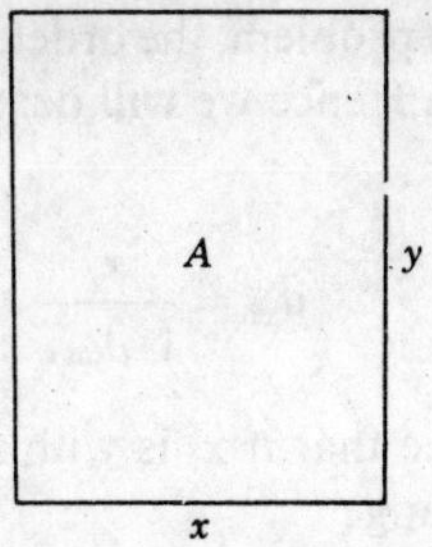

Figure 6.15 Fenced yard of rectangular shape for example 6.6.

6.20 Normalized Equation Systems for AOD

The variables and function expressions of an AOD equation system have to be in proper balance for accurate convergence to the point of optimum design by the digital computer. Some AOD basic format equation systems are acceptable as initially derived for a particular optimization problem. However, in other cases a transformation to an AOD unit based equation system is necessary for a meaningful solution. To be safe in all cases, the simple normalization transformation process might be made for all AOD equation systems.

Unit Based Variables, u_{xi}. As explained later for the algorithm OPTIGO, convergence to the solution point neighborhood is sensed by a measure of the "distance" between successive base points as calculated by equation 6.31.

$$\delta_r = \left\{ \sum_{i=1}^{N} [(x_i)_{Br+1} - (x_i)_{Br}]^2 \right\}^{1/2} \tag{6.31}$$

Hence, if as an extreme example x_1 has an order of magnitude of say 10^{10} and x_2 has an order of magnitude of 10^{-10}, x_2 will lose its significance in equation 6.31 for the convergence test. Thus transformation of the independent variables x_i to unit based variables u_{xi} is required for a proper balance in some problems, and this we will explain next.

For any design variable x_i it will be possible to estimate limit values defining a broad range of interest or feasible variation expressed as follows.

$$(x_i)_{\min} \le x_i \le (x_i)_{\max} \tag{6.32}$$

For most mechanical element problems, any design variable x_i will have only positive values over the range of interest. To have a proper balance among variables in an optimization problem, the order of magnitude for each shoulc be in the same general area. Hence we will define a *unit based variable* u_{xi} ir the following way.

$$u_{xi} = \frac{x_i}{(x_i)_{\max}} \tag{6.33a}$$

From equation 6.33a we see that if x_i is within the range of equation 6.32 then u_{xi} will be within the range

$$0 \leq u_{xi} \leq 1 \tag{6.34}$$

Thus the relative orders of magnitude for the independent unit based variable u_{xi} will be kept in the same general area during the search process, and the measure of convergence expressed by equation 6.35 derived from equatior 6.31 will be meaningful. Each variable will maintain its significance at leas until the neighborhood of the optimum design solution is reached in the search process.

$$\delta_{ur} = \left\{ \sum_{i=1}^{N} \left[(u_{xi})_{Br+1} - (u_{xi})_{Br} \right]^2 \right\}^{1/2} \tag{6.35}$$

Occasionally, a variable x_i is bounded only on its lower end with a limi $(x_i)_{\min}$, and $(x_i)_{\max}$ is effectively infinite. In such cases, the unit based variable u_{xi} for the particular variable x_i described is defined as

$$u_{xi} = \frac{(x_i)_{\min}}{x_i} \tag{6.33b}$$

The assumption is of course that $(x_i)_{\min}$ is a positive value and not zero.

In any case, the definition of a unit based variable u_{xi} is made on the premise that its range of variation should be within the limits $0 \leq u_{xi} \leq 1$, corresponding to the feasible range of variation for x_i. This will keep the relative orders of magnitude of the independent variables in proper balance for accurate convergence control in the execution of the search process by the computer

Unit Based Equations. As will be explained later, penalty functions are introduced in AOD to account for violations of the equality constraints and the regional constraints during the search process. If an equality constraint is violated for it we have $E_j \neq 0$, whereas if a regional constraint is violated we have $R_k < 0$. Hence we define R_{k-} as follows.

$$\begin{aligned} R_{k-} &= R_k \quad \text{for} \quad R_k < 0 \\ R_{k-} &= 0 \quad \text{for} \quad R_k \geq 0 \end{aligned} \tag{6.36}$$

For violations of constraints in the search process of OPTIGO we will use an external parabolic type of penalty function P defined by equations 6.37 and 6.38.

$$P = C_p \eta_p (\text{RE}) \tag{6.37}$$

where

$$(\text{RE}) = \sum_{j=1}^{N_E} E_j^2 + \sum_{k=1}^{N_R} R_{k-}^2 \tag{6.38}$$

In equation 6.37 C_p is a coefficient whose proper value is determined in the self-tuning feature of OPTIGO, as will be explained later. Also, η_p is a parameter initially taken as unity and successively increased in the search process of OPTIGO to steepen the penalty walls as convergence is being made to the neighborhood of the optimum design point.

In the search procedure of OPTIGO, a point may be reached which violates several constraints resulting in several values of $E_j \neq 0$ and/or R_{k-} being applied in equation 6.38. To keep the proper balance between the E_j's and the R_{k-}'s for such a case, in some problems it is recommended to convert the AOD basic format equation system to an *AOD unit based equation system*, also incorporating the unit based variables as previously explained. To accomplish an approximate balance, the varying part of an E_j or R_k equation is made to approach a value between zero and unity at the boundary of the constraint violation. Incidentally, the criterion function Q should be transformed to incorporate the previously defined unit variables. However, transformation of criterion function Q beyond that point should not be made. Proper balance with the penalty function P will be achieved by the self-tuning feature of OPTIGO.

A summary is given below of the basic format of the unit based AOD equation system, which is comparable to the basic format summarized in section 6.19 from which it was derived.

Unit Based AOD Equation System Format.

$Q\downarrow = Q(u_{xi})$	for $i = 1, \ldots, N$	Criterion function
$(E_u)_j = E_{uj}(u_{xi}) = 0$	for $j = 1, \ldots, N_E$	Equality constraints
$(R_u)_k = R_{uk}(u_{xi}) \geq 0$	for $k = 1, \ldots, N_R$	Regional constraints

To illustrate more specifically the conversion of an AOD basic format equation system to the equivalent unit based AOD equation system, we will consider example 6.7 next.

EXAMPLE 6.7

Let us continue with example 6.6 for the fenced yard design problem. The AOD basic format equation system summarized in that example would be adequately balanced as it stands, but for illustrative purposes here we will transform it to a unit based AOD equation system. Also, for many problems it is not known beforehand whether or not a transformation is necessary, and since the conversion is simple it is oftentimes made for all problems to be safe.

To start we will define unit based variables u_{x1} and u_{x2} for $x_1 = x$ and $x_2 = y$ respectively, based on the feasible ranges of $0 \leq x_1 \leq x_{\max}$ and $0 \leq x_2 \leq y_{\max}$ as explained in example 6.6. Thus we define the unit variables as follows, by equation 6.33a.

$$u_{x1} = \frac{x_1}{x_{\max}} \tag{6.39}$$

$$u_{x2} = \frac{x_2}{y_{\max}} \tag{6.40}$$

Regarding the unit based equations we refer to the AOD basic format equations 6.25 through 6.30 as summarized in example 6.6. In these equations we can replace x_1 and x_2 with $x_1 = x_{\max} u_{x1}$ and $x_2 = y_{\max} u_{x2}$, from equations 6.39 and 6.40. In this way we introduce the unit based variables.

Regarding unit based equations, for the equality constraint function E_1 of equation 6.26 we merely divide both sides by L and incorporate the unit based variables by equations 6.39 and 6.40. Then, as desired, the varying part of E_1 during the search approaches unity as the equality constraint E_1 value approaches zero. Similarly, for the regional constraint R_1 of equation 6.27 we introduce the unit based variable by $x_1 = x_{\max} u_{x1}$ from equation 6.39 and then divide the R_1 function through by $x_{\max}$. We follow the same procedure for the other regional constraint functions R_2, R_3, and R_4 of equations 6.28, 6.29, and 6.30. Finally, for the unit based AOD equation system, we transform the criterion function Q of equation 6.25 by incorporating the unit based variables using equations 6.39 and 6.40.

Hence the AOD unit based equation system for the fenced yard problem is summarized as shown below. The subscript u notation has also been introduced for the unit based equations below.

$$Q\downarrow = \frac{1}{(x_{\max} y_{\max} u_{x1} u_{x2})} \tag{6.41}$$

$$(E_u)_1 = 1 - \left(\frac{2}{L}\right)(x_{\max} u_{x1} + y_{\max} u_{x2}) = 0 \tag{6.42}$$

$$(R_u)_1 = 1 - u_{x1} \geq 0 \tag{6.43}$$

$$(R_u)_2 = 1 - u_{x2} \geq 0 \tag{6.44}$$

$$(R_u)_3 = u_{x1} \geq 0 \tag{6.45}$$

$$(R_u)_4 = u_{x2} \geq 0 \tag{6.46}$$

If the preceding unit based AOD equation system were used for solution by OPTIGO, at the conclusion we would be interested in the determined values for x, y, and A, and perhaps other items as well. Thus equations 6.39 and 6.40 would give

$$x = x_1 = x_{\max} u_{x1}$$

$$y = x_2 = y_{\max} u_{x2}$$

and A would be calculated by equation 6.21, giving the value achieved for the original optimization quantity of example 6.6.

6.21 Penalized Optimization Quantity, (QP)

Assume the existence of an optimization problem expressed by an AOD equation system, either as summarized on an absolute basis in section 6.19 or as summarized on a unit basis in section 6.20. In the search process of OPTIGO, equality constraints and regional constraints will be violated and penalty zones will be temporarily invaded. Such violations to any appreciable extent are not acceptable for the solution point, and penalty zone penetrations beyond a small amount should be discouraged in the search process. This is accomplished by penalizing criterion function Q with a penalty function P. There are many types of penalty functions which have been tried, both external and internal. For typical examples see the indices of references 6.2, 6.3, 6.5, and 6.21.

For the algorithm OPTIGO we have chosen an external parabolic type of penalty function as defined by equations 6.37 and 6.38 of the preceding section. The greater the depth of penetration into a penalty zone, the greater will be the magnitude for either E_j or R_{k-} for the violations. Thus for OPTIGO we define a *penalized optimization quantity* (QP) to be minimized as follows.

$$(QP)\downarrow = Q\downarrow + P \tag{6.47}$$

Therefore, to the criterion function Q of the AOD equation system we add the penalty function P, which by equations 6.37 and 6.38 must have a positive value or be equal to zero. The value $P = 0$ exists if there is no constraint

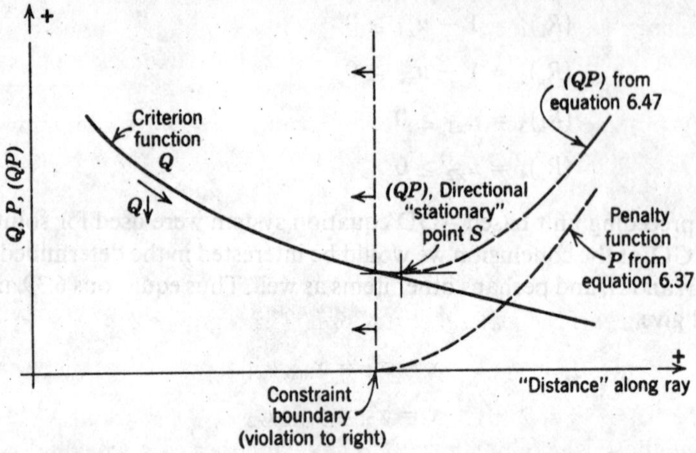

Figure 6.16 "Wall" and directional "stationary" point S created by penalty function when added to Q.

violation for the point in question, since then the E_j's are all zero and th R_{k-}'s are all zero as defined by equation 6.36.

Hence the effect of the penalty function P is to erect a smoothly varyin "wall" discouraging excessive penetration into a penalty zone or region c constraint violation during the search process. This becomes clear if we realiz that the optimization objective controlling the search will be to minimiz penalized optimization quantity (QP). Thus we generally find that the penalt function P creates stationary points of minimum (QP) value as shown i figure 6.16. The algorithm OPTIGO will search for such stationary points i its automated procedure leading to the neighborhood of the solution poir for the optimization problem, as will be explained later.

6.22 Minima–Maxima Equivalence

The standard formats of AOD equation systems are based on minimizatio of optimization quantity, $Q\downarrow$, as summarized in sections 6.19 and 6.20. I fact, for the algorithm OPTIGO the automated search will be controlled b the objective of minimizing penalized optimization quantity, $(QP)\Downarrow$. How ever, as illustrated in section 6.6, for many optimization problems the obje tive is actually to *maximize* a quantity Q_m of interest, denoted as $Q_m\uparrow$. Henc it is necessary to convert the criterion function for $Q_m\uparrow$ to an equivaler criterion function $Q\downarrow$ which is compatible with the standard AOD forma and which can be handled by the algorithm search in AOD by the compute

Incidentally, equivalence for $Q_m\uparrow$ and $Q\downarrow$ by definition is based on the fact that each function must have the same values for the independent variables x_i at the solution point.

Transforming a given $Q_m\uparrow$ function to an equivalent $Q\downarrow$ function can be handled very simply in one of two ways. We can either reciprocate the given $Q_m\uparrow$ function, which gives

$$Q\downarrow = \frac{1}{Q_m} \tag{6.48a}$$

or we can negate the given $Q_m\uparrow$ function, which gives

$$Q\downarrow = -Q_m \tag{6.48b}$$

In either case, we would find that the values of x_i for minimizing Q are the same as those which maximize Q_m. Thus in either case we could say from the optimization problem solution standpoint that $Q\downarrow$ is equivalent to $Q_m\uparrow$.

To the newly created $Q\downarrow$ function we will be adding the penalty function P by equation 6.47 and as illustrated in figure 6.16, and our AOD search will be for minimizing $(QP)\downarrow$. Our control for not significantly violating constraints will be the penalty function P. However, for some problems it has been found that the given $Q_m\uparrow$ function increases very rapidly as a penalty zone is entered. If the rate of increase in Q_m is greater than the rate of increase for P, creating the $Q\downarrow$ function by the negation approach of equation 6.48b can lead to serious problems in the AOD search. For such cases, where Q_m is negated, decreasing Q overpowers the increasing P, and the (QP) function of equation 6.47 decreases the further we enter the penalty zone. Hence, in our AOD search the algorithm leads the computer far into a penalty zone of large constraint violation, giving therefore an erroneous solution to the problem.

To avoid this possible difficulty, we generally choose to define the $Q\downarrow$ function by reciprocation of $Q_m\uparrow$ according to equation 6.48a. Hence, if Q increases very rapidly as a penalty zone is entered, $Q = 1/Q_m$ will merely flatten out approaching the asymptote $Q = 0$. In such cases the penalty function P will continue with its intended effect and will not be overpowered as before. A stopping "wall" as shown in figure 6.16 will be created when entering a constraint violation zone. The reciprocation approach of equation 6.48a is easy to apply and handles maximization problems very well, and for reasons now described we will use it for later applications of OPTIGO.

6.23 Good Start Point Strategy

In algorithms for AOD the determination of a solution point is facilitated by choosing a good start point not too far away. Naturally we would do this if a good start point were known before the AOD search began. However, i such a point was not known we could determine the coordinates of one by a random number generation application, which will now be explained. Incidentally, for problems with many variables x_i such an approach is probably easier than a uniform grid type search which might be followed for finding a good start point instead.

Random Number Generation. This is a subject of mathematics which we will consider briefly. The problem is to generate a sequence of random numbers $(RN)_n$ for $n = 1, 2, 3, 4, \ldots$ within the range $0 \leq (RN)_n \leq 1$. Incidentally, a truly random number generation for x within the range $0 \leq x \leq 1$ would have the following statistical characteristics for an infinite population.

$$x_{avg} = \int_0^1 x\,dx = 0.500 \text{ arithmetic mean value}$$

$$D_x = \left[\int_0^1 (x - 0.5)^2\,dx\right]^{1/2} = 0.2887 \text{ standard deviation}$$

The arithmetic average and standard deviation described above are statistical measures which are often applied to random number generation techniques to see how appropriate they are.

A widely used random number generator is given by the following equation [6.41].

$$(RN)_n = k(RG)_n(\text{mod } p) = (RG)_{n+1} \tag{6.49}$$

For this power residue method proper values compatible with the number base of the computer must be chosen for constant k and modulus p. A start value or seed must be arbitrarily supplied for $(RG)_1$ in the range $0 \leq (RG)_1 \leq 1$, and random number $(RN)_1$ is calculated by the equation, which then becomes the next generator value $(RG)_2$ used for calculating $(RN)_2$. This is repeated successively to generate random numbers $(RN)_n$ for $n = 1, 2, 3, 4, \ldots$.

Another very simple equation for the approximate generation of random numbers is equation 6.50 [6.42].

$$(RN)_n = \text{fractional part of}\{[\pi + (RG)_n]^5\} = (RG)_{n+1} \tag{6.50}$$

This equation is repeated successively as was done with equation 6.49 for the approximate generation of random numbers $(RN)_n$, for $n = 1, 2, 3, 4, \ldots$. From a statistical analysis of 100 numbers generated by equation 6.50 for each of five different arbitrarily chosen seeds, the statistical characteristics summarized in table 6.1 were calculated.

TABLE 6.1

Statistical Characteristics of Random Number Generation by Equation 6.50

Seed $(RG)_1$	Values Generated	Average $(RN)_{avg}$	Standard Deviation
0.136802	100	0.482152	0.291438
0.398123	100	0.515127	0.310585
0.524397	100	0.484207	0.282880
0.729432	100	0.481282	0.296102
0.986374	100	0.504349	0.282435
Overall for 500 numbers	500	0.493423	0.292688

Random Design Point Generation. A random design point will be defined as one having independent coordinates x_i determined by random number generation according to equation 6.51.

$$x_i = (x_i)_{\min} + (\mathrm{RN})_n[(x_i)_{\max} - (x_i)_{\min}] \qquad \text{for} \quad \mathrm{i} = 1, \ldots, N \quad \text{and} \quad n = 1, 2, 3, \ldots \tag{6.51}$$

We assume that each variable x_i has a known feasible range of interest as follows.

$$(x_i)_{\min} \leq x_i \leq (x_i)_{\max} \qquad \text{for} \quad i = 1, \ldots, N \tag{6.52}$$

Each calculation by equation 6.51 would be based on a random number $0 \leq (\mathrm{RN})_n \leq 1$ successively generated by equation 6.50 say. The (x_i)'s correspond to those in the AOD equation system summarized in section 6.19 and values for these uniquely define a design point. After equation 6.51 is applied for $i = 1, \ldots, N$, a random design point is generated. This is repeated successively using the last $(\mathrm{RN})_n$ value for the new seed $(\mathrm{RG})_{n+1}$ in each case. Hence a sequence of N_{SG} random design points can be generated in this way.

For the random design point generation to be meaningful in locating a good start point, the value of N_{SG} must be adequate. On the other hand, if N_{SG} is chosen too large, computation time becomes significant. The following guide may be used for estimating an appropriate value for N_{SG}.

$$N_{SG} \approx (N_d)^N \tag{6.53}$$

In equation 6.53 N is as before, whereas N_d is an index of design point density for the generation in the feasible design space. For good shotgun density, a

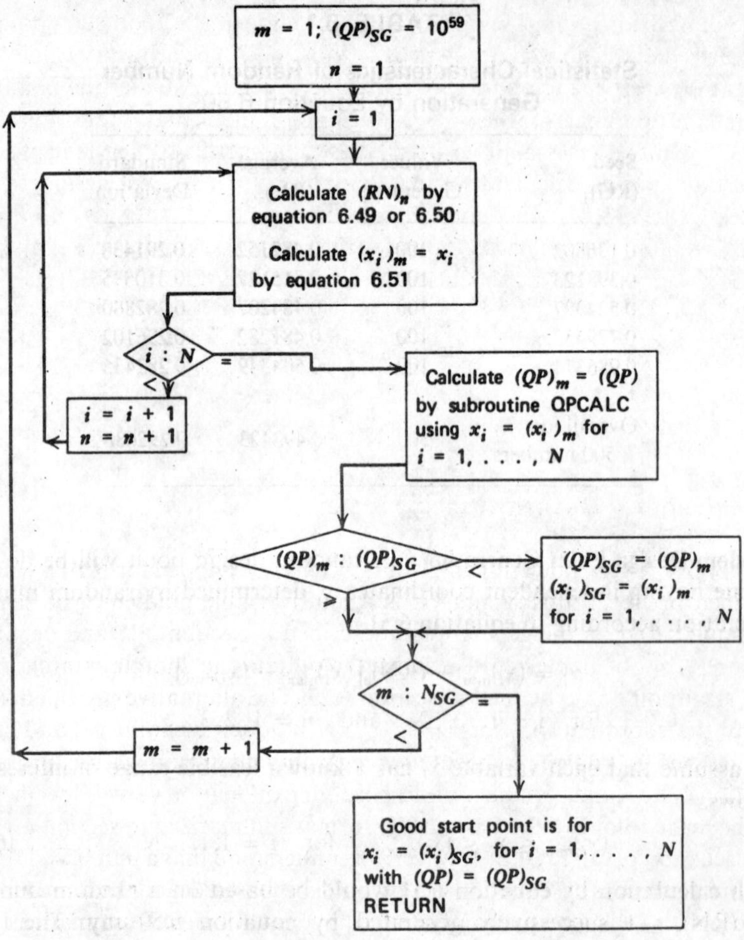

Figure 6.17 Subroutine SG for shotgun search at start if desired in OPTIGO.

value of $N_d = 5$ or more would be desired. However, for problems of larg N a value of N_d as low as 2 may be used to save on computation time, wit acceptable results.

Good Start Point Choice. If a sequence of say N_{SG} random design point are determined as explained, the values of (QP) can be calculated for each b equation 6.47. More specifically, let subscript m designate a random desig point determined as described. Hence we have for each a penalized optimiza tion quantity $(QP)_m$, for $m = 1, \ldots, N_{SG}$. The one of smallest $(QP)_m$ value wi be the good start point chosen for the algorithm OPTIGO.

The described shotgun search strategy for choosing a good start point is ummarized in the flow chart segment of figure 6.17. The arbitrary value 10^{59} iven at the start is a temporary initial value which is lost after the first time hrough the flow chart for the $m = 1$ design point. Incidentally, for some comuters it might be necessary to use a smaller value such as 10^{19} instead of 10^{59}, ecause of specific calculation range limitations. $(QP)_{SG}$ will designate the mallest value of $(QP)_m$ encountered in the shotgun search of figure 6.17. The ood start point determined at the end of the shotgun search as now explained ill be designated with the subscripts SG for the AOD OPTIGO search hich follows.

)PTIGO

.24 General Description

OPTIGO, a very powerful algorithm recently developed for AOD, will be escribed briefly in this section. It is a multivariable gradient type search chnique based on the general plan described in section 2.10 and depicted 1 figure 2.14 of Chapter 2. It includes the option of a shotgun search for a ood start point as explained in section 6.23. The alternative provided is to pecify the coordinates x_i for $i = 1, \ldots, N$ uniquely defining a good start oint if one is known. Incidentally, regarding the optimization problem quation system and variables, to simplify our explanation we will specifically se the terminology of the basic AOD format summarized in section 6.19 in ne discussion of OPTIGO. However, it is understood that a unit based AOD quation system as described in section 6.20 is often used, and the correspondng terms for such an equation system can be interchanged automatically in ne discussion of the algorithm operation.

OPTIGO incorporates features of self-tuning for automatic calculation of oefficient C_p to use in the penalty function P of equation 6.37, automatic erivative determination by finite differences to establish gradient directions t base points in the search, and automatic convergence to within a neighborood of practical size for the solution point. Input for the specific AOD quation system and given constants is simple to incorporate in OPTIGO, nd coordinates x_i including items of analysis are always printed for the olution point at the end of the search. It has the option of printing items of nterest for base points during the search.

The general construction of OPTIGO will be described so the user will nderstand basically how it works and how to use it for solving new optimizaon problems for mechanical elements. Examples of application will be

TABLE 6.2

Main Program and Subroutines for OPTIGO Fortran IV Program Listed in Appendix B

Line Number Range in Program	Statement Number Range in Program	Name	Comments
1–182	1–1450	OPTIGO MAIN PROGRAM	Remains unchanged from one problem to another
183–207	2000–2250	SUBROUTINE PRTOUT	Remains unchanged from one problem to another
208–236	3000–3180	SUBROUTINE SG	Remains unchanged from one problem to another
237–261	4000–4100	SUBROUTINE QPCALC	Remains unchanged from one problem to another
262– ↑	5000–5999*	SUBROUTINE Q5000	Criterion function programmed here to calculate Q. Also, final items programmed here as required; i.e., F_j's for $j = 1, \ldots, N_F$
Depend	6000–6999*	SUBROUTINE EJ6000	Equality constraints programmed here to calculate E_j's, for $j = 1, \ldots, N_E$. Also, final items programmed here as required; i.e., F_j's for $j = 1, \ldots, N_F$
on specific problem	7000–7999*	SUBROUTINE RK7000	Regional constraints programmed here to calculate R_k's, for $k = 1, \ldots, N_R$. Also, final items programmed here as required; i.e., F_j's for $j = 1, \ldots, N_F$
↓	8000–9999*	SUBROUTINE INPUT	Input data entered here. Also calculated constants programmed here to calculate C_m's, for $m = 1, \ldots, N_C$, in terms of given constants G_l, $l = 1, \ldots, N_G$

* Available for user.

presented in this section as well as in Chapter 8. The Fortran IV listing of the computer program for OPTIGO is presented in Appendix B for batch operation with punched card input and printed output in hard copy. Incidentally, conversion of the listed program to terminal operation and/or to the Basic language would be very simple, and this was recently done by the author for use on a microcomputer. As you will note, the algorithm contains a main program plus seven subroutines as summarized in table 6.2, and these will be explained as follows. In the descriptions we will refer to line numbers and statement numbers, which specifically refer to the Fortran IV Program of OPTIGO as listed in Appendix B.

6.25 OPTIGO Main Program

Basic construction and operation of the main program will be outlined, including the equations and some of the decision making process incorporated in the OPTIGO program. A simplified block diagram of the OPTIGO main program operation is summarized in figure 6.18, which will be explained more fully. However, the general outline of the figure should be of help in a better understanding of OPTIGO.

Initial Calculations. Actually, the first calculations made for OPTIGO are for the calculated constants C_m in SUBROUTINE INPUT as described in section 6.32. Next, we calculate finite difference increments $(\Delta x_i)_d$ for each of the independent variables x_i, and these will be needed to surround the base points B_r when calculating gradient or ray directions in the search process.

$$(\Delta x_i)_d = \frac{[(x_i)_{\max} - (x_i)_{\min}]}{\alpha} \qquad \text{for} \quad i = 1, \ldots, N \tag{6.54}$$

In this equation $(x_i)_{\max}$ and $(x_i)_{\min}$ are input items defining the feasible ranges of interest estimated initially for the x_i's of the AOD equation system. The parameter α is normally taken as 10^5 and is not changed from one problem to another. It is already built in as an input item of 10^5 in SUBROUTINE INPUT and must be changed internally if another value is desired.

Another initial calculation is to determine the large step size Δ taken from the first base point B_1 along the first ray in the search routine.

$$\Delta = \frac{\left\{ \sum_{i=1}^{N} [(x_i)_{\max} - (x_i)_{\min}]^2 \right\}^{1/2}}{\beta} \tag{6.55a}$$

Referring to equation 6.55a, we recognize that the greatest diagonal in the feasible range of interest for the independent variables x_i has been divided

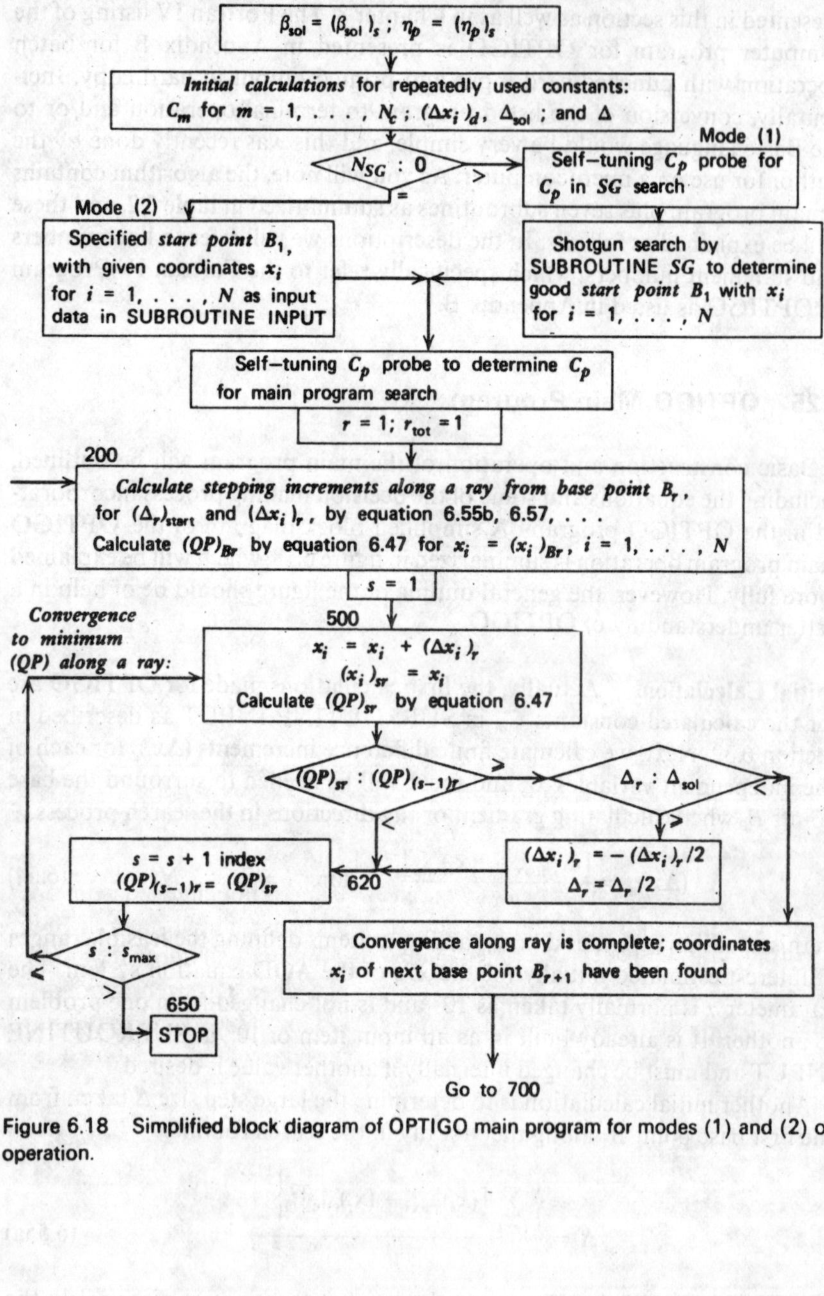

Figure 6.18 Simplified block diagram of OPTIGO main program for modes (1) and (2) of operation.

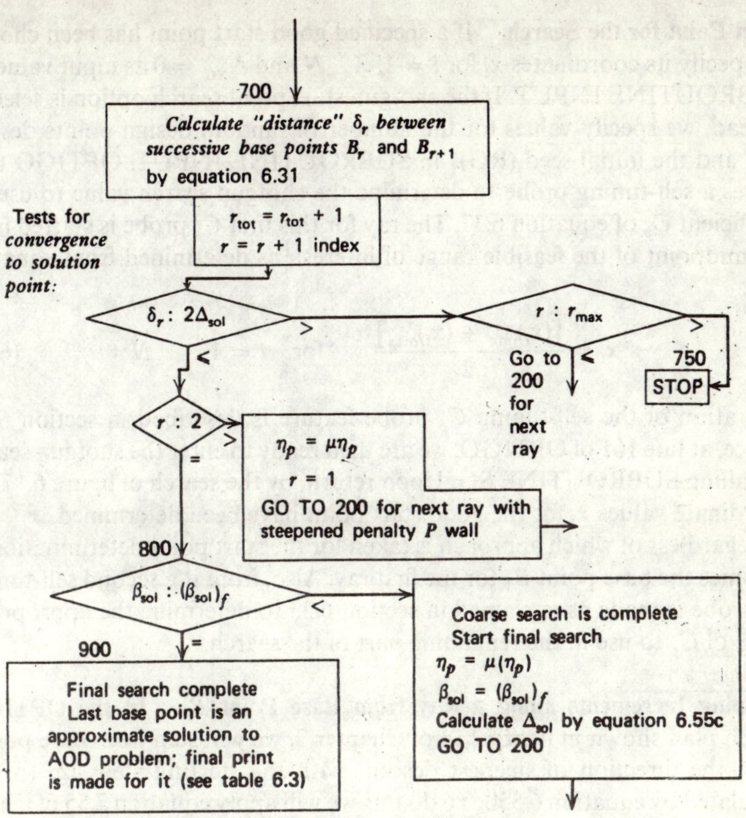

Figure 6.18 *(continued)*.

by a parameter β, which is normally taken as 10 and is not changed from one problem to another. Parameter β is already built in as an input item of 10 in SUBROUTINE INPUT and must be changed internally if another value is desired. For subsequent gradient lines or rays the large step size is gradually reduced by dividing Δ by the total number of rays for the search to that point, r_{tot}, giving equation 6.55b for the starting large step size $(\Delta_r)_{start}$ along a ray in the search process.

$$(\Delta_r)_{start} = \frac{\Delta}{r_{tot}} \tag{6.55b}$$

Also, the solution test increment Δ_{sol} is calculated by

$$\Delta_{sol} = \frac{\beta\Delta}{\beta_{sol}} \tag{6.55c}$$

where β_{sol} is a built in parameter to be explained later.

Start Point for the Search. If a specified good start point has been chosen, we specify its coordinates x_i for $i = 1, \ldots, N$ and $N_{SG} = 0$ as input values in SUBROUTINE INPUT. If the shotgun start point search option is selected instead, we specify values for the number of random design points desired N_{SG} and the initial seed $(\mathrm{RG})_1$ in SUBROUTINE INPUT. OPTIGO then makes a self-tuning probe to determine the shotgun search value to use for coefficient C_p of equation 6.37. The ray for this first C_p probe is started from the midpoint of the feasible range of interest, as determined from equation 6.56.

$$x_i = \frac{[(x_i)_{\min} + (x_i)_{\max}]}{2} \qquad \text{for} \quad i = 1, \ldots, N \tag{6.56}$$

Operation of the self-tuning C_p probe feature is described in section 6.33. Hence, at line 161 of OPTIGO, we are then ready to enter the shotgun search by calling SUBROUTINE SG. Upon return, by the search of figure 6.17 the coordinate values x_i for the good start point have been determined.

Regardless of which approach is taken for the start point determination, it becomes the base point B_1 for the first ray. Also, from it a second self-tuning C_p probe is made as explained in section 6.33 to determine the appropriate value of C_p to use in the remaining part of the search.

Stepping Increments along a Ray from Base Point B_r. In the OPTIGO search plan shown in figure 2.14 of Chapter 2, we will step from base points B_r in the direction of steepest descent with the starting step size $(\Delta_r)_{\text{start}}$ calculated by equation 6.55b. To do this we will apply equation 2.55 of Chapter 2 to the penalized optimization quantity (QP) of equation 6.47 instead of Q. Hence the starting stepping components $(\Delta x_i)_r$ along a ray are calculated by equation 6.57, which is similar to equation 2.55 of Chapter 2.

$$(\Delta x_i)_r = \frac{-(QP)'_{iBr}(\Delta_r)_{\text{start}}}{\left\{\sum_{i=1}^{N} [(QP)'_{iBr}]^2\right\}^{1/2}} \qquad \text{for} \quad i = 1, \ldots, N \tag{6.57}$$

In this equation, $(\Delta_r)_{\text{start}}$ is calculated from equation 6.55b, and the $(QP)'_{iBr}$ terms are the partial derivatives defined as follows.

$$(QP)'_{iBr} = \left[\frac{\partial(QP)}{\partial x_i}\right]_{Br} \qquad \text{for} \quad i = 1, \ldots, N \tag{6.58a}$$

These partial derivatives at the base point B_r are calculated by application of finite difference equation 2.4 of Chapter 2, giving equation 6.58b.

$$(QP)'_{iBr} \approx \frac{[(QP)_{id+} - (QP)_{id-}]}{2(\Delta x_i)_d} \qquad \text{for} \quad i = 1, \ldots, N \tag{6.58b}$$

For equation 6.58b we must calculate the values of (QP) for the appropriate finite difference points surrounding the base point in question, and this we will explain next. This explanation will define what is meant by $(QP)_{id+}$ and $(QP)_{id-}$, as found in equation 6.58b.

First we will calculate the coordinates $(x_i)_{d+}$ adjacent to and positive by $(\Delta x_i)_d$ from the base point B_r by equation 6.59a. Similarly, we calculate the coordinates $(x_i)_{d-}$ adjacent to and negative by $(\Delta x_i)_d$ from the base point B_r by equation 6.59b.

$$(x_i)_{d+} = (x_i)_{Br} + (\Delta x_i)_d \qquad \text{for} \quad i = 1, \ldots, N \tag{6.59a}$$

$$(x_i)_{d-} = (x_i)_{Br} - (\Delta x_i)_d \qquad \text{for} \quad i = 1, \ldots, N \tag{6.59b}$$

In equations 6.59, $(x_i)_{Br}$ are the coordinates for the base point in question, and $(\Delta x_i)_d$ are the finite difference increments calculated by equation 6.54. Next, we calculate for equation 6.58b $(QP)_{id+}$ using equation 6.47 with coordinates $(x_i)_{Br}$ of the base point, except for the i of the partial derivative we use coordinate $(x_i)_{d+}$ from equation 6.59a. Similarly, we calculate for equation 6.58b $(QP)_{id-}$ using equation 6.47 with coordinates $(x_i)_{Br}$ of the base point, except for the i of the partial derivative we use coordinate $(x_i)_{d-}$ from equation 6.59b.

As just explained, by equation 6.57 we calculate the starting component step size increment $(\Delta x_i)_r$ for each of the coordinates for $i = 1, \ldots, N$. If the proportions of these components are maintained in the stepping process, we will stay in the direction of steepest descent along the gradient line or ray from the base point B_r.

Converging to Minimum (QP) Along a Ray. Penalized Optimization quantity (QP) will be characterized by stationary points as illustrated in figure 6.16. Hence, as described in section 2.10 and illustrated in figures 2.13 and 2.14 of Chapter 2, we will apply an interval halving technique to converge to the directional stationary point of minimum (QP) along the ray. This point will then become the next base point for a new ray, and in this iterative manner we converge to the solution stationary point of minimum (QP) similar to figure 2.14.

Coordinates $(x_i)_{sr}$ for a step point s along a ray r are calculated successively as shown in figure 6.18, in an incrementing procedure starting from the base point B_r. Reversal of direction and reduction of step size are obtained by sign change and division by 2 for the stepping increments $(\Delta x_i)_r$ and Δ_r, and this is done upon sensing that the directional stationary point for (QP) along the ray has been reached. When total step size Δ_r has been reduced to the size of Δ_{sol}, we have reached the next base point B_{r+1}. Specific controls not described here have been incorporated in the OPTIGO listing of Appendix B, to avoid excessive stepping in unusual cases. Thus step number limitation

$s \leq s_{max}$ is incorporated, as well as a limitation of number of steps between reversals after the first reversal.

Distance δ_r between Successive Base Points. In convergence to the solution point, the general trend for "distance" δ_r between successive base points is to become smaller and smaller, as illustrated in figure 2.14 for Q instead of (QP). The Euclidean N-dimensional space distance δ_r is determined by the following elementary equation.

$$\delta_r = \left\{ \sum_{i=1}^{N} [(x_i)_{B_{r+1}} - (x_i)_{B_r}]^2 \right\}^{1/2} \tag{6.31}$$

In equation 6.31, the subscripts B_r and B_{r+1} are for successive base points in the searching process.

Convergence to the Solution Point. The final part of the figure 6.18 block diagram is concerned with convergence to the practical neighborhood of the solution point for the optimization problem. The stepping process from base point to base point is continued with increasing number of rays until "distance" $\delta_r \leq (2\Delta_{sol})$. The value of Δ_{sol} is as calculated from equation 6.55c, which is the same value used previously for convergence control in the search along a ray.

When the $\delta_r \leq (2\Delta_{sol})$ smallness check is reached, a local ray number check is made, and if $r > 2$ is the case another round of convergence is started with steeper penalty function walls. This is obtained by increasing coefficient η_p of equation 6.37 by $\eta_p = 8\eta_p$ and starting a new local ray count with $r = 1$. Eventually the "distance" δ_r between the base points on ray $r = 1$ meets the $(2\Delta_{sol})$ test, and what we call the *coarse search* is complete. At this stage of OPTIGO we have found the solution point with reasonably good accuracy. However, in an attempt to improve the accuracy still further we reduce Δ_{sol} by a factor of 10 and continue with what we will call the *final search*.

The coarse and final search stages of OPTIGO are obtained by using two different built-in values for β_{sol} of equation 6.55c. For the coarse search, $\beta_{sol} = 2000$ is incorporated in SUBROUTINE INPUT, whereas for the final search, $\beta_{sol} = 20{,}000$ is incorporated. To change either value, the proper two lines following statement 8050 in the Appendix B listing must be changed. After the final search is complete, the solution point for the optimization problem will have been located quite accurately for all practical purposes. The final printout uniquely defines this point with the x_i values for $i = 1, \ldots, N$, and all final items of interest are printed as calculated for this solution point.

A measure of success for the convergence to an acceptable solution point with no appreciable constraint violations is the value of η_{QP} defined by equation 6.60, which should be unity for an exact solution point. However, at the approximate solution point found by OPTIGO, the value of P should at least be small compared with that of Q. Thus for good convergence to the practical neighborhood of the exact solution point, we generally obtain a value of η_{QP} slightly greater than unity at completion of the final search.

$$\eta_{QP} = \frac{(QP)}{Q} \tag{6.60}$$

The value of (QP) is calculated by equation 6.47, and the value of Q is calculated by the AOD equation system. Hence, for $P = 0$ we see by equation 6.60 that $\eta_{QP} = 1$.

Incidentally, near the end of figure 6.18 a ray limit control $r \leq r_{\max}$ is incorporated in OPTIGO to prevent the search from continuing indefinitely for an unusual (QP) function situation. A value of $r_{\max}$ such as 100 is entered as input data of SUBROUTINE INPUT.

6.26 SUBROUTINE PRTOUT

This subroutine starts at line 183 and statement 2000 of the Appendix B listing for OPTIGO. It remains unchanged from one specific problem to another, and it serves the function of printing out in readable fashion pertinent base point information on call from the main program. The base point printout items are summarized in table 6.3 and are specifically illustrated in part B.2 of Appendix B from example 6.9b.

In modes (1) and (2) of OPTIGO operation (described in section 6.35), the first base point for the main search is always printed just *prior* to the "COARSE SEARCH STARTED . . ." printed statement. This is either the design state point determined from the shotgun search mode (1) or the specified point of mode (2). OPTIGO provides the option of having the table 6.3 information printed for successive base points B_r during the coarse search, if we wish to review how the stepping process is converging. However, for both modes (1) and (2), the first base point printout *after* the "COARSE SEARCH STARTED . . ." printed statement is actually carried over as the point of greatest penetration into a penalty zone from the self-tuning C_p probe of the main search, as explained in section 6.34.

The successive base point print option for the coarse search is chosen by specifying a value of NPR = 1 in the input data of SUBROUTINE INPUT. However, base point printout is suppressed during the search if a value of NPR = 0 is entered instead. Regardless of the NPR input value used, base point printout is always made for the solution point at the end, as well as for the last base point whenever the search is automatically stopped because of

TABLE 6.3

Summary of Printed Items for Base Points from SUBROUTINE PRTOUT of OPTIGO, in Order of Print

General Symbol	Fortran IV Name	Description
r	RA	Ray number count during search between changes of η_p
r_{tot}	TRA	Total number of rays made during search
$(\Delta_r)_{start}$	DRS	Large step size from base point, by equation 6.55b
Last s	S	Total number of steps made along last ray
δ_r	SD	"Distance" between successive base points, by equation 6.31
η_p	EP	Wall steepening coefficient for P in equation 6.37
C_p	CP	Tuning coefficient for P in equation 6.37
x_i	X(I)	Independent variables in AOD equation system, for $i=1, \ldots, N$
E_j	E(J)	Equality constraint function of AOD equation system, value calculated during search, for $j=1, \ldots, N_E$
R_k	R(K)	Regional constraint function of AOD equation system, value calculated during search, for $k=1, \ldots, N_R$
P	P	Penalty function value, by equation 6.37
Q	Q	Optimization quantity value of AOD equation system
(QP)	QP	Penalized optimization quantity value, by equation 6.47
η_{QP}	EQP	Ratio of (QP) to Q, by equation 6.60

exceeding either the ray number limit r_{max} or the step number limit s_{max} on any ray. Also, the last base point in the coarse search is always printed before starting the final search, and usually it will agree quite closely with the final solution point. Base point printout is always suppressed during the final search, except for the solution point at the end.

6.27 SUBROUTINE SG

This subroutine starts at line 208 and statement 3000 of the Appendix B listing for OPTIGO. It has been discussed in section 6.23 and summarized in figure 6.17. It remains unchanged from one specific problem to another and as previously explained, this option serves the function of determining a good start point for the OPTIGO solution search process. Incidentally, some users may wish to modify the subroutine slightly, such as for the use of 10^{1} instead of 10^{59} as explained in section 6.23, or if the particular computer used has an intrinsic random number generation feature.

6.28 SUBROUTINE QPCALC

This subroutine starts at line 237 and statement number 4000 of the Appendix B listing for OPTIGO. Its flow chart is given in figure 6.19. The subroutine remains unchanged from one specific problem to another, and it serves the purpose of calculating by equation 6.47 the value of penalized optimization quantity (QP) at a design state point in the search process. A design state point is uniquely defined by known values of x_i for $i = 1, \ldots, N$ in the AOD equation system.

SUBROUTINE QPCALC always calls on SUBROUTINE Q5000 to calculate the value of Q from the AOD equation system. Also, it calls on SUBROUTINE EJ6000 next to calculate the values of the equality constraint

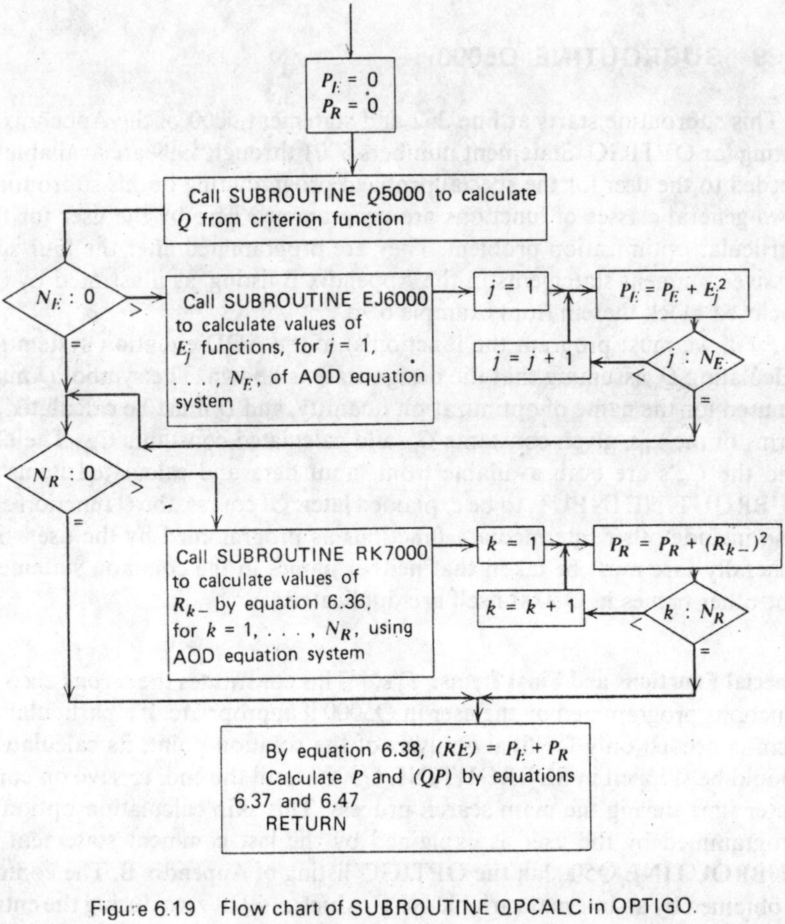

Figure 6.19 Flow chart of SUBROUTINE QPCALC in OPTIGO.

functions E_j for $j = 1, \ldots, N_E$ of the AOD equation system. Similarly, it calls on SUBROUTINE RK7000 to calculate the values of the regional constraint functions R_k for $k = 1, \ldots, N_R$ of the AOD equation system. In the process it compares R_k with zero to determine whether or not a regional constraint is violated, and it assigns the proper value to R_{k-} in accordance with equation 6.36. Through these last two subroutines, the value of (RE) is calculated by equation 6.38 followed by determination of (QP) by equations 6.37 and 6.47.

SUBROUTINE QPCALC is called on command from the main program of OPTIGO. Upon return to the main program, the values of Q, E_j, R_k, (RE), P, (QP), and η_{QP} have been determined and retained by the computer at the design state point in question. This information is needed either in the stepping search process of OPTIGO or for base point printout.

6.29 SUBROUTINE Q5000

This subroutine starts at line 262 and statement 5000 of the Appendix B listing for OPTIGO. Statement numbers 5001 through 5999 are available as needed to the user for the special problem programming in this subroutine. *Two* general classes of functions are programmed here by the user for the particular optimization problem. They are programmed after the four successive comment statements in the Appendix B listing, as illustrated by the enclosed block therein from example 6.9b.

First, we must program the function(s) of the AOD equation system for calculating Q, assuming that the design state is known. The symbol Q must be used for the name of optimization quantity, and Q must be calculated in terms of the x_i's, given constants G_l, and calculated constants C_m. The G_l's and the C_m's are both available from input data and calculated items of SUBROUTINE INPUT, to be explained later. Of course, the Q function can also include other intermediary functions as programmed by the user, but generally care must be taken that neither names in the common statement nor other names in Q5000 itself are duplicated.

Special Functions and Final Items: F_j's. This constitutes the *second* class of functions programmed by the user in Q5000 if appropriate. If a particular F_j item is needed only for final analysis of the solution point, its calculation should be skipped in SUBROUTINE Q5000 until the end, to save on computer time during the main search process. This skip calculation option is programmed by the user as explained by the last comment statement of SUBROUTINE Q5000 in the OPTIGO listing of Appendix B. The control is obtained by an internal variable NFC, which is set at zero during the entire search but is given the value NFC = 1 for analysis of the final solution base point only. The skip F_j calculation option is programmed by *logical if* as specifically illustrated in SUBROUTINE Q5000 of the Appendix B listing.

The entire set of special functions and final items are designated by subscripted variable name notation F_j, for $j = 1, \ldots, N_F$, and their calculation may be programmed by the user for distribution throughout SUBROUTINES Q5000, EJ6000, and RK7000 as needed. The total number of such items, N_F, is entered as an input item in SUBROUTINE INPUT. The F_j's are items of interest for the design analysis and are generally programmed in terms of the available AOD equation system variables, the given constants G_l, and the calculated constants C_m. At the conclusion of the OPTIGO search the entire set of F_j values for $j = 1, \ldots, N_F$ is printed for the solution point.

6.30 SUBROUTINE EJ6000

This subroutine starts at statement number 6000 of the Appendix B listing of OPTIGO. Statement numbers 6001 through 6999 are available to the user as needed for programming specific problem functions in EJ6000. Two general classes of functions are programmed in this subroutine after the five successive comment statements in the Appendix B listing. Specific illustration is enclosed in a block therein, from example 6.9b.

First, we must program the equality constraint function(s) E_j for $j = 1, \ldots, N_E$. These E_j functions are of the AOD equation system, and obviously they are not set equal to zero in the programming as shown in the section 6.19 summary; this is the *requirement* for solution. At any design state point during the search, normally the E_j's will not be zero and thus will contribute to the penalty function P value as seen from equations 6.37 and 6.38. However, the OPTIGO search process should eventually converge to a point where the E_j's are all at least very close to zero, near the solution point. Generally, the E_j functions are programmed in terms of the x_i's, the given constants G_l, and the calculated constants C_m. These have already been discussed under SUBROUTINE Q5000 in section 6.29, and they will be discussed further under SUBROUTINE INPUT in section 6.32. Of course, the E_j functions can also include other intermediary functions as programmed by the user, but generally care must be exercised that neither names in the common statement nor other names in EJ6000 itself are duplicated.

Second, special functions and final items F_j may be programmed here by the user as appropriate. These are designated by the subscripted variable name F_j, and the discussion for the F_j's presented in section 6.29 on SUBROUTINE Q5000 similarly applies here to the F_j's programmed in SUBROUTINE EJ6000. Hence, as previously explained, any particular F_j item needed only in the final analysis for the solution point can be skipped to save on computer time during the main search process. This is done by a logical if statement on NFC as explained in section 6.29.

6.31 SUBROUTINE RK7000

This subroutine starts at statement number 7000 of the Appendix B listing of OPTIGO. Statement numbers 7001 through 7999 are available to the user as needed for programming specific problem functions in RK7000. Two general classes of functions are programmed in this subroutine, and they are programmed after the five successive comment statements in the Appendix B listing. Specific illustration is enclosed in a block therein, from Example 6.9b.

First, we must program the regional constraint function(s) R_k for $k = 1, \ldots, N_R$. These R_k functions are of the AOD equation system. At any design state point during the search, the R_k's may have either positive or negative values. Hence, in accordance with equations 6.36, 6.37, and 6.38, the negative ones violate a constraint and these will contribute to the penalty function P value. However, the OPTIGO search process will eventually converge to the neighborhood of the solution point where all the R_k's are either positive as required by the AOD equation system or negative and very close to zero. Generally, the R_k functions are programmed in terms of the x_i's, the given constants G_l, and the calculated constants C_m. These have already been discussed under SUBROUTINE Q5000 in section 6.29, and they will be discussed further under SUBROUTINE INPUT in section 6.32. As before, the R_k functions can also include other intermediary functions as programmed by the user, but generally care must be taken that neither names in the common statement nor other names in RK7000 itself are duplicated.

Second, special functions and final items of interest may be programmed here by the user as appropriate. The discussion previously made in sections 6.29 and 6.30 applies similarly to SUBROUTINE RK7000, with respect to programming of F_j functions.

6.32 SUBROUTINE INPUT

All the input data supplied to OPTIGO are introduced in this subroutine. First, there are internal parameter values not normally changed from one optimization problem to another. If a change is desired for any of these items, the program statement as listed in SUBROUTINE INPUT of

TABLE 6.4

Input Data for SUBROUTINE INPUT of OPTIGO

General Symbol	Fortran IV Name	Real or Integer	Description
			INTERNAL DATA (normally not changed)
α	AL	R	$1. \times 10^5$ (see equation 6.54)
β	B	R	10. (see equation 6.55)
$(\beta_{sol})_s$	BSS	R	2000. (see figure 6.18 and
$(\beta_{sol})_f$	BSF	R	$2. \times 10^4$ equation 6.55c)
$(n_p)_{max}$	NPM	I	2 (large steps into penalty zone)
$(\eta_p)_s$	EPS	R	1. (see figure 6.18)
μ	EM	R	8. (see figure 6.18)
			FIRST INPUT DATA CARD
N	N	I	Number of independent variables, x_i; for AOD equation system
N_E	NE	I	Number of equality constraints, E_j; for AOD equation system
N_R	NR	I	Number of regional constraints, R_k; for AOD equation system
N_C	NC	I	Number of calculated constants, C_m
N_F	NF	I	Number of final items, F_j
N_G	NG	I	Number of given constants, G_l
r_{max}	RM	I	Ray limit control (100 is typical value); see figure 6.48
s_{max}	SM	I	Step limit control (100 is typical value); see figure 6.18
			SECOND INPUT DATA CARD
N_{SG}	NSG	I	Number of random design points for shotgun search
$(RG)_1$	RSD	R	Seed for start of shotgun search; $0 < (RG)_1 < 1$
N_{pr}	NPR	I	Base point print control (0 for skip; 1 for print)
			NEW INPUT DATA CARDS FOR EACH GROUP
$(x_i)_{min}$	XMN(I)	R	Lower limit estimate for feasible range of x_i; for $I = 1, N$
$(x_i)_{max}$	XMX(I)	R	Upper limit estimate for feasible range of x_i; for $I = 1, N$
G_l	G(L)	R	Given constants for programming use of optimization problem functions; for $L = 1, NG$
x_i	X(I)	R	Independent variable coordinates x_i for specified start point (optional, with $N_{SG} = 0$); for $I = 1, N$

Appendix B must be changed for a different value of the parameter. Second, there are input items supplied by the user for the particular optimization problem at hand, in the form of punched cards for batch mode operation or keyed in data for terminal mode operation. These user supplied input data are with respect to the five READ statements listed in the SUBROUTINE INPUT of OPTIGO in Appendix B. As you will note, all five READ statements refer to the statement number 8150 FORMAT (8G10.3) general field descriptor specification. Thus all the input items will be punched in 8 fields for each data card, divided by columns 1–10, 11–20, 21–30, . . . , 71–80 for each punched card of input data. Summary of pertinent information for the input data items is given in table 6.4, which is self-explanatory.

As you will note in table 6.4 and the listing of SUBROUTINE INPUT in Appendix B, *given constants* are entered and assigned the subscripted variable names G_l, for $l = 1, \ldots, N_G$. Also, you will note that provision is made for calculated constants C_m for $m = 1, \ldots, N_C$. These must be programmed by the user as a function of the given constants G_l, the purpose being to reduce execution time for the computer during the searching process of OPTIGO. The calculated constants C_m are items which are used repeatedly in OPTIGO, and to repeat calculations of these constants in terms of supplied input data each time needed would be wasteful, particularly for lengthy or complicated problems. Specific illustration of calculated constants as programmed by the user in SUBROUTINE INPUT is given in the enclosed block of the Appendix B listing after statement 8500, from example 6.9b.

SUBROUTINE INPUT is called on command early in the main program of OPTIGO, specifically at line 6 of the Appendix B listing. It prints the supplied input data items summarized in table 6.4 as well as all the calculated constants C_m for $m = 1, \ldots, N_C$. Illustration of the input data printout format is presented in part B.2 of Appendix B. Upon return to the main program of OPTIGO, the search procedure is carried through as generally outlined in the flow chart of figure 6.18.

6.33 Size Limitations for OPTIGO

The algorithm OPTIGO is virtually unlimited in the size of optimization problem it can handle. However, in the Fortran IV listing of Appendix B the dimension statements as they stand limit the size of subscripted variables as summarized in table 6.5. For mechanical elements, the provided size should be more than adequate for the vast majority of optimization problems. However, if it is necessary to either increase or decrease any of the table 6.5 size limitations, change in the appropriate dimension statements in the Appendix B listing can readily be made.

TABLE 6.5

Size Limitations for OPTIGO in Appendix B Listing

Variable	Size Limitation	Comments
x_i	$i = 1, \ldots, N \leq 25$	See AOD equation system
E_j	$j = 1, \ldots, N_E \leq 25$	See AOD equation system
R_k	$k = 1, \ldots, N_R \leq 100$	See AOD equation system
G_l	$l = 1, \ldots, N_G \leq 100$	On given constants
C_m	$m = 1, \ldots, N_C \leq 100$	On calculated constants
F_j	$j = 1, \ldots, N_F \leq 100$	On final items and special functions

6.34 Self-Tuning Probes for C_p

For proper operation of an AOD algorithm such as OPTIGO, the iterative calculation procedure must move the design state point in a pattern which converges towards the solution point, as shown in figure 2.14 of Chapter 2. Introduction of the penalty function P defined by equation 6.37 should create a block for the described movement which prevents excessive penetration into penalty zones of constraint violation for the AOD equation system. The block is created in the form of a (QP) function wall as shown in figure 6.16, where penalized optimization quantity (QP) to be minimized is defined by equation 6.47.

It is found that two unacceptable conditions of imbalance can exist between the penalty function P and the criterion function Q. First, if the function P is excessively strong compared with the function Q the created (QP) function wall is too steep and the required movement towards the solution point does not take place. In this event, we might say that the search process "hangs up" at a region appreciably away from the solution point. Second, if the function P is too weak compared with the function Q the created wall for (QP) is too gradual and, in fact, it may be but a minor "hill" which is overpowered by the Q function. In this event the search process might very well "take off" with the movement leading far into penalty zones unchecked and not converging to the solution point. Either one of the described difficulties is caused by a gross imbalance between the P and Q functions and results in failure of the search process. The penalty function P is then not properly tuned to the criterion function Q.

For proper movement as described in the search process converging to the solution point, the penalty function P must be appropriately tuned to the criterion function Q. This is accomplished by having C_p in equation 6.37 of proper value. For acceptable tuning C_p need not be precisely determined, but it must have at least the appropriate order of magnitude for avoidance of the two difficulties described. This is automatically accomplished in OPTIGO by means of a self-tuning feature which will now be described.

When stepping along a ray of steepest descent for (QP) in the OPTIGO search process, the effect of a limit condition will be encountered and a penalty zone will be entered corresponding to a constraint violation, as shown in figure 6.20. If the movement is being made at the large step size increment Δ as determined by equation 6.55a, for proper tuning a block against further movement should be made after about one or two such steps into the penalty zone. According to figure 6.18, the block should be made when an increase in (QP) is sensed.

For a self-tuning C_p probe in OPTIGO, C_p is initially set at zero and the direction of steepest descent for Q is determined from a start point. A stepping process at increment Δ is started along such a ray with counter $n_p = 0$ retained at that value until a constraint violation is sensed by a positive value of (RE) from equation 6.38. The immediately preceding value of Q is retained as $Q_{np=0}$ with the corresponding value of (RE) retained as $(\text{RE})_{np=0}$. The value of n_p is then indexed by one for each succeeding step into the penalty zone until $n_p = (n_p)_{\max}$ is reached. The appropriate value for $(n_p)_{\max}$ is generally 2, which is built into OPTIGO as summarized in table 6.4. For this final step point in the probe the value of Q is retained as $Q_{np\,\max}$ and the corresponding value of (RE) is retained as $(\text{RE})_{np\,\max}$. Hence, for the block to be effective at that point, as seen from figure 6.20, the decrease in Q should equal the

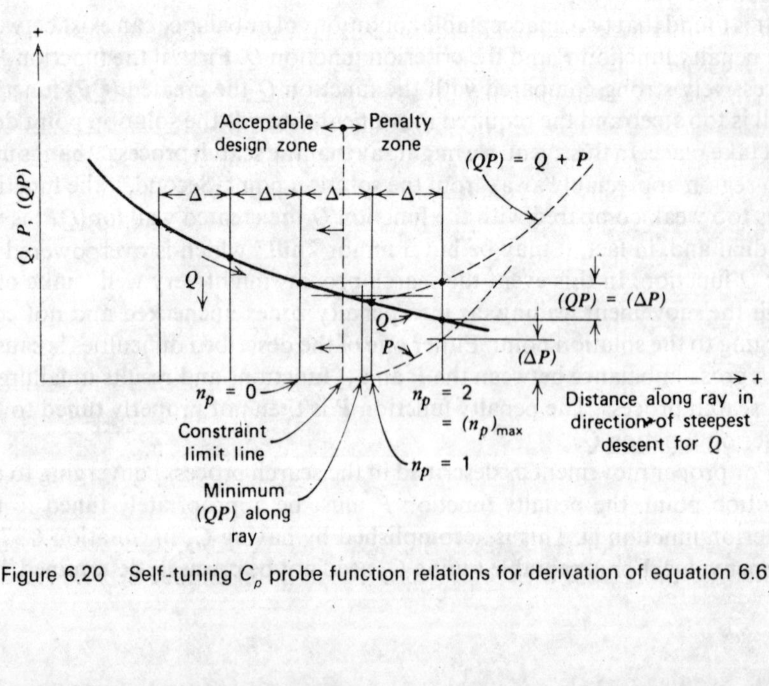

Figure 6.20 Self-tuning C_p probe function relations for derivation of equation 6.61.

increase in P during the $(n_p)_{\max}$ steps. Thus incorporating equation 6.37 we have for a proper block the following relationship.

$$(\Delta Q) = [Q_{np=0} - Q_{np\,\max}] = C_p \eta_p[(\mathrm{RE})_{np\,\max} - (\mathrm{RE})_{np=0}] = (\Delta P)$$

However, for the self-tuning C_p probe we use wall steepening factor η_p at its start value of $(\eta_p)_s = 1.0$, according to table 6.4. Hence, from the preceding equation we obtain equation 6.61, which determines a proper value of C_p for acceptable tuning.

$$C_p = \frac{[Q_{np=0} - Q_{np\,\max}]}{(\mathrm{RE})_{np\,\max} - (\mathrm{RE})_{np=0}} \tag{6.61}$$

The value of C_p from equation 6.61 is determined automatically by OPTIGO in the described stepping process of a self-tuning C_p probe. As used in OPTIGO, this normally determines a proper order of magnitude for C_p which will result in an acceptably tuned condition between P and Q. With this value for C_p, movement should converge close to the solution point during the search process which follows.

As seen near the start of figure 6.18, there are two places in OPTIGO where self-tuning probes are made for determination of C_p. First, a value for C_p is needed at the start of the shotgun search. For this case, the self-tuning C_p probe is made along a ray emanating from the midpoint of the $(x_i)_{\min}$, $(x_i)_{\max}$ feasible range. Second, a properly tuned C_p value is needed at the start of the main program search process. For this case, the self-tuning C_p probe is made along a ray emanating from the start point of the main search. This would be either from the point determined by the shotgun search or from the specified start point if that option is selected.

Finally, in very unusual situations the experienced user of OPTIGO may wish to override the self-tuning feature and specify a particular value of C_p for the main program search. This is readily accomplished by specifying values for $C_p > 0$, the start coordinates x_i for the main search, and $N_{SG} = 0$. Specific details for this option are given under mode (3) in section 6.35.

6.35 Application of OPTIGO

Automatic solution of optimization problems by OPTIGO is usually straightforward and relatively simple. The first step of application is to formulate the optimization problem mathematically according to the basic format for AOD equation systems as described in section 6.19. For most problems, transformation to unit based variables and unit based equations as described in section 6.20 is recommended. The various equations so derived for the specific optimization problem are then programmed by Fortran IV into SUBROUTINES Q5000, EJ6000, RK7000, and INPUT as previousl described in sections 6.29 through 6.32. Then input data for the numerica problem must be decided upon for the items listed in table 6.4. Finally, for th batch mode of computer operation, cards must be punched for the describe problem equations and input data. With these cards in place in the OPTIG deck, we are ready to run the computer for a quick and fully automati solution to the optimization problem at hand.

Before presenting several introductory examples, we outline very briefl the three modes of operation for OPTIGO.

Mode (1): Shotgun Start and Automatic C_p Determinations. This is th normal mode of operation most commonly used for OPTIGO. The numbe of shotgun points $N_{SG} > 0$ must be specified along with a starting seed valu chosen at random in the range $0 < (\mathrm{RG})_1 < 1$. Equation 6.53 may be use as a guide for selection of N_{SG}. The other input data items summarized i table 6.4 must of course be entered as well, except for the last item in the tabl which is omitted for this mode of OPTIGO. Two self-tuning C_p probes ar automatically made by OPTIGO in this mode of operation. As seen fror figure 6.18, these are made prior to the shotgun search and prior to the mai program search.

Mode (2): Specified Start Point and Automatic C_p Determination. This i a seldom used mode of operation for OPTIGO to accommodate the case c specified start point. Selection of this mode is indicated by specifying $N_{SG} =$ and coordinates x_i for the start point as described in the last line of table 6.4 Of course, the other input items summarized in the table must be entered a well for the particular optimization problem. One self-tuning C_p probe i automatically made by OPTIGO in this mode of operation as seen in figur 6.18, prior to the main program search.

Mode (3): Specified C_p Value and Start Point. This is a rarely used mod of operation for OPTIGO selected for the special case where specified valu for C_p is desired. Thus the automatic self-tuning C_p probe feature of OPTIG is suppressed in this mode of operation. Selection of this mode is accom plished specifically as follows, referring to the Appendix B listing of th program.

1. Remove statement number 105 from the main program.
2. Add a program statement similar to the one removed, specifying th value of C_p as "$C_p = \ldots$". However, this specification for C_p must be adde between lines 309 and 310 in SUBROUTINE INPUT, as called for by th comment statements in that location of the Appendix B listing.
3. Enter regular input data items of table 6.4, including a value of $N_{SG} =$

and the coordinates x_i for a specified start point in accordance with the last line of the table.

Let us now consider some very simple examples illustrating some basic characteristics of application for OPTIGO.

EXAMPLE 6.8

The fenced yard design optimization problem of example 6.6 will be solved by OPTIGO for various boundary values and by the three modes of operation summarized in this section. The basic AOD equation system has already been derived and summarized in example 6.6, as equations 6.25 through 6.30. Given constants G_l will be taken as $G_1 = x_{max}$, $G_2 = y_{max}$, and $G_3 = L$. For this example we will have no calculated constants C_m. However, suppose final items F_j of interest for the design solution are (1) enclosed area A and (2) the diagonal D of the figure 6.15 rectangle, as calculated by equation 6.62 from elementary trigonometry.

$$D = [x^2 + y^2]^{1/2} \tag{6.62}$$

Subscripts 1 and 2 will be used for variables x and y respectively, as was done in example 6.6. Hence, referring to equations 6.25 and 6.62, after the four successive comment statements in the SUBROUTINE Q5000 listing of Appendix B, we would program the following for the enclosed block therein.

```
      Q = 1./(X(1)*X(2))
      IF (NFC.EQ. O) GO TO 5100
      F(1) = 1./Q
      F(2) = (X(1)**2 + X(2)**2)**.5
 5100 CONTINUE
```

Similarly, referring to equation 6.26 of example 6.6, after the five successive comment statements in the SUBROUTINE EJ6000 listing of Appendix B we would program the following for the enclosed block therein.

```
      E(1) = G(3) - 2*(X(1) + X(2))
```

Finally, referring to equations 6.27 through 6.30 of example 6.6, after the five successive comment statements in the SUBROUTINE RK7000 listing of Appendix B we would program the following for the enclosed block therein.

```
      R(1) = G(1) - X(1)
      R(2) = G(2) - X(2)
      R(3) = X(1)
      R(4) = X(2)
```

TABLE 6.6

Summary of Some Input–Output Data Values for Example 6.8 Solutions by OPTIGO and Comparison with Exact Solution by MOD

Item	*a*	*b*	*c*	*d*
		INPUT DATA (see table 6.4)		
$x_{max} = (x_1)_{max}$	160.	75.	75.	75.
$y_{max} = (x_2)_{max}$	180.	180.	180.	180.
G_1	160.	75.	75.	75.
G_2	180.	180.	180.	180.
G_3	400.	400.	400.	400.
Start point				
$x = x_1$	—	—	10.	10.
$y = x_2$	—	—	170.	170.
		OUTPUT DATA (see table 6.3), final solution		
r_{tot}	26	74	68	70
δ_r	2.4839E-04	2.1579E-04	9.2525E-03	2.3835E-04
η_p	2.6214E 05	2.6214E 05	4096	4096
C_p	4.9112E-10	2.5801E-09	1.6288E-08	1.0000E-08
$x = x_1$	100.0460	75.04005	75.03546	75.06001
$y = x_2$	99.9701	124.9440	124.9706	124.9523
Q	9.9984E-05	1.0665E-04	1.0664E-04	1.0662E-04
η_{QP}	1.001334	1.016623	1.000879	1.001616
$A = F_1$	10,001.6	9375.80	9377.23	9378.92
$D = F_2$	141.433	145.746	145.767	145.764
		MOD SOLUTION		
x	100.0000	75.0000	75.0000	75.0000
y	100.0000	125.0000	125.0000	125.0000
A	10,000.0	9375.00	9375.00	9375.00

This concludes the special programming required for the optimization problem for application of OPTIGO.

Numerical problems will be solved for different boundary values and the three modes of operation for OPTIGO as described earlier in this section. For each case we will use a specified length of fence $L = 400$ m and the space limit $y_{max} = 180$ m. Also, common input data values for table 6.4 items will e as follows:

$$N = 2, \quad N_E = 1, \quad N_R = 4, \quad N_C = 0, \quad N_F = 2, \quad N_G = 3,$$

$$r_{max} = 100, \quad s_{max} = 100, \quad (RG)_1 = 0.34567, \quad N_{pr} = 1,$$

$$(x_1)_{min} = 0, \quad (x_2)_{min} = 0.$$

he remaining input data values for the four numerical problems are ımmarized in table 6.6, again with reference to table 6.4. A summary of the our numerical problems is given below, including the modes of operation for PTIGO and the boundary values which are changed.

a. Mode (1) operation of OPTIGO (i.e., shotgun start with C_p's determined utomatically) and $x_{max} = 160$ m.
b. Mode (1) operation of OPTIGO and $x_{max} = 75$ m.
c. Mode (2) operation of OPTIGO (i.e., specified start point and C_p etermined automatically). Again, as in part b use $x_{max} = 75$ m. However, a oor start point will be chosen purposely to see how OPTIGO performs. lence, for the poor start point we specify x = X(1) = 10., and y = X(2) = 70.
d. Mode (3) operation of OPTIGO (i.e., specified value of C_p and start oint). Again, as in parts b and c use $x_{max} = 75$ m. Also, as in part c use the oor start point x = X(1) = 10., and y = X(2) = 170. Finally, we will urposely choose a poor value of C_p to see how OPTIGO performs, but at ast it will be of the proper order of magnitude. We specify $C_p = 1.0 \times 10^{-8}$, hich compares with the automatically determined value of $C_p = 1.629 \times$ $)^{-8}$ from the printout solution to part c.

Some pertinent output data from the OPTIGO runs of the four numerical roblems are also summarized in table 6.6. The table also includes results om exact solutions to the problems as determined by the MOD, for comarative purposes. As you will note, the correlation is excellent in each case.

XAMPLE 6.9

The fenced yard design optimization problem of the last example was ansformed to an AOD unit based equation system in example 6.7, summarзed as equations 6.41 through 6.46 therein, which will be solved by OPTIGO s shown below. Hence the given constants G_l will be taken as $G_1 = x_{max}$, $_2 = y_{max}$, and $G_3 = L$. Referring to equations 6.41 and 6.42, the calculated onstants C_m will be chosen as follows for illustrative purposes, in Fortran IV.

```
C(1) = 1./(G(1) * G(2))
C(2) = 2./G(3)
```

They are programmed as shown in the enclosed block in SUBROUTIN INPUT of the Appendix B listing of OPTIGO. Final items F_j of interest fc the design solution are programmed using equations 6.39, 6.40, 6.21, 6.2 and 6.62 as follows.

$$x = \mathrm{F(1) = G(1) * X(1)}$$

$$y = \mathrm{F(2) = G(2) * X(2)}$$

$$A = \mathrm{F(3) = F(1) * F(2)}$$

$$L_{\text{actual}} = 2(x + y) = \mathrm{F(4) = 2. * (F(1) + F(2))}$$

$$D = \mathrm{F(5) = (F(1)**2 + F(2)**2)**.5}$$

Also, optimization quantity Q is programmed from equation 6.41 of exampl 6.7 as follows.

$$Q = \mathrm{C(1) / (X(1) * X(2))}$$

The preceding equations for Q and the F_j's calculated only for the fina solution point are programmed in the enclosed block as shown in SUI ROUTINE Q5000 of the Appendix B listing, including the necessary logica if transfer statement as previously explained in section 6.29.

Similarly, referring to equation 6.42 of example 6.7, the following statemer is included in the enclosed block of SUBROUTINE EJ6000 of the Appendi B listing.

$$\mathrm{E(1) = 1. - C(2) * (G(1) * X(1) + G(2) * X(2))}$$

Finally, referring to equations 6.43 through 6.46 of example 6.7, the followin statements are programmed in the enclosed block of SUBROUTIN RK7000 of the Appendix B listing as shown.

$$\mathrm{R(1) = 1. - X(1)}$$

$$\mathrm{R(2) = 1. - X(2)}$$

$$\mathrm{R(3) = X(1)}$$

$$\mathrm{R(4) = X(2)}$$

This concludes the special programming required for the optimizatio problem for application of OPTIGO.

Two numerical problems will be solved for different boundary values, bot for the mode (1) operation of OPTIGO as described earlier in this sectio For each numerical problem we will use a specified length of fence $L = 400$ r and the space constraint $x_{\max} = 160$ m. Input data values which are commo

TABLE 6.7

Summary of Some Input–Output Data Values for Example 6.9 Solutions by OPTIGO and Comparison with Exact Solution by MOD

Item	a	b
INPUT DATA (see table 6.4)		
G_1	160.	160.
G_2	180.	75.
G_3	400.	400.
OUTPUT DATA (see table 6.3), final solution		
r_{tot}	35	27
δ_r	1.3075E-04	1.2887E-04
η_p	2.6214E 05	4096
C_p	7.8672E-05	1.0168E-03
$u_{x1} = X(1)$	0.6262598	0.7815481
$u_{x2} = X(2)$	0.5543621	0.9995199
Q	1.000136E-04	1.0668E-04
η_{QP}	1.000911	1.000132
$x = F(1)$	100.202	125.048
$y = F(2)$	99.7852	74.9640
$A = F(3)$	9998.63	9374.07
$L_{actual} = F(4)$	399.973	400.023
$D = F(5)$	141.412	145.796
MOD SOLUTION		
x	100.000	125.000
y	100.000	75.000
A	10,000.000	9375.000
L	400.000	400.000

to both numerical problems are as follows, referring to table 6.4 and the preceding summary based on the example 6.7 equation system.

$N = 2,\quad N_E = 1,\quad N_R = 4,\quad N_C = 2,\quad N_F = 5,\quad N_G = 3,$

$r_{max} = 100,\quad s_{max} = 100,\quad N_{SG} = 50,\quad (\text{RG})_1 = 0.34567,$

$N_{pr} = 1,\quad (x_1)_{min} = 0.,\quad (x_2)_{min} = 0.,\quad (x_1)_{max} = 1.,$

$(x_2)_{max} = 1.$

The two numerical problems will differ as follows.

a. Space constraint $y_{max} = 180$ m.
b. Space constraint $y_{max} = 75$ m.

Some of the input and output data for these two numerical problems are summarized in table 6.7. The table also includes results for exact solutions to the problems as determined by the MOD. for comparative purposes. A you will note, the correlation is excellent for all practical purposes in each case. Also, you will find all the input and output printouts to the solution o example 6.9b under part B.2 of Appendix B.

Incidentally, the two introductory examples 6.8 and 6.9 presented in thi section were purposely chosen as very simple equation system optimizatio problems, used to illustrate simply the application details of OPTIGO. How ever, a more complicated and realistic application is given in Chapter 8 for complex mechanical element.

6.36 Some Characteristics of OPTIGO in Operation

For the vast majority of optimization problems with mechanical elements OPTIGO works very well and automatically converges to the solution poin *if one exists.* As described in section 6.13 for the MOD, in numerical problem of design optimization the possibility exists for a case of incompatible speci fications with no acceptable solution to the problem. In this event, an algo rithm such as OPTIGO may converge to a stationary point of minimum (QP) value, which is for an unacceptable design appreciably violating on or more of the constraints. Such a situation is readily recognized by th appreciable constraint violations in the final printout, with one or mor $E_j \neq 0$ and/or one or more $R_k < 0$. Also, the value of η_{QP} in the final printou will be appreciably greater than unity. For the case of an acceptable solutio to the optimization problem, when (I.S.) does not exist, in the final solutio printout *all* the E_j's will be reasonably close to zero *and all* the R_k's will satisf the requirement $R_k \geq 0$ for all practical purposes with but minor infraction at most.

For proper convergence to an acceptable solution by OPTIGO, severa other characteristics will be recognized in the successive base point printout of the coarse search. These are listed below.

Characteristics of Normal Convergence by OPTIGO

1. For any group of constant η_p base point printouts, the general tren should be for decreasing δ_r value. (In review, δ_r is the "distance" betwee successive base points and it is denoted by SD in the printout, per table 6.3

2. Wall steepening factor η_p for the penalty function P of equation 6.37 hould be increasing periodically by factors of μ, at least several times during ıe coarse search. (See table 6.4 and figure 6.18 for μ.)

3. A graphical plot of any two x_i's from the base point printouts should how a general trend of convergence towards a particular point, which would e the solution point.

4. Also, as described earlier, for the final solution point to be acceptable ve should have $\eta_{QP} \approx 1.00$ and reasonably good satisfaction of the AOD quation system constraint requirements $E_j = 0$ and $R_k \geq 0$ for all of the j's nd k's.

The normal application of OPTIGO is to start with the mode (1) operation escribed in section 6.35. On rare occasions it is possible that the above haracteristics may not be evident, indicating that convergence to an acceptble solution point is not taking place. For example, the search may be oscilating between stationary points of minimum (QP) which are located very lose together. Incidentally, this was the reason for gradually reducing the tarting step size along the rays according to equation 6.55b, which should ake care of most such situations. However, if the preceding summarized haracteristics are not evident in the successive base point printouts with no onvergence to an acceptable solution obtained, the user might take action s described below.

emedial Action for Unusual Cases

1. Try operating OPTIGO with a greater number of shotgun points to tart, that is, a larger value for N_{SG}. Hence the feasible design range will be overed more densely in the selection of a good start point.

2. Change appropriate boundary values to open up the feasible design pace and alleviate constraints. This is especially appropriate if (I.S.) is uspected.

3. Try operating OPTIGO with various start points specified, in either ıode (2) or mode (3) as described in section 6.35.

4. If at all possible, try solving the problem by the MOD, which will reveal learly what the situation is and what must be done to obtain an acceptable olution to the optimization problem.

ienerally the preceding course of action will not have to be taken, since the ast majority of optimization problems will be readily solved automatically y OPTIGO in the first attempt.

XAMPLE 6.10

Pertinent base point printout data during the coarse search in the OPTIGO olution of example 6.8b are summarized in table 6.8, which illustrates

TABLE 6.8

Pertinent Items from Base Point Printouts During Convergence b OPTIGO to Solution of Example 6.8b Optimization Problem

η_p	r_{tot}	r	δ_r	x	y	Comments
1				37.50	90.00	Midrange point
	1	1		57.57	161.98	Start point from SG searc
			20.79			
	2	2		78.04	165.61	Coarse search started
			23.54			
	3	3		82.28	142.46	
			8.785			
	4	4		90.92	144.02	
			9.826			
	5	5		92.64	134.34	
			4.326			
	6	6		96.89	135.14	
			4.570			
	7	7		97.61	130.63	
			2.002			
	8	8		99.58	130.96	
			3.275			
	9	9		100.72	127.89	
			0.812			
	10	10		101.49	128.15	
			1.462			
	11	11		102.59	127.19	
			0.166			
8	12	1		102.46	127.08	Penalty walls steepened
			18.03			
	13	2		88.47	115.71	
			7.219			
	14	3		83.61	121.04	
			0.261			
	15	4		83.81	121.21	
			0.163			
64	16	1		83.69	121.10	Penalty walls steepened
			3.504			
	17	2		80.80	119.11	
			3.513			
	18	3		79.18	122.23	
			0.880			
	19	4		78.40	121.82	
			0.641			
	20	5		78.31	122.46	
			0.4267			
	21	6		77.89	122.37	
			0.406			
	22	7		77.93	122.77	
			0.3325			
	23	8		77.60	122.73	
			0.318			
	24	9		77.64	123.04	
			0.254			
	25	10		77.39	123.01	
			0.244			
	26	11		77.42	123.25	
			0.0938			
512	27	1		77.33	123.24	Penalty walls steepened
			0.452			
	28	2		76.92	123.05	
			0.522			
	29	3		76.43	123.23	
			0.168			
4096	30	1		76.47	123.39	Penalty walls steepened
			1.056			
	31	2		75.60	123.99	
			0.157			
	32	1		75.69	124.12	Coarse search completed
	⋮			⋮	⋮	Final search started
262,144	74			75.04	124.94	Solution found

specifically the previously summarized characteristics of normal convergence for the algorithm. Also, for the final solution to that problem we have from the last printout the following values.

$$E_1 = 0.03189$$

and

$$R_1 = -0.040054$$

represent only small violations of AOD constraints.

$$R_2 = +55.056$$

$$R_3 = +75.040$$

$$R_4 = +124.944$$

where $R_k \geq 0$ is acceptable for AOD equation system requirement.

$\eta_{QP} = 1.0166$, which is reasonably close to unity.

Finally, the graphical plot of (x_1, x_2) during the convergence process is given in figure 6.21, which satisfies the general convergence criterion described. Hence, from a review of the summarized characteristics, we conclude that the OPTIGO solution to example 6.8b converges normally to an acceptable optimization solution for all practical purposes. This is confirmed by comparison with the exact solution as follows, which was determined independently by the MOD.

$$x = 75.000 \text{ m}$$

$$y = 125.000 \text{ m}$$

A final characteristic of importance should be mentioned for solution of optimization problems by algorithms. Different design solution points of the same minimum Q value may be obtained under the following conditions of operation for OPTIGO in some numerical problems.

1. In mode (1) operation, using different values for N_{SG} and/or $(RG)_1$.
2. In mode (2) or (3), using different start points.

This situation is generally the result of a particular optimization problem having constant minimum Q along a boundary curve segment or locus in the feasible design range. Possibility of such a situation is readily revealed by a MOD solution to the optimization problem. Example 2.10 of Chapter 2 specifically illustrates the possibility of such a case.

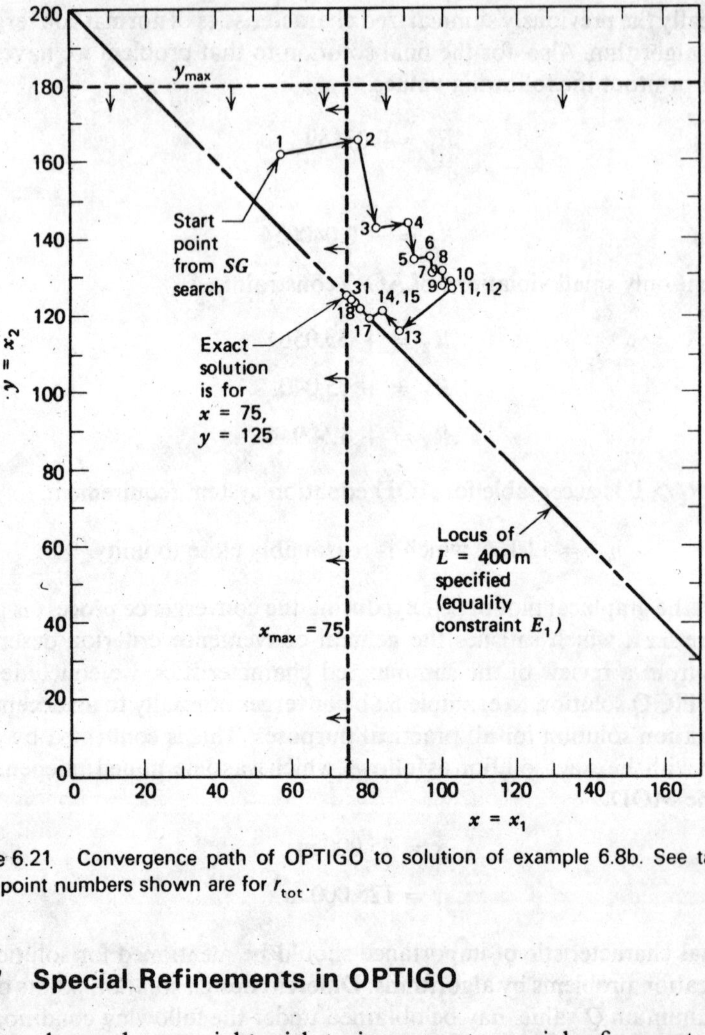

Figure 6.21 Convergence path of OPTIGO to solution of example 6.8b. See table 6.8 data; point numbers shown are for r_{tot}.

6.37 Special Refinements in OPTIGO

The advanced user might wish to incorporate special refinements in the OPTIGO listing of Appendix B. As an example, for relatively simple optimization problems with few stationary points in the QP function, a very simple modification can give greater accuracy in the solution found. Specifically this is to replace the line 136 statement with the following statement.

$$136 \quad EP = EPS * EM$$

This very minor modification only affects the final search of OPTIGO. It effect it uses the base point determined from the coarse search solution as a good start point for the final search with a reduced value for $EP = \eta_p$. This

TABLE 6.9

Comparison of Solution Accuracies for Example 6.9b Optimization Problem

Line 136 of OPTIGO	x	y	A	L
Line 136 of OPTIGO Modified: $EP = EPS * EM$	125.05	74.96	9374	400.02
Line 136 of OPTIGO Standard: $EP = EP * EM$	125.25	74.74	9361	399.81
True solution by MOD	125.00	75.00	9375	400.00

expands the search space for the final search, which for relatively simple problems can give a more accurate final solution at the expense of increased computation time. Solutions to examples 6.8 and 6.9 as previously presented were obtained using this minor modification. A comparison of final solutions for example 6.9b is given in table 6.9.

However, a word of caution should be given for the preceding modification in very complicated optimization problems, such as the example of section 8.3 in Chapter 8. For such problems, the expanded final search will appreciably increase the computation time. Also, difficulties may be encountered if the complicated QP function has many stationary points. Such difficulties can be readily recognized if the coarse search solution differs appreciably from the final search solution.

Recommendation for General Application of OPTIGO. The general recommendation is to use the standard version of OPTIGO as listed in Appendix B, with $EP = EP * EM$ as given therein. For all practical purposes, the accuracy achieved will be good enough and computation time will be reduced. If solution difficulties are recognized on rare occasions, it is generally more practical to follow the remedial action described in section 6.36, using the standard version of OPTIGO as listed in Appendix B.

REFERENCES

6.1 D. J. Wilde, *Globally Optimal Design*, Wiley-Interscience, New York, 1978, 288 pp.

6.2 G. R. Walsh, *Methods of Optimization*, Wiley, New York, 1975, 200 pp.

6.3 D. M. Burley, *Studies in Optimization*, Wiley, New York, 1974, 228 pp.

6.4 D. R. Smith, *Variational Methods in Optimization*, Prentice-Hall, Englewood Cliffs N.J. 1974, 378 pp.

6.5 R. L. Fox, *Optimization Methods for Engineering Design*, Addison-Wesley, Reading, Mass 1971, 270 pp.

6.6 R. L. Zahradnik, *Theory and Techniques of Optimization for Practicing Engineers*, Barne & Noble, New York, 1971, 326 pp.

6.7 W. H. Middendorf, *Engineering Design*, Allyn and Bacon, Boston, 1969, Chapter 9.

6.8 M. J. Forray, *Variational Calculus in Science and Engineering*, McGraw-Hill, New York 1968, 221 pp.

6.9 R. J. Duffin, E. L. Peterson, and C. Zener, *Geometric Programming—Theory and Applicatio* Wiley, New York, 1967, 278 pp.

6.10 D. J. Wilde and C. S. Beightler, *Foundations of Optimization*, Prentice-Hall, Englewoo Cliffs, N.J., 1967, 480 pp.

6.11 J. R. Dixon, *Design Engineering: Inventiveness, Analysis, and Decision Making*, McGraw Hill, New York, 1966, Chapter 12..

6.12 A. Lavi and T. P. Vogl, *Recent Advances in Optimization Techniques*, Wiley, New York 1966, 656 pp.

6.13 T. T. Woodson, *Introduction to Engineering Design*, McGraw-Hill, New York, 196 Chapter 15.

6.14 S. E. Dreyfus, *Dynamic Programming and the Calculus of Variations*, Academic Press, Ne York, 1965, 248 pp.

6.15 G. Leitmann, *Optimization Techniques with Applications to Aerospace Systems*, Academi Press, 1962, 453 pp.

6.16 R. C. Johnson, "How to Design High-Speed Cam Mechanisms for Optimum Cam an Follower Proportions," *Machine Design*, Cleveland, Jan. 26, 1956, pp. 85–89.

6.17 R. C. Johnson, "How to Design Geneva Mechanisms to Minimize Contact Stresses an Torsional Vibrations," *Machine Design*, Cleveland, March 22, 1956, pp.107–111.

6.18 R. C. Johnson, *Mechanical Design Synthesis—Creative Design and Optimization*, 2nd ec Krieger, Huntington, N.Y., 1978, 448 p.

6.19 R. C. Johnson, "Three Dimensional Variation Diagrams for Control of Calculations Optimum Design," *Transactions of the ASME*, *Journal of Engineering for Industry*, Au 1967, pp. 391–398.

6.20 J. P. Vidosic, *Elements of Design Engineering*, Ronald, New York, 1969, pp. 223, 228–23

6.21 J. N. Siddall, *Analytical Decision-Making in Engineering Design*, Prentice-Hall, Englewoc Cliffs, N.J., 1972, pp. 94, 192–195.

6.22 R. C. Johnson, "Design Synthesis," *Machine Design*, Penton, Cleveland; Part 1, Oct. 1 1973, pp. 149–155; Part 3, Dec. 13, 1973, pp. 164–169; Part 4, Dec. 27, 1973, pp. 52–57.

6.23 R. C. Johnson, "Evolution of the Method of Optimum Design," Keynote Address Symposium on Mechanical Design, University of Mexico, Mexico City, D. F., Aug., 2 1975, 17 pp.

6.24 R. C. Johnson, "A Method of Optimum Design," invited paper presented at ASME Desi Automation Conference in Chicago on Sept. 27, 1977; *ASME Journal of Mechanical Desi* in press.

6.25 J. Ellis, "Johnson's Method of Optimum Design Applied to a Problem with Simp Functional Relationships," *Computer Aided Design*, Vol. 8, No. 1, Jan. 1976, pp. 9–12.

6.26 J. Ellis, "An Introductory Problem in Mechanical Design Optimization," *Internation Journal of Mechanical Engineering Education*, Vol. 5, No. 2, 1977, pp. 121–126.

6.27 J. Ellis, "A Note on Small-Scale Design Optimization," *Transactions of the ASME, Journal of Engineering for Industry*, Ser. B, Vol. 99, No. 4, Nov. 1977, pp. 915–916.

6.28 J. Ellis, "Small Scale Design Optimization Using an Interactive Minicomputer," accepted for presentation at 1978 Winter Annual Meeting of the ASME in San Francisco in December 1978 and publication in *ASME Transactions*.

6.29 M. Pappas and J. Y. Moradi, "An Improved Direct Search Mathematical Programming Algorithm," *Transactions of the ASME, Journal of Engineering for Industry*, Nov. 1975, pp. 1305–1310.

6.30 W. T. Zangwill, *Nonlinear Programming: A Unified Approach*, Prentice-Hall, Englewood Cliffs, N.J., 1969.

6.31 D. M. Himmelblau, *Applied Nonlinear Programming*, McGraw-Hill, New York, 1972.

6.32 E. D. Eason and R. G. Fenton, "A Comparison of Numerical Optimization Methods for Engineering Design," *Transactions of the ASME, Journal of Engineering for Industry*, Ser. B, Vol. 96, No. 1, Feb. 1974, pp. 196–201.

6.33 M. Pappas, "Use of Direct Search in Automated Optimal Design," *Transactions of the ASME, Journal of Engineering for Industry*, Ser. B, Vol. 94, No. 2, May 1972, pp. 395–401.

6.34 J. N. Siddall, '*OPTISEP*' *Designers' Optimization Subroutines*, ME/71/DSN, REP 1, Faculty of Engineering, McMaster University, Hamilton, Ontario, Canada, 1971.

6.35 J. N. Siddall, *Analytical Decision-Making in Engineering Design*, Prentice-Hall, Englewood Cliffs, N.J., 1972, Appendices A–O.

6.36 J. L. Kuester and J. H. Mize, *Optimization Techniques with Fortran*, McGraw-Hill, New York, 1973, 500 pp.

6.37 D. E. Whitney, "Pattern Search," Joint Mechanical and Civil Engineering Computer Facility, M.I.T., Cambridge, Mass., April 1974, 7 pp.

6.38 R. Hooke and T. A. Jeeves, "Direct Search Solution of Numerical and Statistical Problems," *Journal of the Association for Computing Machinery*, Vol. 8, No. 2, April 1961, pp. 212–229.

6.39 G. E. Johnson and M. A. Townsend, "Nonoptimal Termination Properties of Quadratic Interpolation Univariate Searches," *The Journal of the Franklin Institute*, Vol. 306, No. 3, Sept. 1978, pp. 257–266.

6.40 G. E. Johnson and M. A. Townsend, "Numerical Investigation of Some Potential Problems of Univariate Minimization Methods," *Transactions of ASME, Journal of Mechanical Design*, in press.

6.41 J. T. Golden, *Fortran IV Programming and Computing*, Prentice-Hall, Englewood Cliffs, N.J., 1965, pp. 210–213.

6.42 *HP-25 Applications Programs*, Hewlett-Packard Co., Cupertino, Calif., 1975, p. 116.

EXERCISES

6.1. In example 6.4 of section 6.15 a tensile bar was designed for minimization of material cost C_m, and the MOD initial formulation (I.F.) was summarized therein as equations 6.5, 6.6, 6.7, 6.10, 6.13, 6.14, and 6.15. The specified items were also summarized for the optimization problem in that example.

(a) Express the preceding MOD (I.F.) as a basic AOD format equation system of the type described in section 6.19, with a criterion function, equality constraint, and regional constraints. Be sure to account for the fact that none of the variables can be negative, which was merely implied in the MOD (I.F.)

of example 6.4. (You as an engineer would automatically assume this requirement in the MOD solution process. However, computers must have requirements such as positive values for the design variables spelled out in the AOD equation system. Otherwise, according to Murphy's law, they surely would wander far into unacceptable negative value zones during the search process and a totally unacceptable solution if any might be arrived at.)

(b) Transpose the basic AOD equation system of part a into a unit based AOD equation system format of the type in section 6.20.

(c) For the unit based AOD equation system of part b, summarize the given constants expressed as G_l for $l = 1, \ldots, N_G$. Also, decide on what you will have for the calculated constants C_m and the final items F_j. Summarize the equations C_m for $m = 1, \ldots, N_C$ and F_j for $j = 1, \ldots, N_F$. Program all the functions in Fortran IV for the optimization problem and show where they go in the Appendix B listing of OPTIGO.

(d) Solve the optimization problem by OPTIGO in three ways, that is, by modes (1), (2), and (3), for the following numerical boundary values.

1. AISI 2340 steel oil quenched and drawn at 400°F (see table 4.1 for properties).
2. $c = \$0.40/\text{lbm}$; $P = 20{,}000\ \text{lbf}$; $N_y = 1.5$; $A_{\text{min}} = 0.120\ \text{in}^2$; $\Delta_{\text{max}} = 0.050\ \text{in}$; $L_{\text{min}} = 8.000\ \text{in}$; $L_{\text{max}} = 12.000$ in. Compare your solutions from OPTIGO with the MOD solution to the same problem. (See figure 6.13 for exact solution flow chart derived by MOD in example 6.4.)

6.2. A rectangular fenced yard is to be designed for maximization of enclosed area A with length L specified. The problem is solved by OPTIGO in examples 6.8 and 6.9 of section 6.35. The (I.F.) for solution by the MOD is summarized at the start of example 6.6 in section 6.19.

(a) Apply the exploratory equations of section 6.13 and derive the (F.F.) for two different solution approaches to the optimization problem by the MOD. For each case sketch the (V.S.) variation diagram and derive the calculation flow chart. Solve both of the following numerical problems by the MOD, with given boundary values as follows.

1. $x_{\text{max}} = 160$ m, $y_{\text{max}} = 180$ m, and $L = 400$ m.
2. $x_{\text{max}} = 75$ m., $y_{\text{max}} = 180$ m, and $L = 400$ m.

Compare your MOD solutions with those obtained by OPTIGO in example 6.8 of section 6.35.

(b) Change the optimization problem of part a to include a requirement $L_{\text{min}} \le L \le L_{\text{max}}$ instead of L specified.

1. Summarize the (I.F.), apply the exploratory equations, derive the (F.F.

by an approach of your choice, sketch the (V.S.) variation diagram, and derive the calculation (F.C.), all by the MOD.

2. Derive the AOD equation system normalized to the unit based format of section 6.20 and program the necessary equations in Fortran IV for solution by OPTIGO.

3. Apply the results from 1 and 2 to the solution of the following numerical problem, by both the MOD and AOD OPTIGO techniques.

$$x_{\max} = 120\ \text{m}, \qquad y_{\max} = 150\ \text{m}, \qquad L_{\min} = 300\ \text{m}, \qquad L_{\max} = 600\ \text{m}$$

Compare the numerical problem design solutions obtained by the two techniques.

6.3. Read Chapter 7 and solve the exercises at its end to gain experience with both the MOD and AOD optimization techniques.

7

Optimal Design of Simple Mechanical Elements

The purpose of this chapter is to present some illustrative examples of design optimization for some simple mechanical elements and at the same time to introduce some basic concepts and techniques of general value in applications of the MOD. A variety of optimization objectives and realistic boundary conditions will be illustrated. However, the equation systems will be simple and practical, and emphasis will be on the MOD as the basic solution technique. Also, many exercises will be presented at the end of the chapter, thereby providing ample opportunity for developing a working facility with the MOD as well as in transformation of equation systems to AOD formats for solution by an algorithm such as OPTIGO. Solution of more complicated optimization problems will be illustrated in Chapter 8 for several complex mechanical elements, with emphasis on both the MOD and AOD techniques.

7.1 Tensile Bar in Dynamic Loading

In this section we shall consider the design of a coupling stud which is to be subjected to repeated energy loadings which cannot be predicted precisely for the intended usage of the bar. Hence, the basis for optimum design will be maximum energy absorption *capability* per cycle of repeated loading, and the design must be accomplished under the existence of certain space limitations and from a limited list of available materials.

For the simple tensile bar shown in figure 7.1, specific amounts of *energy*

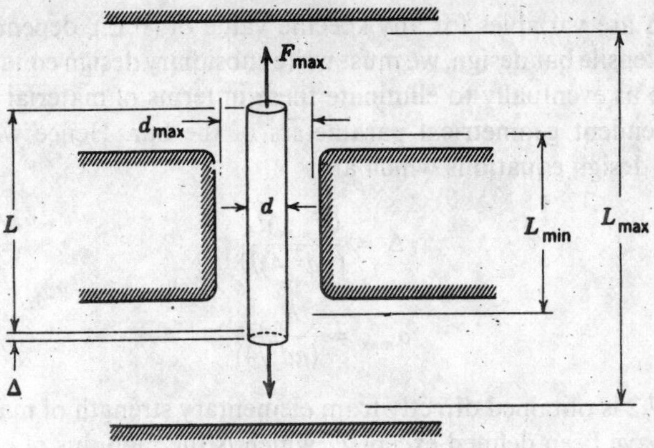

Figure 7.1 Simple tensile bar shown at instant of greatest energy absorption.

from an external source are to be absorbed, and the value for the maximum axial force F_{max} will be a variable depending on the particular design. Hence, for any specific design, assume the variation of loading typically depicted in figure 7.2.

The primary design equation will express the maximum energy absorbed per cycle of repeated loading. Therefore, if the maximum axial elongation Δ corresponds to the maximum axial force F_{max} for any specific design, from elementary mechanics the primary design equation will be equation 7.1. The reasonable assumption is made that stresses will be below the proportional limit for any good design.

$$(\text{P.E.}) = \tfrac{1}{2}(F_{max})(\Delta) \tag{7.1}$$

The objective for this problem is to arrive at a final optimum design for the tensile bar so that allowable (P.E.) will be as great as possible. Since both

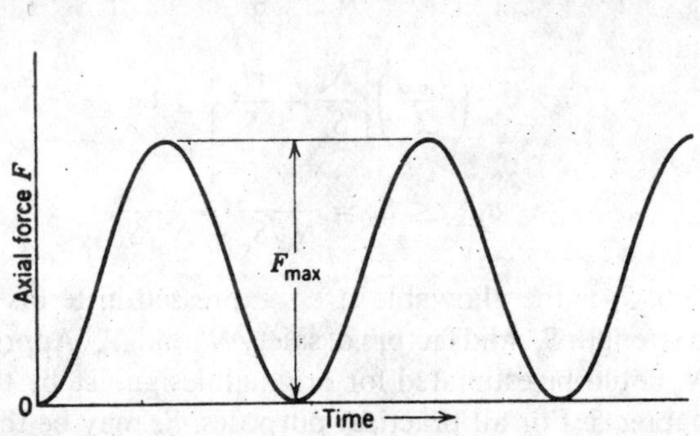

Figure 7.2 Typical variation of axial force F with respect to time for a specific feasible tensile bar design.

F_{max} and Δ are variables for any specific value of (P.E.), depending on the particular tensile bar design, we must write subsidiary design equations which will enable us eventually to eliminate these in terms of material parameters and independent geometrical parameters of the bar. Hence we have two subsidiary design equations which are

$$\Delta = \frac{(F_{max})L}{(\pi d^2/4)E} \tag{7.2}$$

$$\sigma_{max} = \frac{F_{max}}{(\pi d^2/4)} \tag{7.3}$$

Equation 7.2 is obtained directly from elementary strength of materials, and all terms have been defined except E, which is the modulus of elasticity for the bar material. Equation 7.3 expresses the maximum principal normal stress from elementary strength of materials, which is needed in the criterion for fatigue failure.

Limit equations exist on feasible values of diameter and length for the tensile bar, as depicted in figure 7.1. The limit equations are

$$d \leq d_{max} \tag{7.4}$$

$$L_{min} \leq L \leq L_{max} \tag{7.5}$$

Let us now write the limit equation for avoidance of fatigue failure. Thus we apply directly equation 4.17 of Chapter 4, using the following relations for σ which for any design will vary in a manner similar to F as shown in figure 7.2. As defined in Chapter 4, σ_a, σ_M, and σ_{max} are the amplitude, mean value, and maximum value respectively for uniaxial stress σ from a time variation standpoint of the figure 7.2 shape.

$$\sigma_a = \sigma_M = \frac{\sigma_{max}}{2}$$

$$\therefore \quad \left(\frac{\sigma_{max}}{2}\right)\left[\frac{N_e}{S_e} + \frac{N_y}{S_y}\right] \leq 1$$

$$\therefore \quad \sigma_{max} \leq S_a = \frac{2}{N_e/S_e + N_y/S_y} \tag{7.6}$$

In equation 7.6, S_a is the allowable stress expressed in terms of endurance limit S_e, yield strength S_y, and factors of safety N_e and N_y. Appropriate values for N_e and N_y could be estimated for original design use by the basic techniques of Chapter 5. For all practical purposes, S_a may be thought of as a material parameter term based on S_e and S_y generally available for materials of interest as from table 4.1 of Chapter 4.

The method of optimum design (MOD) covered in Chapter 6 will be used to solve the optimization problem now described and mathematically defined by the initial formulation (I.F.) (equations 7.1 through 7.6). For this problem we have two free variables (Δ and F_{max}) and two subsidiary design equations (7.2 and 7.3). Thus as explained in section 6.13 of Chapter 6, with $n_f = N_s = 2$ in the (I.F.), we have a case of normal specifications. Hence, in the MOD the (I.F.) will be reduced to a single developed (P.D.E.) from which we can draw the design conclusions for maximizing allowable (P.E.) of the tensile bar.

First, we shall eliminate the unspecified and unlimited variable Δ in primary design equation 7.1 by combining it with subsidiary design equation 7.2.

$$(\text{P.E.}) = \tfrac{1}{2}(F_{max})(\Delta) = \tfrac{1}{2}(F_{max})\left[\frac{(F_{max})L}{(\pi d^2/4)E}\right]$$

$$\therefore \quad (\text{P.E.}) = \frac{2F_{max}^2 L}{\pi d^2 E} \tag{a}$$

Next, we shall eliminate the unspecified variable F_{max} in primary design equation a by combining it with subsidiary design equation 7.3.

$$(\text{P.E.}) = \frac{2[(\pi d^2/4)\sigma_{max}]^2 L}{\pi d^2 E}$$

$$\therefore \quad (\text{P.E.}) = \frac{\pi}{8}\left[\frac{\sigma_{max}^2}{E}\right](d^2 L) \tag{b}$$

At this point it is important to realize that in equation b, which is the developed (P.D.E.), L, d, and σ_{max} are *independent* parameters which can now be placed equal to their upper limits from equations 7.4, 7.5, and 7.6. From equation b, this is desirable in order to obtain the greatest possible energy storage *capability* for the tensile bar. Thus equation b becomes equation 7.7, which follows; it is purposely grouped into functional requirement parameters and material parameters.

$$\max\,(\text{P.E.}) = \frac{\pi}{8}\, d_{max}^2 L_{max} \left(\frac{S_a{}^2}{E}\right) \tag{7.7}$$

Summarizing, the optimum design of the tensile bar based on maximum energy absorption *capability* per cycle of repeated loading is as follows.

Optimum Material. From the list of available materials select the material having the highest value for the material selection factor, which from equation 7.7 is $(S_a{}^2/E)$. S_a is defined by equation 7.6 in terms of readily available material properties S_e and S_y as typically found in table 4.1 of Chapter 4.

Optimum Geometry. $d = d_{max}$; $L = L_{max}$.

An important point should be observed with respect to the general method of optimum design. This point concerns the placing of variables at their limit values in the primary design equation development procedure. Hence in equation a we might erroneously say that the geometrical parameter d should be made as small as possible in order to have (P.E.) as large as possible. However, this would be a fallacy since, in equation a, subsidiary design equation 7.3 and limit equation 7.6 have not been accounted for. In other words, in equation a the variables F_{max} and d are dependent on each other according to equations 7.3 and 7.6, and we cannot arbitrarily let either variable go to a specific value in the (P.D.E.), equation a, at that stage. Similarly, it would be erroneous to select the optimum material from equation a based on smallest value for E. This would be a fallacy also, since in equation a the variable F_{max} for any specific value of (P.E.) and for any geometry would be limited by properties of the material related by equations 7.3 and 7.6 which cannot be ignored in selecting the optimum material. However, in equation b, σ_{max}, d, and L can be varied *independent* of each other and it is permissible at that stage to place them equal to their appropriate limit values.

7.2 Torsion in Simple Cylindrical Bars

We shall begin a study of the optimum design of torsion shafts by considering a simple cylindrical bar in this section and the next. Then we shall extend the study in later sections to include the effects of threaded portions, stepped shafts, keyways, and other real conditions which are encountered in practical torsion shafts.

A simple cylindrical bar, shown in figure 7.3, is transmitting a twisting moment M_t, which would be caused by an applied external couple. The geometry of the simple cylindrical bar would be uniquely defined by the independent geometrical parameters l and d shown in the figure. From elementary studies in the theory of elasticity [7.1] it can be shown that the circular bar at any point is in a biaxial state of pure shearing stress, which varies in magnitude directly as the distance from the geometric axis of the bar. The

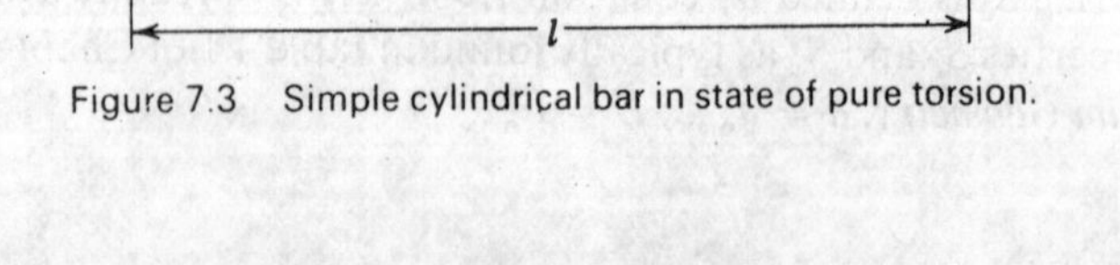

Figure 7.3 Simple cylindrical bar in state of pure torsion.

maximum shearing stress occurs at the outermost fibers of the bar, and its magnitude would be

$$\tau_{\max} = \frac{16M_t}{\pi d^3} \tag{7.8}$$

Also, from elementary studies in the theory of elasticity [7.1] it can be shown that cross sections of the circular bar remain plane and rotate without distortion during loading. Hence straight lines scribed on the end planes of the bar remain straight after loading, and the angle of twist θ for the entire bar would be given by

$$\theta = \frac{32lM_t}{\pi d^4 G} \tag{7.9}$$

For equations 7.8 and 7.9, $\tau_{\max}$ is the maximum shearing stress in pounds per square inch; M_t is the applied twisting moment, in inch-pounds; d is the shaft diameter, in inches; l is the shaft length, in inches; G is the modulus of rigidity for the material, in pounds per square inch; and θ is the angle of twist for the shaft, in radians.

For a biaxial state of pure torsion at any point it is well known that the principal normal stresses are of opposite sign and they are both of the same magnitude as the shearing stress. Also, the planes upon which the principal normal stresses act are inclined at 45° to the planes upon which the pure shearing stresses act. Hence compressive stress trajectories would hypothetically be wrapped around the cylindrical bar at a helix angle of 45°, as illustrated in figure 7.3. From the theory of elastic stability [7.2], it can be shown that a cylindrical bar will buckle if the twisting moment exceeds a critical value, depending on the end conditions. This is analogous to the buckling of slender columns where the compressive stress trajectories are in the direction of the column axis. Thus, if the cylindrical bar in pure torsion is assumed to have pinned ends which are free for rotation about axes perpendicular to the shaft but fixed against translation in directions normal to the shaft, buckling will be avoided if equation 7.10 is satisfied. Of course, this equation is conservative if the ends are not entirely free for rotation about axes perpendicular to the shaft axis.

$$\frac{32}{\pi}\frac{M_t}{Ed^4} \leq \frac{\pi}{l}$$

or

$$M_t \leq \frac{\pi^2 Ed^4}{32l} \tag{7.10}$$

For most practical cases, the danger of buckling because of twisting moment in a shaft is insignificant. However, it is an undesirable effect which occasionally is significant and it should not be completely ignored for all problems.

7.3 Simple Torsion Bar for Minimum Weight

In this section we shall consider the optimum design of a simple cylindrical torsion bar on the basis of minimum weight, for the case of normal specifications. We shall assume a constant twisting moment in the shaft, and hence we shall use a static criterion for failure. In later sections of this chapter we shall progress to practical shafts having irregularities such as threaded sections and keyways which are subjected to variable twisting moments requiring a criterion for failure based on fatigue. For these later studies we shall consider various bases of optimum design and cases of redundant specifications.

First, considering the problem of this section in general, suppose that the cylindrical bar in figure 7.3 is to transmit a virtually constant twisting moment M_t which is a specified functional requirement value. Also, suppose an analysis of the entire machine has shown that the shaft should have a specified value for its torsional rigidity k, which by definition is merely M_t/θ. For the design of the bar, assume that our choice of materials is confined to those whose properties are given in tables 4.1 and 4.2, and assume that the independent geometrical parameters d and l are *not* limited to specific ranges. The torsion bar is to be designed for use in a machine where weight is very significant. Hence the problem is to design the bar on the basis of *minimum weight*.

Following the procedural steps in section 6.14 of Chapter 6, having sketched the bar in figure 7.3 we shall write the primary design equation, which for this problem expresses the weight of the bar. Hence the (P.D.E.) is

$$W = w\left(\frac{\pi d^2}{4}\right)l \tag{7.11}$$

On the right side of equation 7.11, there are no functional requirement parameters, the specific weight w is the only material parameter, and the independent geometrical parameters are d and l.

The next step will be to write the subsidiary design equations, and for this problem we have two. Equation 7.12, derived from equation 7.9, expresses the specified functional requirement k, and equation 7.8 expresses the implied undesirable effect τ_{max}. Hence the subsidiary design equations are

$$\frac{M_t}{\theta} = k = \frac{\pi d^4 G}{32l} \tag{7.12}$$

$$\tau_{max} = \frac{16M_t}{\pi d^3} \tag{7.8}$$

The next step will be to write *all* significant limit equations, and for this problem we shall assume that the only one is for stress. Since the number of cycles of load application are considered to be relatively small, a static criterion of failure will be appropriate. Based on the review in section 4.7 of Chapter 4, three failure criteria will be of interest for application. First, for pure torsion in relatively brittle materials, the maximum normal stress theory would be most appropriate but least conservative. Hence, equation 4.6 would give the following requirement for an acceptable design.

$$\tau_{max} = \sigma \leq \frac{S_y}{N_y} \tag{7.13a}$$

Second, for pure torsion in relatively ductile materials, the distortion energy theory would probably be most accurate according to figure 4.4. Hence equation 4.11 would give the following requirement for an acceptable design, since for pure torsion we have $\tau_{max} = \sigma_1 = -\sigma_2$.

$$\tau_{max} \leq 0.577 \frac{S_y}{N_y} \tag{7.13b}$$

Finally, if we wish to be most conservative and still reasonably accurate, for ductile materials in pure torsion the maximum shearing s ress theory would be appropriate. Thus by equation 4.10 we would require the following relationship for an acceptable design.

$$\tau_{max} \leq 0.5 \frac{S_y}{N_y} \tag{7.13c}$$

Hence, from review of equations 7.13a, 7.13b, and 7.13c, in general we can write the following requirement for an acceptable design for the case of pure torsion.

$$\tau_{max} \leq \frac{C_f S_y}{N_y} \tag{7.13}$$

In equation 7.13, τ_{max} is the maximum shearing stress in the bar, S_y is the tensile yield strength readily available for most materials such as in table 4.1, N_y is the factor of safety selected as in Chapter 5, and C_f is a failure theory coefficient depending primarily on the type of material. From the preceding equations, the appropriate values for C_f would be as follows.

C_f	Type of Material
1.00	Brittle
0.577	Ductile (generally most accurate)
0.5	Ductile (most conservative)

Incidentally, a measure of ductility for a material is its elongation, as given in table 4.1. For elongation greater than 10% the material is generally classified as ductile, whereas for elongation less than about 5% the material is considered to be in the brittle zone. For the transitional region of 5 to 10% elongation, for this application it would be most conservative to use the ductile classification.

The next step in the MOD will be to combine all of the (S.D.E.)'s with the (P.D.E.) by eliminating an unlimited and unspecified independent geometrical parameter from the (P.D.E.) for each combination. Hence, using the (S.D.E.), equation 7.12, to eliminate l in the (P.D.E.), equation 7.11, we obtain

$$W = w\left(\frac{\pi d^2}{4}\right)\left(\frac{\pi d^4 G}{32k}\right) = \left(\frac{\pi^2}{128}\right)\left(\frac{wG}{k}\right)(d^6)$$

Now using the other (S.D.E.), equation 7.8, it is possible to eliminate d in the (P.D.E.) just derived. Hence

$$W = \left(\frac{\pi^2}{128}\right)\left(\frac{wG}{k}\right)\left(\frac{16^2 M_t^{\,2}}{\pi^2 \tau_{\max}^2}\right)$$

Therefore

$$W = 2\left(\frac{M_t^{\,2}}{k}\right)(wG)\left(\frac{1}{\tau_{\max}}\right)^2 \tag{a}$$

Since it was possible to combine *all* the (S.D.E.)'s with the (P.D.E.) without either eliminating or ignoring a limited or specified parameter, we have the case of normal specifications discussed in Chapter 6.

At this point we have developed the (P.D.E.) to equation a, and in that equation we see that the optimization quantity W varies inversely as the square of the independent parameter $\tau_{\max}$. Hence it is obvious that weight W is minimized if $\tau_{\max}$ has its greatest permissible value, which is $C_f S_y/N_y$ according to equation 7.13. Hence, with $\tau_{\max}$ at that value in equation a, the final developed (P.D.E.) is derived as equation 7.14. It has been grouped purposely into the two bracketed terms of functional requirement parameters and material parameters. Incidentally, it just happens that there are no geometrical parameters in the final (P.D.E.).

$$W = 2\left[\frac{(M_t N_y)^2}{k}\right]\left[\frac{wG}{(C_f S_y)^2}\right] \tag{7.14}$$

The final (P.D.E.), equation 7.14, includes the effects of all subsidiary design equations and all significant limit equations, and from it we can at once extract our optimum design. We shall assume that our functional requirement parameters M_t, N_y, and k are specified values which cannot be changed. However, if it were possible to change them, the main objective would be to reduce the value of the *group* of functional requirement parameters, which is the first bracket in equation 7.14. The independent *group* of material parameters is the second bracket in equation 7.14, and this is the *material selection factor* (MSF). Hence, in selecting the *optimum* material, the main objective is to select the *available feasible* one having the *smallest* value for this factor, which from equation 7.14 is $[wG/(C_f S_y)^2]$. Since the MSF group is independent of the functional requirement parameter group in equation 7.14, the optimum material will be the same *regardless* of specific values for M_t, k, and N_y.

Once the optimum material has been selected from the list of available feasible materials using the MSF, the optimum values for the independent geometrical parameters can be calculated directly from the original subsidiary design equations. Hence the diameter d for the bar is determined by using equation 7.8 with $\tau_{max} = C_f S_y/N_y$. Then equation 7.12 would be used to determine the length l for the bar. After the validity of the assumption of no buckling has been verified by using equation 7.10, the optimum design of the torsion bar can be considered complete. The design so obtained will satisfy the specified functional requirements of transmitting the torque M_t with a torque gradient of k, the important undesirable effect stress will be within tolerable limits, and the bar will have the *lowest possible* weight. If the buckling check is not satisfied, a case of redundant specifications will exist and a second approach should be tried with the MOD.

Regarding the buckling possibility, it is interesting to notice that the following equation 7.10a is obtained by combining equation 7.12 with equation 7.10, which if satisfied would assure us of no buckling.

$$\pi\left(\frac{k}{M_t}\right) \geq \left(\frac{G}{E}\right) \tag{7.10a}$$

This equation is a relationship between an independent functional requirement parameter group (k/M_t) and a material parameter group (G/E). Furthermore, the material parameter group can be simplified more by recalling the relationship between modulus of rigidity G and modulus of elasticity E from elementary strength of materials. This relationship is merely $G = E/2(1 + \nu)$, where ν is Poisson's ratio for the material. Combining this with equation 7.10a, we obtain the following equation, which, if satisfied, will assure us of no buckling.

$$\left(\frac{k}{M_t}\right) \geq \frac{1}{2\pi(1 + \nu)} \tag{7.10b}$$

However, for any metal Poisson's ratio is approximately $^1/_4$ or possibly $^1/_3$, and this range of variation would change the right side of equation 7.10b only very little. Hence, from equation 7.10b, we can conclude that the possibility of buckling of a torsion bar for *any* metal is primarily a function of the value of the ratio (k/M_t), a functional requirement parameter group. Also, for problems such as this one where both M_t and k are specified values, we can apply equation 7.10b to check for buckling at the very start. If buckling will occur, we must either change the design to provide for lateral bracing or change the functional requirements to obtain an acceptable value for (k/M_t) in equation 7.10b.

7.4 Practical Deviations from Simple Geometry

In practice, one of the fundamental purposes of a torsion shaft is to *transmit* a twisting moment *between* other mechanical elements. Hence all *real* torsion shafts are characterized by *regions of connection* to other mechanical elements, as with splines, keys, taper pins, and threaded sections. Also, a torsion shaft must be compatible with the entire mechanical structure wherein it exists; hence the necessity for stepped shafts, grooves, and holes is introduced. Thus real torsion shafts *cannot* be simple cylindrical bars as discussed in section 7.2.

The introduction of geometrical deviations to a simple cylindrical shaft greatly complicates the distribution of stresses. Geometrical deviations are characterized by appreciable increases of stress at highly localized regions, the increases being above the nominal values as calculated with elementary theory for a simple cylinder. Relatively simple deviations from simple cylinders, such as general *prismatical* bars and shafts with variable *diameter*, have been studied in great detail using the theory of elasticity. Fortunately, for many specific geometrical deviations significant results for *design* usage are available in the form of published stress concentration factors, as in references 7.3 through 7.7.

The term *stress concentration factor* is defined as the ratio of the actual *maximum stress* to a corresponding *nominal stress*, where the nominal stress

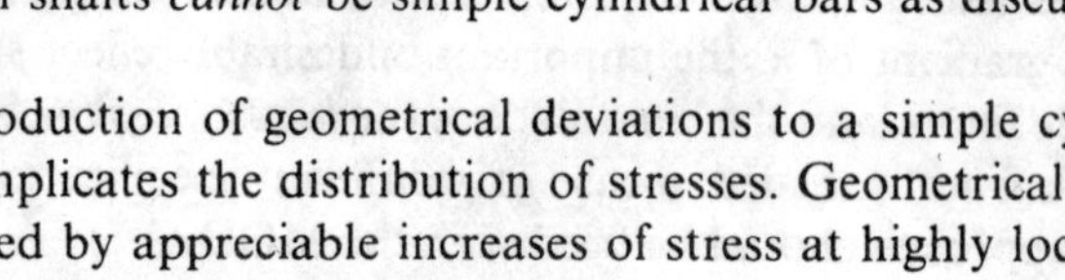

Figure 7.4 Typical part of a real torsion shaft showing filleted shoulder and stepped shaft introduced in design as a geometrical deviation to simple cylinder shown in phantom. $(r/d) = 0.20$; $(D/d) = 2$; $K_{ts} = 1.26$; $K_t = 10.1$.

is referred to a particular arbitrarily chosen section. The value for the nominal stress is calculated on the assumption of a hypothetical elementary stress distribution with respect to the particular specified reference area. Thus, using the published stress concentration factor, the particular value for the *actual* maximum stress of significance can be calculated. Hence considering a stepped torsion shaft having the specific proportions given in figure 7.4, the stress concentration factor K_{ts} would be 1.26, from published graphs such as figure 67 of reference 7.3, and the nominal stress would be calculated using equation 7.8. Therefore, for this particular example, the actual maximum shearing stress would be

$$\tau_{max} = 1.26\left(\frac{16M_t}{\pi d^3}\right) \tag{a}$$

From this simple example in *analysis*, the great value for published stress concentration factors is quite apparent.

In original *design* work, the engineer will find it necessary to *introduce* geometrical deviations to mechanical elements for reasons such as those mentioned in the first paragraph of this section. Hence, in order to make a *fair* evaluation for the *absolute effect* of the *introduced* geometrical deviation, the resulting actual maximum shearing stress should be compared with the nominal maximum shearing stress which would exist in a *hypothetical* simple cylindrical shaft.

For general *design* usage, we shall define the term *stress increase factor* K_i as the ratio of actual *maximum stress* in the mechanical element having the geometrical deviation to a corresponding *nominal stress*, where the nominal stress is calculated using elementary theory referred to a hypothetical mechanical element having *no* geometrical deviation. Thus the resulting *penalty* paid for *introducing* the geometrical deviation in *design* will be truly evaluated in the proper light. Also, when choosing the particular *type* of geometrical deviation to introduce in design for a mechanical element for satisfying certain functional requirements, a fair comparison will be made by considering *stress increase factors*, which are referred to the same hypothetical simple element.

For the specific example in figure 7.4, suppose that in *design* we must introduce the stepped portion because of existing space restrictions in the surrounding structure. Hence, as depicted in the figure, the hypothetical simple element would be the cylinder of diameter D. The actual maximum shearing stress is given by equation a, and the nominal maximum shearing stress is $(16M_t/\pi D^3)$ by equation 7.8. Thus, since $D = 2d$ for this specific example, by definition the *stress* increase factor K_i is

$$K_i = \frac{1.26(16M_t/\pi d^3)}{(16M_t/\pi D^3)} = 1.26(2)^3 = 10.1$$

Therefore, for original *design* work, when considering the effect of introducing the geometrical deviation shown in figure 7.4, we might better think of the resulting *stress increase factor* $K_i = 10.1$ rather than the *stress concentration factor* $K_{ts} = 1.26$.

There are many cases where the *stress increase factor* has exactly the same value as the *stress concentration factor*. This is so whenever the nominal stress upon which each is based *happens* to be the same. However, the importance of thinking in terms of *stress increase factors* for original *design* work cannot be overemphasized, and we shall now illustrate this further with another example.

In figure 7.5 is a sketch of a typical practical shaft showing several geometrically deviated regions where stress concentration effects are significant. Suppose that the transverse hole in figure 7.5 is to be introduced to the shaft in order to satisfy some functional requirement, but that we have some freedom in selecting the value for the diameter *a*. Fortunately, for this kind of a problem stress concentration factors have been determined and published, and the curve in figure 7.6 is repeated from reference 7.3. From a cursory look at figure 7.6, the design engineer might erroneously decide to select the value for *a* so that the ratio (a/d) is as large as practical, because in so doing the *stress concentration factor* will be as small as possible. However, the fallacy is quite evident when the *stress increase factor* for this case is taken into account. By definition the *stress increase factor* $(K_i)_B$, because of the hole in figure 7.5, would relate the actual maximum shearing stress τ_B, at a point in the shaft

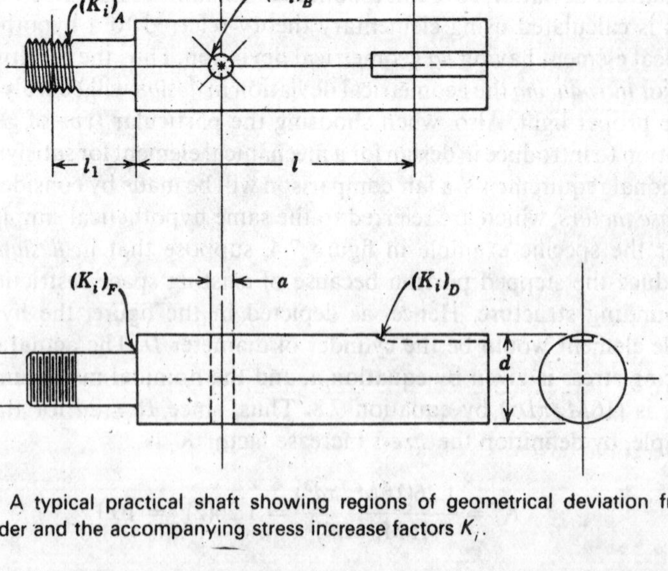

Figure 7.5 A typical practical shaft showing regions of geometrical deviation from a simple cylinder and the accompanying stress increase factors K_i.

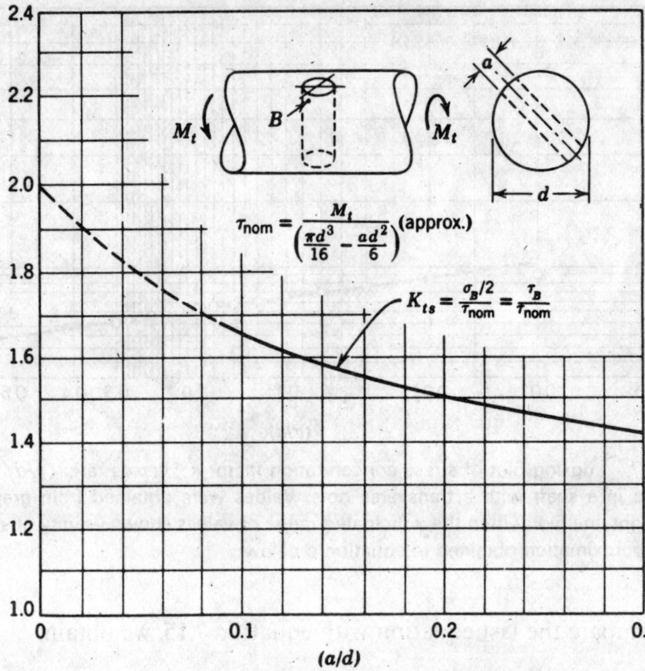

Figure 7.6 Shear stress concentration factor K_{ts} for the case of torsion in a shaft with a transverse hole (determined by strain method of Thum and Kirmser). (Taken from figure 89 of reference 7.3. By permission of John Wiley and Sons, Inc.)

corresponding to B in figure 7.6, to the nominal maximum shearing stress as given by equation 7.8 in a hypothetical simple cylindrical shaft of diameter d. The relationship is

$$\tau_B = (K_i)_B\left(\frac{16M_t}{\pi d^3}\right) \tag{7.15}$$

But, from figure 7.6,

$$\tau_B = (K_{ts})\tau_{nom} = K_{ts}\left(\frac{M_t}{\pi d^3/16 - ad^2/6}\right)$$

where K_{ts} is the theoretical *stress concentration factor*. If we multiply numerator and denominator of the right side of the last equation by $16/\pi d^3$ and regroup some terms, we obtain

$$\tau_B = \frac{K_{ts}}{[1 - (8/3\pi)(a/d)]}\left(\frac{16M_t}{\pi d^3}\right)$$

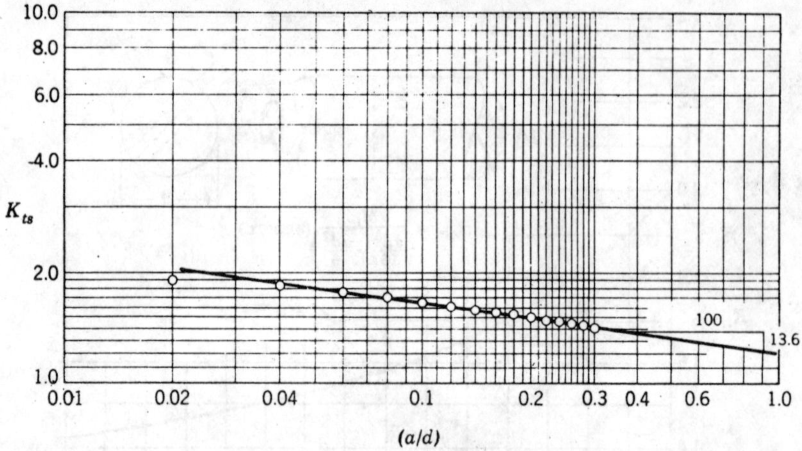

Figure 7.7 Log-log plot of stress concentration factor K_{ts} versus ratio (a/d) for the case of torsion in a shaft with a transverse hole. Values were obtained from graph in figure 7.6. Straight line plot within the anticipated range of values shows validity of simple exponential approximation obtained in equation c below.

If we compare the last equation with equation 7.15, we obtain

$$(K_i)_B = \frac{K_{ts}}{[1 - (8/3\pi)(a/d)]} \tag{b}$$

To determine explicitly how the stress increase factor $(K_i)_B$ varies with (a/d), it is necessary to express the stress concentration factor K_{ts} as a mathematical function of (a/d). We shall do this by a simple exponential equation, as discussed in Chapter 2, from the published graph of figure 7.6. A plot of K_{ts} versus (a/d) from figure 7.6 on log-log graph paper resulted in figure 7.7, which obviously is nearly a perfect straight line. Hence the simple exponential approximation is appropriate and for (K_{ts}) it results in

$$K_{ts} = \frac{1.20}{(a/d)^{0.136}} \tag{c}$$

Thus our stress increase factor $(K_i)_B$ can be expressed explicitly by combining equation b with c:

$$(K_i)_B = \frac{1.20}{(a/d)^{0.136}[1 - (8/3\pi)(a/d)]} \tag{7.16}$$

Without the hole, the maximum shearing stress in the shaft would be given by equation 7.8. Because of the hole, the actual maximum shearing stress is

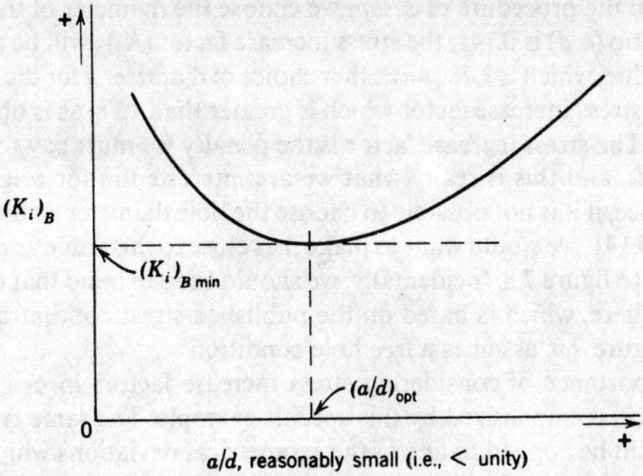

Figure 7.8 Rough free-hand sketch of stress increase factor $(K_i)_B$ versus a/d obvious from equation 7.16 for a shaft with a transverse hole as in figure 7.5, assumed to be subjected to pure torsion load.

greater than such a nominal value by the *stress increase factor* $(K_i)_B$, according to equation 7.15. Hence our objective in *selecting* the value for the hole diameter a should be, if possible, to minimize the *stress increase factor* $(K_i)_B$. From equation 7.16, we see that the stress increase factor $(K_i)_B$ has a minimum value at a certain optimum value for the ratio (a/d). This is obvious from a sketch of the variation of $(K_i)_B$ versus (a/d) from equation 7.16, as shown in figure 7.8. To find the minimum value of $(K_i)_B$ in figure 7.8, we shall take the derivative of $(K_i)_B$ with respect to (a/d) in equation 7.16 and equate it to zero; for the sake of simplicity we need to consider only the denominator in that equation. Hence

$$\frac{\partial\{(a/d)^{0.136}[1-(8/3\pi)(a/d)]\}}{\partial(a/d)} =$$

$$\left(\frac{a}{d}\right)^{0.136}\left(-\frac{8}{3\pi}\right)+\left[1-\frac{8}{3\pi}\left(\frac{a}{d}\right)\right](0.136)\left(\frac{a}{d}\right)^{0.136-1} = 0$$

$$\therefore \quad \left(\frac{a}{d}\right)_{\text{opt}} = 0.141$$

For this optimum value of (a/d), the stress increase factor $(K_i)_B$ from equation 7.16 would be

$$(K_i)_{B\,\text{min}} = \frac{1.20}{(0.141)^{0.136}[1-(8/3\pi)(0.141)]}$$

$$\therefore \quad (K_i)_{B\,\text{min}} = 1.78$$

Hence, if in the procedure of *design* we choose the diameter of the hole such that the ratio (a/d) is 0.141, the stress increase factor $(K_i)_B$ will be at its *lowest possible* value, which is 1.78. *Any* other choice of diameter a for the hole would result in a stress increase factor which is greater than 1.78, as is obvious from figure 7.8. The stress increase factor is the penalty we must pay to put a hole in the shaft, and this is really what we are interested in for original design work. Hence, if it is not possible to choose the hole diameter so the ratio (a/d) is exactly 0.141, we would want to make it as close to this value as permissible according to figure 7.8. Incidentally, we should keep in mind that the analysis presented here, which is based on the published stress concentration factor curve in figure 7.6, assumes a *free* hole condition.

The importance of considering stress increase factors in original design work has been emphasized by this specific example. The same type of viewpoint should be applied to any of the geometrical deviations which are commonly necessary in the *design* of real mechanical elements. The primary objective when selecting a connection or introducing any geometrical deviation to an ideal cylindrical torsion shaft should be minimization of the *stress increase factor*. Hence, with this objective in mind, we would find that it would be worth while to select an Aero Stud type of screw thread instead of the U.S. Standard. Also, it would be better to choose a sled runner type of keyway instead of the profile type.

For many cases, stress increase factors will be functions really of *ratios* which are *independent* of other geometrical parameters in the design. This was the case for the transverse hole in the cylindrical bar, where the stress increase factor $(K_i)_B$ was a function of the *independent ratio* (a/d) (equation 7.16). Obviously the ratio (a/d) is independent of the other geometrical parameters for the remainder of the torsion shaft, such as the diameter d. This is also generally true for standard proportion screw threads, keys and other practical geometrical deviations. For such cases, it will be possible for us to introduce merely the symbol K_i for *stress increase factor* where appropriate in our design equations and its value will depend on the proportions chosen for the geometric deviation.

7.5 Practical Torsion Bar for Minimum Weight

In this section we shall study the optimum design of a real torsion shaft on the basis of *minimum weight*. Assume that the shaft is to connect pure torque-transmitting elements in a machine setting as typically depicted in figure 7.9. From considerations of space available as depicted in the figure we have estimated the limits L_{max}, L_{min}, and d_{max} on feasible values of length l and diameter d respectively. Also, from an analysis of the entire machine,

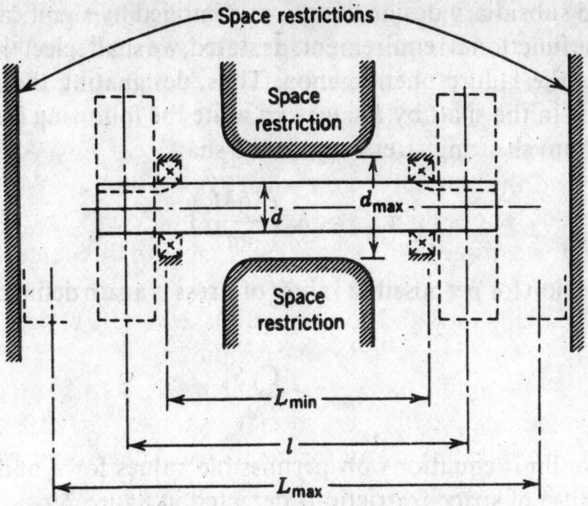

Figure 7.9 Schematic diagram of a feasible torsion shaft design shown in machine setting with typical space restrictions. Pure torque-transmitting elements and mounting bearings are shown in phantom.

it is known that the shaft must transmit a virtually constant twisting moment M_t and have a torque gradient k, both of which must be specific values for proper functioning of the system. We shall assume that the shaft will have a keyed connection at each end as depicted in figure 7.9, and, for the particular application, yielding of the material must be avoided because loosening of the keys cannot be tolerated. Also, assume that the available list of materials is that given as table 4.1 in Chapter 4.

First, we shall write the primary design equation, which will be the equation for expressing shaft weight, and for all practical purposes it will be identical with equation 7.11:

$$W = w\left(\frac{\pi}{4}d^2\right)l \tag{7.11}$$

One of the subsidiary design equations will express torque gradient k, since it is assumed to be a specified value. Again, for all practical purposes it will be identical with equation 7.12:

$$k = \frac{\pi d^4 G}{32l} \tag{7.12}$$

It will be permissible to ignore stress concentration effects from the torque gradient equation since the stress concentrations are assumed to be highly localized and the integrated effect on torque gradient for the whole shaft would be small.

The second subsidiary design equation is implied by significant stress. On the basis of the functional requirements as stated, we shall select the occurrence of yielding as the failure phenomenon. Thus, designating the *largest* stress increase factor in the shaft by K_i, we can write the following (S.D.E.) for the actual maximum shearing stress τ_{max} in the shaft.

$$\tau_{max} = K_i\left(\frac{16M_t}{\pi d^3}\right) \tag{7.17}$$

The limit equation for permissible values of stress is again defined by equation 7.13.

$$\tau_{max} \leq \frac{C_f S_y}{N_y} \tag{7.13}$$

Next, we have limit equations on permissible values for d and l which are imposed because of space restrictions depicted in figure 7.9.

$$d \leq d_{max} \tag{7.18}$$

$$L_{min} \leq l \leq L_{max} \tag{7.19}$$

We shall assume that the possibilities of buckling can be ignored in the optimization study, based on satisfaction of equation 7.10b for the specified M_t and k values.

Solution to the optimization problem now described will be determined by the MOD. Reviewing the (I.F.) equation system summarized above, we readily determine the following. (See Chapter 6 and Appendix A for nomenclature.)

$$n_f = 0; \qquad N_s = 2$$

Thus, with $n_f < N_s$ we have the (R.S.) type of optimization problem.

$$n_c = 3 \text{ (for } d, l, \text{ and } \tau_{max});$$

$$\therefore \quad n_v = n_f + n_c = 3$$

Hence, applying exploratory equations 6.1 through 6.4 from Chapter 6, we predict for execution of the MOD the following characteristics.

$$D_{vs} = n_v - N_s + 1 = 3 - 2 + 1 = 2$$

$$A_t = \frac{n_c!}{[n_e!(n_c - n_e)!]} = \frac{3 \times 2}{[2 \times (1)]} = 3$$

where $n_e = N_s - n_f = 2 - 0 = 2$.

$$n_r = n_c - n_e = 3 - 2 = 1$$

Thus very simply we have learned that the (V.S.) will be two dimensional in character, with one related parameter and two eliminated parameters. Also, there are three approaches theoretically possible for derivation of final formulations, namely choosing either d, l, or τ_{max} as the related parameter with the other two being the eliminated parameters.

As illustrated in examples 6.2 and 6.3 of Chapter 6, for two dimensional problems involving simple exponential equations, we can execute the MOD by a sequential set of two dimensional log-log straight line graphs, which clearly presents the final solution on the last one. In a brief and general review of this procedure, the first step is to derive the (F.F.) for the selected approach in the MOD. Then, the first log-log graph will present optimization quantity Q versus the related parameter from equation I of the (F.F.), imposing the related parameter constraints which determine limits on Q to be carried over to the next graph. Then the related parameter is traded for first eliminated parameter using equation II combined with I. Hence a second log-log graph is determined with effects of constraints on the particular eliminated parameter imposed and the corresponding limits on Q determined. The procedure is successively applied until all eliminated parameters are expended, with equations II, III, and so on having been used successively in combination with equation I for trading of parameters by equation combination. Therefore, with the corresponding limits on Q carried over from one graph to another in the successive procedure, the final graph clearly presents the overall solution to the numerical optimization problem. The final presentation includes all feasible materials of interest as well as the effects of all specifications and limits in the (I.F.). Let us illustrate the general technique applied to the practical torsion bar problem for weight minimization as described earlier.

Suppose for the specific problem described at the start of this section we choose τ_{max} as the related parameter with d and l as the eliminated parameters. Thus by the MOD we derive the (F.F.) for equations I, II, and III.

Start with the (P.D.E.), equation 7.11, and combine a (S.D.E.), equation 7.12, with it by eliminating geometrical parameter l. In so doing, we shall temporarily ignore limit equation 7.19 on permissible values for l. Hence, we obtain

$$W = w\left(\frac{\pi d^2}{4}\right)l = w\left(\frac{\pi d^2}{4}\right)\left(\frac{\pi d^4 G}{32k}\right)$$

$$W = \left(\frac{\pi^2}{128}\right)\left(\frac{wG}{k}\right)d^6 \qquad \text{(a)}$$

Next, combine the other (S.D.E.), equation 7.17, with the (P.D.E.), equation a, by eliminating d, which will be ignoring temporarily limit equation 7.18

existing on permissible values for d. Hence we have

$$W = \left(\frac{\pi^2}{128}\right)\left(\frac{wG}{k}\right)\left(\frac{K_i^2 256 M_t^2}{\pi^2 \tau_{\max}^2}\right)$$

$$W = 2\left(\frac{M_t^2}{k}\right)(wG)\left(\frac{K_i}{\tau_{\max}}\right)^2 \tag{I}$$

Equation I as it now stands is the developed (P.D.E.) of the ideal problem as defined in Chapter 6. For that equation, realize that both M_t and k are specified values of functional requirements, (wG) is a material characteristic and, as discussed in the previous section, K_i is dependent on proportional design for any value of d, as evident from figure 7.10. The two relating equations of the (F.F.) are derived from equations 7.17 and 7.12 as follows.

$$d = \left[\frac{16K_i M_t}{(\pi\tau_{\max})}\right]^{1/3} \tag{II}$$

$$l = \frac{\pi d^4 G}{(32k)}$$

$$= \frac{\pi[16K_i M_t/(\pi\tau_{\max})]^{4/3} G}{(32k)}$$

$$l = \frac{[(16/\pi)^{1/3}(K_i M_t)^{4/3}](G/2k)}{(\tau_{\max})^{4/3}} \tag{III}$$

By the described procedure of successive graphs, we would first look at equation I for W versus $\tau_{\max}$, expressed in logarithmic form as follows.

$$\log W = \log\left(\frac{2M_t^2 K_i^2}{k}\right) + \log(wG) - 2\log\tau_{\max} \tag{Ia}$$

Equation Ia obviously will result in a series of straight parallel lines of slope -2 on a log-log sketch as shown by the typical plot in figure 7.11. For each of these lines the lower limit for feasible designs is established by $\tau_{\max} = S_y/2N_y$ from equation 7.13. We shall retain these lower limits on W as determined by stress for future reference. Incidentally, the maximum shearing stress failure criterion of equation 7.13c was chosen, with $C_f = 0.5$, and the materials of tables 4.1 and 4.2 in Chapter 4 were assumed to be of interest for the specific example.

For the next graph, we combine equation II with I, eliminating $\tau_{\max}$ and thereby reverting to equation a expressed in logarithmic form as follows.

$$\log W = \log\left(\frac{\pi^2}{128k}\right) + \log(wG) + 6\log d \tag{Ib}$$

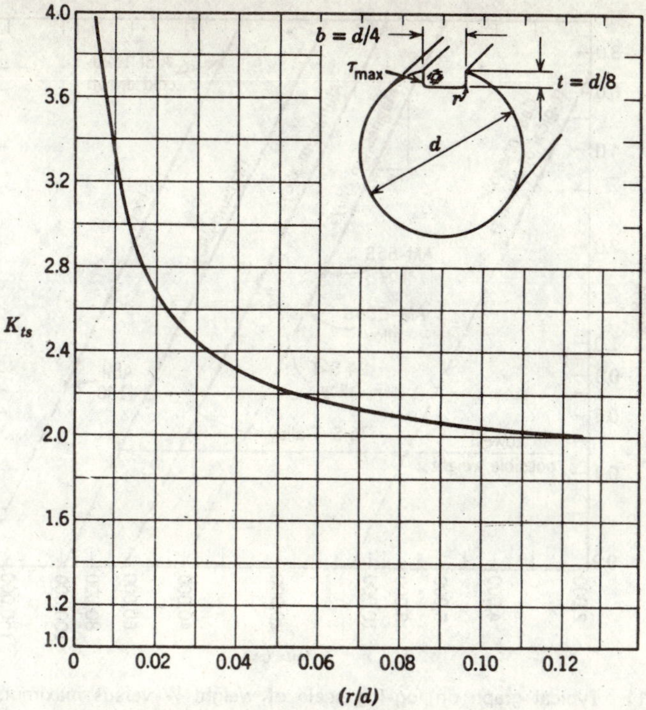

Figure 7.10 Stress concentration factor K_{ts} for the straight portion of a keyway in a shaft in torsion. $K_{ts} = \tau_{max}/\tau_{nom}$, where $\tau_{nom} = 16M_t/\pi d^3$. This figure was taken directly from figure 100 of reference 7.3. By permission of John Wiley and Sons, Inc.

In equation Ib, $(\pi^2/128k)$ is a specified value common for all materials, material parameters (wG) are obtained from tables 4.1 and 4.2, and the straight lines will all have a slope of $+6$ for a log-log graph. Hence we obtain figure 7.12 as a typical plot, and the lower limits on W imposed by stress are carried over from figure 7.11 for each material. Notice, in figure 7.12, that the ordering of the materials in a vertical direction is exactly the same as it was in figure 7.11, since both are based on a log (wG) term in equations Ia and Ib.

In figure 7.12 we readily see the implication of the limit equation on permissible values for diameter d, equation 7.18. Hence, if the limit d_{max} falls as typified by the type 1 region in figure 7.12, we will have a case of incompatible specifications and no feasible design will exist. For a feasible design to exist we would have to increase limit d_{max}, or possibly seek additional materials having a higher yield strength. On the other hand, if the limit d_{max} falls as typified by the type 2 region in figure 7.12, a feasible and optimum design would exist although two of the materials would be automatically eliminated

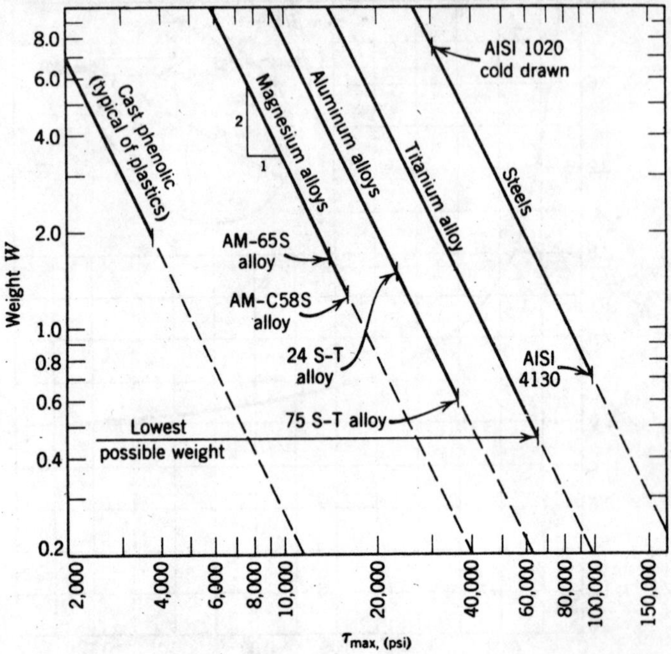

Figure 7.11 Typical graph on log-log scale of weight W versus maximum shearing stress τ_{max} for all feasible designs of practical torsion shaft as typified in figure 7.9, temporarily ignoring limits on d and l. Lower limits on W are established by upper limits on τ_{max} by equation 7.13. Refer to equation Ia in section 7.5.

from further consideration, since all feasible designs must lie to the left of the vertical line d_{max} in that figure.

It would be possible for the value of d_{max} to eliminate all materials but steel and in that event steel would be the optimum and only available material. However, for such a condition, a very small increase in the value of d_{max} could result in a very large reduction in W, since then the titanium alloy might be used, as is evident from figure 7.12. Finally, if the limit d_{max} were in the type 3 region of figure 7.12, all available materials would be feasible and obviously the titanium alloy would be the optimum choice, unless modified by limits on l.

Let us now assume that we have the type 2 region specifically shown in figure 7.12, and proceed to the next step, remembering that in that figure for this stage all feasible designs are represented by *solid* lines to the *left* of the line d_{max}. With this assumption, it is obvious that the magnesium alloy and cast phenolic have been eliminated from further consideration. Hence we

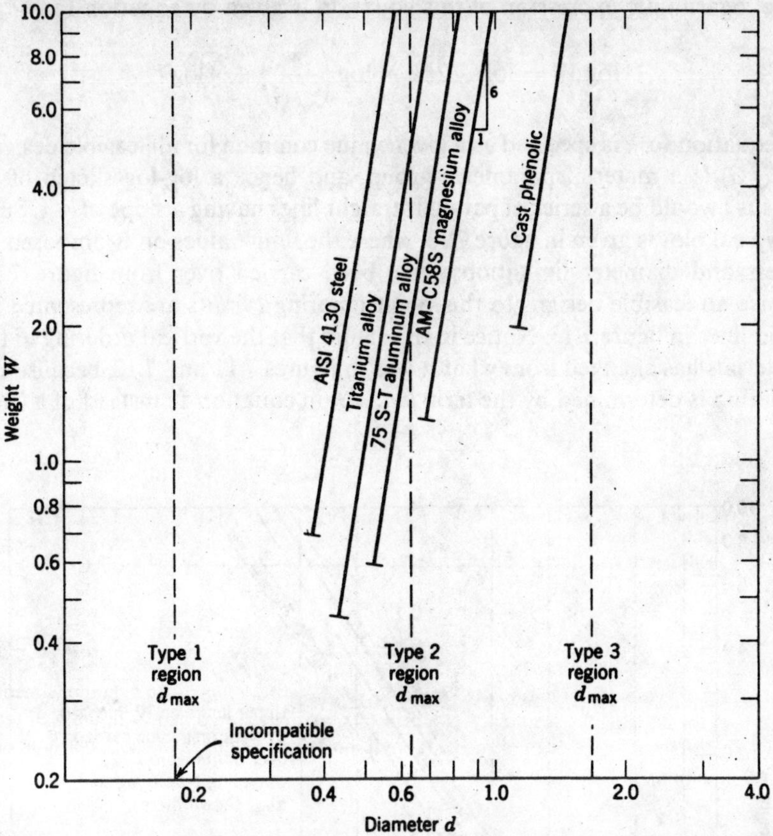

Figure 7.12 Log-log graph of weight W versus diameter d showing all feasible designs of practical torsion shaft as typified in figure 7.9, temporarily ignoring limits on l. Lower limits on W as governed by stress limitation equation 7.13 are carried over from figure 7.11. Typical limit values d_{max} on diameter d as given by equation 7.18 are shown. Refer to equation Ib in section 7.5.

shall retain the information obtained to this point by carrying over the proper limits on W for each remaining feasible material to our next sketch.

Let us now proceed by eliminating τ_{max} from equation I using equation III, which we see is equivalent to a simple combination of equations 7.12 and 7.11. Hence we obtain equation b as follows, which defines the variation of W with respect to l for the graph in figure 7.13.

$$W = \sqrt{2\pi}\left(\frac{w}{\sqrt{G}}\right)\sqrt{k}\, l^{3/2} \tag{b}$$

The logarithmic expression of this equation is given by equation Ic.

$$\log W = \log\sqrt{2\pi k} + \log\left(\frac{w}{\sqrt{G}}\right) + \tfrac{3}{2}\log l \qquad \text{(Ic)}$$

In equation Ic, k is specified as a given value common for all feasible designs $(w/\sqrt{G})$ is a material parameter group, and hence a log-log sketch of W versus l would be a series of parallel straight lines having a slope of $+\frac{3}{2}$. Such a typical plot is given in figure 7.13, where the limit values on W imposed by stress and diameter limitations have been carried over from figure 7.12 Hence all feasible designs to this point ignoring l limits are represented by solid lines in figure 7.13. Notice in this figure that the vertical ordering of the materials has changed from what it was in figures 7.11 and 7.12, because the ordering is determined by the term $(w/\sqrt{G})$ in equation Ic instead of a (wG)

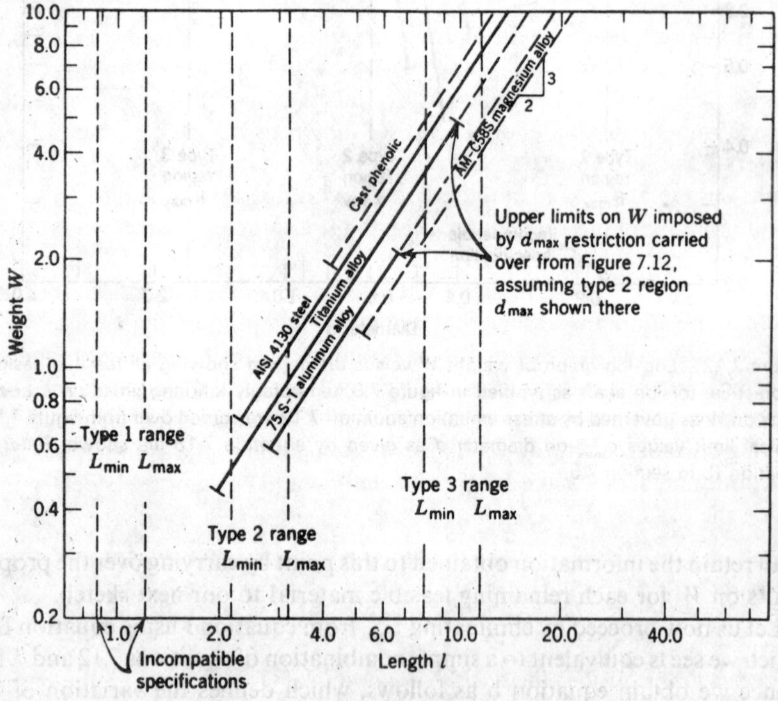

Figure 7.13 Log-log graph of weight W versus length l showing all feasible designs by solid lines, assuming practical torsion shaft as typified in figure 7.9, available materials in table 4.1, and type 2 region d_{max} limitation shown in figure 7.12. Typical limit range L_{min} and L_{max} on length l are shown. Lower limits on W carried over from figure 7.11 are imposed by stress limitation equation 7.13. Refer to equation Ic above.

term as in equations Ia and Ib. Hence the dotted line shown for cast phenolic has gone from the bottom to the top.

Some startling facts are evident from figure 7.13. Three types of ranges exist for feasible values of l. For type 1 in figure 7.13, no feasible design exists and we have a case of incompatible specifications. For a type 2 range in figure 7.13, many interesting points can be brought out. For instance, with the type 2 range as specifically shown, the optimum material would be the titanium alloy, but the stress in the bar would *not* be at the upper limit for the material. However, for the type 2 range, if the L_{min} value should be increased to correspond with the lower limit for the aluminium alloy, from figure 7.13 the optimum design would be with the aluminium alloy and the stress in the bar would be at its upper limit. On the other hand, for the type 2 region situation specifically shown in figure 7.13, the limit value for L_{max} is insignificant in the optimum design study. However, a type 2 region situation could exist where the value of L_{max} would be significant. For instance, if L_{min} and L_{max} were very close in figure 7.13 and both slightly to the left of the lower limit on aluminium, titanium would be the optimum material with the stress not at its upper limit. Now, if L_{max} were increased slightly, the optimum material might change to the aluminium alloy with the stress at its upper limit. If a type 3 range situation should exist as shown in figure 7.13, the optimum material would be the titanium alloy with stress in the bar not at its upper limit. On the other hand, if the entire type 3 range should be located slightly to the right in figure 7.13, eliminating the titanium alloy, steel would be the optimum material and the stress in the bar would not be at its upper limit. Also, if the type 3 range in figure 7.13 should be located slightly to the left to include the aluminium alloy, that would be the optimum material.

From this study it is evident that optimum design for cases of redundant specifications has many interesting aspects. For any of the possible situations described, the final optimum design would be uniquely determined after having considered figure 7.13. Thus, after selection of the optimum material, the optimum geometry of the torsion shaft would be completely determined, using the original (S.D.E.)'s if necessary.

7.6 Torsion Shaft for Minimum Cost

Let us now consider the design of a practical torsion shaft on the basis of minimum manufacturing cost. Assume that the general shape of the shaft has been decided upon from basic functional considerations of the machine and that this general shape is typically illustrated in figure 7.14. Assume that the shaft is to transmit a specified virtually constant twisting moment M_t, and it is to have a certain torque gradient k. Assume a factor of safety N_y based on

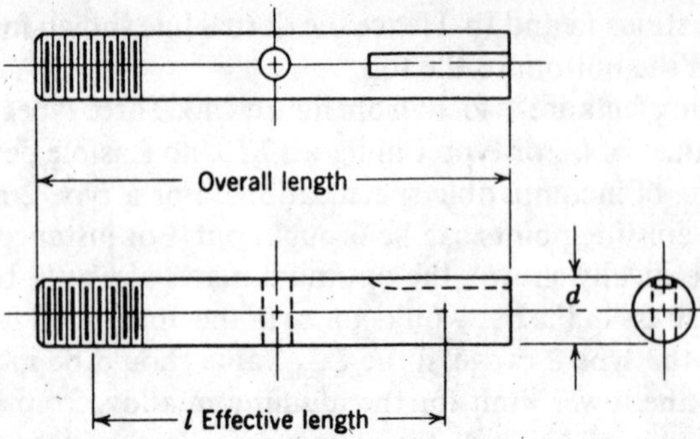

Figure 7.14 Typical basic geometry of shaft for minimum cost problem.

the occurrence of significant yielding, and to be reasonable and conservative assume the maximum shearing stress theory of failure. Also, assume in figure 7.14 that the twisting moment is transmitted from the keyway to the threaded portion.

Following the MOD, the first step will be to write the primary design equation, which is for total manufacturing cost C_T of the shaft. The total cost can be resolved into three significant components. The first will be the same regardless of what our final design is, and we shall call it fixed cost, C_F. It will be an invariant component which is insignificant as far as optimum design is concerned, and it will include charges for such services as handling and paper work, which cannot be avoided. The second component will be the cost of the material. Generally a price can be determined for materials in usable stock form on a unit weight cost basis, and hence we shall let the material parameter c_1 represent the cost of the material in dollars per pound. The value for the parameter c_1, of course, will vary from material to material and possibly even with the shape of stock available. The third component will be the cost for machining the shaft. This is a difficult part to account for precisely in many cases, but it is primarily a function of material machining characteristics and the amount of machining which has to be done on the shaft. For any specific material we could probably say that this part of the cost would be proportional to the size of the shaft. To be specific, since cutting speeds and feeds are generally limited to certain magnitudes for each material because of machining characteristics, it would be reasonable to say for the shaft in figure 7.14, that the machining cost would be approximately proportional to both diameter d and length l. If information is available from past machining experience on similarly shaped shafts, possibly a simple exponential equation could be derived for expressing anticipated machining cost as a function of diameter d and length l. However, for the sake of simplicity in this example, we shall assume a linear relationship which will adequately illustrate the general procedure. The constant of proportionality for the product ld will depend on the specific material, and we shall denote its value by the machinability parameter c_2. Hence the total manufacturing cost for the shaft could be expressed by

$$C_T = C_F + c_1 w\left(\frac{\pi d^2}{4}\right)l + c_2 ld \tag{7.20}$$

Equation 7.20 is our primary design equation, and w is the weight density of the material, c_1 and c_2 are other material parameters already defined and d and l are independent geometrical parameters. We shall make no attempt to give specific values to the various cost factors in this book. These factors change too greatly from year to year and from locality to locality. Also, they depend on factors peculiar to specific companies, such as overhead, types and quality of available machinery and equipment, etc. An experienced engineer will be able to assess *reasonable* values for c_1 and c_2 suitable for conduct of the design study for his specific problem for each of the available feasible materials, and the value for the constant C_F need not be known for determining the optimum shaft design.

The next step in the MOD will be to write the subsidiary design equations. The equation for torque gradient k will be merely equation 7.12:

$$k = \frac{\pi d^4 G}{32l} \tag{7.12}$$

This assumes that the integrated effect of the thread, small hole, and keyway will be negligible as far as torsional rigidity of the entire shaft is concerned. This is quite reasonable, since the effects of these stress concentrations are highly localized. Incidentally, if the shaft had been stepped, it would be advisable to consider an *effective* shaft length referred to a single diameter, and this would be a straightforward procedure even though it would add some complexity to the problem.

The equation for maximum shearing stress in the shaft would be identical with equation 7.17:

$$\tau_{\max} = K_i\left(\frac{16M_t}{\pi d^3}\right) \tag{7.17}$$

In this equation K_i is the value of the *largest* or governing stress increase factor for the shaft, in view of the various geometrical deviations depicted typically in figure 7.14. Incidentally, if the shaft had been stepped, it would again be feasible to refer all stress increase factors to a single diameter of a hypothetical simple cylinder, and this too would be a straightforward procedure even though it would add some complexity to the problem.

Next we shall write the only significant limit equation, which is that for permissible values of stress, assuming the reasonably conservative maximum shearing stress theory of failure.

$$\tau_{max} \le \frac{S_y}{2N_y} \tag{7.13c}$$

Assume that buckling is of no further concern, based on satisfaction of equation 7.10b by the specified values for M_t and k.

From a quick review of the (I.F.) now presented, we note that there are two free variables (d and l) and two (S.D.E.)'s (equations 7.12 and 7.17). Hence, as explained in section 6.13 of Chapter 6, this problem is of the (N.S.) type since $n_f = N_s = 2$. Thus following the MOD, we shall combine the (S.D.E.)'s with the (P.D.E.) eliminating the independent geometrical parameters d and l:

$$C_T = C_F + 2\left(\frac{M_t^2}{k}\right)\left(\frac{c_1 wG}{\tau_{max}^2}\right)(K_i)^2 + \frac{1}{2}\left(\frac{16}{\pi}\right)^{2/3}\left(\frac{M_t^{5/3}}{k}\right)\left(\frac{c_2 G}{\tau_{max}^{5/3}}\right)(K_i)^{5/3} \tag{a}$$

In the developed (P.D.E.), equation a, it is obvious that the total cost C_T is minimized if the independent parameter τ_{max} is made as large as possible, so we shall let τ_{max} be equal to its upper limit by equation 7.13c. Hence, placing $\tau_{max} = S_y/2N_y$ in equation a, we obtain the final primary design equation

$$C_T = C_F + 8\left(\frac{M_t^2 N_y^2}{k}\right)\left(\frac{c_1 wG}{S_y^2}\right)(K_i)^2 + \left(\frac{32}{\pi}\right)^{2/3}\left(\frac{M_t^{5/3} N_y^{5/3}}{k}\right)\left(\frac{c_2 G}{S_y^{5/3}}\right)(K_i)^{5/3} \tag{7.21}$$

The final (P.D.E.), equation 7.21, includes all the design equations, and it is purposely bracketed according to significant groups. For optimum design we are concerned with the last two parts of equation 7.21, and for each notice that the first term is merely a number. However, the second group for each part is the functional requirement parameter group. We have assumed that values for M_t and k have been specified, and the lowest permissible value for N_y has been selected as discussed in Chapter 5. The third group for each part is the material parameter group. The last term in each part contains only the largest or governing stress increase factor of the shaft, K_i, which might be considered an independent geometrical parameter affected primarily by the proportional design of the thread, hole, or keyway in figure 7.14. Of course, the objective would be to have the proportions such that the governing value for K_i would be as small as possible.

For equation 7.21 we have a condition of *inseparable* functional, material and geometrical parameters. Such cases are sometimes encountered in real problems, and then we cannot explicitly select the optimum material from a single material selection factor. Hence we must be careful to consider the effects of the functional and geometrical terms in equation 7.21 when selecting the optimum material.

Once we have decided on specific values for M_t, N_y, k, and K_i, it would be simple to select the optimum material through equation 7.21. We would be concerned with only the last two parts in that equation, and we would calculate values for them, using characteristics of the available feasible materials from tables 4.1 and 4.2. We would assign values for c_1 and c_2 based on prevailing costs at our particular locality according to the preceding discussion of these factors. As a suggestion for the order of making the optimum material selection calculations, from equation 7.21 it would be wise to consider materials having the lowest values for $(c_1 wG/S_y{}^2)$ and $(c_2 G/S_y^{5/3})$ first.

Finally, after the optimum material has been selected, the last step will be to calculate the optimum values for d and l. We would first calculate d, using the stress equation 7.17 with $\tau_{max} = S_y/2N_y$, where the value of S_y is for the optimum material. Then we would calculate the optimum value for l, using the torque gradient equation 7.12 with the value of G for the optimum material and the optimum value of d just calculated. The optimum design at this point is complete, since the thread, hole, and keyway have already been selected by proportional design in the step for minimizing K_i.

The lowest possible total cost C_T could be calculated from either equation 7.20 or 7.21. Obviously, whatever we can do to reduce the independent fixed cost C_F will be beneficial also.

7.7 Torsion Shaft for Minimum Dynamic Torque

Let us consider now the design of a torsion bar spring which is to absorb and release a specified amount of energy (P.E.) for each cycle of repeated loading, and the shape of the torque variation is generally depicted in figure 7.15. The bar is to have a desired life of 500 million cycles, and M_t will denote the maximum twisting moment in the bar corresponding to a maximum angle of twist which will be denoted by θ. The maximum twisting moment in the bar, M_t, will also be the maximum torque felt by the frame of the machine. The problem will be to design the bar so that the value of torque M_t transmitted to the machine frame for each cycle of repeated loading will be as small as possible. It is assumed that this intermittent torque is an undesirable effect which we are trying to minimize by proper design of the torsion bar spring. Regarding space restrictions, we shall say that d_{max}, L_{min}, and L_{max} exist as depicted in figure 7.9. In addition, we shall assume that the value for maximum angle of twist θ cannot exceed a limit value θ_{max}, which is a functional requirement of the machine.

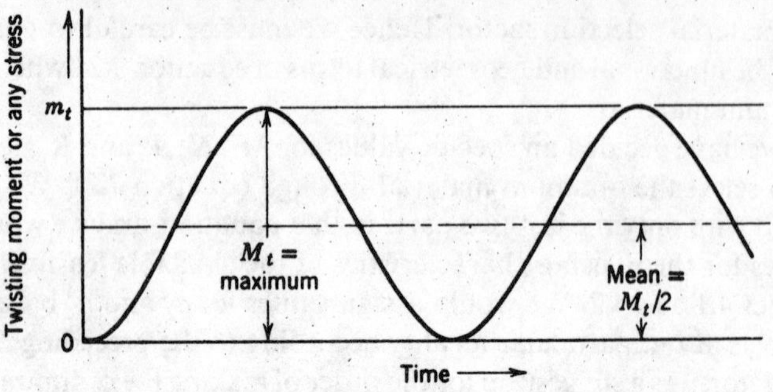

Figure 7.15 Assumed general type of cyclical loading for torsion bar spring in absorbir and releasing energy.

Initial Formulation. Let us now derive the (I.F.) equation system for th optimization problem. The first step will be to derive the criterion functio for M_t to be minimized, and referring to figure 7.16 by elementary mechanic we obtain the following equation.

$$(\text{P.E.}) = \tfrac{1}{2}M_t\theta$$

$$\therefore \quad M_t = \frac{2(\text{P.E.})}{\theta} \qquad (7.2$$

In primary design equation 7.22, the energy to be stored (P.E.) is a specifie functional requirement value, and the maximum angle of twist, θ, is a limite parameter whose value will be determined by optimal design.

Next, writing the subsidiary design equations, equation 7.9 for angle c twist can be used directly, the effects of stress concentrations being neglecte

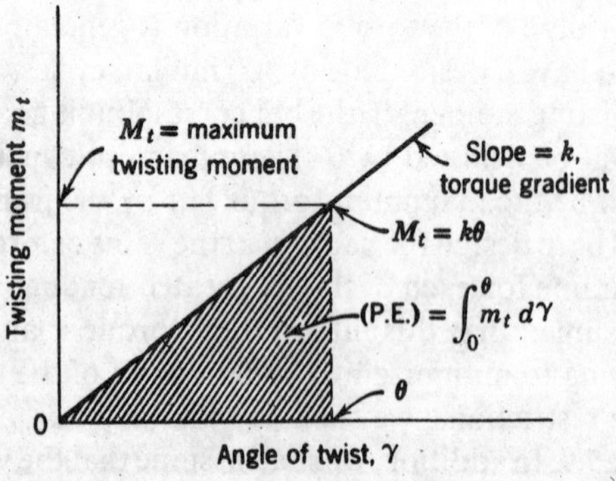

Figure 7.16 Twisting moment versus angle of twist for a torsion bar, showing ener stored.

This is a reasonable assumption since the effects of stress concentrations are generally highly localized, and the integrated effect for the whole shaft with respect to angle of twist would be small. Hence the first (S.D.E.) is

$$\theta = \frac{32lM_t}{\pi d^4 G} \tag{7.9}$$

The other (S.D.E.) is equation 7.17, expressing the greatest timewise value for the maximum shearing stress, at the critical point in the torsion bar, designated τ_{max}.

$$\tau_{max} = K_i\left(\frac{16M_t}{\pi d^3}\right) \tag{7.17}$$

Again K_i is the *largest* or governing stress increase factor of all the geometrical deviations, such as keyways or holes which cause stress concentrations in the real bar.

Regarding limit equations, the space restrictions on permissible values of d and l given by equations 7.18 and 7.19 will be retained, since such limitations are generally encountered in real problems. They are repeated here:

$$d \leq d_{max} \tag{7.18}$$

$$L_{min} \leq l \leq L_{max} \tag{7.19}$$

The limit equation on permissible values for greatest angle of twist, θ, can be written

$$\theta \leq \theta_{max} \tag{7.23}$$

Regarding the criterion for fatigue failure, we shall again assume the reasonable but conservative maximum shearing stress theory as reviewed in Chapter 4 and apply equation 4.16 derived therein. Critical shearing stress will vary with time as shown in figure 7.15, so for the equation we use $(\tau_G)_{max} = \tau_{max}$ and $(\tau_G)_M = \tau_{max}/2$. Thus the fatigue failure criterion is expressed by equation 7.24 as follows, which must be satisfied for an acceptable design.

$$\tau_{max} - (1 - p)\frac{\tau_{max}}{2} \leq \frac{S_e}{2N_e}$$

$$\therefore \quad \tau_{max} \leq \frac{S_e}{(1 + p)N_e} \tag{7.24}$$

For any specific material, equation 7.24 defines the range of *permissible values* for the greatest value of the maximum shearing stress τ_{max} as defined for equation 7.17; S_e is the published fatigue strength of the material determined by a standard reversed bending type of test based on 500 million cycles:

$p = S_e/S_y$, where S_y is the published yield strength of the material from a simple tensile test, and N_e is the appropriate factor of safety depending on the significance of the occurrence of a fatigue failure before 500 million cycles as discussed in Chapter 5.

Solution to the optimization problem now formulated will be determined in two ways in this book. First, the MOD technique will be applied for the solution in this section. Second, the same problem will also be solved by automated optimal design as the illustrative example of Appendix C. Correlation between the numerical solutions is excellent for the two approaches.

Exploratory Calculations. Regarding the MOD solution, upon review of the (I.F.) presented above, we readily determine that $n_f = 0$ and $N_s = 2$. Hence, with $n_f < N_s$ we have the (R.S.) type of problem, as explained in section 6.13 of Chapter 6. Also, from the (I.F.) we see that $n_v = n_c = 4$ (for θ, τ_{max}, d, and l). Applying exploratory equations 6.1 through 6.4 from Chapter 6, we predict the following characteristics for execution of the MOD.

$$D_{vs} = n_v - N_s + 1 = 4 - 2 + 1 = 3$$

$$A_t = \frac{n_c!}{[n_e!(n_c - n_e)!]} = \frac{4 \times 3 \times 2}{2 \times (4 - 2)!} = 6$$

where $n_e = N_s - n_f = 2 - 0 = 2$.

$$n_r = n_c - n_e = 4 - 2 = 2$$

Thus by the simple exploratory equations we have learned that the (F.F.) and (V.S.) will be three dimensional in character, with two related parameters and two eliminated parameters. Also, there are six approaches theoretically possible for derivation of (F.F.)'s. The eliminated parameter possibilities corresponding to the six approaches would be the combinations of the four constrained parameters taken two at a time disregarding the order. Graphically, they are represented by the terminal points for the six lines of the circle diagram shown in figure 7.17. For any eliminated parameter combination, the related parameters of course would be the remaining two of the four constrained parameters.

Final Formulation. In transforming the (I.F.) to (F.F.), equation reduction is not possible since there are no free variables. However, as described in section 6.13 of Chapter 6, some equation manipulation by combination will be necessary for organization to the desired format of the (F.F.). Approach 1 of the figure 7.17 circle diagram is a relatively simple determinate approach,

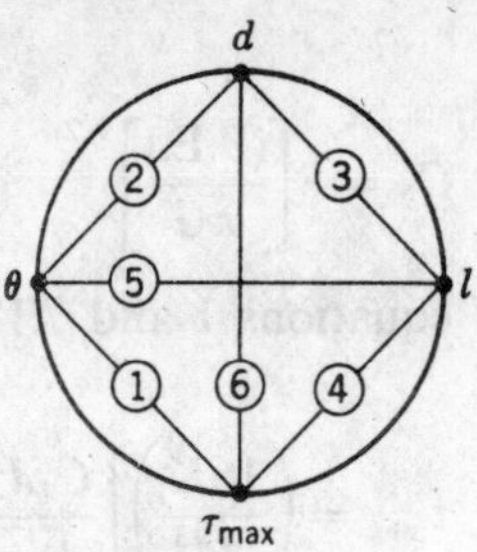

Figure 7.17 Circle diagram for section 7.7 problem.

as can be seen from review of the (I.F.). Its (F.F.) general format is given below. Incidentally, as explained in Chapter 6, material parameters will be held constant for the (V.S.).

$$M_t = f_1(d, l) \tag{I}$$

$$\theta = f_2(d, l) \tag{II}$$

$$\tau_{max} = f_3(d, l) \tag{III}$$

where θ, τ_{max} = eliminated parameters

d, l = related parameters.

With the preceding general format as a guide for the necessary equation combination manipulation, we specifically derive the (F.F.) for approach 1 starting with the (I.F.) for the optimization problem.

By combination of equations 7.22 and 7.9 thus eliminating θ we obtain equation I.

$$M_t = \frac{2(\text{P.E.})}{\theta} = \frac{2(\text{P.E.})\pi d^4 G}{32 l M_t}$$

$$\therefore M_t = \frac{C_1 d^2}{l^{1/2}} \tag{I}$$

where

$$C_1 = \left[\left(\frac{\pi}{16}\right)(\text{P.E.})G\right]^{1/2} \tag{7.25}$$

Similarly, by combination of equations 7.9 and 7.22 thus eliminating M_t we obtain relating equation II.

$$\theta = \left(\frac{32l}{\pi d^4 G}\right)\left[\frac{2(\text{P.E.})}{\theta}\right]$$

$$\therefore \theta = \frac{C_2 l^{1/2}}{d^2} \tag{II}$$

where

$$C_2 = 8\left[\frac{(\text{P.E.})}{\pi G}\right]^{1/2} \tag{7.26}$$

Finally, by combination of equations I and 7.17 thus eliminating M_t we obtain relating equation III.

$$\tau_{\max} = \left(\frac{16K_i}{\pi d^3}\right)\left[\frac{C_1 d^2}{l^{1/2}}\right]$$

$$\tau_{\max} = \frac{C_3}{(dl^{1/2})} \tag{III}$$

where

$$C_3 = \frac{16K_i C_1}{\pi} \tag{7.27}$$

In review of the (F.F.) equations I, II, and III, some items of importance should be noted. First, the coefficient for each of these equations contains the material parameter G (i.e., modulus of rigidity), as found in equations 7.25, 7.26, and 7.27. Hence the values of these coefficients, C_1, C_2, and C_3, will change from one material to another. Also, they will change from one numerical problem to another, since they contain the special functional requirement (P.E.).

Second, we note that stress increase factor K_i appears only in relating equation III of the (F.F.), which expresses eliminated parameter $\tau_{\max}$ in terms of the related parameters d and l. The constraint on $\tau_{\max}$ for an acceptable design is expressed by equation 7.24. In general, the effect of a constraint on an eliminated parameter is to restrict the feasible design space on the related parameter plane. Hence choosing optimum geometric proportions for the design of the critical stress concentration regions would reduce K_i and thus $\tau_{\max}$ for any design. (As a specific example, see exercise 7.9 at the end of this chapter.) Therefore, minimizing K_i would have the same effect as opening the design space on the (d, l) related parameter plane for our 3-D variation diagram which follows. Hence we should strive for optimum proportions for minimization of K_i, since this can only improve the solution for minimization of M_t if equation III of the (F.F.) is active in a numerical application.

Variation Study. The typical three dimensional variation diagram for this problem is sketched for a specific material of interest in figure 7.18. Its construction will now be explained in the same sequence as it was drawn. To start, the related parameter axes for d and l are drawn and typical limit lines are imposed for $d_{\max}$, $L_{\min}$, and $L_{\max}$. These axes and the limit lines define the

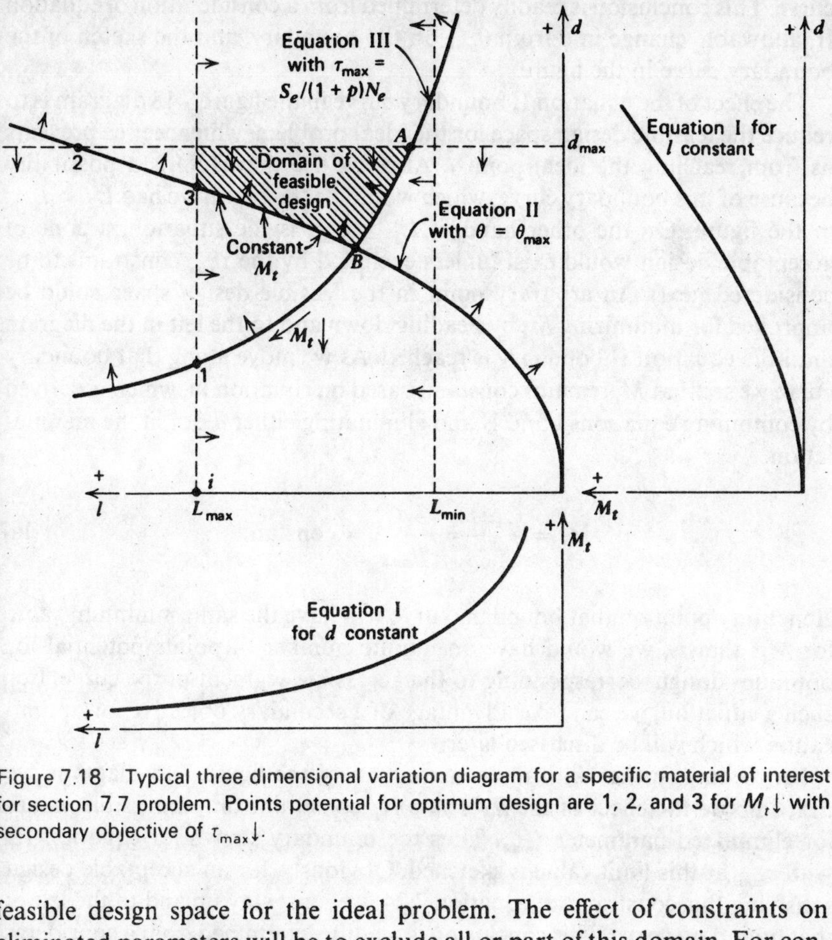

Figure 7.18 Typical three dimensional variation diagram for a specific material of interest for section 7.7 problem. Points potential for optimum design are 1, 2, and 3 for $M_t\downarrow$ with secondary objective of $\tau_{max}\downarrow$.

feasible design space for the ideal problem. The effect of constraints on eliminated parameters will be to exclude all or part of this domain. For construction of the diagram, side and bottom cuts are now sketched from equation I for optimization quantity M_t, showing how it varies independently with respect to the related parameters d and l. Thus we conclude that the ideal point of optimum design is i in the figure, but obviously this is not a realistic possibility for design. The eliminated parameter constraints will prevent us from ever reaching that point, and i will not be potential for optimum design in the total picture.

Proceeding with the construction of the figure 7.18 diagram, we next impose the effect of the constraint θ_{max} on permissible values for eliminated parameter θ. Thus the boundary shown for equation II with $\theta = \theta_{max}$ is sketched. Obviously, for an acceptable design satisfying the constraint of equation 7.23, we must stay up and to the right of the described boundary curve. This conclusion is readily determined from a consideration of equation II, allowable change in θ from θ_{max} on the boundary, and the sketch of the boundary curve in the figure.

The effect of the equation II boundary curve in the figure 7.18 diagram is to reduce the feasible design space for the ideal problem, which we see prevents us from reaching the ideal point *i*. Also, we see that (I.S.) is a possibility because of this boundary curve, which would be the case if we had $L_2 < L_{min}$ in the figure. On the other hand, if $L_2 \geq L_{min}$ is the situation, a zone of acceptable design would exist (unless excluded by the τ_{max} constraint to be considered next). An arbitrary point in the feasible design space could be improved for minimizing M_t by heading down and to the left in the diagram until the equation II boundary is reached. As we move along that boundary curve we see that M_t remains constant, based on equation Ia, which is derived by combining equations I and II and eliminating either *d* or *l* in the manipulation.

$$M_t = \frac{C_1 d^2}{l^{1/2}} = \frac{C_1 C_2}{\theta_{max}} = \text{constant} \tag{Ia}$$

Hence any point on that boundary curve will have the same minimum value for M_t; that is, we would have an infinite number of points potential for optimum design, corresponding to the accessible segment of the curve. For such a situation, we can take advantage of a secondary objective for optimization which will be discussed later.

Proceeding further with the construction of the figure 7.18 diagram, we next impose the effect of the $S_e/(1 + p)N_e$ constraint on permissible values for eliminated parameter τ_{max}. Thus the boundary shown for equation III with τ_{max} at this limit value is sketched. Obviously, for an acceptable design satisfying this constraint of equation 7.24, we must stay up and to the left of that boundary curve. This conclusion is readily determined from a consideration of equation III, allowable change in τ_{max} from its limit value on the boundary, and the sketch of the boundary curve in the figure.

The possible effect of the equation III boundary curve in the figure 7.18 diagram is to reduce further the feasible design space in the *d*, *l* plane. In fact, two additional possibilities for (I.S.) are introduced, which would be for $L_A > L_2$ and for $d_1 > d_{max}$ as seen in the figure. However, if this equation III boundary is active along a feasible design space, starting from an arbitrarily chosen interior point we would obviously move towards that curve for reduction in M_t. Then, from the side and bottom views of the diagram, moving down and to the left along the boundary would provide additive components which would reduce M_t still further. Specifically, we would move along the boundary to get as close as possible to point 1, which is thereby revealed as a singular point potential for optimum design. This would be the case in a numerical problem if $d_1 \leq d_{max}$ and $d_3 \leq d_1$.

On the other hand, if the constraints are such for a numerical problem that $d_1 \leq d_{max}$, and $d_B < d_2$, and $d_3 > d_1$, we would have to stop at point B in our attempt to reach point 1. Hence equation II with $\theta = \theta_{max}$ would become active, and as learned earlier minimum M_t would be constant along that boundary curve. In this event it is possible to take advantage of a worthwhile secondary objective for optimum design with no increase in M_t. In this particular problem we might choose minimization of τ_{max} or maximization of safety factor N_e as the secondary objective. With this choice, two additional points are seen to be potential for optimum design, namely, points 2 and 3 in figure 7.18. They are the points on the equation II boundary which are furthest from the τ_{max} limit curve, and from the diagram we see that the point of smaller L value would govern.

Construction of the figure 7.18 variation diagram has now been explained, as have the significant conclusions drawn from it. In summary, we have learned that there are three possibilities for (I.S.). Also, points 1, 2, and 3 are potential for optimum design in minimizing M_t, where the secondary objective of minimizing τ_{max} becomes active when point 2 or 3 is chosen. If the figure 7.18 variation diagram is thoroughly understood as explained, the flow chart of figure 7.19 logically falls together. However, assuming such an understanding of the (V.S.), some general guidelines for flow chart construction in the MOD should be outlined next.

General Guidelines for Flow Chart Synthesis in MOD. The construction of a calculation (F.C.) is one of the last steps in solution of an optimization problem by the MOD. The (F.C.) logically falls together from a thorough understanding of the (V.S.), which in turn depends on a thorough understanding of the normalized (F.F.) equation system as derived from the (I.F.). Being the last item in the derivation process for the MOD, synthesis of a calculation (F.C.) is often a stumbling block to those first learning the optimization technique. Hence the following guidelines are summarized for general application and are illustrated specifically in the (F.C.) of figure 7.19.

1. Start the (F.C.) by input of the boundary values, in the reading of specified items including limit values.

2. Screen out the possibilities of (I.S.) at the start, based on simple comparisons between pertinent points. The necessary comparisons become evident from a study of the (V.S.) variation diagram, including a consideration of the possible ways in which the boundaries can fall relative to each other.

3. Determine which point of the potential ones is specifically the one of optimum design in numerical application, based on simple comparisons

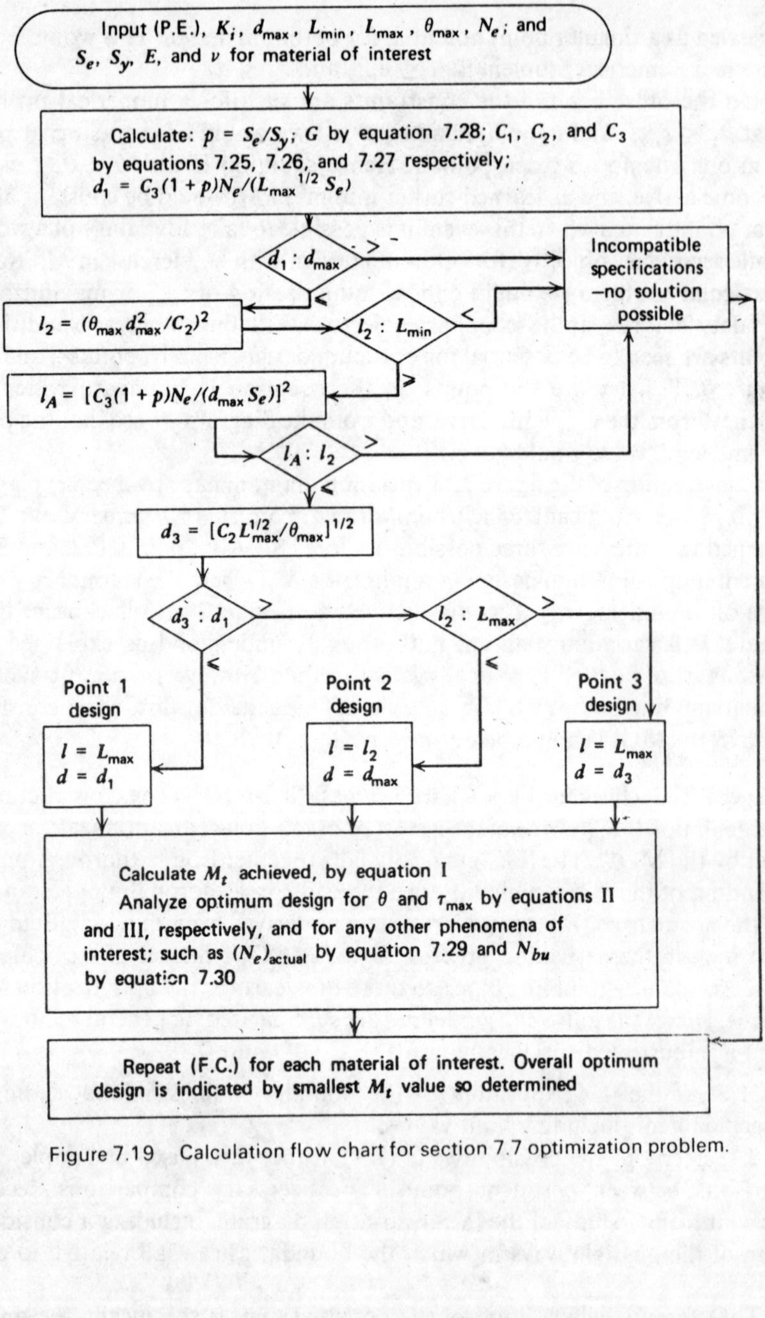

Figure 7.19 Calculation flow chart for section 7.7 optimization problem.

between pertinent points. The necessary comparisons become evident from a study of the (V.S.) variation diagram, including a consideration of the possible ways in which the boundaries can fall relative to each other. It is assumed that variation of optimization quantity has been determined in the (V.S.) along any pertinent boundary curve of the variation diagram.

4. Determine all pertinent points for steps 2 and 3 from a review of the (V.S.) variation diagram, using appropriate equations from the (F.F.), as well as specified values from the (I.F.) in the process. To execute this step simply, it is important to label all boundaries and curves on the variation diagram in terms of the (F.F.) equation numbers, including constraint conditions where appropriate, as shown in figure 7.18. Incidentally, a design state point is always uniquely determined by the values for the related parameter coordinates alone.

5. Analyze the point of optimum design determined in step 3 for all phenomena of interest. For example, we are always interested in the value achieved for the optimization quantity, which can always be calculated at this stage using equation I of the (F.F.). Next, analyze for values of any free variables using the (I.F.) equation system by merely reversing the equation combination procedure used for derivation of the (F.F.). Finally, analyze the determined optimum design for all other phenomena of interest, such as eliminated parameter values using the relating equations of the (F.F.). Final items of analysis can usually be calculated sequentially at this time for the design state point which has been uniquely determined for solution of the optimization problem.

The (F.C.) synthesized by the preceding steps can be a very valuable tool in solution of numerical applications for the optimization problem. This is especially true for cases where digital computers or even programmable calculators are available for repeated solutions of the problem type in the long range future, such as for mechanical elements often encountered in practice.

EXAMPLE 7.1

Let us now solve a numerical example of the section 7.7 optimization problem. Specifically, we will apply the figure 7.19 flow chart as explained and derived by the MOD. Input data for the boundary values of the numerical problem will be as follows:

(P.E.) = 500 in-lb; $K_i = 1.41$; $d_{\max} = 1.500$ in; $L_{\min} = 20.0$ in; $L_{\max} = 85.0$ in; $\theta_{\max} = 0.200$ rad; and $N_e = 1.30$ as required for design to be safe against fatigue breakage.

The values of 500×10^6 cycle endurance limit S_e, yield strength S_y, modulus of elasticity E, and Poisson's ratio ν are determined for various materials of interest in tables 4.1 and 4.2 of Chapter 4. Regarding the steels, as called for in table 4.1, we use equation 4.1 for S_e, and surface factor k_a is taken from figure 3.8 of Chapter 3, assuming an appropriate surface finish for the type of steel. The values of S_e for the other materials, as well as the values of S_y and E for all the materials, are taken directly from table 4.1. The values of Poisson's ratio ν are obtained from table 4.2, and these are needed to calculate the modulus of rigidity G by equation 7.28 from elementary strength of materials.

$$G = \frac{E}{2(1 + \nu)} \tag{7.28}$$

The best design for minimization of M_t is determined for each of the materials from the figure 7.19 (F.C.), as derived from the figure 7.18 variation diagram to which we will now refer. If either point 2 or 3 in that diagram is the point of optimum design for a particular material, we will certainly be interested in calculating the actual factor of safety, since the secondary objective of maximizing it then becomes active. It is calculated as follows, based on the equation 7.24 criterion using the equality sign.

$$(N_e)_{\text{act}} = \frac{S_e}{(1 + p)\tau_{\max}} \tag{7.29}$$

Finally, as mentioned earlier, we will be interested in analyzing each of the best designs at the end to determine whether or not the possibility of buckling is of concern. Thus in reference to the criterion equation 7.10 discussed earlier, we derive equation 7.30 for factor of safety against buckling, N_{bu}. It is in agreement with the basic definition of factor of safety by equation 5.3 of Chapter 5, expressed in general as the ratio of load capability to actual load with respect to a failure phenomenon such as buckling.

$$N_{bu} = \frac{\pi^2 E d^4}{32 l M_t} \tag{7.30}$$

For an acceptable design having no chance of buckling, we should calculate in the end analysis a value for N_{bu} appreciably greater than unity. If this is not the case we must include a buckling constraint in our (I.F.), but we will do this only if necessary. Based on our initial judgment, our strategy for simplification reasons in deriving the (I.F.) was that buckling was of no concern and we must validate this assumption by analysis at the end.

A computer program was written from the figure 7.19 (F.C.), and the results for the numerical problem are summarized in table 7.1. From the table we note that all three points potential for optimum design have been uncovered, and all the (F.C.) tracks leading to (I.S.) have been followed for the cases of $d_1 > d_{\max}$, $l_A > l_2$, and $l_2 < L_{\min}$. In review of the results in table 7.1, the

TABLE 7.1

Summary of Best Designs for Various Materials as Calculated by Figure 7.19 Flow Chart for Example 7.1 Torsion Bar Optimization Problem (Primary Objective is $M_t\downarrow$)

Material (see tables 4.1 and 4.2)	S_e, kpsi	S_y, kpsi	G, Mpsi	d, in	l, in	M_t, in-lb	θ rad	$(N_e)_{act}$	N_{bu}	Comments (see figures 7.18 and 7.10)
AISI 1020 HR steel	28.5	35	11.417	—		—		—	—	(I.S.): $d_1 = 2.158$ in $> d_{max} = 1.500$ in
AISI 1020 CR steel	37.2	60	11.417	1.476	85.0	7915	0.126	1.300	63.16	Point 1 design (figure 7.18)
AISI 2340 steel Oil Q & D 800°F	60.0	167	11.417	1.173	85.0	5000	0.200	1.986	39.90	Point 3 design (figure 7.18)
AISI 2340 steel Oil Q & D 400°F	63.0	232	11.417	1.173	85.0	5000	0.200	2.229	39.90	Point 3 design (figure 7.18)
Titanium alloy	85.0	130	6.015	1.377	85.0	5000	0.200	3.740	41.78	Point 3 design (figure 7.18)
7075-T6 aluminium alloy	23.0	73	3.910	1.500	77.73	5000	0.200	1.644	41.78	Point 2 design (figure 7.18)
AZ61A magnesium alloy	17.0	33	2.407	—	—	—	—	—	—	(I.S.): $l_A = 72.72$ in $> l_2 = 47.86$
Acetal homopolymer plastic	3.8	9	0.148	—	—	—	—	—	—	(I.S.): $l_2 = 2.945$ in $< L_{min} = 20.0$ in

smallest possible value for M_t is 5000 in-lb, and this is achieved for the best designs of four of the eight materials. From final considerations of cost, availability, machinability, and such practical factors, we probably would choose the AISI 2340 steel oil Q & D at 800 F with $d = 1.173$ in and $l = 85.0$ in for the overall optimum design. Analysis of this design gave from the table $M_t = 5000$ in-lb, $\theta = 0.200$ rad, $(N_e)_{act} = 1.986$, and with $N_{bu} = 39.90$ we see that our initial assumption of no buckling was a valid one.

7.8 Dynamic Shaft in Combined Loading (N.S.)

In this section we shall consider a problem in optimum design which exemplifies two important characteristics common to many real mechanical elements. First, in the optimum design study we shall encounter certain important factors which are accurately indeterminable *at the time of design.* However, we shall circumvent this seemingly formidable obstacle and, with the general method of optimum design, determine explicitly the optimum material and the optimum values for geometrical parameters. Second, by following the MOD we shall explicitly determine an optimum geometry which is somewhat unorthodox in nature and which will result in a significant improvement in product performance. Thus application of the MOD will determine explicitly an improvement in basic geometry which otherwise would be limited by traditional practices.

The specific problem in this section will be the optimum design of a practical torsion bar spring on the basis of maximum energy absorption *capability*, for the case of fatigue loading. However, the MOD can be similarly applied to *practical shafts* in combined loading and for cases having other bases of optimum design. For certain unique problems additional limit equations might exist, as on maximum permissible deflection of a shaft in combined loading, thereby resulting in (R.S.).

Statement of the Problem. The specific problem which confronts us in this section is to design a practical torsion bar spring, as depicted in figure 7.20, on the basis of maximum energy absorption *capability* per cycle of repeated loading. We shall assume that energy is introduced to the torsion bar by means of force F applied to the end of a load arm, where F varies cyclically as generally depicted in figure 7.21. Since given amounts of *energy* of unpredictable magnitude in the form of shock loading from an external source are to be absorbed and released by the spring, the particular value for F_{max} is dependent upon the particular final design which will be determined. The objective in design will be to select the material and the geometry which will result in the greatest possible energy absorption *capability* per cycle of shock

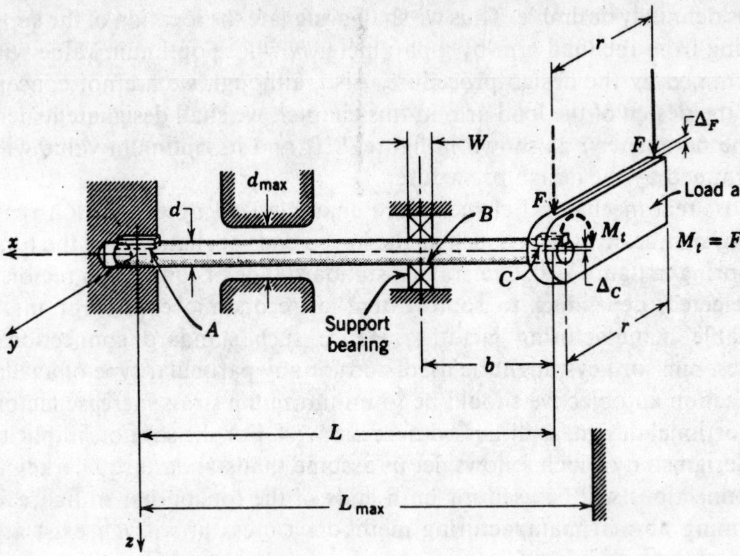

Figure 7.20 Torsion bar spring showing practical connection regions and load arm. (Sled-runner type of keyways at both ends.)

loading. We shall assume that space restrictions d_{max} and L_{max} exist on feasible values of diameter d and length L respectively, because of functional requirements of the entire structure.

In figure 7.20, normally the support bearing would be placed as close to the load arm as possible, merely on the basis of intuition or by following traditional practice. However, at this stage of design we have not proved that

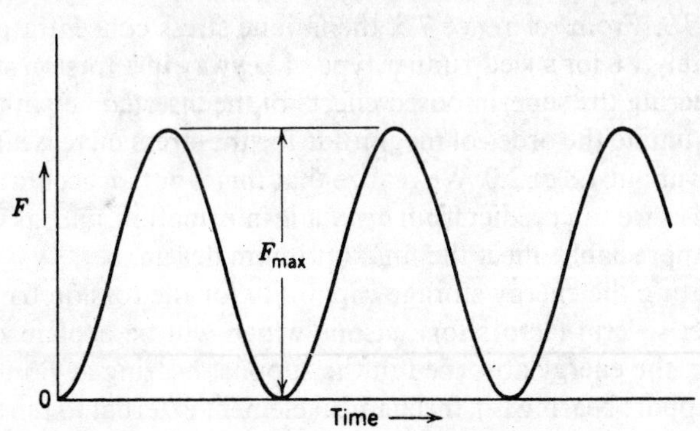

Figure 7.21 General variation of force F in figure 7.20, as introduced in shock loading of torsion bar spring.

this is definitely desirable. Thus we shall designate the location of the support bearing from the load arm by a parameter b, whose optimum value will be determined by the design procedure. Also, although we are not concerned with the design of the load arm in this chapter, we shall designate its length by the parameter r as shown in figure 7.20, and its optimum value will be determined by the design procedure.

Most real mechanical elements are characterized by connection regions having significant stresses, such as the keyed ends at A and C for the torsion bar spring in figure 7.20. Generally a standard type of connection region will be selected, depending to some extent on economic considerations and available manufacturing facilities. Hence such standard connections as splines, pins, or keys might be used, and for any particular type of available connection an objective should be to minimize the stress increase factor by proportional design, as discussed in section 7.4. For the sake of simplicity in the design study which follows, let us assume that standard square key type of connections will be used for both ends of the torsion bar in figure 7.20. Assuming normal manufacturing methods, a press fit will *not* exist at the keyed regions, and therefore we can assume with reasonable confidence that virtually a pure twisting moment will exist at sections A and C of the torsion bar, with the bending moment essentially zero at those regions. We shall be interested in using the *stress increase factors*, $(K_i)_A$ and $(K_i)_C$, in the bar at sections A and C, respectively. Although precise values for these stress increase factors are indeterminable because of boundary conditions and the presence of the key in the keyway, we can estimate their orders of magnitude from available information in the literature [7.3–7.5]. Also, it is well known from experimental evidence that the commonly used sled-runner keyway results in an appreciably lower stress concentration factor than the profiled type of keyway. Thus we would obviously select the sled-runner type of keyway having good design proportions as explicitly determined by a study such as in exercise 7.9. From reference 7.5, the fatigue stress concentration factor is approximately 1.6 for a sled-runner type of keyway in a torsion shaft. Therefore, considering the superimposed effects of the inserted key in the keyway, we could estimate the order of magnitude for the stress increase factors $(K_i)_A$ and $(K_i)_C$ as about 1.8 or 2.0. We realize that this is not an accurate value, but it is as good as we can predict from *available* information, and, as we shall see, it will not appreciably affect the final optimum design.

In calculating the energy storage capability for the torsion bar spring, we shall neglect several factors for reasons which will be explained. First, we shall neglect the energy absorbed in the support bearing at B in figure 7.20, since the support bearing is a mechanical element external to the torsion bar. Second, we shall neglect the energy absorbed in the load arm, since it is also a mechanical element which is really external to the torsion bar and its optimum design would be a separate problem. Finally, we shall neglect energy absorbed in the torsion bar because of beam shear deflections; this is reasonable for metallic materials having a high span to depth ratio. The primary amount of energy absorbed by the bar in figure 7.20 will be the result of twisting and bending deformations.

Optimum Design of the Torsion Bar Spring. Following the MOD, we shall write the primary design equation expressing the energy stored in the torsion bar spring. Hence, from basic mechanics, we can resolve the load arm force F at any instant into a directly applied force F and a twisting moment couple $M_t = Fr$, as shown in figure 7.20 by dotted line designation. The directly applied force F causes bending of the bar, acting as a beam, whereas the twisting moment couple M_t causes twisting of the bar, acting as a pure torsion shaft. Therefore the total energy stored in the torsion bar spring at any instant consists of a bending component superimposed on a torsion component. Thus we can write

$$(\text{P.E.}) = (\text{P.E.})_t + (\text{P.E.})_b \tag{a}$$

In equation a, (P.E.) is the total energy stored in the torsion bar spring at any instant, $(\text{P.E.})_t$ is the energy stored because of twisting, and $(\text{P.E.})_b$ is the energy stored because of bending. Since the effects of the keyway stress concentrations are localized to relatively small vicinities in the connection regions, we can ignore their influence on the total energy (P.E.), which is integrated over the entire torsion bar spring.

From a knowledge of elementary strength of materials, we can express the energy stored because of twisting in the bar as follows:

$$(\text{P.E.})_t = \tfrac{1}{2}M_t\theta = \tfrac{1}{2}M_t\left(\frac{M_tL}{JG}\right)$$

where

$$G = \frac{E}{2(1 + \nu)}$$

$$J = \frac{\pi d^4}{32}$$

$$M_t = Fr$$

$$\therefore \quad (\text{P.E.})_t = \tfrac{1}{2}(Fr)\left[\frac{(Fr)L}{(\pi d^4/32)}\left(\frac{1 + \nu}{E/2}\right)\right]$$

$$\therefore \quad (\text{P.E.})_t = \frac{32}{\pi}\frac{(1 + \nu)}{E}\left(\frac{F^2r^2L}{d^4}\right) \tag{b}$$

In equation b, ν is Poisson's ratio and E is the modulus of elasticity, both for the bar material; F is the force depicted in figure 7.20; and r, L, and d are the independent geometrical parameters shown in that figure.

From elementary strength of materials we can also express the energy stored because of bending in the bar as follows:

$$(\text{P.E.})_b = \tfrac{1}{2}F(\Delta_C)$$

where Δ_C is the vertical deflection of the axis of the bar at section C as depicted in figure 7.20. Using the second area-moment proposition from elementary strength of materials applied to the beam in figure 7.22, we obtain

$$\Delta_C = \frac{FLb^2}{3EI}$$

where $I = \pi d^4/64$.

$$\therefore \quad (\text{P.E.})_b = \tfrac{1}{2}F(\Delta_C) = \tfrac{1}{2}F\left(\frac{FLb^2 64}{3E\pi d^4}\right)$$

$$\therefore \quad (\text{P.E.})_b = \left(\frac{32}{3\pi}\right)\left(\frac{F^2b^2L}{Ed^4}\right) \tag{c}$$

In equation c, E is the modulus of elasticity for the bar material, F is the force depicted in figure 7.20, and b, L, and d are the independent geometrical parameters shown in that figure.

Let us now combine equations a, b, and c:

$$(\text{P.E.}) = (\text{P.E.})_t + (\text{P.E.})_b$$

$$= \left(\frac{32}{\pi}\right)\left(\frac{F^2L}{Ed^4}\right)\left[(1+\nu)r^2 + \frac{b^2}{3}\right]$$

$$\therefore \quad (\text{P.E.}) = \left(\frac{32}{3\pi}\right)\left(\frac{F^2r^2L}{Ed^4}\right)\left[3(1+\nu) + \left(\frac{b}{r}\right)^2\right] \tag{7.31}$$

Equation 7.31 is the (P.D.E.) for this problem, and all terms have already been defined in conjunction with equations a, b, and c.

Subsidiary design equations will be written for significant stresses in the torsion bar spring. Since we shall be concerned with fatigue loading of the bar, we shall select the maximum shearing stress theory of fatigue failure for reasons explained in Chapter 4. Thus we shall write equations for the maximum shearing stresses at significant places in the bar. Hence, since virtually pure twisting exists in the bar at sections A and C in figure 7.20, we can write

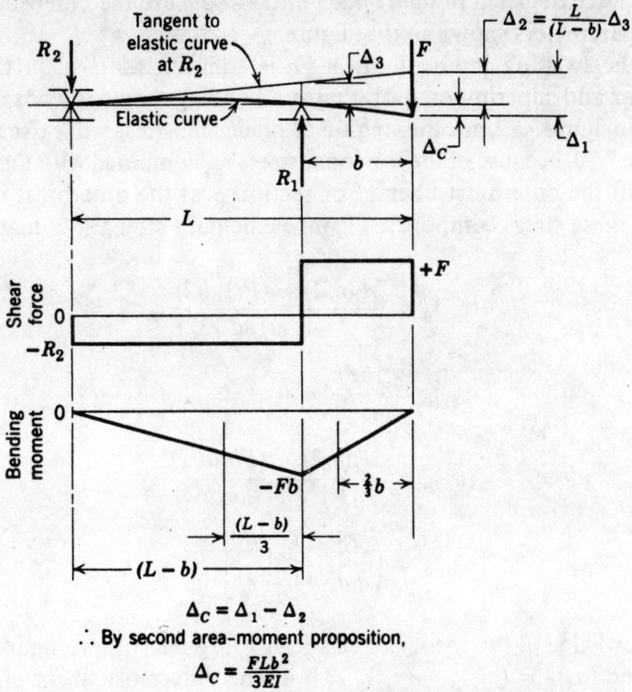

Figure 7.22 Beam from figure 7.20 showing loads, reactions, shearing forces, bending moments, and elastic curve.

an eqation for $(\tau_{max})_A$ and $(\tau_{max})_C$ similar to equation 7.17. Therefore we obtain

$$(\tau_{max})_A = (K_i)_A\left(\frac{16M_t}{\pi d^3}\right)$$

$$\therefore\quad (\tau_{max})_A = (K_i)_A\left(\frac{16Fr}{\pi d^3}\right) \tag{7.32}$$

and

$$(\tau_{max})_C = (K_i)_C\left(\frac{16Fr}{\pi d^3}\right) \tag{7.33}$$

In equations 7.32 and 7.33, $(\tau_{max})_A$ and $(\tau_{max})_C$ are the maximum shearing stresses at regions A and C respectively in figure 7.20, $(K_i)_A$ and $(K_i)_C$ are the *stress increase factors* for the key connection regions at A and C, F is the load arm force depicted in figure 7.20, and r and d are the independent geometrical parameters shown in that figure.

Since the twisting moment $M_t = Fr$ is transmitted *through* the entire torsion bar and superimposed on that is a bending moment whose diagram is shown in figure 7.22, another region of *significant* stress will exist at section B in figure 7.20, because of the torsional stress τ_{xy} combined with the bending stress σ_x at the outermost fibers. For section B at the outermost fibers, we can write these stress components from elementary strength of materials:

$$(\tau_{xy})_B = \frac{M_t(d/2)}{J} = \frac{(Fr)(d/2)}{(\pi d^4/32)}$$

$$\therefore \quad (\tau_{xy})_B = \frac{16Fr}{\pi d^3} \tag{d}$$

$$(\sigma_x)_B = \frac{M_b(d/2)}{I} = \frac{(Fb)(d/2)}{(\pi d^4/64)}$$

$$\therefore \quad (\sigma_x)_B = \left(\frac{16Fr}{\pi d^3}\right)\left[2\left(\frac{b}{r}\right)\right] \tag{e}$$

From knowledge of the theory of elasticity [7-1], referring to figure 7.20 we can assume $(\sigma_y)_B = (\tau_{xz})_B = (\tau_{yz})_B = 0$ at the outermost fibers of the bar. Also, we can show by appropriate calculations that compressive stress $(\sigma_z)_B$ will be less than one-tenth of $(\tau_{xy})_B$ for any reasonable bearing support. Thus by combined stress theory as reviewed in section 4.5 of Chapter 4 for the triaxial state, we conclude that the governing principal shearing stress will be as follows, giving equation 7.34a for $(\tau_{max})_B$. The derivation incorporates equations d and e for $(\tau_{xy})_B$ and $(\sigma_x)_B$ respectively.

$$\tau_{1,2} = \sqrt{\left(\frac{\sigma_x - \sigma_y}{2}\right)^2 + \tau_{xy}^2}$$

$$= \sqrt{\left[\left(\frac{16Fr}{\pi d^3}\right)\left(\frac{b}{r}\right)\right]^2 + \left(\frac{16Fr}{\pi d^3}\right)^2}$$

$$\therefore \quad (\tau_{max})_B = \frac{16Fr}{\pi d^3}\sqrt{\left(\frac{b}{r}\right)^2 + 1} \tag{7.34a}$$

In comparing equation 7.34a with equations 7.32 and 7.33, we might consider $\sqrt{(b/r)^2 + 1}$ a stress increase factor, $(K_i)_B$, at section B of the bar because of the effect of bending. This conformation with equations 7.32 and 7.33 will be advantageous in the conduct of the method of optimum design. Thus we can write the (S.D.E.), equation 7.34a, as follows:

$$(\tau_{max})_B = (K_i)_B \frac{16Fr}{\pi d^3} \tag{7.34b}$$

where

$$(K_i)_B = \sqrt{\left(\frac{b}{r}\right)^2 + 1} \tag{7.34c}$$

Hence the significant stresses $(\tau_{max})_A$, $(\tau_{max})_B$, and $(\tau_{max})_C$ are all expressed as the product of a stress increase factor (K_i) and the nominal maximum torsion stress $16Fr/(\pi d^3)$. Notice that stress increase factors $(K_i)_A$ and $(K_i)_C$ will depend on the proportional design of the key connection regions, and as previously discussed their order of magnitude will be about 1.8. On the other hand, the stress increase factor $(K_i)_B$ will be a function of the ratio (b/r) in figure 7.20. Considering the significant stress equations 7.32, 7.33, and 7.34b in comparable form will enable us to conduct a single stage of optimum design instead of two or three.

Before starting the procedure of optimum design we shall write the limit equations, of which there are three for this problem. Equations 7.35 and 7.36 can be written directly because of the existing space limitations in the particular structure.

$$d \leq d_{max} \tag{7.35}$$

$$L \leq L_{max} \tag{7.36}$$

A third limit equation will exist on permissible values for any critical shearing stress τ_{max}, and it is written from the selected criterion for failure. Since the time variation of loading for the torsion bar is as generally shown in figure 7.21, the limit equation on stress could be derived exactly as was equation 7.24. Thus we have the following equation, which if satisfied at every point in the torsion bar will assure us of no undesirable fatigue failures.

$$\tau_{max} \leq \frac{S_e}{(1 + p)N_e} \tag{7.24}$$

In equation 7.24, τ_{max} is the greatest value for the maximum shearing stress at any point of significance in the bar; S_e is the published fatigue strength from a standard reversed bending type of fatigue test for the desired number of cycles of life; p is the ratio S_e/S_y, where S_y is the published yield strength of the material from a standard tensile test; and N_e is a factor of safety based on the significance attached to the occurrence of a fatigue failure and which is an estimated reasonable and conservative value, as discussed in Chapter 5.

Following the MOD, we shall now combine the significant stress equation of the form $\tau_{max} = K_i[16Fr/(\pi d^3)]$, from the (S.D.E.)s, equations 7.32, 7.33, or 7.34b, with the (P.D.E.), equation 7.31, by eliminating the unspecified and unlimited variable F. Thus we obtain the developed (P.D.E.), equation 7.31a, as follows:

$$\text{(P.E.)} = \left(\frac{32}{3\pi}\right)\left(\frac{\tau_{max}\pi d^3}{16rK_i}\right)^2\left(\frac{r^2L}{Ed^4}\right)\left[3(1 + \nu) + \left(\frac{b}{r}\right)^2\right]$$

$$\therefore \quad \text{(P.E.)} = \left(\frac{\pi}{24}\right)\left(\frac{\tau_{max}^2 Ld^2}{EK_i^2}\right)\left[3(1 + \nu) + \left(\frac{b}{r}\right)^2\right] \tag{7.31a}$$

From this (P.D.E.), we see that, for any feasible geometry and for any feasible material, energy absorption in the bar will be as great as permissible if the independent parameter τ_{max} is equal to its upper limit as specified by equation 7.24. Thus we obtain the (P.D.E.), equation 7.31b, as follows, where (P.E.)_c is the energy absorption *capability* of the bar.

$$\text{(P.E.)}_c = \left(\frac{\pi}{24}\right)\left[\frac{S_e}{(1 + p)N_e}\right]^2\left(\frac{Ld^2}{EK_i^2}\right)\left[3(1 + \nu) + \left(\frac{b}{r}\right)^2\right]$$

$$\therefore \quad \text{(P.E.)}_c = \left(\frac{\pi}{24N_e^2}\right)\left[\frac{S_e^2}{(1 + p)^2E}\right]\left(\frac{Ld^2}{K_i^2}\right)\left[3(1 + \nu) + \left(\frac{b}{r}\right)^2\right] \tag{7.31b}$$

From equation 7.31b, we would consider the independent factor $[S_e^2/(1 + p)^2E]$ as our material selection factor, neglecting the effect of Poisson's ratio ν in that equation. This is valid because Poisson's ratio has virtually the same numerical value for all available feasible metals, such that $\frac{1}{4} \leq \nu \leq \frac{1}{3}$, and hence variations of ν in this range would have only a *negligible* effect on energy absorption *capability* according to equation 7.31b. Thus, from the discrete list of available materials in Table 4.1, we would select the optimum material as the one having the highest value for the material selection factor $[S_e^2/(1 + p)^2E]$.

From the (P.D.E.), equation 7.31b, it is obvious that, for any material, energy absorption *capability* (P.E.)_c is maximum if the independent geometrical parameters L and d are placed at their upper limits as specified by equations 7.35 and 7.36. Hence we obtain the (P.D.E.), equation 7.31c, as follows:

$$\text{(P.E.)}_c = \left(\frac{\pi L_{max} d_{max}^2}{24N_e^2}\right)\left[\frac{S_e^2}{(1 + p)^2E}\right]\left[\frac{3(1 + \nu) + (b/r)^2}{K_i^2}\right] \tag{7.31c}$$

In this developed (P.D.E.), the only unknown geometrical parameter is the ratio (b/r). The stress increase factor K_i is a constant of approximately 1.8 for the keyway regions, and $(K_i)_B$ is a function of the ratio (b/r) according to equation 7.34c. Thus, regardless of whether we are considering region A, B, or C in figure 7.20, our remaining task in optimum design is to determine the optimum value for the independent parameter (b/r). Hence from the (P.D.E.), equation 7.31c, in order to complete the optimum design study we need merely look at the variation of the independent factor

$$\left[\frac{3(1+\nu)+(b/r)^2}{K_i^2}\right] \sim (\text{P.E.})_c$$

versus the ratio (b/r), for the three regions A, B, and C.

First, consider the variation of $(\text{P.E.})_c$ versus the independent geometrical parameter ratio (b/r) as governed by the keyway regions A and C, where $(K_i)_A$ and $(K_i)_C$ are *constants* which are independent of (b/r). Hence the variation of $(\text{P.E.})_c$ with respect to (b/r) is parabolic, having increasing energy absorption *capability* $(\text{P.E.})_c$ with increasing (b/r). Thus energy absorption *capability* as governed by stress at keyway regions A and C is as shown in the parabolic sketch of figure 7.23. The interesection of this curve with the $(b/r) = 0$ axis is obviously $3(1+\nu)/K_i^2$, where K_i is either $(K_i)_A$ or $(K_i)_C$ whose orders of magnitude are both about 1.8, assuming sled-runner keyways with good geometrical proportions.

Second, consider the variation of $(\text{P.E.})_c$ versus the independent geometrical parameter ratio (b/r) as governed by combined stresses at region B in figure

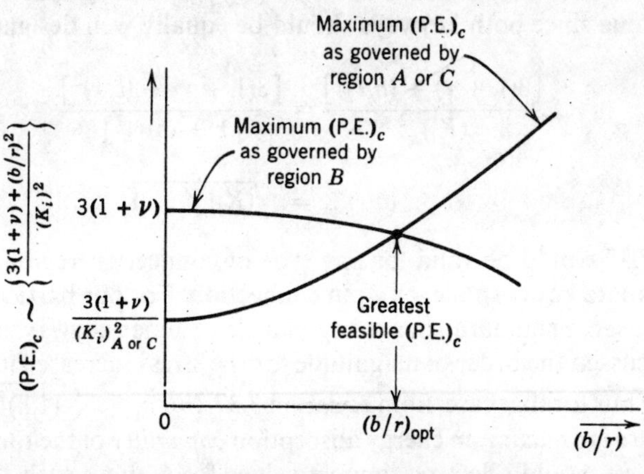

Figure 7.23 Variations of energy absorption capability for torsion bar spring with respect to independent geometrical parameter (b/r) as governed by region A, B, or C of figure 7.20.

7.20. Hence, by equations 7.31*c* and 7.34 *c*, we are interested in studying the following variation, which is proportional to energy absorption *capability* as governed by combined stresses at region *B*.

$$(\text{P.E.})_c \sim \frac{3(1+\nu)+(b/r)^2}{(K_i)_B{}^2} = \frac{3(1+\nu)+(b/r)^2}{[(b/r)^2+1]} \tag{f}$$

Therefore

$$\frac{\partial(\text{P.E.})_c}{\partial(b/r)} \sim \frac{[(b/r)^2+1]2(b/r) - [3(1+\nu)+(b/r)^2]2(b/r)}{[(b/r)^2+1]^2}$$

$$\therefore \quad \frac{\partial(\text{P.E.})_c}{\partial(b/r)} \sim -\frac{2(b/r)(2+3\nu)}{[(b/r)^2+1]^2} \tag{g}$$

Thus the variation of energy absorption *capability* versus the ratio (b/r) as governed by stress at region *B* is as shown in the sketch of figure 7.23. This curve is roughly drawn by taking into account its interesection with the $(b/r) = 0$ axis, using equation f, and the facts that its slope is always negative in the region $0 < (b/r) < \infty$ from equation g, and that its slope is zero at $(b/r) = 0$ and at $(b/r) \to \infty$ from that equation.

Thus the relative positions of the energy absorption *capability* curves are as shown in figure 7.23. Obviously, the optimum value for the ratio (b/r) will be at the intersection of the two curves in that figure. This value will be designated $(b/r)_{opt}$, and for that particular ratio the significant stresses would be equally important at regions *A*, *B*, and *C*. Hence, equating the energy factors from equation 7.31c, let us determine $(b/r)_{opt}$ as follows. For the keyed regions we will denote the stress increase factor by $(K_i)_A$, assuming that $(K_i)_C$ will be the same value since both keyways should be equally well designed.

$$\therefore \quad \frac{[3(1+\nu)+(b/r)^2]}{(K_i)_A{}^2} = \frac{[3(1+\nu)+(b/r)^2]}{[1+(b/r)^2]}$$

$$\therefore \quad (b/r)_{opt} = \sqrt{(K_i)_A{}^2 - 1} \tag{7.37}$$

Equation 7.37 would be valid for any type of connection region, such as a standard square key, a spline, or a pin connection. For our particular design we have chosen a standard square key in a sled-runner keyway, and as previously discussed the order of magnitude for the stress increase factor will be about 1.8. Thus for that case, from equation 7.37, $(b/r)_{opt} \approx \sqrt{(1.8)^2 - 1} = 1.5$, and therefore for maximum energy absorption *capability* of the torsion bar in figure 7.20 we would select reasonable values for *b* and *r* such that (b/r) is approximately 1.5. This will be illustrated in the numerical example below.

Example 7.2

Suppose that it is desired to design a torsion bar as shown in figure 7.20 on the basis of maximum energy absorption *capability* per cycle of repeated loading, using the available materials from table 4.1. The geometrical limitations imposed by the structure are

$$L \leq 0.75 \text{ m}$$

$$d \leq 35 \text{ mm}$$

Hence, from the study covered in the section, assuming well-designed square key connections, the optimum design would be as described below.

Optimum Material. Select the material having the highest value for the material section factor, $[S_e^2/(1 + p)^2E]$ from equation 7.31c. Thus for the titanium alloy of table 4.1, we calculate

$$\text{MSF} = \frac{S_e^2}{(1 + p)^2E} = \frac{(586 \times 10^6)^2}{[1 + (586/896)]^2(110 \times 10^9)} = 1.141 \text{ MPa}$$

Of the steels in table 4.1, for illustrative purposes we will consider only the AISI 2340 Oil Q & D 400°F one. Suppose the surface finish will be about 63×10^{-6} in AA, which gives $k_a \approx 0.6$ from figure 3.8 of Chapter 3. Hence, assuming $k_b = k_c = 1.0$, equation 3.17 gives $S_e \approx k_a k_b k_c S'_e \approx 0.6 \times 100{,}000 = 60{,}000 \text{ psi} \times 6.895 \times 10^3 = 414$ MPa. (For a summary of some English and SI unit relations, see Appendix D of reference 7.8.) Therefore, for this particular steel we calculate

$$\text{MSF} = \frac{S_e^2}{(1 + p)^2E} = \frac{(414 \times 10^6)^2}{[1 + (414/1600)]^2(200 \times 10^9)} = 0.541 \text{ MPa}$$

With the MSF values now calculated, we must make a final decision on the material selection. Thus if all practical considerations are acceptable, such as cost, availability, and machinability, we would choose the titanium alloy as the better choice for maximization of $(\text{P.E.})_c$. However, if from such final considerations we believe that the steel would be more practical and accepted by our particular company, we would choose it as the second best material from the standpoint of maximizing $(\text{P.E.})_c$. With the MSF values at hand, we can at least compare the materials from the standpoint of the optimization objective and decide whether or not the gain is worth the price which might have to be paid because of the practical considerations. Tempering final decisions of design in this way is often necessary in realistic situations of professional practice.

Optimum Geometry. From the results of our optimization study, for maximizing $(P.E.)_c$ we set $d = d_{max} = 35$ mm and $L = L_{max} = 0.75$ m. Also, we select reasonable values for b and r such that $b/r = 1.5$. Good proportions would be obtained by letting $r = 110$ mm and $b = 165$ mm. This would complete the optimum design of the torsion bar shown in figure 7.20.

If the torsion bar had been designed with $b \approx 0$ based on intuition, or traditional practice, an appreciable decrease in energy absorption *capability* would result even if the optimum material were used and if $d = d_{max}$ and $L = L_{max}$ were selected. Hence, using equation 7.31c we can see the value of using $(b/r)_{opt}$ instead of $b \approx 0$. Thus

$$\frac{(P.E.)_{(b/r)opt}}{(P.E.)_{(b/r)\sim 0}} = \frac{[3(1 + \nu) + (b/r)^2_{opt}]/(K_i)_A{}^2}{[3(1 + \nu) + 0]/(K_i)_A{}^2}$$

$$= 1 + \frac{(b/r)^2_{opt}}{3(1 + \nu)} = 1 + \frac{(1.5)^2}{3(1 + \frac{1}{4})}$$

$$= 1 + 0.60$$

Therefore a 60% increase in energy absorption *capability* is realized by choosing $(b/r) = (b/r)_{opt} = 1.5$ for this problem, assuming Poisson's ratio $\nu = \frac{1}{4}$. Thus, in actual design of the torsion bar, it would be profitable to select reasonable values of b and r so that (b/r) will be as close to the optimum value as possible.

7.9 Dynamic Shaft in Combined Loading, (R.S.)

Let us modify the torsion bar design problem of section 7.8 as follows, retaining the configuration of figure 7.20 and the optimization objective of maximizing strain energy capacity, $(P.E.)_c$. Because of functional requirements of the entire system in which the bar must operate, r is a specified value and diameter d must be a standard size value d_s of which there are N. Also, from a vibratory and dynamic analysis of the entire system, force gradient $k = F/\Delta_F$ must be within an acceptable range, $k_{min} \leq k \leq k_{max}$. Limit values k_{min} and k_{max} have been determined by the dynamic analysis, for acceptable performance of the system.

For derivation of the k equation, we apply the theorem of Castigliano [7.9] to equation 7.31.

$$\Delta_F = \frac{\partial(P.E.)}{\partial F} = \left(\frac{64Fr^2L}{3\pi Ed^4}\right)\left[3(1 + \nu) + \left(\frac{b}{r}\right)^2\right]$$

Thus since $k = F/\Delta_F$ we obtain equation 7.38 from the preceding.

$$k = \frac{3\pi E d^4}{64 r^2 L[3(1 + \nu) + (b/r)^2]} \tag{7.38}$$

All terms for equation 7.38 have now been defined above or in equation 7.31 of the previous section.

Suppose for a practical design that we wish to constrain b of figure 7.20 in two ways. For one, the proportions should be practical and the geometric ratio b/L should not be excessive. Hence we define a new dimensionless variable x by

$$x = \frac{b}{L} \tag{7.39}$$

and impose the regional constraint

$$x \leq x_{max}$$

A typical value for x_{max} might be, say, 0.3. Second, suppose we wish to limit b of figure 7.20 to an acceptable range because of mounting considerations in the machine setting not shown. Thus we will impose the regional constraint $b \leq b_{max}$ in our (I.F.).

In summary, the constraints for the modified problem will be as follows.

$d = d_s \leq d_{max}$, for $s = 1, \ldots, N$; $k_{min} \leq k \leq k_{max}$; $x \leq x_{max}$; $b \leq b_{max}$; and limit equations 7.36 and 7.24 are carried over from the previous problem, requiring $L \leq L_{max}$ and $\tau_{max} \leq S_e/(1 + p)N_e$ for an acceptable design.

Specified values are known for the standard sizes d_s, for $s = 1, \ldots, N$; and all of the preceding limit values are known at this stage of design. Also, the value of $(K_i)_A$ has been determined by geometric proportions for the connection region, and as in the previous section we will assume a value of about 1.8 for K_i at A and C of figure 7.20. Also, as stated earlier, r is a given constant. Finally, a list of feasible available materials is assumed to exist with known properties, such as those in tables 4.1 and 4.2 of Chapter 4.

Final Formulation. A simple approach for derivation of the (F.F.) is quite obvious from a review of the optimization problem as stated, with the objective of maximizing strain energy capacity, (P.E.). Suppose we choose k and x as the eliminated parameters for this (R.S.) type of optimization problem.

Then, equations 7.31a, 7.38, and 7.39 will constitute the (F.F.) as they stand. Thus

$$(\text{P.E.}) = \left(\frac{\pi}{24}\right)\left(\frac{\tau_{max}^2 Ld^2}{EK_i^2}\right)\left[3(1+\nu) + \left(\frac{b}{r}\right)^2\right] \quad \text{(I)}$$

$$k = \frac{3\pi Ed^4}{64r^2L[3(1+\nu)+(b/r)^2]} \quad \text{(II)}$$

$$x = \frac{b}{L} \quad \text{(III)}$$

where k, x = eliminated parameters

d, L, b = related parameters.

In the (F.F.) we note that τ_{max} exists only in the developed (P.D.E.) of the ideal problem, equation I. Hence it is a truly independent parameter, as defined in Chapter 6. Thus for maximization of (P.E.) we place τ_{max} at its upper limit value $S_e/(1+p)N_e$, the decision being reached from a consideration of equation I alone. Therefore, from now on the developed (P.D.E.) of the ideal problem becomes as follows, expressing energy storage *capacity* $(\text{P.C.})_c$ as a function of geometry for a material of interest.

$$(\text{P.E.})_c = \left(\frac{\pi}{24N_e^2}\right)\left[\frac{S_e^2}{(1+p)^2E}\right]\left(\frac{Ld^2}{K_i^2}\right)\left[3(1+\nu) + \left(\frac{b}{r}\right)^2\right] \quad \text{(Ia)}$$

However, the fact that we have material parameters E and ν in the relating equations II and III means that there is no MSF for this problem, which is generally the case for the (R.S.) type of problem.

With three related parameters in the (F.F.), our variation study is basically four dimensional in character. However, a practical approach for the (V.S.) would be to hold d constant at a standard size d_s value, which will then make our variation diagram three dimensional. A typical variation diagram is sketched in figure 7.24 for $d = d_s$ constant and for a specific material of interest. From this variation diagram, a (F.C.) will be derived which would be the basis for writing a computer program. The (F.C.) would be executed by the computer for each value of d_s for each material of interest. Thus the overall optimum design would be the one of greatest $(\text{P.E.})_c$ value so determined, and this would be the index for our specification of the optimum material and the optimum geometry.

Variation Study. The sequence of steps for sketching the typical 3-D variation diagram of figure 7.24 will now be explained, including interpretation of its meaning as the basis for logical derivation of the calculation (F.C.) which follows. First we sketch the related parameter axes b, L and impose

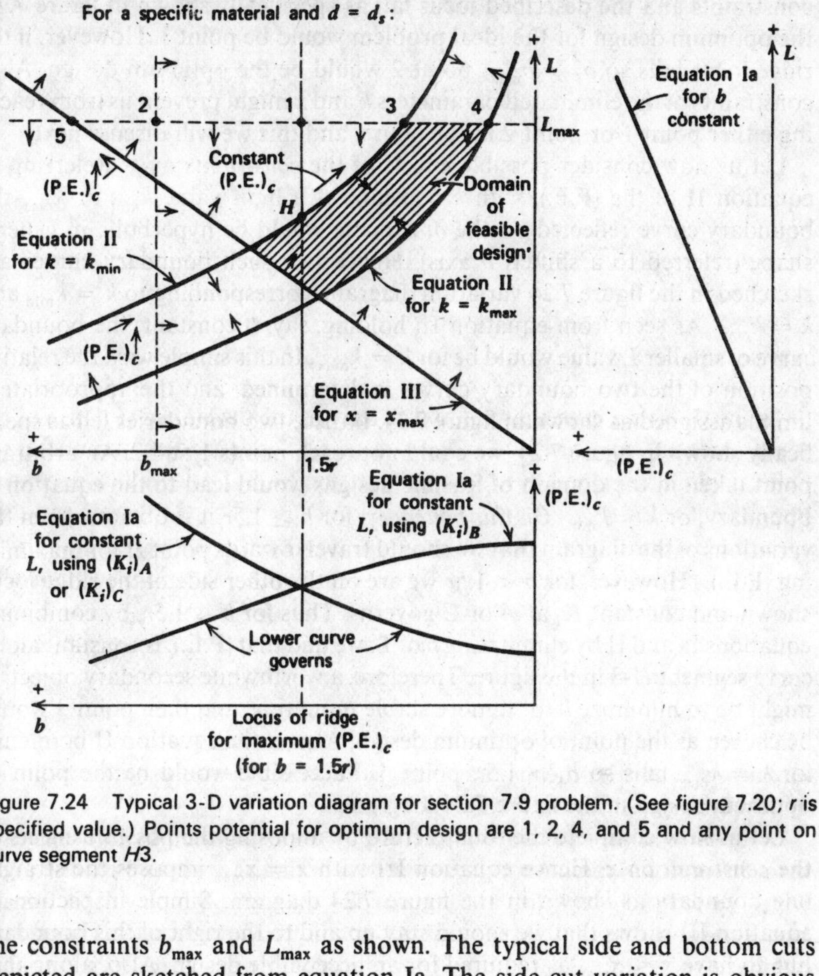

igure 7.24 Typical 3-D variation diagram for section 7.9 problem. (See figure 7.20; r is pecified value.) Points potential for optimum design are 1, 2, 4, and 5 and any point on urve segment $H3$.

he constraints b_{max} and L_{max} as shown. The typical side and bottom cuts lepicted are sketched from equation Ia. The side cut variation is obvious ince $(P.E.)_c$ varies linearly with L in equation Ia. The bottom cut is more lifficult, but its variation is exactly the same as the composite one already erived for figure 7.23. The same explanation presented near the end of the ast section applies directly here, since both d and L are constant for the bo - om cut of figure 7.24 as they were for figure 7.23. Hence, in the 3-D diagram of figure 7.24 there is a locus in the form of a ridge for the criterion function P.E.)$_c$, defined by $b = 1.5r$. Since r is a specified value for this problem the ocus so defined is of constant b value, as shown in figure 7.24.

Let us now consider the *ideal problem* by itself, temporarily ignoring the onstraints on the eliminated parameters k and x. The feasible design space s the rectangle bounded by the b, L axes and the b_{max}, L_{max} limit lines. If these constraints and the described locus fall as specifically shown in figure 7.24 the optimum design for the ideal problem would be point 1. However, if the ridge locus falls so $b_1 > b_{max}$, point 2 would be the optimum design. Also constraints of the eliminated parameters k and x might prevent us from reaching either point 1 or point 2 in the figure, and this we will discuss next.

Let us now consider possible effects of the *constraints on k*. Referring to equation II of the (F.F.), with k constant at a limit value k_{min} or k_{max} the boundary curve reflected on the b, L plane would be hyperbolic in general shape (referred to a shifted b axis). Hence two such boundary curves are sketched in the figure 7.24 variation diagram, corresponding to $k = k_{min}$ and $k = k_{max}$. As seen from equation II, holding, say, b constant, the boundary curve of smaller L value would be for $k = k_{max}$. In this simple way, the relative position of the two boundary curves is determined, and the appropriate k limit is assigned as shown in figure 7.24. If these two boundaries fell as specifically shown in figure 7.24, we could not reach points 1 and 2. An arbitrary point taken in the domain of feasible designs would lead to the equation II boundary for $k = k_{min}$. *On this boundary*, for $b \geq 1.5r$ it is obvious from the variations of the diagram that we should travel towards point H for maximizing $(\text{P.E.})_c$. However, for $b < 1.5r$ we are on the other side of the ridge locus shown and constant K_i at A or C governs. Thus for $b < 1.5r$, by combining equations Ia and II by eliminating b or L, we find that $(\text{P.E.})_c$ is constant along curve segment H-3 in the figure. Therefore, a worthwhile secondary objective might be to minimize b for a more stable mounting, and then point 3 would be chosen as the point of optimum design. Also, if the equation II boundary for $k = k_{max}$ falls so $b_4 > 1.5r$, point 4 if accessible would be the point of optimum design in the figure 7.24 diagram.

Let us now complete the total picture by imposing the possible effects of the *constraint on x*. Hence equation III with $x = x_{max}$ imposes the straight line boundary as shown in the figure 7.24 diagram. Simple inspection of equation III shows that we should stay up and to the right of this boundary line to have $x \leq x_{max}$ as required for an acceptable design. Also, along that boundary line we could prove mathematically by combining equations Ia and III by eliminating b or L that $(\text{P.E.})_c$ increases as we move away from the origin regardless of whether we use constant K_i for A or C or vary K_i for B by equation 7.34c. Based on the important characteristics now summarized we conclude that point 5 is potential for optimum design in figure 7.24. It would be the point specifically for the case where $b_5 < b_1, b_5 < b_{max}, b_5 < b_3$ and $b_5 > b_4$.

From our sketching of the eliminated parameter boundary curves in the variation diagram of figure 7.24, we note that there are two possibilities for incompatible specifications. First, (I.S.) would be the case if we had $b_4 > b_{max}$ Second, (I.S.) would also be the case if we had $b_5 < b_4$.

Flow Chart Derivation. Based on the variation study just explained, the (F.C.) is logically derived as presented in figure 7.25. After input of specified values, the two possibilities of (I.S.) are screened out. If the tests for existence of a feasible design are passed in a numerical application, the actual point of optimum design is determined from potential points 1 through 5 of the figure 7.24 diagram. Incidentally, as previously explained, point 3 is selected if curve segment H-3 is accessible for the case where $b_3 < b_1$, and

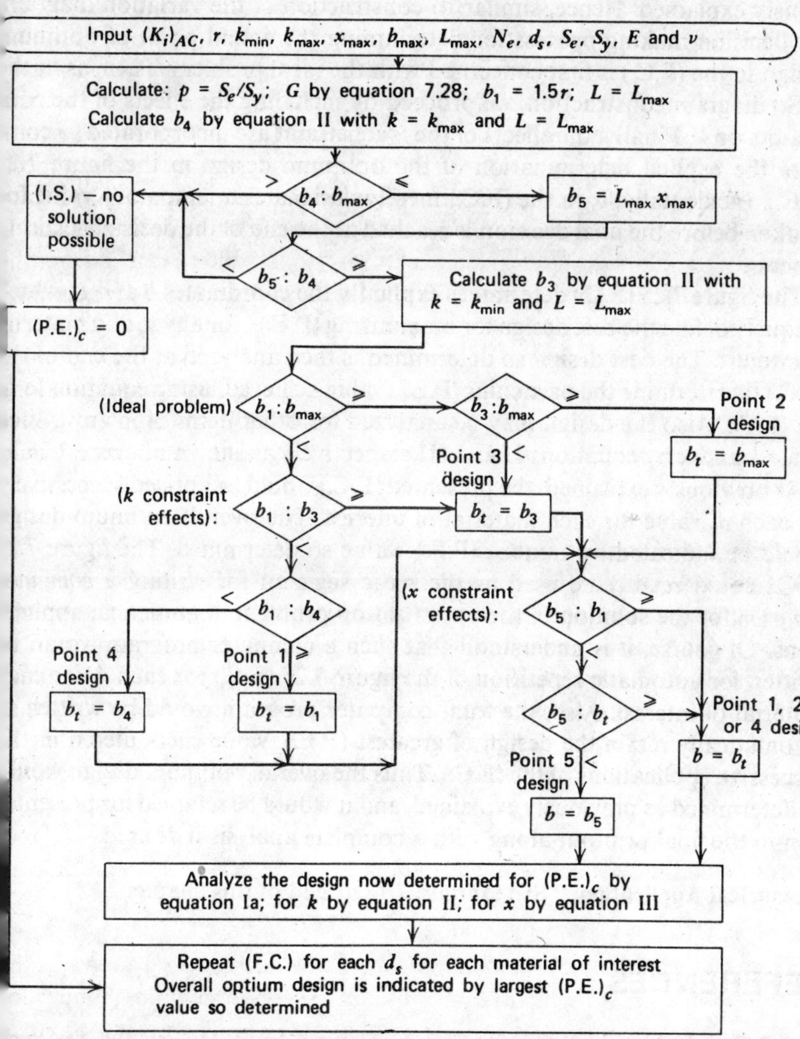

Figure 7.25 Basic flow chart segment for solution of section 7.9 optimization problem.

minimization of b is chosen as the secondary objective along that curve segment of constant peak $(P.E.)_c$ value.

Coordinates for the various points needed for the comparisons in the decision-making process of the (F.C.) are readily determined from the figure 7.24 diagram, using the equations involved as summarized therein and the particular boundary conditions of the point in question. The decision making procedure of the (F.C.) is synthesized from an understanding of how the variation diagram was constructed and the variations involved, as previously explained. Hence, similar to construction of the variation diagram the decision-making process for determining the actual point of optimum design in the (F.C.) is first concerned with the ideal problem. Then, as in the (V.S.) diagram construction, we proceed by including the effects of the constraints on k. Finally, the effects of the x constraint are incorporated to complete the explicit determination of the optimum design in the figure 7.25 (F.C.). Incidentally, b_t in the (F.C.) merely designates a temporary value for b, taken before the final decision is reached at the end of the decision-making process.

The figure 7.25 (F.C.) determines explicitly the coordinates b and L, which uniquely define the best design for maximizing $(P.E.)_c$, for any d_s and material of interest. The best design so determined is then analyzed at the end of the (F.C.) to determine the particular $(P.E.)_c$ value achieved, using equation Ia of the (F.F.). Also the design may be analyzed for other items of interest, such as for k and x by equations II and III respectively as shown in figure 7.25.

As previously explained, the presented (F.C.) would be applied successively for each d_s value for each material of interest. The overall optimum design would be indicated by greatest $(P.E.)_c$ value so determined. The figure 7.25 (F.C.) could readily be used as the basic segment for writing a *computer program* for the solution of the optimization problem in numerical applications. Of course, it is understood that such a computer program would be written for automatic repetition of the figure 7.25 (F.C.) for each d_s for each material of interest. Also, the total computer program would be written to automatically retain the design of greatest $(P.E.)_c$ value encountered in the successive applications of the (F.C.). Thus the overall optimum design would be determined as previously explained, and it would be retained for presentation in the final printout along with a complete analysis if desired.

Numerical Application. See exercise 7.12 at end of this chapter.

REFERENCES

7.1 S. P. Timoshenko and J. N. Goodier, *Theory of Elasticity*, 3rd ed., McGraw-Hill, New York, 1970, 567 pp.

.2 S. Timoshenko, *Theory of Elastic Stability*, McGraw-Hill, New York, 1936, p. 169.

.3 R. E. Peterson, *Stress Concentration Design Factors*, Wiley, New York, 1953, 155 pp.

.4 R. E. Peterson, *Stress Concentration Factors*, Wiley-Interscience, New York, 1974, 317 pp.

.5 C. Lipson, G. C. Noll, and L. S. Clock, *Stress and Strength of Manufactured Parts*, McGraw-Hill, New York, 1950, 257 pp.

.6 R. J. Roark and W. C. Young, *Formulas for Stress and Strain*, 5th ed., McGraw-Hill, New York, 1975, pp. 590–606.

.7 J. H. Faupel, *Engineering Design*, Wiley, New York, 1964, pp. 767–795.

.8 R. C. Johnson, *Mechanical Design Synthesis—Creative Design and Optimization*, 2nd ed., Krieger, Huntington, New York, 1978, 448 pp.

.9 J. P. Den Hartog, *Advanced Strength of Materials*, McGraw-Hill, New York, 1952, Chapter VII.

EXERCISES

Exercises for the design optimization of some simple mechanical elements re described in the remaining part of the chapter. As called for in the particuar exercises, the general problem statement given below applies. Solve all or art of it for the particular exercise as appropriate and to the extent assigned y the instructor.

General Problem Statement.

(a) Solve the described optimization problem by the MOD as follows. See Appendix A for nomenclature.)

1. Summarize the (I.F.), including the (P.D.E.), the (S.D.E.)'s, the (L.E.)'s, nd a list of the specified items. Derive equations where necessary, or when ossible use equations already derived elsewhere in the literature.

2. Review the (I.F.) of step 1, and from the values of n_f and N_s determine hether the type of problem is (N.S.) or (R.S.). If the problem is of the (R.S.) ype, determine values for n_c, n_v, D_{vs}, n_e, n_r, A_t, and N_{ff} . (See section 6.13 of hapter 6 for exploratory equations.)

3. If the problem is of the (N.S.) type, derive the developed (P.D.E.) and ndicate the MSF if separable. Also, outline how the optimum geometry is niquely determined, based on your (V.S.). Finally, solve the numerical roblem if given. Specify the material from table 4.1, the dimensions for the eometry, and the value achieved for optimization quantity.

f the problem is of the (R.S.) type as determined in step 2, then steps 4 through 0 apply if appropriate for the specific exercise.

4. Summarize the (F.F.) equation systems in general format expressions, or the various approaches theoretically possible, as described in section 6.13 f Chapter 6.

5. Specifically choose one of the determinate approaches from step 4 after eviewing the (I.F.) from step 1. Derive and summarize the specific (F.F.) for

the approach you select. Incorporate the general terminology for the MOD where appropriate, such as equations I, II, III, and so on, eliminated parameter(s), related parameter(s), and truly independent parameter(s).

6. Based on steps 4 and 5, determine which of the theoretical approaches are determinate and which, if any, are indeterminate.

7. Sketch the (V.S.) typical variation diagram from the (F.F.) of step 5 for a specific material of interest. Clearly label all curves and limits in the diagram and uniquely define all points which are potential for optimum design. Also from the variation diagram sketched, determine the possibilities, if any, for (I.S.).

8. From the (V.S.) now made in step 7, derive the calculation (F.C.) which could be applied for solution of a numerical problem.

9. Using the (F.C.) from step 8, write a computer program (programmable calculator type is all right if you wish). Apply your computer program to solve the numerical problem if one is given. For your numerical answer specify the optimum material from tables 4.1 and 4.2 of Chapter 4 and specify the dimensions uniquely defining the optimum design. Also, specify the value for optimization quantity achieved and any other items of interest for analysis of the optimum design.

10. Choose one of the other determinate approaches from steps 4 and 6. Starting from the (F.F.) of step 5, derive the specific (F.F.) for the other approach selected. Then, from this (F.F.) sketch the (V.S.) variation diagram and determine all points potential for optimum design. Also, determine the possibilities, if any, for (I.S.). (Your conclusions determined independently in this step should be entirely compatible with those from step 7.)

(b) Solve the described optimization problem by AOD as follows.

1. Starting with the (I.F.) from step 1 of part a, derive the basic AOD equation system which could be used for solution of the optimization problem by an algorithm such as OPTIGO. (For background see sections 6.1 and 6.20 in particular of Chapter 6.)

2. Program your equation system from step 1 into an algorithm for AOD solution, such as OPTIGO. (For background on OPTIGO, see sections 6.2 through 6.37 of Chapter 6 and Appendix B for Fortran IV listing.)

3. Solve the numerical optimization problem, if given, by an AOD algorithm such as OPTIGO, using your programming from step 2. Compare your results with those from step 9 of part a. Explain any discrepancies in the two numerical solutions, based on the (V.S.) from step 7 of part a. (For example, the algorithm might find a solution point violating one or more constraints in a situation of (I.S.). Or the algorithm might find a single solution point in a situation where there are an infinite number of points for optimum design along a boundary curve segment on the variation diagram

7.1. A tensile bar threaded at each end is to be subjected to shock loading as shown in figure 7.26. For each cycle of repeated loading the tensile bar must absorb and release a specified amount of *energy*, designated (P.E.). The length L of the tensile bar to be designed must lie within the specified range $L_{\text{min}} \leq L \leq L_{\text{max}}$. Designate the stress concentration factor at the threaded sections by K_r and the factor of safety based on the occurrence of a fatigue failure by N_e. Suppose that the desired life for the tensile bar is 500 million cycles of repeated loading.

Assume that the diameter d must be equal to or less than a maximum value d_{max} imposed by space restrictions and that axial force gradient k of the tensile bar must be equal to or greater than a minimum value k_{min} imposed by vibrational considerations of the entire mechanical system. Thus we have two additional limit equations:

$$d \leq d_{\text{max}}$$

$$k \geq k_{\text{min}},$$

Optimal design of the tensile bar is to be conducted on the basis of minimizing the force transmitted to the machine frame; therefore the objective in the procedure of *design* will be to obtain minimum F_{max}.

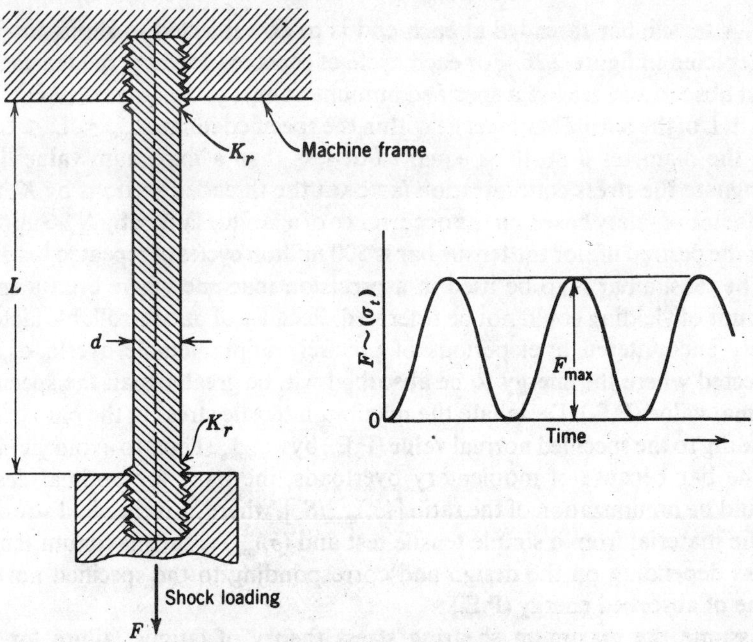

Figure 7.26 For exercises 7.1 and 7.2.

See the general problem statement given at the beginning of the exercises and solve the particular steps as assigned.

For a numerical application assume the following boundary values.

(P.E.) = 100.0 in-lb to be absorbed and released by the tensile bar per cycle of repeated loading

$$N_e = 1.2$$
$$L_{max} = 10 \text{ in}$$
$$L_{min} = 6 \text{ in}$$
$$d_{max} = 1 \text{ in}$$
$$k_{min} = 0.80 \times 10^6 \text{ lb/in}$$
$$K_r = 3.0$$

Assume that the available materials are those in tables 4.1 and 4.2, and completely specify the optimum design of the tensile bar. Finally, calculate the value of F_{max} felt by the machine frame for the optimum tensile bar design.

From the study using the method of optimum design in previous steps, discuss what changes should be considered for each of the numerical specifications in order to alleviate F_{max} still more.

7.2. A tensile bar threaded at each end is to be subjected to shock loading as depicted in figure 7.26. For each cycle of repeated loading, the tensile bar must absorb and release a specified amount of energy designated (P.E.). The length L of the tensile bar must lie within the specified range $L_{min} \leq L \leq L_{max}$, and the diameter d must be equal to or less than a maximum value d_{max}. Designate the stress concentration factor at the threaded sections by K_r and the factor of safety based on the occurrence of a fatigue failure by N_e. Suppose that the desired life for the tensile bar is 500 million cycles of repeated loading.

The tensile bar is to be used in a precision machine where even a small amount of yielding could not be tolerated. Because of uncontrollable factors, rarely encountered brief periods of precisely unpredictable overloads are expected where the energy to be absorbed will be greater than the specified normal value (P.E.). Designate the maximum tensile stress in the bar corresponding to the specified normal value (P.E.) by $(\sigma_t)_{max}$. Thus, to avoid yielding of the bar because of momentary overloads, the basis for optimal *design* should be minimization of the ratio $[(\sigma_t)_{max}/S_y]$, where S_y is the yield strength of the material from a simple tensile test and $(\sigma_t)_{max}$ is the maximum tensile stress depending on the design and corresponding to the specified normal value of absorbed energy (P.E.).

Assume the maximum shearing stress theory of fatigue failure for the general shape of cyclical loading depicted in figure 7.26, based on the specified normal energy (P.E.), since overloads will occur only very rarely. Assume that the available materials are those given in tables 4.1 and 4.2.

See the general problem statement given at the beginning of the exercises and solve the particular steps as assigned.

For a numerical application assume the following boundary values.

$$(P.E.) = 100 \text{ in-lb}$$
$$d_{max} = 1 \text{ in}$$
$$L_{min} = 6 \text{ in}$$
$$L_{max} = 10 \text{ in}$$
$$N_e \geq 1.5$$
$$K_r \approx 3.0 \text{ for Aero stud thread system}$$

7.3. A tensile bar is to be designed to withstand a specified static axial force of value P. The bar is to have a length L of 10 in, and the axial elongation Δ caused by the force P must not exceed a value $\Delta_{max} = 0.010$ in. The bar is to be manufactured in large quantities; therefore cost is a very important consideration. Assume that two significant materials are available for consideration, and they will be designated material a and material b. The properties of interest for these materials follow.

Material a: $c = \$5/10$ lb; $w = 0.100$ lb/in^3; $S_y = 15{,}000$ psi; $E = 10 \times 10^6$ psi

Material b: $c = \$3/10$ lb; $w = 0.300$ lb/in^3; $S_y = 20{,}000$ psi; $E = 30 \times 10^6$ psi

where c is the unit weight cost of the material; w is the weight density of the material; S_y is the yield strength of the material determined by a simple tensile test; and E is its modulus of elasticity.

Assume that, for manufacturing any reasonable tensile bar design, fixed costs and tooling costs are essentially the same for both of the available feasible materials in this problem. Hence the objective in optimal design will be to minimize total material cost, which is $C = cwLA$, where all terms have been defined except A, which is the unspecified cross sectional area of the bar. Also assume a theory of failure with a factor of safety N_y equal to unity based on the occurrence of yielding in the bar.

See the general problem statement given at the beginning of the exercises and solve the particular steps as assigned.

For a numerical application assume the given specifications, with $P = 30{,}000$ lbf, and $N_y = 1.2$.

7.4. A torsion shaft is to transmit a virtually pure twisting moment whose static value is 15,000 in-lb. However, vibrational considerations of the entire machine are of great importance, since it is known that a small varying moment will be superimposed on the theoretical static moment. The entire mechanical system has several degrees of freedom and resonance must be avoided at any of its natural frequencies. From a vibrational study of the entire mechanical system, it has been determined that the torsional rigidity of the shaft should be 500 in-lb per degree of twist. The length of the shaft must be 10 in, and space limitations restrict any feasible diameter to a maximum value of 2 in. Assume the maximum shearing stress theory of failure with a factor of safety $N_y = 2.0$ based on the occurrence of significant yielding. Assume that the only available materials are those given in tables 4.1 and 4.2.

The torsion shaft is to be used in a machine where weight should be as low as possible. Hence optimal design will be conducted on the basis of minimizing the shaft weight.

(a) With the specifications given, design the torsion shaft on the basis of minimum weight.

(b) If a case of incompatible specifications exists for this problem, suggest, on the basis of studies using the method of optimum design, what might be done to the functional requirements or the space limitations in order to obtain a feasible and optimum design of the shaft.

7.5. A torsion shaft to be subjected to a virtually pure twisting moment, (whose magnitude varies) is to be used as a spring member in a machine for absorbing and releasing a specified amount of energy designated $(P.E.)_s$ during repeated cyclical loading. For this type of shock loading assume that the twisting moment varies unidirectionally and cyclically from zero to a maximum and back to zero, where the maximum value is initially unspecified. Because this is a practical torsion shaft, assume that there will be geometrical deviations such as keyways, transverse holes, or threaded sections with accompanying stress increase factors which are functions of independent proportions. Designate the largest or governing one of these stress increase factors by (K_i), referred to the shaft diameter d. Space restrictions for feasible values of diameter d and length L are $d \leq d_{max}$ and $L_{min} \leq L \leq L_{max}$. Also, from vibrational studies of the entire machine, the torsional rigidity of the shaft designated k must lie within the range $k_{min} \leq k \leq k_{max}$. Suppose that the desired life for the torsion shaft is 500 million cycles of repeated loading. Assume that the only available materials are those given in tables 4.1 and 4.2.

The normal amount of energy which the torsion shaft spring must absorb per cycle of repeated loading is the specified amount $(P.E.)_s$. However, on very rare occasions an indeterminate amount of energy overload can be expected. The torsion shaft spring is to be used in a precision machine where even a small amount of yielding cannot be tolerated. Hence the torsion shaft spring is to be designed on the basis of maximizing the energy absorption *capability* for the rarely encountered overloads.

Two significant criteria for failure exist for this problem. First, during the rarely encountered cycles of *overload*, yielding of the material must not occur. Hence, to account for this possible failure phenomenon, select the maximum shearing stress theory of failure for static loads as discussed in Chapter 4. Second, during *normal* operation the periodically absorbed energy $(P.E.)_s$ must not cause a fatigue failure. To account for this possible failure phenomenon select the maximum shearing stress theory of fatigue failure as discussed in Chapter 4.

Optimal design is to be conducted on the basis of maximizing energy absorption *capability* for the rarely encountered overload cycles.

See the general problem statement given at the beginning of the exercises and solve the particular steps as assigned.

For a numerical application assume the following boundary values.

$(P.E.)_s = 300$ in-lb

$d_{max} = 2$ in

$L_{min} = 10$ in

$L_{max} = 15$ in

$(K_i) = 3.0$

$k_{min} = 5 \times 10^5$ in-lb/rad

$k_{max} = 2 \times 10^6$ in-lb/rad

$N_e \geq 1.2$, permissible values for factor of safety based on the occurrence of a fatigue failure.

7.6. A torsion shaft is to be used as a spring member subjected to repeated loading in a machine as shown in figure 7.27. For each cycle of repeated loading the spring must absorb a specified amount of *energy* designated $(P.E.)_s$. The twisting moment M_t felt by the machine frame will vary cyclically as shown in the figure between zero and a maximum value $(M_t)_{max}$ which depends on the design. Assume that space restrictions exist on feasible values of diameter d and length L such that $d \leq d_{max}$ and $L_{min} \leq L \leq L_{max}$. Also, because of functional requirements of the machine, a limitation exists on permissible values of the greatest angle of twist θ for the spring, such that $\theta \leq \theta_{max}$. Assume that standard proportion square keys will be used for both connections on the spring as shown in the figure.

Assume the maximum shearing stress theory of fatigue failure with a specified value for factor of safety N_e, as discussed in Chapter 5. Suppose that

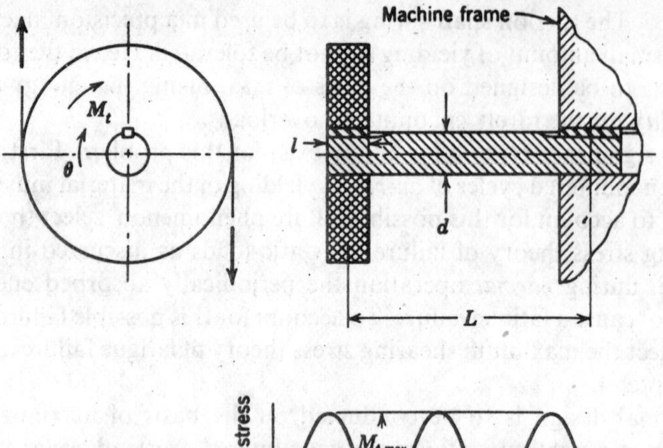

Figure 7.27. For exercises 7.6 and 7.8.

the desired life for the torsion shaft spring is 500 million cycles of repeated loading. Assume that the only available materials are those given in tables 4.1 and 4.2.

For the particular machine under consideration, the torque $(M_t)_{max}$ felt by the machine frame is a significant undesirable effect. Hence optimal design of the torsion shaft spring is to be conducted on the basis of minimizing $(M_t)_{max}$.

See the general problem statement given at the beginning of the exercises and solve the particular steps as assigned.

For a numerical application assume the following boundary values.

$$(\text{P.E.})_s = 200 \text{ in-lb}$$
$$d_{max} = 2 \text{ in}$$
$$L_{min} = 6 \text{ in}$$
$$L_{max} = 20 \text{ in}$$
$$\theta_{max} = 30°$$
$$N_e = 1.2$$

7.7. A torsion shaft is to be subjected to a specified twisting moment of precisely indeterminable magnitude M_t, which is a virtually static load with respect to the rotating shaft. The shaft is to be used in a precision machine where absolutely no yielding can be tolerated. From the study in Chapter 5, realize that it would be desirable for maximum reliability to have the factor of safety N_y as large as possible for this problem. Assume the maximum shearing stress theory of failure, and write the primary design equation directly from equation 4.10 in Chapter 4 as follows, by using the equality sign:

$$N_y = \frac{S_y}{2\tau_{max}}$$

Space restrictions in the machine limit feasible values for diameter d and length L to the following ranges:

$$d \leq d_{max}$$

$$L_{min} \leq L \leq L_{max}$$

Also, vibrational considerations of the entire machine show that the torsional rigidity k for the shaft should be equal to or greater than a certain minimum permissible value. Hence another limit equation is

$$k \geq k_{min}$$

The shaft is to be designed with good proportions for inherently unavoidable geometrical deviations, and the greatest or governing stress increase factor for all deviations in the shaft will be designated K_i. Assume that the only available materials are those whose properties are given in tables 4.1 and 4.2.

Optimal design is to be conducted on the basis of maximizing factor of safety N_y for the torsion shaft. See the general problem statement given at the beginning of the exercises and solve the particular steps as assigned.

For a numerical application assume the following boundary values.

$$M_t = 20{,}000 \text{ in-lb}$$

$$d_{max} = 1.5 \text{ in}$$

$$L_{min} = 10 \text{ in}$$

$$L_{max} = 30 \text{ in}$$

$$k_{min} = 50 \times 10^3 \text{ in-lb/rad}$$

$$K_i = 2.0$$

7.8. A torsion shaft is to be used as a spring member for the repeated absorption and release of *energy* in a machine. The value for the energy to be absorbed per cycle is not accurately determinable because of practical factors beyond your control. The twisting moment M_t transmitted by the spring will vary cyclically between zero and a maximum value $(M_t)_{max}$ which depends on the design, similarly to the variation shown in figure 7.27. Assume that space restrictions exist on feasible values for diameter d and length L such that

$d \leq d_{max}$ and $L_{min} \leq L \leq L_{max}$. Also, because of functional requirements of the machine, a limitation exists on the permissible angle of twist θ for the spring such that $\theta \leq \theta_{max}$. From a dynamic analysis of the entire machine it has been determined that the torque gradient k for the shaft should be a certain specific value.

In solving this problem, use the maximum shearing stress theory of fatigue failure with a specified value for factor safety N_e. The desired life for the torsion bar spring is 500 million cycles of repeated loading. Assume that the only available materials are those given in tables 4.1 and 4.2.

The designed torsion shaft spring is to be used in a machine where reliability of operation is of utmost importance. Since the energy to be absorbed per cycle of repeated loading is not precisely determinable, the design engineer can foresee the possibility of premature spring failures, which should be avoided as much as possible. Hence conduct the optimal design of the torsion shaft spring on the basis of maximum energy absorption *capability* per cycle of repeated loading.

See the general problem statement given at the beginning of the exercises and solve the particular steps as assigned.

For a numerical application assume the following boundary values.

$k = 10{,}000$ in-lb/rad, required torque gradient of shaft

$K_i = 1.5$, largest or governing stress increase factor in shaft

$d_{max} = 1.25$ in

$L_{min} = 15$ in

$L_{max} = 30$ in

$\theta_{max} = 15$

$N_e = 1.2$, desired factor of safety

7.9. A torsion shaft of any specified diameter d is to have a chamfered squar key, as shown in figure 7.28. The shaft is to transmit a twisting moment M which will vary cyclically from zero to an unspecified maximum value an back to zero, with the maximum twisting moment designated by $(M_t)_{max}$ Hence a fatigue failure initiating in the vicinity of the keyway will be c primary interest to the design engineer.

The shaft and the key are to be made of the same ductile type of materia and hence we shall assume the maximum shearing stress theory of fatigu failure as reviewed in Chapter 4. Two significant maximum shearing stresse exist in the shaft; one is $(\tau_{max})_f$ at the fillet surface of the keyway. The stres concentration factor K_{ts} for $(\tau_{max})_f$ is given graphically in figure 7.10. Th other significant maximum shearing stress exists at the bearing surface shown in figure 7.28, acting on planes at 45° to the mating surfaces betwee

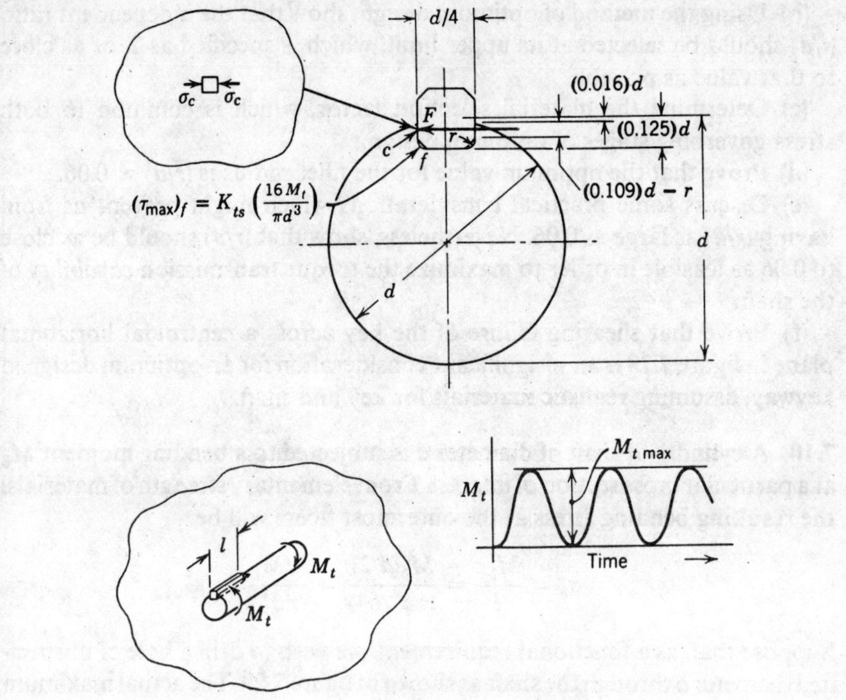

Figure 7.28 For exercise 7.9. Shaft with filleted keyway showing square chamfered key of length l. (See figure 7.10 for K_{ts}.)

key and keyway. This stress is determined approximately by the equation $(\tau_{max})_c = \sigma_c/2 \approx F/2l[(0.109)d - r]$.

Standard proportions are to be used for the key size with respect to the shaft diameter, so the width of the square will be $d/4$ and the depth of the keyway in the shaft will be $d/8$ as shown in the figure. These conform to the standard proportions assumed in figure 7.10. Also assume from practical experience that the value for the ratio (l/d) in the figure should not be greater than 2, in order to be assured of having a good bearing contact for the entire length l of the key.

Optimal design is to be conducted on the basis of maximizing the torque transmission *capability* of the shaft.

(a) From figure 7.10 derive the following simple exponential equation, which is valid over the entire given graphical range, using techniques from Chapter 2.

$$K = \frac{1.392}{(r/d)^{0.165}}$$

(b) Using the method of optimum design, show that the independent ratio (l/d) should be selected at its upper limit, which is specified as 2, or as close to that value as possible.

(c) Determine the material selection factor, which is common to both stress governing stages of optimum design.

(d) Prove that the optimum value for the fillet radius is $(r/d) \approx 0.06$.

(e) Discuss some practical considerations which might prevent us from having (r/d) as large as 0.06. Nevertheless, show that (r/d) should be as close to 0.06 as feasible in order to maximize the torque transmission *capability* of the shaft.

(f) Prove that shearing failure of the key across a centroidal horizontal plane in figure 7.28 is an insignificant consideration for an optimum designed keyway, assuming realistic materials for key and shaft.

7.10. A cylindr •ıl shaft of diameter d is subjected to a bending moment M_b at a particular cross section of interest. From elementary strength of materials, the resulting bending stress at the outermost fibers will be

$$\sigma_b = \frac{M_b c}{I} = \frac{M_b(d/2)}{(\pi d^4/64)} = \frac{32M_b}{\pi d^3}$$

Suppose that, as a functional requirement, we wish to drill a hole of unspecified diameter a through the shaft as shown in figure 7.29. The actual maximum normal stress σ_{max} will be *greater* than the nominal value σ_b; the following equation is obtained from figure 88 of reference 7.3:

$$\sigma_{max} = K_t\left[\frac{M_b}{(\pi d^3/32) - (ad^2/6)}\right]$$

where K_t is a published theoretical *stress concentration factor* referred to the net section of the shaft. A graphical expression for K_t versus (a/d) is given in figure 88 of reference 7.3, which is also repeated in the insert of figure 7.29 herein.

On the basis of the discussion in section 7.4, we could define the *stress increase factor*, K_i, by the equation

$$\sigma_{max} = K_i \sigma_b$$

In selecting the value of the diameter a for the transverse hole, the objective will be to minimize the *stress increase factor* K_i. If this is done, the undesirable effect of introducing the transverse hole in *design* will be minimized because of minimization of significant stresses. Thus, assuming that bending moment M_b and diameter d are independent parameters, the objective in optimum design will be to determine the *ratio* (a/d) for which the *stress increase factor* K_i is minimum.

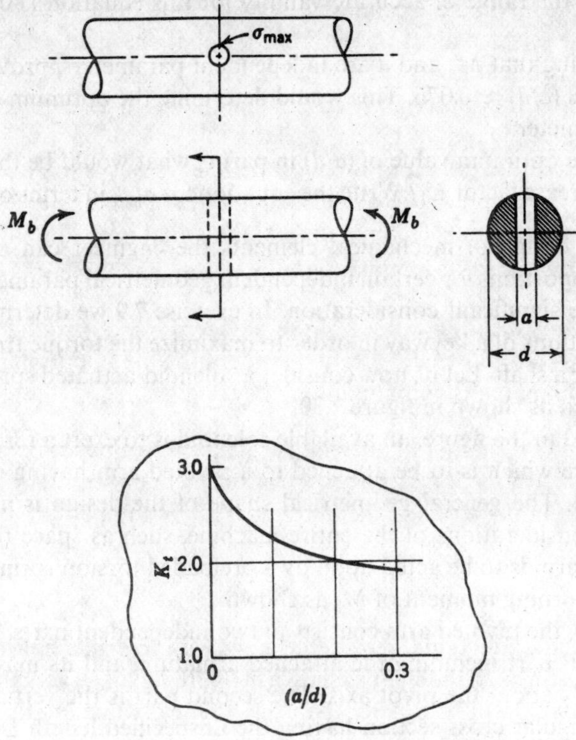

Figure 7.29 For exercise 7.10.

(a) Derive the following equation for the stress increase factor K_i in terms of the published theoretical stress concentration factor K_t, both of which are as previously defined:

$$K_i = \left[\frac{K_t}{1 - (16/3\pi)(a/d)}\right]$$

This is the primary design equation for this problem. Incidentally, this equation is valid only over the range $0 \leq (a/d) \leq 0.3$, which is the given graphical range for K_t in figure 7.29.

(b) Using the simple exponential approximation for curve fitting as described in Chapter 2, derive the following equation from the graph in figure 7.29:

$$K_t = \frac{1.61}{(a/d)^{0.148}}$$

Incidentally, the range of accurate validity for this equation is $0.03 \leq (a/d) \leq 0.3$.

(c) Assuming that M_b and d are independent parameters, prove that σ_{max} is minimum if $(a/d) \approx 0.076$. This would determine the optimum size for the free hole diameter.

(d) For the optimum value of (a/d) in part c, what would be the value for the stress increase factor K_i? Write the equation for σ_{max} in terms of M_b and d.

7.11. In the design of mechanical elements the engineer can often select optimum proportions for certain independent geometrical parameters on the basis of some significant consideration. In exercise 7.9 we determined optimum proportions of a keyway in order to maximize the torque transmission capability for a shaft. Let us now consider a solenoid-actuated spring-loaded contact switch as shown in figure 7.30.

As depicted in the figure, an available solenoid is to exert a force F_s on an iron armature which is to be attached to a pivoted arm having contacts at its other end. The general geometrical shape of the design is imposed by functional considerations of the entire machine, such as space restrictions. The pivoted arm is to be acted upon by a preloaded torsion spring, which is to exert a returning moment of M_s as shown.

Essentially, the pivoted arm consists of two independent parts. The first is the horizontal part including the attached armature, and its mass moment of inertia is I_2 about the pivot axis. The second part is the vertical contact arm of rectangular cross section having the unspecified length L as shown. Its mass moment of inertia about the pivot axis is designated I_1, which can be closely approximated by the following equation from basic mechanics:

$$I_1 \approx \frac{1}{3} ML^2 = \frac{1}{3}\left(\frac{wbh}{g}\right) L^3$$

Assume that it is desired to break the contacts by a specified distance s in as short a period of time as possible. Hence, in selecting the value for the arm length L, the engineer realizes that having it either too large or too small would result in a poor design. If the arm length L is too large, the contact breaking time will be large because of the large inertia of the vertical arm. On the other hand, if the arm length L is too small, the contact breaking time will be large because of the relatively large angular rotation of the arm which is required for obtaining the specified linear motion s.

In figure 7.30 the force W_2 is the gravitational force on the solenoid part of the pivot arm, including the weight of the armature. Assume that the electrical response of the solenoid circuit is much faster than the mechanical response of the pivot arm, so that the force F_s is virtually constant as the contacts separate through the very small distance s. Also because of its torque

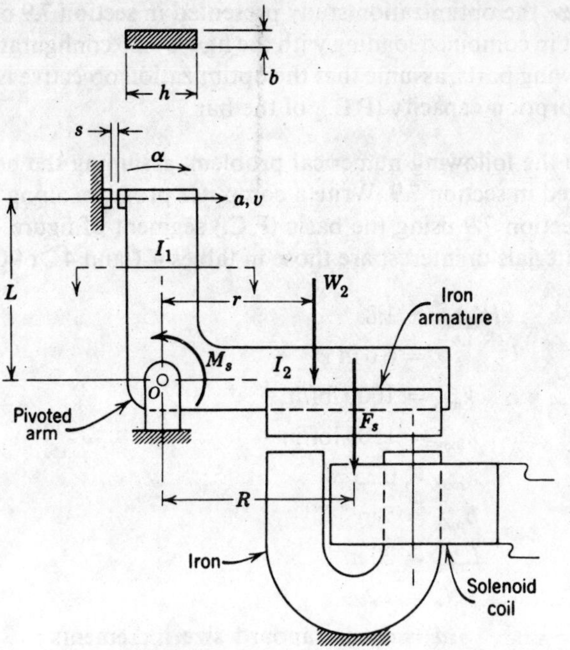

Figure 7.30 For exercise 7.11.

gradient, assume that the torsion spring moment M_s is virtually constant as the contacts separate through the specified distance s. Assume that M_s, W_2, r, F_s, R, I_2, b, and h are independent of the length L whose optimum value is to be determined. Optimal design is to be based on minimization of the time t required to cause the specified separation s of the contacts starting at rest from a closed contact position.

(a) Prove that the optimum arm length L is such that $I_1 = I_2/2$.

(b) Explain why use of the approximation $I_1 \approx \frac{1}{3}(wbh/g)L^3$ for determining the optimum arm length results in an insignificant error with respect to the manufactured product performance. Cumbersome calculations are somewhat reduced by using this approximation.

(c) Consider how you might approximately take into account in the design study the variation in F_s, if it is appreciable, as the arm swings through the small angle required for the specified separation gap s.

(d) In addition to part c, consider how you would take into account in the design study the variation of M_s, if it is appreciable.

(e) How might we modify the basic shape of the vertical arm to obtain a still better improvement in design?

7.12. Review the optimization study presented in section 7.9 of this chapter for the shaft in combined loading with the figure 7.20 configuration. For each of the following parts, assume that the optimization objective is to maximize energy absorption capacity $(P.E.)_c$ of the bar.

(a) Solve the following numerical problem, assuming the boundary conditions stated in section 7.9. Write a computer program along the lines described in section 7.9 using the basic (F.C.) segment of figure 7.25. Assume that the materials of interest are those in tables 4.1 and 4.2 of Chapter 4.

$$
\begin{aligned}
(K_i)_{AC} &\approx 1.8 \\
r &= 6.0 \text{ in} \\
k_{min} &= 1000 \text{ lbf/in} \\
k_{max} &= 1500 \text{ lbf/in} \\
x_{max} &= 0.35 \\
b_{max} &= 5 \text{ in} \\
L_{max} &= 20 \text{ in} \\
N_e &= 1.2 \\
d &= d_s \text{ in standard size increments} \\
&\quad \text{of } 0.05 \text{ in between } 0.5 \text{ and } 1.5 \text{ in}
\end{aligned}
$$

For your solution, specify the optimum material, the optimum values for r, L, b, and d, and the value achieved for $(P.E.)_c$. Also, analyze the optimum design for x and k.

(b) Modify the boundary conditions for the section 7.9 problem as follows:

1. Assume that d is unconstrained (i.e., it is a free variable).
2. Assume that r is not a specified value, but that it is constrained to the region $r_{min} \leq r \leq r_{max}$ with the limit values specified.

Then see the general problem statement given at the beginning of the exercises and solve the particular steps as assigned. Use the same numerical problem values as for part a of this exercise, except here d is unconstrained and 4 in $\leq r \leq 8$ in.

(c) Modify the boundary conditions for the section 7.9 problem as follows:

Assume that r is not specified but is constrained to the region $r_{min} \leq r \leq r_{max}$ with the limit values specified.

Then see the general problem statement given at the beginning of the exercises and solve the particular steps as assigned. Use the same numerical problem values as for part a of this exercise, except 4 in $\leq r \leq 8$ in.

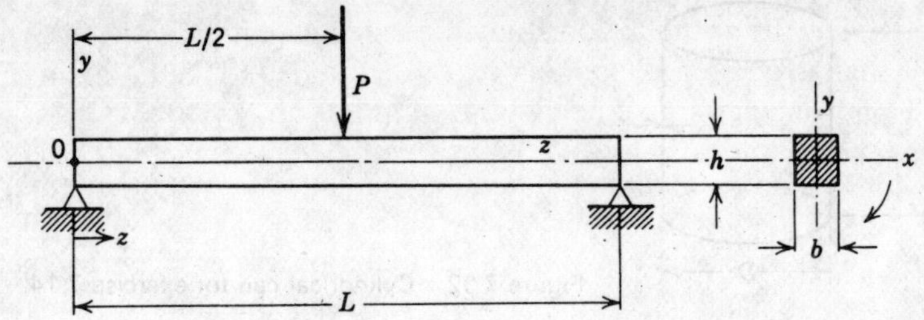

Figure 7.31 Rectangular beam for exercise 7.13.

7.13. A rotating beam of given span L is subjected to a specified central load P oriented relative to the beam as shown in figure 7.31. Force P is of constant magnitude and rotates with the beam. The cross sectional shape is to be that of a solid rectangle, with sides b and h to be determined by optimal design. Yielding is to be avoided in the ductile material to be selected from table 4.1 of Chapter 4. For an acceptable design, functional requirements and space constraints require satisfaction of the following regional constraints on b and h.

$$b_{min} \le b \le b_{max}$$
$$h \le h_{max}$$

Also, mass moment of inertia J_z of the rotating beam is limited as follows for an acceptable design.

$$J_z \le J_{max}$$

Assume that all limit values are known at this stage of design.

The application is one where transverse deflection Δ of the beam is critical. Hence optimal design will be conducted for the objective of minimizing Δ. See the general problem statement given at the beginning of the exercises and solve the particular steps as assigned.

For a numerical application assume the following boundary values.

$$P = 12{,}000 \text{ lbf}$$
$$L = 116 \text{ in}$$
$$b_{min} = 5 \text{ in}$$
$$b_{max} = 8 \text{ in}$$
$$h_{max} = 11.5 \text{ in}$$
$$J_{max} = 84.0 \text{ lb-sec}^2\text{-in}$$

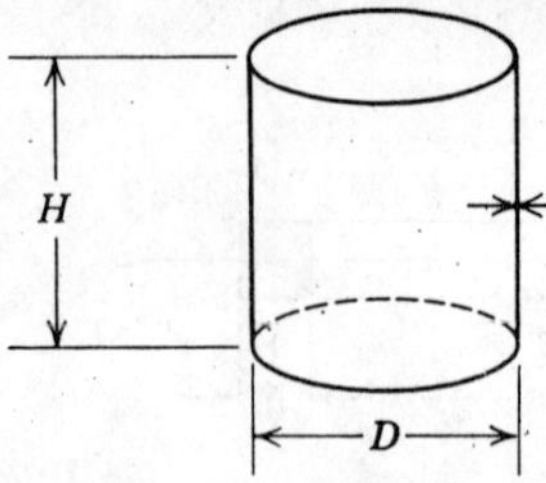

Figure 7.32 Cylindrical can for exercise 7.14.

7.14. A thin-walled cylindrical can with both ends closed as shown in figure 7.32 is to be designed for holding a specified volume V of liquid. Hence, in your derivations, assume that $t \ll D$ and $t \ll H$ will be true for any feasible design. The optimization objective is to minimize the amount of metal required for the can, and wall thickness t is a specified value. From space constraints we require $D \leq D_{max}$ and $H \leq H_{max}$ for an acceptable design. Also, from considerations of practical proportions, we define $x = H/D$ and require $x \leq x_{max}$ for an acceptable design. Assume that all limit values are specified.

See the general problem statement given at the beginning of the exercises and solve the particular steps as assigned.

For a numerical application assume the following boundary values.

$$V = 0.1 \text{ m}^3$$
$$D_{max} = 0.3 \text{ m}$$
$$H_{max} = 1.5 \text{ m}$$
$$x_{max} = 2.0$$

7.15. A rectangular parallelepiped thin-walled can with all sides closed as shown in figure 7.33 is to be designed for holding a specified volume V of a granulated material. For any feasible design, assume in your derivations that $t \ll x$, $t \ll y$, and $t \ll h$. Wall thickness t is a specified value, and for an acceptable design we require $x \leq x_{max}$, $y \leq y_{max}$, and $h \leq h_{max}$. Assume that all limit values are specified.

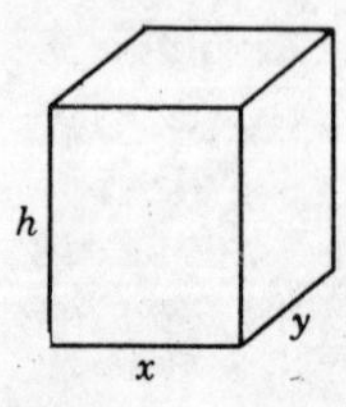

Figure 7.33 Rectangular can for exercise 7.15.

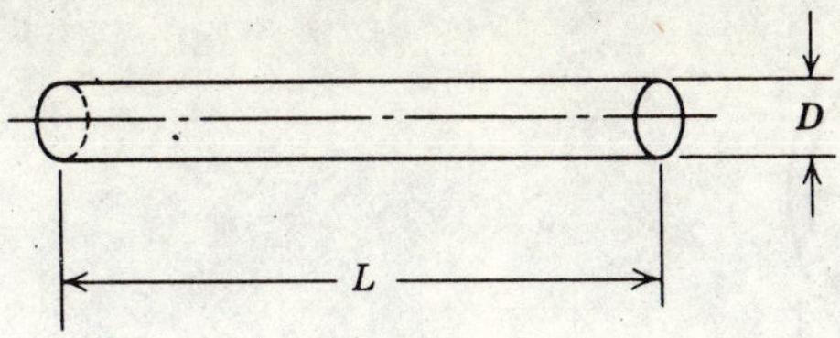

Figure 7.34 Cylindrical roll for exercise 7.16.

The conduct of optimal design is to be on the basis of minimizing the amount of metal required for the can. See the general problem statement given at the beginning of the exercises and solve the particular steps as assigned.

For a numerical application assume the following boundary values.

$$V = 1.0 \text{ m}^3$$
$$x_{max} = 1.5 \text{ m}$$
$$y_{max} = 0.8 \text{ m}$$
$$h_{max} = 1.35 \text{ m}$$

7.16. A cylindrical roll as shown in figure 7.34 is to be designed for an application where lateral surface area $A_s = \pi DL$ must meet minimum requirements. Therefore, for an acceptable design we must satisfy the regional constraint $A_s \geq (A_s)_{min}$. The rolls are to be sold in a highly competitive field, and the selected objective for optimal design is minimization of cost C. Hence the following criterion function is derived from a cost analysis study, which includes cost components for tooling, machining, and material.

$$C = c_1 L + c_2 DL + c_3 D^2 L$$

Space constraints impose the regional constraints $D \leq D_{max}$ and $L \leq L_{max}$ for an acceptable design. At this stage of design all limit values are known, as are c_1, c_2, and c_3. See the general problem statement given at the beginning of the exercises and solve the particular steps as assigned.

7.17. A thin-walled spherical container of metal as shown in figure 7.35 is to be designed for storage of a gas under internal pressure p. Assume in your derivations that $t \ll D$ is true for any feasible design. Mass M of metal for the container is to be limited by the constraint $M \leq M_{max}$, and volume V of gas stored must be within the range $V_{min} \leq V \leq V_{max}$. Several materials of interest have known properties for tensile strength S_t and density ρ. Possibility of pressure surging is anticipated, and reliability of the design must be maximized against the possibility of bursting failure. Hence the optimization objective is to maximize factor of safety N_t, which is defined

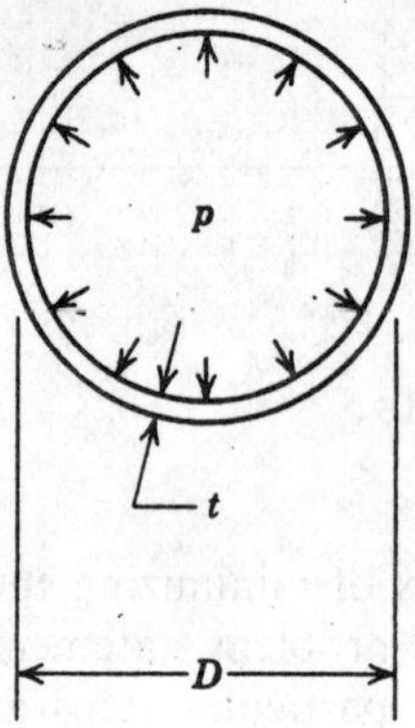

Figure 7.35 Spherical container for exercises 7.17 and 7.18

as $N_t = S_t/\sigma_t$, where σ_t is the tensile stress in the container wall for nomin pressure p, expressed by the following equation derived from elementa strength materials.

$$\sigma_t = \frac{K_t pD}{4t}$$

In this equation, K_t is a stress concentration factor for the inlet and outl fittings, and wall thickness t and diameter D are shown in figure 7.35. For a acceptable design we must satisfy the regional constraints $D \leq D_{max}$ an $t_{min} \leq t \leq t_{max}$. All limit values are specified, as well as nominal pressure stress concentration factor K_t, and S_t and ρ for several materials of interes

See the general problem statement given at the beginning of the exercise and solve the particular steps as assigned.

7.18. The spherical container shown in figure 7.35 for gas storage unde given pressure p is to be designed for an application where we wish to max mize diameter D. Tensile stress σ_t is expressed by the following equation, whic is derived from elementary strength of materials.

$$\sigma_t = \frac{K_t pD}{4t}$$

In this equation, K_t is a stress concentration factor introduced because c inlet and outlet fittings required in the wall. For an acceptable design w require $\sigma_t \leq S_t/N_t$, where S_t is the tensile strength of the material and N_t is factor of safety selected according to the guidelines described in Chapter

Mass M of the container wall is limited as follows for an acceptable desig

$$M \leq M_{max}$$

In deriving your equation for wall mass M as a function of material density ρ, thickness t, and diameter D, assume that $t \ll D$ will be true for any feasible design.

Annual cost C_p of painting the outer surface of the container is limited as follows for an acceptable design.

$$C_p \leq (C_p)_{max}$$

Annual painting cost C_p is expressed by the following equation, where c_1 is the cost per unit area.

$$C_p = c_1 \pi D^2$$

Treat thickness t as a free variable with no constraints directly imposed on it. Assume that all limit values are specified, as well as pressure p, stress concentration factor K_t, unit area painting cost c_1, and S_t, N_t, and ρ for several materials of interest.

The conduct of optimal design is to be based on maximization of diameter D. See the general problem statement given at the beginning of the exercises and solve the particular steps as assigned.

7.19. The flywheel shown in figure 7.36 is to be designed for a special machine, and the solid cylindrical shape shown is desired for simplification in manufacturing. For an acceptable design, width b, diameter D, weight W, and critical bursting stress σ_θ must satisfy the following regional constraints.

$$b \leq b_{max}$$

$$D_{min} \leq D \leq D_{max}$$

$$W \leq W_{max}$$

$$\sigma_\theta \leq \frac{S_t}{N_t}$$

It is believed that the most critical tensile stress σ_θ will be on the surface of a small pilot hole of diameter d_p shown in the figure, which is used for purposes of centering the cylinder in its mounting. Other details of the mounting are not shown in the figure. Assuming for any feasible design that $d_p \ll D$ will be true, the bursting stress σ_θ is obtained as follows, from equation 58 on page 82 of reference 7.1.

$$\sigma_\theta \approx \frac{(3 + \nu)}{16} \rho \omega^2 D^2$$

In this equation ν is Poisson's ratio from table 4.2 of Chapter 4, and density $\rho = w/g$, where w is the weight density from the same table. Also, g is the

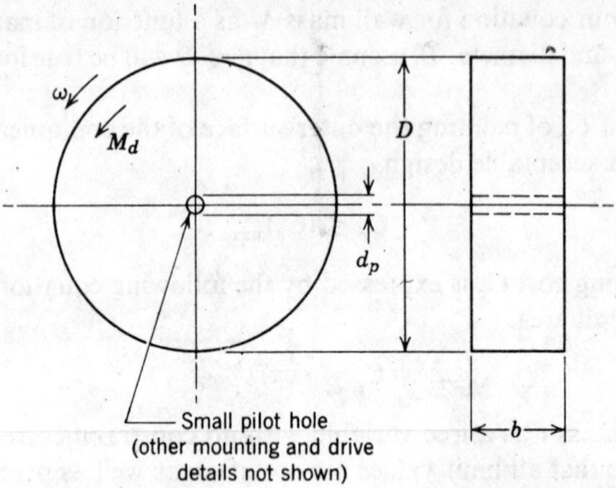

Figure 7.36 Cylindrical flywheel for exercises 7.19 and 7.20

gravitational constant of 386 in/sec^2, and ω is the unspecified maximum speed of rotation for the cylinder in radians per second.

Through electromagnetic clutching not shown, the flywheel is acted upon by a specified drive torque M_d for a specified interval of time Δt. The flywheel starts from rest and achieves speed ω determined by the design, after given M_d assumed constant acts for the given period Δt. Assume that all limit values are specified for the regional constraints, as well as w, S_t, and ν for several materials of interest from tables 4.1 and 4.2 of Chapter 4. Also, the acceptable value for factor of safety N_t has been estimated based on the techniques discussed in Chapter 5.

The purpose of the flywheel is to store kinetic energy E_k; and we wish to have this item as large as possible for best performance of the machine. Hence optimal design of the flywheel will be conducted on the basis of maximizing E_k. See the general problem statement given at the beginning of the exercises and solve the particular steps as assigned.

For a numerical application the boundary values are as follows.

$$b_{\max} = 1 \text{ in}$$
$$D_{\min} = 4 \text{ in}$$
$$D_{\max} = 8 \text{ in}$$
$$W_{\max} = 10 \text{ lb}$$
$$N_t = 1.5$$
$$M_d = 30 \text{ in-lb}$$
$$\Delta t = 1.0 \text{ sec}$$
$$(d_p = 0.5 \text{ in}).$$

7.20. A solid cylindrical flywheel as shown in figure 7.36 is to be designed for a tight space situation with the objective of minimizing diameter D. See exercise 7.19 for descriptive background and definition of all terminology given below. Constraints for an acceptable design are as follows.

$$b \leq b_{\max}$$

$$W \leq W_{\max}$$

$$\sigma_\theta \leq \frac{S_t}{N_t}$$

$$E_k \geq (E_k)_{\min}$$

Assume that all limit values are specified, as well as N_t, and material properties S_t, w, and v from tables 4.1 and 4.2 of Chapter 4. As in exercise 7.19, given drive torque of constant magnitude M_d is to act for a given period of time Δt with the flywheel starting from rest. Maximum speed ω achieved depends on the design. Assume for any feasible design that $d_p \ll D$ will be true.

Optimal design is to be conducted on the basis of minimizing diameter D. See the general problem statement given at the beginning of the exercises and solve the particular steps as assigned.

For a numerical application the boundary values are as follows.

$$b_{\max} = 1 \text{ in}$$

$$W_{\max} = 6 \text{ lb}$$

$$N_t = 1.5$$

$$(E_k)_{\min} = 2000 \text{ in-lb}$$

$$M_d = 30 \text{ in-lb}$$

$$\Delta t = 1.0 \text{ sec}$$

8

Optimal Design of Complex Mechanical Elements

In this chapter we shall be concerned with optimal design of some comple mechanical elements, such as gears, springs, and bearings. The equatio systems and basic design information will be more complicated than thos for the simple mechanical elements of Chapter 7. Hence emphasis will be o practical techniques of application for handling such optimization problem Specifically, these problems will be solved in two ways: (1) by the method c optimum design, and (2) by automated optimal design using the algorithr OPTIGO presented in Chapter 6.

Much work of analysis has been published for the various comple mechanical elements, and in various sources you will find different specifi information given for design use of the same element. Hence no attempt wi be made to cover all the published information for the complex mechanica elements. In fact, there may be places in this chapter where the reader woul choose a different equation or graph, or different tabulated data for th detailed application, and this is a prerogative of the design engineer whic we will not attempt to change. Thus the emphasis will be on handling com plicated information representative of complex mechanical elements fo realistic design optimization problems using the MOD and AOD.

Because of the preceding facts and the need to keep the book to a reason able size, a minimal amount of space will be devoted to explanation of back ground information for the various complex mechanical elements. Thu references will be cited to which the interested reader can go for specifi explanations of background information. However, for most cases thi should be unnecessary if the engineer has some background in basic machine design [8.1–8.9].

In section 8.1 we will be concerned with the optimal design of helical and spur gearsets. For the example on gears, the particular optimization problem will be for maximization of power transmission capability, with the solution obtained by the MOD. In section 8.2 we will be concerned with the optimal design of helical springs, and in a unique approach three optimization objectives will be considered simultaneously by the MOD. This leads to the derivation of a common flow chart, which may be used for numerical application when any one of the three objectives is appropriate. In section 8.3 we will be concerned with the optimal design of hydrostatic journal bearings, for the objective of maximizing mounting rigidity. The equation system is a very complicated one, and solution will be obtained only by AOD using OPTIGO. Finally, at the end of the chapter many exercises are presented for various complex mechanical elements with realistic optimization objectives and boundary conditions. Thus the interested reader can gain basic experience for handling problems of a more complicated nature than the simple mechanical elements of Chapter 7. At the same time, a catalog of pertinent design optimization solutions can be derived, which may be applied directly to numerical problems encountered frequently in practice for various complex mechanical elements. This can be especially valuable if the solutions to the optimization problems are computerized for future applications.

8.1 Optimal Design of Helical and Spur Gears

Helical gears and spur gears are often needed to transmit power, as shown in figure 8.1, in situations where gear tooth loads cannot be predicted accurately. Uncontrollable factors, such as those introduced by operator habits and output loading uncertainties, may result in such a situation. Hence, suppose the design engineer wishes to specify a helical or spur gearset having greatest possible power transmission capability, and the design must satisfy all practical constraints of significance as discussed below. A typical helical gearset is shown in figure 8.2, and a spur gearset is merely the special case of pitch helix angle ψ equal to zero. In fact, the optimization study which follows will be made specifically for helical gears, but it may be directly applied to the more commonly used case of spur gears merely by constraining ψ to 0° as will be explained later.

The design of the gearset in figure 8.2 is uniquely defined by a specification of the materials and the geometry. Materials such as those in table 4.1 of Chapter 4, which are also suitable for gearset use, would be available for consideration. Regarding geometry, we will assume that precision gears of

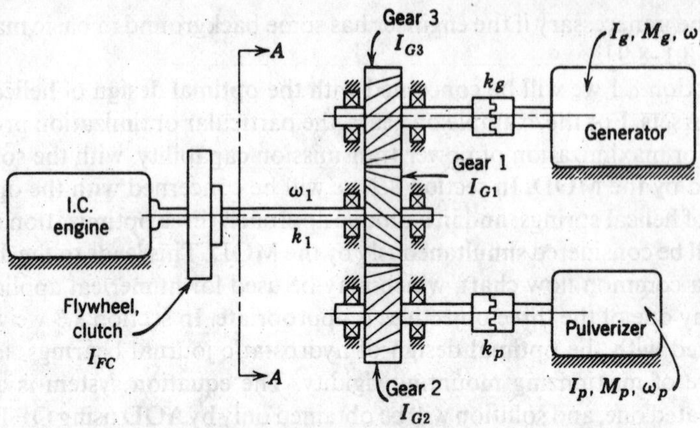

Figure 8.1 Typical gears in a machine setting of uncertain load.

20° involute stub tooth proportions will be used. Hence, with gear ratio a specified, there are four remaining geometrically independent parameters to decide on in design : (1) pitch diameter D_p, (2) face width b, (3) number of teeth N_p, and (4) pitch helix angle ψ. In the text that follows, our focus will be on the determination of the material combination and the geometry for the figure 8.2 gearset which has the greatest power transmission capability.

Initial Formulation. We start by writing the primary design equation 8.1a for horsepower transmitted, $(\mathrm{HP})_t$, as derived from elementary mechanics referring to figure 8.2.

$$(\mathrm{HP})_t = \frac{F_t D_p \omega_p}{126{,}051} \tag{8.1a}$$

In equation 8.1a, F_t is the average energy-transmitting component of force in pounds, D_p is the pinion pitch diameter in inches, and ω_p is the specified value for pinion speed in revolutions per minute.

Power transmission capability of any helical gearset will be governed by one of several possible failure phenomena, such as scoring, tooth wear, and tooth breakage [8.10–8.13]. To be specific, we will choose tooth wear in the form of surface fatigue as the governing failure phenomenon. We would find this to be appropriate for well-designed precision gears of hardened steel. Incidentally, we also would find that the optimum geometry would be exactly the same if tooth breakage had been chosen for the governing failure phenomenon, independent of the material used.

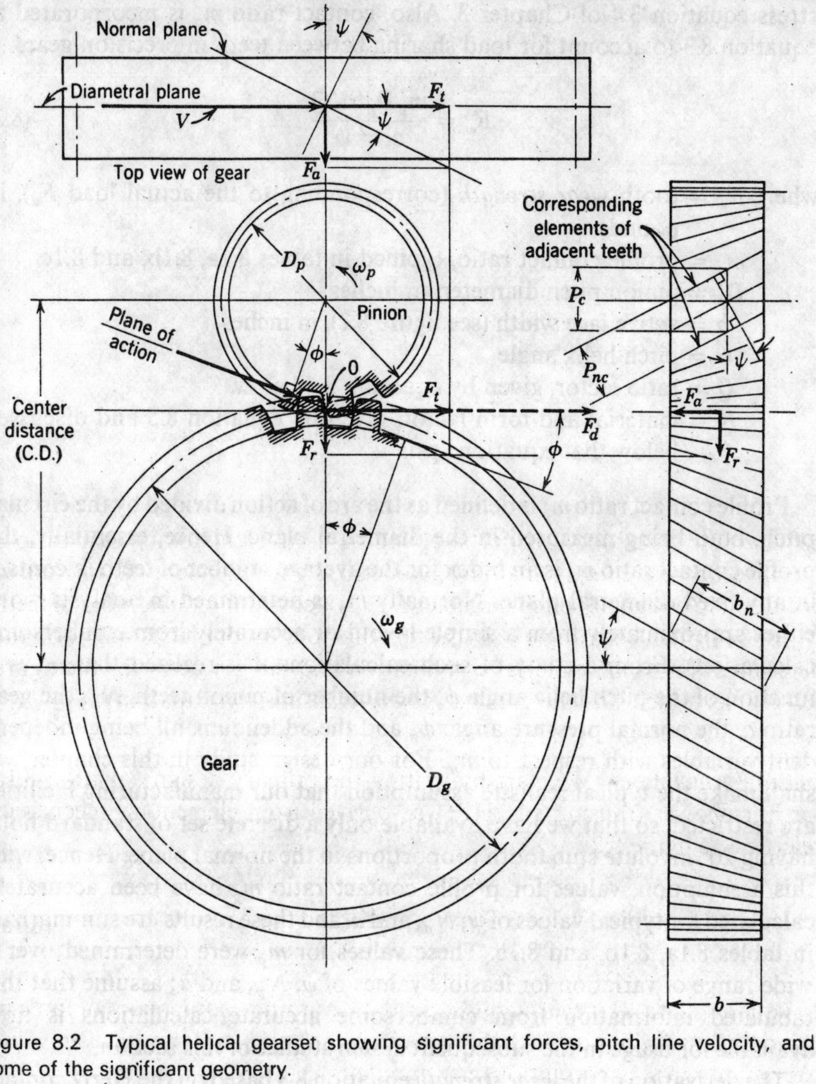

igure 8.2 Typical helical gearset showing significant forces, pitch line velocity, and ome of the significant geometry.

For an acceptable gearset design we will assume that dynamic tooth load $_d$ must satisfy the following regional constraint.

$$F_d \leq F_w \tag{8.2}$$

ooth wear strength F_w can be expressed in terms of the gearset design parameters by means of equation 8.3, which is derived from the Hertz contact stress equation 3.4 of Chapter 3. Also, contact ratio m_p is incorporated in equation 8.3 to account for load sharing between teeth in precision gears.

$$F_w = \frac{m_p D_p b K Q}{\cos^2 \psi} \tag{8.3}$$

where F_w = tooth *wear strength* (corresponding to the actual load F_d), in pounds
m_p = profile contact ratio, typified in tables 8.1a, 8.1b, and 8.1c.
D_p = pinion pitch diameter, in inches
b = active face width (see figure 8.2), in inches
ψ = pitch helix angle
Q = ratio factor, given by equation 8.4 below
K = material and form factor, given by equation 8.5 and discussed below that equation; psi.

Profile contact ratio m_p is defined as the arc of action divided by the circular pitch, both being measured in the diametral plane. Hence, essentially, the profile contact ratio m_p is an index for the average number of teeth in contact in any fixed diametral plane. Normally m_p is determined in analysis work either approximately from a simple layout or accurately from cumbersome calculations. From a study of such calculations it is realized that m_p is a function of the pitch helic angle ψ, the number of pinion teeth, N_p, the gear ratio a, the normal pressure angle ϕ_n and the addendum, all being independent variables with respect to m_p. For our design study in this chapter, we shall make the typical realistic assumption that our manufacturing facilities are restricted, so that we have available only a discrete set of standard hobs having 20° involute stub tooth proportions in the normal plane. Hence, with this assumption, values for profile contact ratio m_p have been accurately calculated for typical values of ψ, N_p, and a, and these results are summarized in tables 8.1a, 8.1b, and 8.1c. These values for m_p were determined over a wide range of variation for feasible values of ψ, N_p, and a; assume that this tabulated information from cumbersome accurate calculations is now available for usage in the subsequent *design* studies of this section.

The derivation of the *wear strength* equation 8.3 based on the Hertz contact stress introduces the ratio factor Q, defined by

$$Q = \frac{2N_g}{N_g + N_p} = \frac{2a}{a + 1} = \frac{2D_g}{D_g + D_p} \tag{8.4}$$

where N_g, N_p = number of teeth in gear and pinion respectively
a = angular velocity ratio of gearset (N_g/N_p)
D_g, D_p = pitch diameters of gear and pinion respectively.

TABLE 8.1a

Typical Values of Profile Contact Ratio m_p as a Function of Pitch Helix Angle ψ for Helical Gears

($N_p = 30$, $a = \omega_p/\omega_g = 3$, $\phi_n = 20°$ involute stub tooth proportions in normal section)

ψ,	m_p
0	1.426
10	1.394
20	1.299
30	1.148
40	0.955

TABLE 8.1b

Typical Values of Profile Contact Ratio m_p as a Function of Number of Pinion Teeth N_p for Helical Gears

($a = \omega_p/\omega_g = 3$, $\psi = 20°$ helix angle, $\phi_n = 20°$ involute stub tooth proportions in normal section)

N_p	m_p
10	1.156
20	1.255
30	1.299
40	1.324
50	1.367

TABLE 8.1c

Typical Values of Profile Contact Ratio m_p as a Function of Gear Ratio $a = \omega_p/\omega_g$ for Helical Gears

($N_p = 30$, $\psi = 20°$ helix angle, $\phi_n = 20°$ involute stub tooth proportions in normal section)

a	m_p
1	1.247
2	1.284
3	1.299
4	1.307
5	1.312

Also, derivation of the *wear strength* equation 8.3 based on the Hertz contact stress introduces the material and form factor K, defined by

$$K = \frac{S_c^2 \sin \phi_n}{1.4}\left(\frac{1}{E_p} + \frac{1}{E_g}\right) \tag{8.5}$$

where ϕ_n = pressure angle of the tooth form in the normal plane
E_p, E_g = moduli of elasticity for the pinion and gear materials respectively, in pounds per square inch
S_c = surface fatigue strength as determined from a test of mating, sliding curved surfaces in contact, in pounds per square inch.

Values of K have been determined experimentally and they are summarized for some materials of interest in table 8.2, for direct use in equation 8.3.

However, for many other material combinations of possible interest, experimental load-stress factors K_1 as described in section 4.9 of Chapter 4 may be used in equation 8.5a to calculate the equivalent K factor for use in gearset F_w equation 8.3. Suppose we solve equation 4.24 of Chapter 4 for load-stress factor K_1 and use $v = 0.3$ for Poisson's ratio. If we divide this equation by equation 8.5, using $S_{ca} = S_c$, we obtain the following relationship.

$$K = \frac{K_1 \sin \phi_n}{4.005} \tag{8.5a}$$

In equation 8.5a, K_1 is the experimental load-stress factor obtained from a published source such as Chapter 4 of reference 8.14, and K and ϕ_n are as previously defined. Incidentally, the values of K_1 used should include the effects of a small amount of sliding, such as would be encountered in meshing gear teeth. Hence the load-stress factors K_1 for pure rolling in table 4.4 of Chapter 4 herein may be used in equation 8.5a for gearsets, but the values should be derated by a factor of about 0.7 or 0.8 to account for the effect of the small amount of sliding in gear teeth.

As a numerical example for equation 8.5a, suppose we were interested in AISI 4340 steel through hardened to R_c 50–55 ($\sim$540 BHN) in contact with carbon steel through hardened to R_c 60–62 ($\sim$640 BHN). Hence, using table 4.4 of Chapter 4, we estimate $K_1 \approx 0.7 \times 13{,}000 = 9100$ psi for this material combination. Thus with $\phi_n = 20°$, we calculate by equation 8.5a $K \approx 9100 \sin 20°/4.005 = 777$ psi. This value agrees quite well with those of table 8.2 for the steels of this hardness range.

To satisfy the tooth wear criterion of equation 8.2 with application in our design optimization study given below, we must also express *dynamic tooth load* F_d as a function of the design parameters. Primarily because of errors in tooth profile and in tooth spacing, a rapidly varying component of force wi

TABLE 8.2

Values of K for Use in Equation 8.3

Material in Pinion	Brinell Hardness Number	Material in Gear	Brinell Hardness Number	S_c, psi	K, $14\frac{1}{2}^\circ$ psi	K, 20° psi
Steel	150	Steel	150	50,000	30	41
Steel	200	Steel	150	60,000	43	58
Steel	250	Steel	150	70,000	58	79
Steel	200	Steel	200	70,000	58	79
Steel	250	Steel	200	80,000	76	103
Steel	300	Steel	200	90,000	96	131
Steel	250	Steel	250	90,000	96	131
Steel	300	Steel	250	100,000	119	162
Steel	350	Steel	250	110,000	144	196
Steel	300	Steel	300	110,000	144	196
Steel	350	Steel	300	120,000	171	233
Steel	400	Steel	300	125,000	186	254
Steel	350	Steel	350	130,000	201	275
Steel	400	Steel	350	140,000	233	318
Steel	500	Steel	350	145,000	250	342
Steel	400	Steel	400	150,000	268	366
Steel	500	Steel	400	175,000	364	497
Steel	600	Steel	400	180,000	385	526
Steel	500	Steel	500	190,000	430	588
Steel	600	Steel	600	230,000	630	861
Steel	150	Cast iron	—	50,000	44	60
Steel	200	Cast iron	—	70,000	87	119
Steel	250	Cast iron	—	90,000	144	196
Steel	>250	Cast iron	—	90,000	144	196
Steel	150	Ph. bronze	—	50,000	46	62
Steel	200	Ph. bronze	—	70,000	91	124
Steel	250	Ph. bronze	—	85,000	135	204
Steel	>250	Ph. bronze	—	85,000	135	204
Cast iron	—	Cast iron	—	90,000	193	284

be superimposed repeatedly on the average tooth force during actual operation of a gearset. Much experimental and analytical work has been done with respect to estimating dynamic tooth forces. As an example, reference 8.15 gives the results of strain gage studies conducted on steel spur gears. With respect to analytical studies, Buckingham has considered the problem of dynamic tooth loads quite extensively in reference 8.16, and his solution technique is also summarized in reference 8.10. The approach presented is quite involved and cumbersome, although it can be applied practically to analysis work for dynamic tooth load. A typical relationship between dynamic load F_d and average energy-transmitting component of force F_t is depicted in figure 8.2. Incidentally, axial and radial components of force are also shown in the figure, and for well-lubricated gears these average tooth force components are related as follows.

$$F_a = F_t \tan \psi \tag{8.6a}$$

$$F_r = F_t \tan \phi \tag{8.6b}$$

where ψ, ϕ, and F_t are as before; F_a = axial force component, in pounds; F_r = radial force component, in pounds.

However, in our design optimization study for gearsets, we will need a reasonable equation for expressing dynamic load F_d as a function of the significant design parameters. Some very simple equations can be found in some of the machine design textbooks expressing F_d as a function only of pitch line velocity V and force F_t. A more complicated and generally conservative expression is equation 8.7, based on the work of Buckingham applied to typical geared systems [8.3, 8.5, 8.6, 8.9]. Even though it assumes average mass and elastic properties for the mechanical system, it does take into account the speed of operation, elastic properties of the pinion and gear materials, the effects of tooth form and pressure angle, the effects of probable tooth errors, the pitch helix angle ψ, the actice face width b, and the effects of the energy-transmitting force F_t.

$$F_d = F_t + \frac{0.05V(bC_1 \cos^2 \psi + F_t) \cos \psi}{0.05V + (bC_1 \cos^2 \psi + F_t)^{1/2}} \tag{8.7}$$

where F_d = dynamic tooth load, in pounds
F_t = energy-transmitting force, in pounds
V = pitch line velocity, in feet per minute
b = active face width, in inches
ψ = pitch helix angle
C_1 = elasticity and tooth error factor, which is determined by equation 8.8.

For equation 8.7 the elasticity and tooth error factor C_1 is determined by

$$C_1 = C_3 e \tag{8.8}$$

where e = the error in tooth action, in inches
C_3 = a constant which is a function of material properties and the tooth system, given by equation 8.9:

$$C_3 = \frac{C_2}{(1/E_p + 1/E_g)} \tag{8.9}$$

TABLE 8.3

Values of Constant C_2 for Equation 8.9

Tooth System	C_2
$14\frac{1}{2}°$ full depth involute	0.107
20° full depth involute	0.111
20° stub involute	0.115

where E_p, E_g = moduli of elasticity of pinion and gear materials respectively, in pounds per square inch

C_2 = a constant determined by tooth proportions and pressure angle, given in table 8.3.

To facilitate the estimation of the probable error in action resulting from inaccuracies in tooth profile and in tooth spacing, table 8.4 gives the usually published information for spur gears. From this table we see that the probable error in action, e, is a function of the tooth size, as measured by the diametral pitch P_{nd} and the degree of accuracy obtained in manufacturing. Class 1 gears are general commercial gears manufactured by form cutting or hobbing in one pass. Class 2 gears are carefully cut by hobbing or generating with a finishing pass. Class 3 gears are precision-made gears which are shaved, ground, or lapped after careful cutting.

At this stage of development, our (I.F.) can be considered to consist of the (P.D.E.) 8.1a, the (S.D.E.)'s 8.3 and 8.7 (including the term-defining equations

TABLE 8.4

Probable Error in Action Resulting from Inaccuracies in Tooth Profile and Tooth Spacing

Diametral Pitch, P_{nd}	Class 1, e, in	Class 2, e, in	Class 3, e, in
1	0.0048	0.0024	0.0012
2	0.0040	0.0020	0.0010
3	0.0032	0.0016	0.0008
4	0.0026	0.0013	0.0007
5	0.0022	0.0011	0.0006
6 and finer	0.0020	0.0010	0.0005

8.4, 8.5, 8.8, and 8.9), and (L.E.) 8.2. However, at this point we might not be certain as to what other equations should be included in the (I.F.) to properly complete tying the optimization problem together mathematically. The basic key for completion of the (I.F.) lies in a recognition of the constraints which we consider to be significant for the problem. This is really how we introduced the (S.D.E.)'s 8.3 and 8.7, based on a recognition of (L.E.) 8.2 as being a significant constraint. Hence, in review of our machine setting including manufacturing and practical considerations, suppose we recognize the significance of the following additional regional constraints, referring to figure 8.2 where appropriate.

$$(D_p)_{min} \leq D_p \leq (D_p)_{max} \tag{8.10}$$

$$0 < b \leq b_{max} \tag{8.11}$$

$$0 \leq \psi \leq \psi_{max} \tag{8.12}$$

$$\infty > N_{np} \geq (N_{np})_{min} \tag{8.13}$$

$$0 < C_{nF} \leq (C_{nF})_{max} \tag{8.14}$$

For the preceding constraints, we assume that all limit values are known based on the discussion which follows. (L.E.)'s 8.10 and 8.11 on pitch diameter D_p and face width b respectively were introduced to account for size and space limitations in the particular machine setting. Incidentally, since center distance (C.D.) in figure 8.2 is related to D_p by equation 8.15a, (C.D.) constraints in the machine setting could be reflected as corresponding pitch diameter constraints by equations 8.15b and 8.15c.

$$(\text{C.D.}) = \frac{D_p + D_g}{2} = \frac{D_p(a + 1)}{2} \tag{8.15a}$$

$$(D_p)_{min} = \frac{2(\text{C.D.})_{min}}{a + 1} \tag{8.15b}$$

$$(D_p)_{max} = \frac{2(\text{C.D.})_{max}}{a + 1} \tag{8.15c}$$

Thus we could account for a center distance constraint $(\text{C.D.})_{min} \leq (\text{C.D.}) \leq (\text{C.D.})_{max}$ by means of the corresponding (L.E.) 8.10 on D_p.

Next, (L.E.) 8.12 on ψ was introduced to assure a helical gearset of reasonable magnitude for pitch helix angle, based on practical judgement as primarily related to axial thrust effects revealed by equation 8.6a. (L.E.) 8.13 is introduced to avoid undercutting of the pinion tooth in the hobbing process which is a condition we wish to avoid for obtainment of involute action in our gearset and for not weakening the tooth. Also, satisfaction of (L.E.) 8.13

will automatically guarantee that we will not have involute interference in our meshing gearset. Incidentally, typical values for the limits on ψ and N_{np} are $\psi_{\max} = 35^\circ$, and $(N_{np})_{\min} = 13.7$ teeth for the 20° involute stub tooth system.

Finally, for a gearset of practical tooth proportions, tooth length to thickness ratio should not be excessive. Otherwise, small errors introduced in manufacturing and mounting might cause our tooth contact condition to bear heavy at one end. As a measure of such length to thickness tooth proportions, we define *face width factor* C_{nF} in the normal plane as follows.

$$C_{nF} = \frac{b_n}{P_{nc}} = \frac{b_n}{\pi/P_{nd}} \tag{8.16a}$$

In equation 8.16a, b_n, P_{nc}, and P_{nd} are the face width, circular pitch, and diametral pitch respectively, all referred to the hypothetical spur gearset in the normal plane. Referring to figure 8.2, we see that b_n is related to b and ψ as shown below, which leads to equation 8.16b, which expresses C_{nF} in terms of constrained parameters b, P_{nd}, and ψ.

$$b_n = \frac{b}{\cos\psi}$$

$$\therefore \quad C_{nF} = \frac{bP_{nd}}{\pi\cos\psi} \tag{8.16b}$$

The terms in equation 8.16b have already been defined. In conclusion, satisfaction of (L.E.) 8.14 on C_{nF} will result in a gearset of good tooth proportions, which should result in a good contact condition for the meshing teeth of gear and pinion. Incidentally, for normal manufacturing and mounting accuracies with precision gears, a typical value for the limit on C_{nF} is $(C_{nF})_{\max} = 4$.

In addition to the preceding regional constraints, we have two discrete value constraints to adhere to for a practical gearset. They are expressed as follows.

$$P_{nd} = (P_{nd})_j \qquad \text{for} \quad j = 1\ldots, N_{sc} \tag{8.17}$$

$$N_p = (N_p)_i \qquad \text{for} \quad i = 1, 2, 3, \ldots \tag{8.18}$$

In equation 8.17, $(P_{nd})_j$ represents a standard size constraint on diametral pitch because of our desire to use standard size available hobs as standard cutting tools. Typical standard size values for P_{nd} are given in table 8.5, of which there are $N_{sc} = 46$ values given for $(P_{nd})_j$.

In equation 8.18, we are merely recognizing that numbers of teeth N_p and

TABLE 8.5

Typical Values of P_{nd} Obtained from Available Standard Hobs

$\frac{1}{2}$	$1\frac{3}{4}$	3	7	12	17	24	34	44	60
$\frac{3}{4}$	2	$3\frac{1}{2}$	8	13	18	26	36	46	—
1	$2\frac{1}{4}$	4	9	14	19	28	38	48	—
$1\frac{1}{4}$	$2\frac{1}{2}$	5	10	15	20	30	40	50	—
$1\frac{1}{2}$	$2\frac{3}{4}$	6	11	16	22	32	42	56	—

N_g must be positive whole numbers which must also satisfy the specified gear ratio a, expressed by equation 8.19.

$$a = \frac{N_g}{N_p} = \frac{D_g}{D_p} = \frac{\omega_p}{\omega_g} \tag{8.19}$$

More specifically, we can also recognize that a lower limit is introduced on N_p because of the constraints on ψ and N_{np}. In a helical gearset, the number of teeth N_{np} for the hypothetical spur gearset in the normal plane is given by equation 8.20 [8.2, 8.4].

$$N_{np} = \frac{N_p}{\cos^3 \psi} \tag{8.20}$$

From this equation we readily see that the limits ψ_{max} and $(N_{np})_{min}$ of (L.E.)'s 8.12 and 8.13 impose a *lower limit* on permissible values for N_p expressed as follows.

$$N_p \geq (N_{np})_{min} \cos^3 \psi_{max} \tag{8.21}$$

Also, for a helical gearset, we have the geometric relationship given by equation 8.22 derived below, using the subscript n to designate the hypothetical spur gearset in the normal plane.

$$P_{nd} = \frac{N_{np}}{D_{np}}$$

However, based on radius of curvature of the pitch ellipse at the pitch point in the normal plane, we have the following relationship [8.2, 8.4].

$$D_{np} = \frac{D_p}{\cos^2 \psi}$$

Hence, also incorporating equation 8.20, we obtain

$$P_{nd} = \frac{(N_p/\cos^3 \psi)}{D_p/\cos^2 \psi}$$

$$\therefore \quad P_{nd} = \frac{N_p}{D_p \cos \psi} \tag{8.22}$$

Therefore, for a standard size value of $P_{nd} = (P_{nd})_j$ by (L.E.) 8.17, and considering the limits $(D_p)_{max}$ and ψ_{min} in (L.E.)'s 8.10 and 8.12, we note that there is an *upper limit on* N_p imposed by these constraints. From equation 8.22, this is expressed as follows.

$$N_p \leq (P_{nd})_j (D_p)_{max} = (N_p)_{max,\,j} \tag{8.23}$$

Thus we see that our whole number constraint on N_p is bounded by the lower limit of equation 8.21 and the upper limit of equation 8.23. To be practical, we could readily calculate the first acceptable whole number for N_p satisfying (L.E.) 8.21, which also gives a whole number for N_g by equation 8.19 for the specified gear ratio a. This value of N_p we designate $(N_p)_{min}$ as an input item for our design specification data. Then we could increment N_p by an amount (ΔN_p), as an additional input item, which must be compatible with the specified gear ratio a giving whole numbers for both N_p and N_g. This practical approach now explained for handling the constraints on N_p will be illustrated in the numerical example presented at the end of the section. In summary, N_p has a discrete value constraint expressed by the whole number requirement explained for equation 8.18, and it is also restricted by the regional constraints of equations 8.21 and 8.23 as reflected by the specified constraints ψ_{max}, $(N_{np})_{min}$, $(P_{nd})_j$, and $(D_p)_{max}$.

At this point in development of the (I.F.), we have incorporated the effects of significant constraints for the optimization problem. Also, our derivation has revealed that some of the constrained parameters are not independent of each other, and appropriate geometric equations must be included as (S.D.E.)'s. Specifically, we must include (S.D.E.)'s 8.16b, 8.20, and 8.22, since all the terms in these equations are constrained by specified limits as explained.

At this stage of development for our (I.F.), we should review the equations to see whether or not there are other equations which should be included because of free variable relations. To start, we note that free variable F_t of the (P.D.E.) 8.1 is related to limited F_d by means of (S.D.E.) 8.7, so the tie has already been completed in the (I.F.). Next, in equation 8.3 for F_w, we note that free variable m_p is related to ψ, N_p, and a, as seen from tables 8.1a, 8.1b, and 8.1c. Hence we need a specific equation for expressing the following relationship.

$$m_p = f_1(\psi, N_p, a) \tag{8.24a}$$

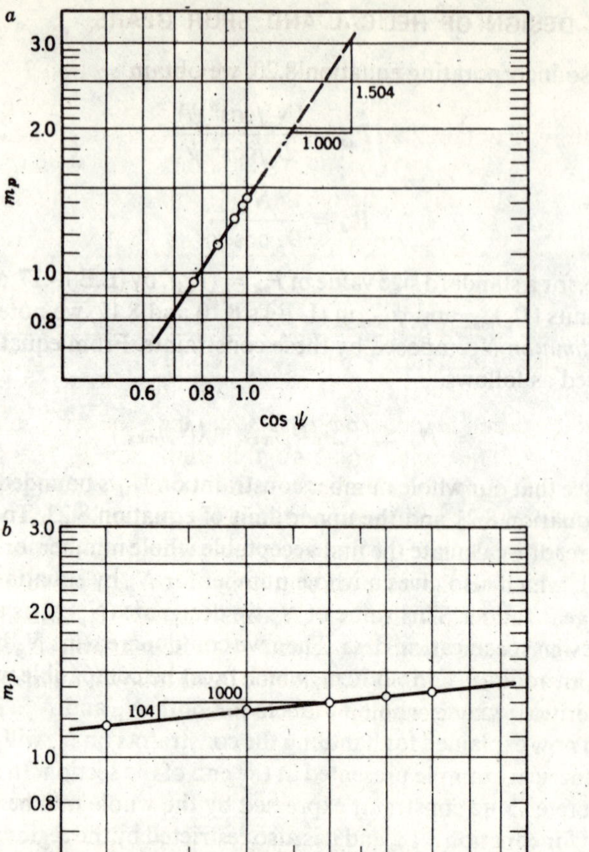
a
3.0
2.0
1.0
0.8
m_p
1.504
1.000
0.6
0.8
1.0
cos ψ
b
3.0
2.0
1.0
0.8
m_p
104
1000
10
20
30
40
50
60
70
N_p
c
3.0
2.0
1.0
0.8
m_p
32
1000
1
2
3
4
5
6

Therefore, referring to our study of curve fitting in section 2.7 of Chapter 2, we first plot the data of tables 8.1a, 8.1b, and 8.1c on log-log graph paper. Specifically, we obtain the straight line plots shown in figure 8.3, which indicates that a simple exponential equation of the equation 2.6 type is appropriate. In this way we derive the following equation for m_p.

$$m_p = 0.966\, a^{0.032} N_p^{\ 0.104} (\cos \psi)^{1.504} \tag{8.24b}$$

Equation 8.24b expresses profile contact ratio m_p in terms of specified gear ratio a, number of pinion teeth N_p, and pitch helix angle ψ, and it is valid for the full ranges of practical interest covered by tables 8.1a, 8.1b, and 8.1c. Accuracy checks on equation 8.24b reveal that the error is less than 1 % with respect to the accurate values in these tables, and the error might be several percent for extreme range value combinations for ψ, N_p, and a. Hence equation 8.24b should be acceptably accurate for our intended design optimization usage. Incidentally, the equation is also limited to 20° involute stub tooth proportions in the normal plane, as originally assumed.

Next, in equation 8.3 we note that ratio factor Q is covered by equation 8.4, and K is a material parameter as seen from table 8.2. Finally, in equation 8.7 for F_d we note that pitch line velocity V (a free variable) is related to specified pinion speed ω_p and pitch diameter D_p as follows, referring to figure 8.2 for the derivation by elementary mechanics.

$$V = \frac{\pi}{12} D_p \omega_p \tag{8.25}$$

In equation 8.25, V = pitch line velocity, in feet per minute; D_p = pitch diameter of pinion, in inches; ω_p = angular velocity of pinion, in revolutions per minute. The pitch line velocity lies in the diametral plane, and it is vectorially shown in figure 8.2.

Finally, in equation 8.7 we note that free variable C_1 is a function of the tooth system chosen, material properties E_p and E_g (i.e., moduli of elasticity), and error in action e. For precision gears, free variable e is a function of diametral pitch P_{nd}, as we see from the class 3 column of table 8.4. As we will see later, curve fitting of these data will not be necessary. Hence we might merely

---

Figure 8.3 Log–log plots for typical helical gearsets used for deriving simple exponential equation 8.24b. (For ϕ_n = 20° involute stub tooth proportions in normal section.) (a) Log–log plot of profile contact ratio m_p versus cosine of pitch helix angle ψ, from available accurately calculated values in table 8.1a. N_p = 30; a = 3. (b) Log–log plot of profile contact ratio m_p versus number of pinion teeth N_p, from available accurately calculated values in table 8.1b. ψ = 20°; a = 3. (c) Log–log plot of profile contact ratio m_p versus gear ratio a, from available accurately calculated values in table 8.1c. N_p = 30; ψ = 20°.

recognize the one-to-one correspondence between e and P_{nd} as follows, in general terminology.

$$e = f_2(P_{nd}) \tag{8.26}$$

As a concluding item in our review of the (I.F.), we note that (L.E.) 8.2 establishes a limit on F_d in terms of F_w, but by equations 8.3 and 8.7 both F_d and F_w are functions of the design parameters. In cases like this, it is often less confusing to define a new design variable such as x, which is limited in terms of a constant. Hence, suppose we define x by equation 8.27 and replace (L.E.) 8.2 with (L.E.) 8.28.

$$x = \frac{F_d}{F_w} \tag{8.27}$$

$$x \leq 1 \tag{8.28}$$

At this point our (I.F.) is complete; the optimization problem has been mathematically coordinated. For a problem of this complexity, we should summarize the (I.F.) prior to application of the exploratory calculations. The summary is given in table 8.6.

Exploratory Calculations. As explained in section 6.13 of Chapter 6, exploratory calculations can be made from a quick review of the (I.F.). This foretells what lies ahead for the (F.F.) and the (V.S.), and it can be a valuable guide in executing the next steps in the MOD. Specifically, in review of the gearset optimization problem (I.F.) of table 8.6, we make the following exploratory calculations using the terminology defined in Chapter 6 and also summarized in Appendix A.

From table 8.6 we count the number of free variables n_f, the number of constrained parameters n_c, their sum n_v, and the number of (S.D.E.)'s as follows.

$$n_f = 8 \text{ (for } F_t, F_w, m_p, F_d, V, C_1, e, \text{ and } Q)$$

$$n_c = 8 \text{ (for } D_p, b, \psi, C_{nF}, P_{nd}, N_{np}, N_p, \text{ and } x)$$

$$n_v = n_f + n_c = 8 + 8 = 16$$

$$N_s = 11$$

Thus, as explained in section 6.13 of Chapter 6, we have the relation $n_f = 8 < N_s = 11$ which foretells a case of redundant specifications with the MOD.

The number of dimensions required for the variation study can be foretold by equation 6.1 as follows.

$$D_{vs} = n_v - N_s + 1 = 16 - 11 + 1 = 6$$

TABLE 8.6

Summary of Initial Formulation for Helical Gearset Problem

$$(HP)_t \uparrow = \frac{F_t D_p \omega_p}{126{,}051} \qquad \text{(P.D.E.) (8.1a)}$$

$$F_w = \frac{m_p D_p b K Q}{\cos^2 \psi} \qquad \text{(S.D.E.) (8.3)}$$

$$Q = \frac{2N_g}{N_g + N_p} = \frac{2a}{a+1} = \frac{2D_g}{D_g + D_p} \qquad \text{(S.D.E.) (8.4)}$$

$$F_d = F_t + \frac{0.05V(bC_1 \cos^2 \psi + F_t)\cos \psi}{0.05V + (bC_1 \cos^2 \psi + F_t)^{1/2}} \qquad \text{(S.D.E.) (8.7)}$$

$$C_1 = C_3 e \qquad \text{(S.D.E.) (8.8)}$$

$$C_{nF} = \frac{bP_{nd}}{\pi \cos \psi} \qquad \text{(S.D.E.) (8.16b)}$$

$$N_{np} = \frac{N_p}{\cos^3 \psi} \qquad \text{(S.D.E.) (8.20)}$$

$$P_{nd} = \frac{N_p}{D_p \cos \psi} \qquad \text{(S.D.E.) (8.22)}$$

$$m_p = 0.966\, a^{0.032} N_p^{\,0.104} (\cos \psi)^{1.504} \qquad \text{(S.D.E.) (8.24b)}$$

$$V = \frac{\pi}{12}\,(D_p \omega_p) \qquad \text{(S.D.E.) (8.25)}$$

$$e = f_2(P_{nd}), \text{ by table 8.4} \qquad \text{(S.D.E.) (8.26)}$$

$$x = \frac{F_d}{F_w} \qquad \text{(S.D.E.) (8.27)}$$

$(D_p)_{min} \le D_p \le (D_p)_{max}$; $0 < b \le b_{max}$; $0 \le \psi \le \psi_{max}$; $\infty > N_{np} \ge (N_{np})_{min}$; $0 < C_{nf} \le (C_{nF})_{max}$; $0 < x \le 1$; $P_{nd} = (P_{nd})_j$, for $j = 1, \ldots, N_{sc}$; $N_p = (N_p)_i$, acceptable whole numbers from $(N_p)_{min}$ in (ΔN_p) increments to (L.E.) 8.23

Specified or known: a; ω_p; $(N_p)_{min}$, (ΔN_p); $(N_{np})_{min}$; $(D_p)_{min}$, $(D_p)_{max}$; ψ_{max}; $(C_{nF})_{max}$; b_{max}; $(P_{nd})_j$, for $j = 1 \ldots, N_{sc}$; and K, C_3 for materials of interest; and table 8.4 for $e = f_2(P_{nd})$

Find: Design of helical gearset which maximizes $(HP)_t$

With a six dimensional variation study indicated for the problem, at this point we might be apprehensive about the possibility of solving the problem by the MOD. However, we may have truly independent parameters in our (F.F.), which in effect will reduce the value for D_{vs}. Also, there are other techniques which we might employ to make the solution practical, especially with a digital computer available. Hence we decide to proceed further and apply exploratory equations 6.3, 6.2, and 6.4 respectively as follows.

$$n_e = N_s - n_f = 11 - 8 = 3$$

$$A_t = \frac{n_c!}{n_e!(n_c - n_e)!} = \frac{8 \times 7 \times 6 \times 5 \times 4 \times 3 \times 2}{3 \times 2(5 \times 4 \times 3 \times 2)} = 56$$

$$n_r = n_c - n_e = 8 - 3 = 5$$

In summary, from the preceding, for the (F.F.) there will be 3 eliminated parameters, 5 related parameters, and 56 different approaches theoretically possible for derivation of (F.F.)'s. The circle diagram for the problem is given in figure 8.4. In it there are 56 triangles, and the vertices for each represent the eliminated parameter combinations for the 56 different approaches. The complementary set of 5 constrained parameters for each case represents the corresponding related parameter set for each approach.

Approach 1 (F.F.). In light of the preceding, from a review of the (I.F.) in table 8.6 we recognize a relatively simple determinate approach for deriva-

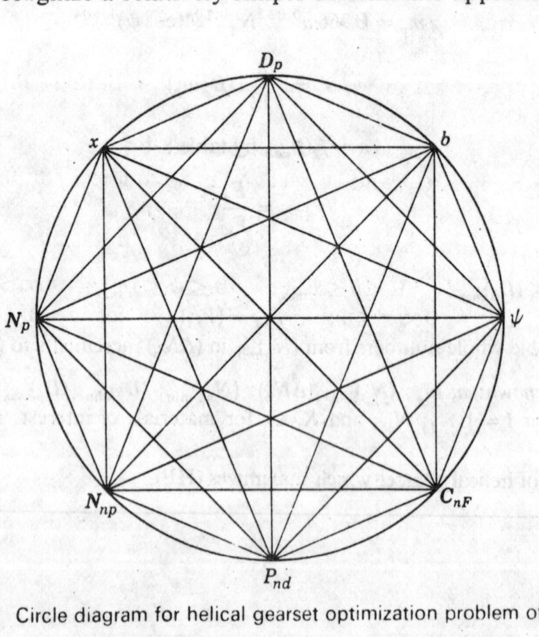

Figure 8.4 Circle diagram for helical gearset optimization problem of table 8.6.

tion of a (F.F.). It would be to choose P_{nd}, N_p, and b as the eliminated parameters, with D_p, ψ, C_{nF}, N_{np}, and x as the related parameters. The equation system general format for this (F.F.) by approach 1 is as follows. The letter a is added to the equation numbers, since another (F.F.) will be derived later for a simplified approach.

$$(\mathrm{HP})_t = f_3(D_p, \psi, C_{nF}, N_{np}, x) \tag{Ia}$$

$$P_{nd} = f_4(D_p, \psi, C_{nF}, N_{np}, x) \tag{IIa}$$

$$N_p = f_5(D_p, \psi, C_{nF}, N_{np}, x) \tag{IIIa}$$

$$b = f_6(D_p, \psi, C_{nF}, N_{np}, x) \tag{IVa}$$

Incidentally, as explained in Chapter 6, there are no free variables in the (F.F.). They are always eliminated in the equation combination procedure of the MOD, with only constrained parameters remaining in the (F.F.). The preceding general format (F.F.) equation system can serve as a valuable guide for the equation manipulations specifically required in transforming from the (I.F.).

From the (I.F.) summary of table 8.6, we note that it will be necessary in the MOD to combine equations 8.1a and 8.7 by eliminating free variable F_t.

Using the (S.D.E.) 8.7 to eliminate F_t from the (P.D.E.) 8.1a; still maintaining an explicit solution for power transmission $(\mathrm{HP})_t$, at first glance appears to be a formidable task. The difficulty lies in obtaining an explicit solution for F_t from equation 8.7. However, this can be accomplished quite readily with good accuracy by converting the available complex equation 8.7 to a simplified approximation which is valid over a limited range of anticipated variation for the variables, as explained in section 2.7 of Chapter 2. Hence, if we consider F_d in equation 8.7 a differentiable function of the independent parameters F_t, V, b, C_1, and ψ, we can approximate F_d quite accurately over the anticipated feasible ranges for the independent parameters by a simple exponential relationship of the general form

$$F_d = (C_4)(F_t)^{C_5}(V)^{C_6}(b)^{C_7}(C_3)^{C_8}(P_{nd})^{C_9}(\cos\psi)^{C_{10}} \tag{a}$$

Notice that the parameter C_1 in equation 8.7 has been replaced in equation a by terms of C_3 as defined by equation 8.9 and P_{nd}. The validity of this replacement is implied by equations 8.8 and 8.26. The theory behind the use of the simple exponential approximation, equation a above, is thoroughly discussed in Chapter 2, and we shall present here only some salient results for this particular application.

We assume that the anticipated feasible ranges for the independent parameters in the original equation 8.7 and its simple exponential approximation equation a are as given in table 8.7. These are reasonable broad assumptions based on a knowledge of previous gear design work, and the anticipated

TABLE 8.7

Approximate Anticipated Ranges for Independent Parameters in Dynamic Tooth Load Equations

Anticipated Range	Approximate Average
1000 lb $\leq F_t \leq$ 12,000 lb	6000 lb
500 fpm $\leq V \leq$ 6000 fpm	3000 fpm
$\frac{1}{2}$ in $\leq b \leq$ 3 in	$1\frac{1}{2}$ in
$0.8 \times 10^6 \leq C_3 \leq 1.72 \times 10^6$	1.26×10^6
0.0005 in $\leq e \leq$ 0.0012 in	0.00085 in
$1 \leq P_{nd} \leq 60$	—*
$0° \leq \psi \leq 45°$	$22\frac{1}{2}°$

* From table 8.4, class 3 gears, $P_{nd} \approx 2.8$ corresponds to the average value of 0.00085 for e.

ranges are not critical, as we will see shortly. The approximate average of each range is also given in table 8.7.

Placing the average values from table 8.7 in equation 8.7, we obtain a dynamic tooth load F_d of 10,330 lb. If we now use equation 8.7 to vary each independent parameter *separately* over its anticipated range, keeping the other parameters at their constant average values, we obtain from such calculations the linear log-log plots of F_d versus each variable as shown in figures 8.5 through 8.10. Slope measurements on these plots give the values for the exponents C_5, C_6, C_7, C_8, C_9, and C_{10} to use for the *simple exponential approximation* equation a. The value for the coefficient constant C_4 in equation a is finally obtained from that equation by using the average value of 10,330 lb for F_d, the exponent values now known for C_5, C_6, C_7, C_8, C_9, and C_{10}, and the appropriate average values for the other terms from table 8.7. The final results of this derivation are

$$F_d = (2.72)\frac{F_t^{0.795}V^{0.169}b^{0.054}(C_3')^{0.053}(\cos\psi)^{0.482}}{P_{nd}^{0.012}} \tag{8.29a}$$

where F_d = dynamic tooth load, in pounds
F_t = energy-transmitting force, in pounds
b = face width, in inches
ψ = helix angle at pitch surface
P_{nd} = diametral pitch referred to normal plane, in teeth per inch
$C_3' = C_3/10^6$, where C_3 is obtained from equation 8.9.

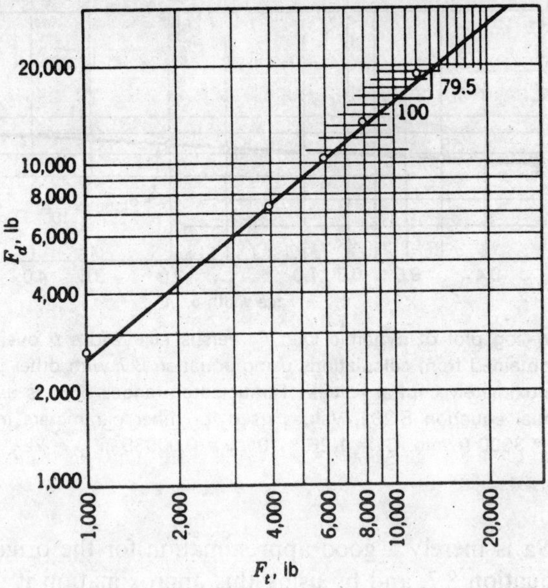

Figure 8.5 Log–log plot of dynamic load F_d versus transmitted force F_t over its anticipated feasible range, obtained from calculations using equation 8.7 with other parameters held constant at approximately central values of anticipated ranges. Plot is used for deriving simple exponential equation 8.29a. Values used for other parameters in equation 8.7: $V = 3000$ ft/min, $b = 1.50$ in, $C_3 = 1.26 \times 10^6$, $e = 0.00085$ in, $\psi = 22\frac{1}{2}°$.

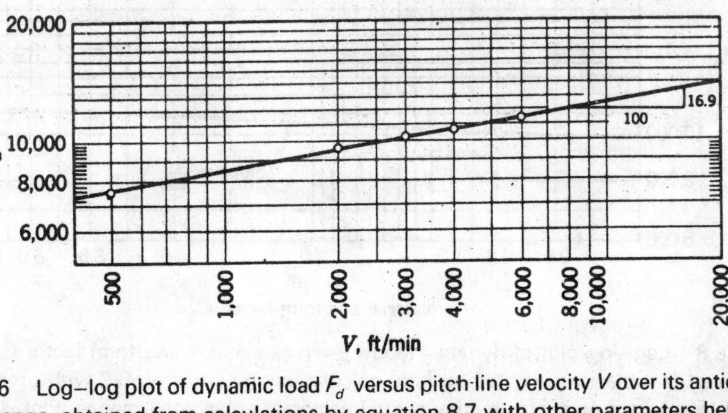

Figure 8.6 Log–log plot of dynamic load F_d versus pitch-line velocity V over its anticipated feasible range, obtained from calculations by equation 8.7 with other parameters held constant at approximately central values of anticipated ranges. Plot is used for deriving simple exponential equation 8.29a. Values used for other parameters in equation 8.7: $F_t = 6000$ lb, $b = 1.5$ in, $C_3 = 1.26 \times 10^6$, $e = 0.00085$ in, $\psi = 22\frac{1}{2}°$

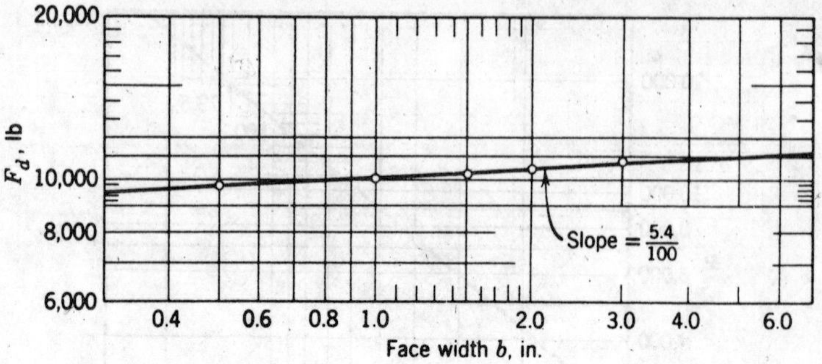

Figure 8.7 Log–log plot of dynamic load F_d versus face width b over its anticipated feasible range, obtained from calculations using equation 8.7 with other parameters held constant at approximately central values of anticipated ranges. Plot is used for deriving simple exponential equation 8.29a. Values used for other parameters in equation 8.7: F_t = 6000 lb, V = 3000 ft/min, C_3 = 1.26 × 10^6, e = 0.00085 in, $\psi = 22\frac{1}{2}°$.

Equation 8.29a is merely a good approximation for the original dynamic tooth load equation 8.7, and by using this approximation it is possible to obtain an explicit solution for F_t, which was our original difficulty. Before actually using equation 8.29a in the design procedure, we shall probably be interested in checking its accuracy. Several random accuracy checks made between equations 8.7 and 8.29a for extreme values of the independent parameters resulted in the percentage difference calculations given in table 8.8.

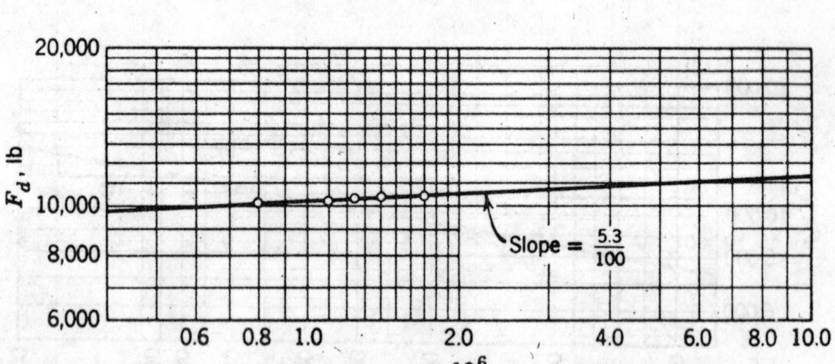

Figure 8.8 Log–log plot of dynamic load F_d versus material and form factor C_3 over its anticipated feasible range, obtained from calculations by equation 8.7 with other parameters held constant at approximately central values of anticipated ranges. Plot is used for deriving simple exponential equation 8.29a. (See table 8.3 and equation 8.9.) Values used for other parameters in equation 8.7: F_t = 6000 lb, b = 1.5 in, V = 3000 ft/min, e = 0.00085 in, $\psi = 22\frac{1}{2}°$.

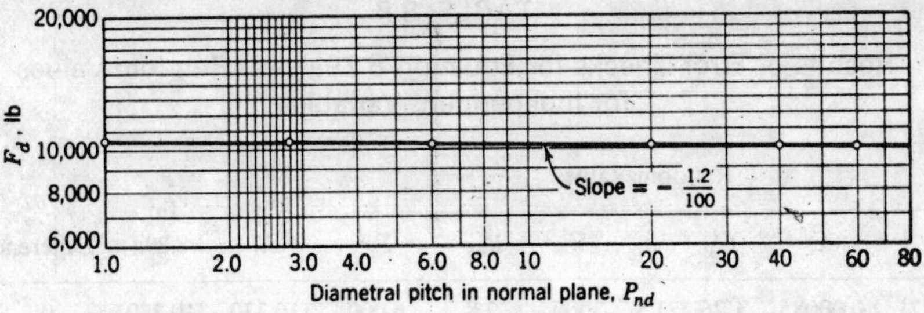

Figure 8.9 Log–log plot of dynamic load F_d versus diametral pitch in normal plane P_{nd} over its anticipated feasible range, giving effect of expected error in action e (see table 8.4). Plot was obtained from calculations using equation 8.7 and table 8.4 with other parameters held constant at approximately central values of anticipated ranges. Plot is used for deriving simple exponential equation 8.29a. One-to-one correspondence between expected error in action e and diametral pitch is given in table 8.4. Values used for other parameters in equation 8.7: F_t = 6000 lb, b = 1.5 in, V = 3000 ft/min, $C_3 = 1.26 \times 10^6$, $\psi = 22\frac{1}{2}°$.

Notice that the 7.7% difference is for a case where the value of F_t is far outside the anticipated range upon which equation 8.29a was derived.

The reasonably accurate results from the random checks in table 8.8 indicate that it will be satisfactory to use equation 8.29a in the optimum design procedure. After the completion of the optimum design study for the helical gearset, the design engineer can always revert to the original equations

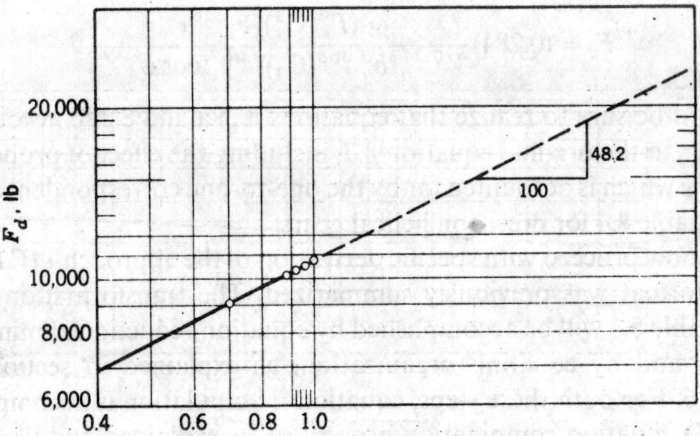

Figure 8.10 Log–log plot of dynamic load F_d versus cosine of helix angle ψ over anticipated feasible range, obtained from calculations using equation 8.7 with other parameters held constant at approximately central values of anticipated ranges. Plot is used for deriving simple exponential equation 8.29a. Values used for other parameters in equation 8.7: F_t = 6000 lb, b = 1.5 in, V = 3000 ft/min, e = 0.00085 in, $C_3 = 1.26 \times 10^6$.

TABLE 8.8

Results of Error Checks for Equation 8.29a with Random Values for Independent Variables

Random Values							F_d,	F_d,	
$\psi,^0$	e	C'_3	b	V	P_{nd}	F_t	Eq. 8.7	Eq. 8.29a	% Difference
$22\frac{1}{2}$	0.00085	1.26	1.5	3000	2.8	6,000	10,330	10,380	0.5
45	0.0005	1.72	3	6000	60	12,000	18,760	18,120	3.4
0	0.0005	0.80	3	500	60	12,000	14,360	13,580	5.4
30	0.0012	1.72	3	4000	1	20,000	31,950	29,500	7.7
30	0.0005	1.72	2	4000	6	8,000	13,410	13,630	1.6

to conduct an accurate *analysis* as a final check. The important point is that the introduction of the approximation by equation 8.29a eventually will make it possible to extract explicitly the optimum available material and the optimum values for the independent geometrical parameters, which we will find to be *limit values* not affected by use of the approximation in the design study. Hence the introduction of the approximation by equation 8.29a is merely a powerful design tool which will make it possible to arrive at an explicit solution to the optimum design problem. Solving equation 8.29a for F_t, we now obtain

$$F_t = (0.284)\frac{(P_{nd}^{0.015})(F_d^{1.26})}{V^{0.213}b^{0.068}(C'_3)^{0.067}(\cos\psi)^{0.607}} \tag{8.29b}$$

We should be sure to realize that equations 8.29a and 8.29b account for *all* the factors in the original equation 8.7, including the effect of probable error in action e which is accounted for by the one-to-one correspondence of e with P_{nd} from table 8.4 for precision helical gears.

Let us now proceed with specific derivation of the approach 1 (F.F.), whose general format was previously summarized. The transformation from the (I.F.) of table 8.6 will be accomplished by equation reduction eliminating free variables and by equation organization, as explained in section 6.13 of Chapter 6. For both these steps, equation manipulation is accomplished by the MOD equation combination procedure, as explained and illustrated in Chapters 6 and 7.

We start deriving equation Ia of the (F.F.) by combining equation 8.29b [equivalent to (S.D.E.) 8.7] with (P.D.E.) 8.1a, eliminating free variable F_t. Then we develop the (P.D.E.) further to the format desired by eliminating the remaining free variables and the "eliminated parameters" of the (F.F.).

Specifically, free variables F_d, V, F_w, m_p, and Q are eliminated using (S.D.E.)'s 8.27, 8.25, 8.3, 8.24b, and 8.4 respectively. Also, organization of equation Ia to the format desired requires further equation combination eliminating P_{nd}, N_p, and b (i.e., the "eliminated parameters"), using (S.D.E.)'s 8.16b, 8.20, and 8.22 in the process. Similarly, we use (S.D.E.)'s 8.16b, 8.20, and 8.22 to derive the relating equations IIa, IIIa, and IVa, applying the MOD equation combination procedure to achieve the desired format previously summarized.

The approach 1 (F.F.) equation system is summarized in table 8.9. In it we note that x is a truly independent parameter, as explained in section 6.15 of Chapter 6. Hence, from equation Ia of the table 8.9 (F.F.), for maximization of $(HP)_w$ we reach the decision to set $x = 1$, the upper limit on x in the table 8.6 (I.F.). This is a valid and final decision to make at this stage of design and will be carried through for the remaining part of the optimization study in this section. Thus from now on x is not treated as a variable but merely a constant of value $x = 1$.

Also, from the (F.F.) of table 8.9 we note three terms enclosed in brackets for developed (P.D.E.) Ia. In the first we have only specified constants ω_p and a. Hence this bracketed term does not change from one feasible design to another. In the second bracket of equation Ia we have the material properties K and C_3'. The K term would normally be determined from table 8.2, but it also could be calculated from either equation 8.5a using published load-stress

TABLE 8.9

Approach 1 (F.F.) Equation System

$$(HP)_w = \left[\frac{\omega_p^{0.787} a^{1.30}}{37{,}200(a+1)^{1.26}}\right]\left[\frac{K^{1.26}}{(C_3')^{0.067}}\right]\left[\frac{D_p^{3.224} C_{nF}^{1.192} x^{1.26}}{N_{np}^{1.046}(\cos\psi)^{2.004}}\right] \qquad \text{(Ia)}$$

$$P_{nd} = \frac{N_{np}\cos^2\psi}{D_p} \qquad \text{(IIa)}$$

$$N_p = N_{np}\cos^3\psi \qquad \text{(IIIa)}$$

$$b = \frac{\pi C_{nF} D_p}{N_{np}\cos\psi} \qquad \text{(IVa)}$$

where P_{nd}, N_p, and b = eliminated parameters

x = truly independent parameter

D_p, ψ, C_{nF}, and N_{np} = related parameters

$MSF = K^{1.26}/(C_3')^{0.067}$

factors or equation 8.5. The term C_3' is defined as $C_3/10^6$ (after equation 8.29a), where C_3 from equation 8.9 is a material property with tooth system already chosen. The third bracket in equation Ia of table 8.9 contains the design variables other than the material properties, which are the related parameters and the truly independent parameter previously discussed.

As an item of importance, we note that there are no material properties in the relating equations IIa, IIIa, and IVa of the table 8.9 (F.F.). Hence the second bracketed term of (P.D.E.) Ia is an independent material selection factor (MSF). From the list of feasible available materials, we would choose the combination having greatest value for the MSF $= K^{1.26}/(C_3')^{0.067}$. In essence, for well-lubricated gears as assumed, we would choose the suitable gearset of greatest hardness value. Incidentally, most optimization problems of the (R.S.) class do not have an independent MSF term, and optimum materials are determined by comparison of best designs for each material of interest as previously illustrated in Chapters 6 and 7.

With our decision to place truly independent parameter x at its upper limit of one, the number of variables n_v has been reduced by one. Hence we see from equation 6.1 of Chapter 6 that the number of dimensions for our variation study has been reduced by one. Thus, with $x = 1$, the value of D_{vs} has been reduced from 6, as previously calculated, to 5. However, a five dimensional variation study is still formidable, and we must now decide on a practical strategy of attack.

Strategy for Simplification of the Variation Study. We start by reviewing the approach 1 (F.F.) of table 8.9 to determine which, if any, of the remaining constrained parameters could be held constant for the variation study. This set consists of the related and eliminated parameters (i.e., D_p, ψ, C_{nF}, N_{np}, P_{nd}, N_p, and b), and the ones with discrete value constraints finite in number are ideal for holding constant during the (V.S.). From the summary of constraints in the (I.F.) of table 8.6, we note that this type of constraint exists on both P_{nd} and N_p. Hence we conclude that it would be practical to hold these parameters constant for our (V.S.), which would reduce the number of dimensions from a formidable 5 to a simple 3. Thus the strategy of attack will be to find the best gearset design for a specific material of interest and for a specific set of P_{nd} and N_p values. From such a (V.S.) a flow chart can be explicitly derived for successive application to the finite number of possible combinations for the P_{nd} and N_p values, utilizing a digital computer in actual execution of the plan. In this way we determine the best design for each of the feasible P_{nd}, N_p combinations. Therefore, by simple comparison of best design $(HP)_w$ values, the overall optimum design will be determined explicitly for maximization of $(HP)_w$.

Simplified (F.F), Approach 2. Considering x, P_{nd}, and N_p as being constants for the (V.S.), we can apply the *exploratory equations* 6.1 through 6.4 of Chapter 6 to the equation system of table 8.9. This is simpler than reverting to the original (I.F.) of table 8.6, since the table 8.9 (F.F.) has already been developed by equation combination to eliminate the free variables. It would be a waste of effort to repeat this equation reduction step in the transformation process. Hence, referring to the table 8.9 equation system and using it as the (I.F.), we make the following exploratory calculations for the approach 2 (F.F.) to be derived.

$$n_f = 0; \qquad n_v = n_c = 5 \text{ (for } D_p, b, \psi, C_{nF}, \text{ and } N_{np})$$

$$N_s = 3 \text{ (for equations IIa, IIIa, and IVa)}$$

Thus by equations 6.1, 6.3, 6.2, and 6.4 respectively we calculate the following:

$$D_{vs} = n_v - N_s + 1 = 5 - 3 + 1 = 3$$

$$n_e = N_s - n_f = 3 - 0 = 3$$

$$A_t = \frac{n_c!}{n_e!(n_c - n_e)!} = \frac{5 \times 4 \times 3 \times 2}{3 \times 2(2)} = 10$$

$$n_r = n_c - n_e = 5 - 3 = 2$$

In summary, we see that the three dimensional variation study given below will have 3 eliminated parameters and 2 related parameters, and theoretically there will be 10 different approaches possible for deriving the simplified (F.F.). The circle diagram for the simplified (F.F.) possibilities is shown in figure 8.11. In it there are 10 lines, and the terminal points for each represent the related parameter combinations for the 10 approaches. The complementary set of three constrained parameters for each case represents the corresponding eliminated parameter set for each approach. Incidentally, these eliminated parameter sets correspond to the vertices for the 10 triangles in the circle diagram of figure 8.11.

Of the 10 theoretically possible approaches for (F.F.)'s, we recognize the fact that some of these might be indeterminate. This would be the case if the proposed eliminated parameter combination is totally absent from at least one of equations IIa, IIIa, and IVa in table 8.9. Hence we recognize that choosing C_{nF}, b, and D_p as the eliminated parameter combination would be an indeterminate approach, since these parameters are totally absent from equation IIIa of table 8.9.

In light of the preceding, and from a review of the table 8.9 equation system, we recognize a relatively simple determinate approach for derivation of the simplified (F.F.), approach 2. It would be to choose C_{nF} and ψ as the related

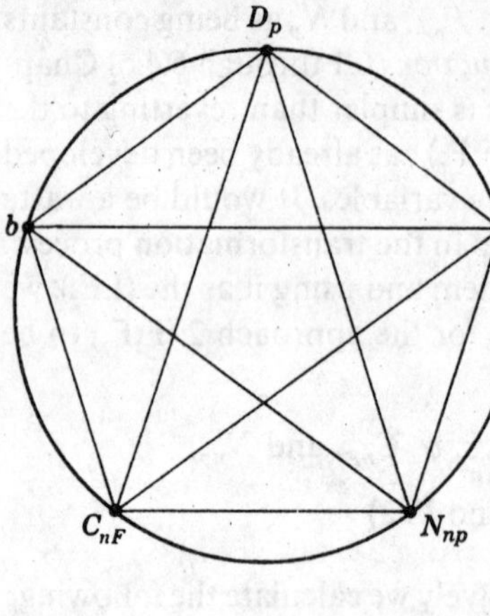

Figure 8.11 Circle diagram for helical gearset optimization problem simplified (F.F.).

parameters, with N_{np}, b, and D_p then being the eliminated parameters. The general format for this simplified (F.F.) approach 2 is summarized as follows.

$$(\text{HP})_w = f_7(\psi, C_{nF}) \quad \text{(I)}$$

$$D_p = f_8(\psi, C_{nF}) \quad \text{(II)}$$

$$N_{np} = f_9(\psi, C_{nF}) \quad \text{(III)}$$

$$b = f_{10}(\psi, C_{nF}) \quad \text{(IV)}$$

The preceding general format equation system will serve as a valuable guide for the equation combination manipulation required in transforming from the table 8.9 approach 1 (F.F.) to the simplified (F.F.) of approach 2.

To simplify notation, we will define C_{sm} by equation 8.30a, which merely denotes the specified item and material parameter brackets of equation Ia.

$$C_{sm} = \left[\frac{\omega_p^{0.787} a^{1.30}}{37{,}200(a+1)^{1.26}}\right]\left[\frac{(K^{1.26})}{(C_3')^{0.067}}\right] \quad \text{(8.30a)}$$

Hence, if we also incorporate the previous decision of setting $x = 1$, equation Ia of table 8.9 becomes

$$(\text{HP})_w = C_{sm}\left[\frac{D_p^{3.224} C_{nF}^{1.192}}{N_{np}^{1.046}(\cos\psi)^{2.004}}\right] \quad \text{(8.30b)}$$

Thus to transform this equation to the simplified (F.F.) format of equation I, we combine equations IIa and IIIa of table 8.9 with it, eliminating D_p and N_{np} in that order. Also, for equation II we combine equations IIa and IIIa, eliminating N_{np} and solving explicitly for D_p. Next, for equation III we merely solve equation IIIa for N_{np}. Finally, for equation IV we combine equations

TABLE 8.10

Simplified (F.F.), Approach 2

$$(HP)_w = C_{sm}\left(\frac{N_p^{2.178}}{P_{nd}^{3.224}}\right)\left[\frac{C_{nF}^{1.192}}{(\cos\psi)^{2.090}}\right] \quad \text{(I)}$$

$$D_p = \frac{N_p}{P_{nd}\cos\psi} \quad \text{(II)}$$

$$N_{np} = \frac{N_p}{\cos^3\psi} \quad \text{(III)}$$

$$b = \frac{\pi}{P_{nd}} C_{nF}\cos\psi \quad \text{(IV)}$$

where D_p, N_{np}, and b = eliminated parameters

ψ and C_{nF} = related parameters

P_{nd} and N_p = constants for the (V.S.) to follow

IIa and IIIa in that order with equation IVa, eliminating D_p and N_{np} respectively. The simplified (F.F.) for approach 2 so derived is summarized in table 8.10.

In the equation system of table 8.10 all terms are as previously defined, with the constraints on the eliminated and related parameters summarized in the (I.F.) of table 8.6. C_{sm} is a constant for the (V.S.) to follow, as defined by equation 8.30a previously explained. In it we note the same MSF as before, for the more complex (F.F.) of table 8.9. Since there are no material parameters in the relating equations of the table 8.10 simplified (F.F.), the MSF of table 8.9 would still apply. Hence we would choose the available gearset material combination of greatest MSF value, being in essence as previously explained the suitable steels of greatest feasible hardness value.

Variation Study with P_{nd} and N_p Constant. Based on the (F.F.) of table 8.10 and the constraints summarized as (L.E.)'s in the (I.F.) of table 8.6, we sketch the typical 3-D variation diagram shown in figure 8.12, which is similar to what we did in previous studies in Chapters 6 and 7. Specifically, for the ideal problem, we start by sketching the variations for optimization quantity $(HP)_w$ by equation I of table 8.10, giving the side and bottom cuts as defined and shown in figure 8.12. We also impose the limits 1, $\cos\psi_{max}$, and $(C_{nF})_{max}$ on the related parameters $\cos\psi$ and C_{nF}. These initial steps reveal that point 4 in figure 8.12 is the point of optimum design for the ideal problem.

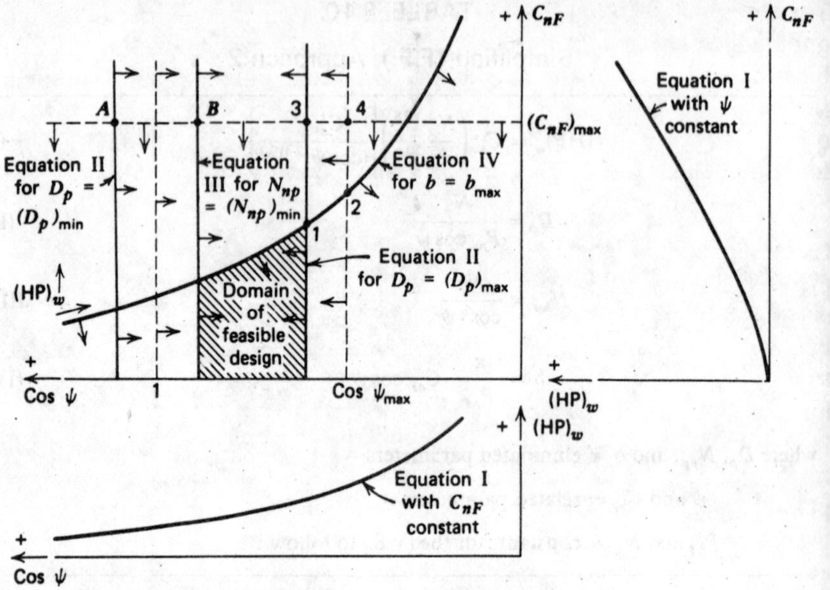

Figure 8.12 Typical 3-D variation diagram for helical gearset problem, for a material combination of interest, and for P_{nd} and N_p constant. See table 8.10 for (F.F.) equations referred to in diagram. Points potential for optimum design are 1, 2, 3, and 4.

Next, we impose typical boundaries in figure 8.12 as imposed by the constraints $(D_p)_{min}$, $(D_p)_{max}$, and $(N_{np})_{min}$ on the eliminated parameters D_p and N_{np}. This also requires a consideration of relating equations II and III of the table 8.10 (F.F.). From the imposition of these boundaries in figure 8.12, we see that point 3 is also potential for optimum design, and there are four possibilities for incompatible specifications. They are for the relations $(\cos\psi)_A < \cos\psi_{max}$, $(\cos\psi)_B < \cos\psi_{max}$, $(\cos\psi)_3 > 1$, and $(\cos\psi)_3 > (\cos\psi)_B$.

Finally, we impose the constraint b_{max} on eliminated parameter b and refer to relating equation IV of the table 8.10 (F.F.) to sketch the hyperbolic shaped boundary shown and defined in figure 8.12. Hence two more points potential for optimum design are revealed, points 1 and 2 in the figure. We also see from figure 8.12 that the hyperbolic shaped boundary in the plan view does not introduce any more possibilities for (I.S.) because of b_{max}.

Incidentally, if the constraints should happen to fall as specifically shown in figure 8.12 for a numerical application, point 1 would represent the optimum design. Similarly, for other numerical applications the boundaries could fall so either point 2, 3, or 4 would be the one of optimum design. In any event, the coordinates ($\cos\psi$, C_{nF}) would be uniquely determined by the intersecting curves as shown and defined in the figure 8.12 variation diagram.

These coordinates would uniquely define the state of design for the point of optimum design for maximization of $(\mathrm{HP})_w$, for the particular (P_{nd}, N_p) combination under consideration. For this best design of the (P_{nd}, N_p) combination, we could then calculate the $(\mathrm{HP})_w$ value by equation I of table 8.10, and other characteristics of interest such as D_p, N_{np}, and b, could be calculated for that point by equations II, III, IV, and so on.

Calculation Flow Chart. From an understanding of the variation study now explained for figure 8.12, the calculation (F.C.) segment of figure 8.13 is readily derived. By simple calculations and comparisons based on the functions and boundaries of figure 8.12, it explicitly determines the best design with respect to maximization of $(\mathrm{HP})_w$, for a specific combination of (P_{nd}, N_p) values and for a specific material combination of interest.

The figure 8.13 flow chart segment is the heart of the overall design optimization (F.C.) presented in figure 8.14. In this total (F.C.), cycling to cover all possible (P_{nd}, N_p) combinations is shown, as well as determination of the overall optimum design and its analysis at the end. The (F.C.)'s of figures 8.13 and 8.14 should be easily understood with some study, based on the discussions above.

At the conclusion of the figure 8.14 (F.C.) the overall optimum design has been uniquely determined for the objective of maximizing power transmission capability $(\mathrm{HP})_w$ for the particular material combination of interest. Generally we would wish to analyze the optimum design further for other phenomena of interest, to be certain that the design is acceptable from all standpoints. Hence, as suggested in the (F.C.), we may wish to analyze the gearset by other approaches, such as by the AGMA techniques of references 8.12 and 8.13. As another illustration, to be assured of full control and smooth action in our meshing gearset, we probably would want to analyze the design for both profile contact ratio m_p and face contact ratio m_F. We defined m_p earlier in the section following equation 8.3. Let us now briefly consider face contact ratio m_F and its relation to smoothness of operation.

Face contact ratio m_F is defined as the ratio of helical advance of a tooth element across the active face width to the circular pitch, both being measured in the diametral plane. Expressed mathematically, face contact ratio m_F is defined by

$$m_F = \frac{b \tan \psi}{P_c} \tag{8.31a}$$

In equation 8.31a, m_F is the unitless face contact ratio, b is the active face width, ψ is the pitch helix angle, and P_c is the circular pitch referred to the diametral plane. Hence, since $P_c = \pi D_p/N_p$ from elementary kinematics, the

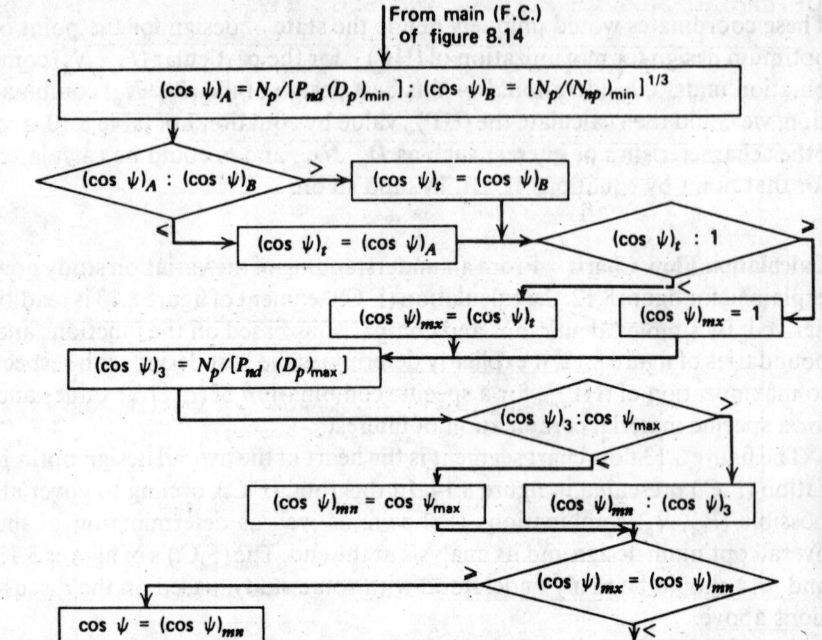

Figure 8.13 (F.C.) segment to determine best design for (P_{nd}, N_p) combination. [See tables 8.6 and 8.10 and figure 8.12 for basis of (F.C.).]

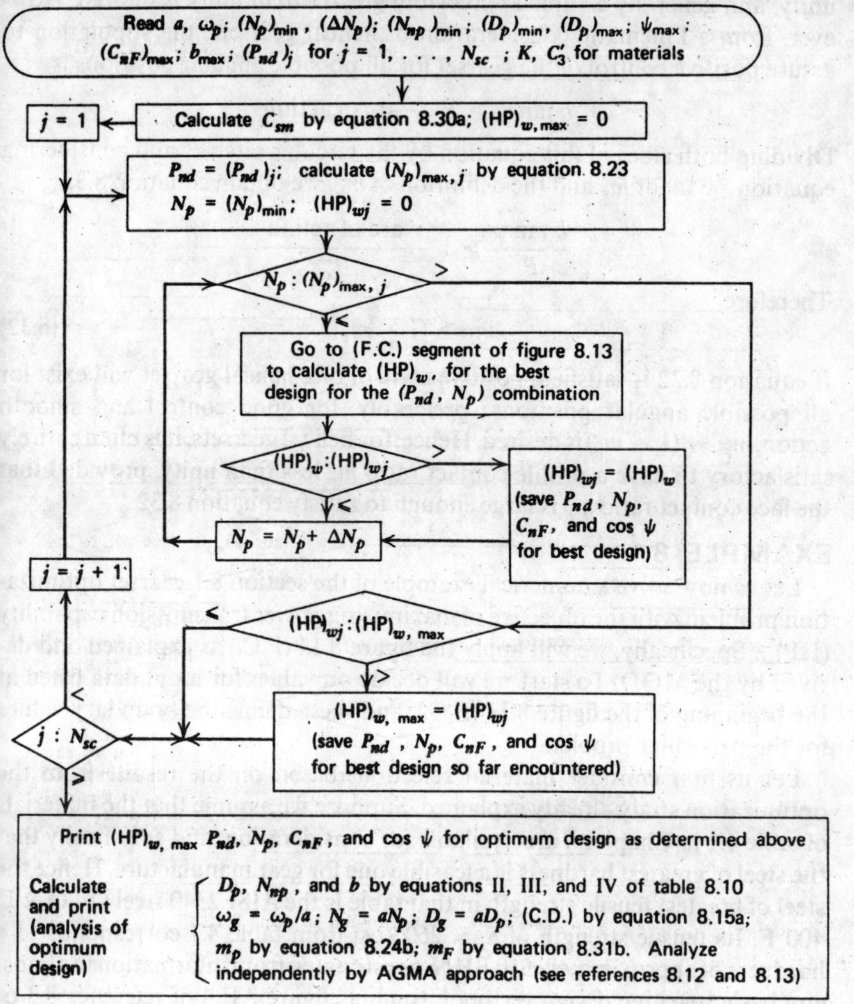

Figure 8.14 Main (F.C.) for optimum design of helical gearset for section 8.1 problem for maximization of $(HP)_w$. Repeat (F.C.) for various material combinations of interest.

equation for m_F becomes

$$m_F = \frac{bN_p \tan \psi}{\pi D_p} \tag{8.31b}$$

To assure positive control of the gearset for all possible angular positions. definite relationship between the contact ratios must be satisfied. Obviously, for a spur gearset, the profile contact ratio m_p must be equal to or greater than unity, and generally a value appreciably greater than unity is desired. However, from a kinematic consideration of helical gearsets, the stipulation to assure positive control of the gearset for all possible angular positions is

$$b \tan \psi \geq P_c - \text{arc of action}$$

Dividing both sides of this equation by the circular pitch P_c and considering equation 8.31a for m_F and the definition of m_p, we obtain equation 8.32:

$$\frac{b \tan \psi}{P_c} \geq 1 - \frac{\text{arc of action}}{P_c}$$

Therefore,

$$m_F \geq (1 - m_p) \tag{8.32}$$

If equation 8.32 is satisfied, positive control of a helical gearset will exist for all possible angular positions; preferably, for good control and smooth action, $m_F \gg (1 - m_p)$ is desired. Hence, for helical gearsets, it is often entirely satisfactory to have a profile contact ratio m_p less than unity, provided that the face contact ratio m_F is large enough to satisfy equation 8.32.

EXAMPLE 8.1

Let us now solve a numerical example of the section 8.1 gearset optimization problem with the objective of maximizing power transmission capability $(HP)_w$. Specifically, we will apply the figure 8.14 (F.C.) as explained and derived by the MOD. To start we will decide on values for input data listed at the beginning of the figure 8.14 (F.C.), and these define the boundary values for the particular problem.

Let us first consider material selection, based on the results from the optimization study already explained. Suppose we assume that the materials of table 4.1 in Chapter 4 are available for consideration, and specifically that the steel of greatest hardness is a feasible one for gear manufacture. Hence the steel of greatest tensile strength in that table is the AISI 2340 steel Oil Q & D 400°F. Its tensile strength of $S_t = 260$ kpsi from table 4.1 corresponds to a hardness of approximately 510 BHN, as estimated from information in almost any basic machine design textbook (such as figure AF 4 of reference 8.3 or figure 3-17 of reference 8.4). Thus according to table 8.2, assuming the proportionalities $S_c \sim S_t \sim$ BHN, we estimate S_c for the steel and calculate the corresponding K value from equation 8.5 as follows.

$$S_c = 190 + (230 - 190) \times \frac{(510 - 500)}{(600 - 500)} = 194 \text{ kpsi}$$

$$\therefore \quad K = \frac{S_c^2 \sin \phi_n}{1.4}\left(\frac{1}{E_p} + \frac{1}{E_g}\right) = \frac{[(194{,}000)^2 \sin 20°/1.4](1/29 + 1/29)}{10^6}$$

$$= 634 \text{ psi}$$

This value for K agrees quite well with that in table 8.2, considering the chosen steel with 510 BHN for pinion and gear.

Next, for steel having $E_p = E_g = 29 \times 10^6$ psi by table 4.1, we calculate C_3' as follows by equation 8.9 using the C_2 value of table 8.3 for the 20° involute stub tooth system already chosen at the beginning of the section.

$$C_3' = \frac{C_3}{10^6} = \frac{C_2}{10^6(1/E_p + 1/E_g)} = \frac{0.115}{(1/29 + 1/29)} = 1.668$$

Thus, we have now determined the material properties K and C_3', which will be used together with the specified items of gear ratio a and pinion speed ω_p to calculate the value for C_{sm} by equation 8.30a at the start of the figure 8.14 (F.C.).

Five numerical problems will be solved in this example, and a general description for these problems is given in table 8.11. A summary of the common input data for these five problems is given below, whereas the input items which differ are summarized in table 8.12.

Common Input Values for Problems A through E.

$K = 634$ psi and $C_3' = 1.668$ (for the chosen steel as previously determined); gear ratio $a = 3$; pinion speed $\omega_p = 1800$ rpm; $\Delta N_p = 1$ tooth; $(N_{np})_{\min} =$ 13.7 teeth, to avoid undercutting in hobbing; $(C_{nF})_{\max} = 4$, for good tooth length to thickness proportions; $N_{sc} = 46$, number of standard hobs available (see table 8.5); $(P_{nd})_j$ for $j = 1, \ldots, N_{sc}$ are the ones in table 8.5.

From table 8.11 we note that problems A, B, and D are for helical gearsets, whereas problems C and E are for spur gearsets. For the helical gears we use equation 8.21 to decide on $(N_p)_{\min}$, using $\psi_{\max} = 35$ as given in table 8.12. Thus we require $(N_p)_{\min} \geq (N_{np})_{\min} \cos^3 \psi_{\max} = (13.7)(\cos 35\ )^3 = 7.53$.

TABLE 8.11

General Description of Problems for Example 8.1

Problem	General Description
A	Helical gearset with liberal $b_{\max}$ limit
B	Helical gearset with reduced $b_{\max}$ limit
C	Same as for problem B, except for spur gearset
D	Helical gearset with specified (C.D.)
E	Same as for problem D, except for spur gearset

TABLE 8.12

Input–Output Data for Gearset Optimization Problems of Example 8.1 (See Figure 8.14 Flow Chart)

Item	Units	A	B	C	D	E
				Problem		
				INPUT DATA*		
$(N_p)_{min}$	—	8	8	14	8	14
$(D_p)_{min}$	inches	2.000	2.000	2.000	2.500	2.500
$(D_p)_{max}$	inches	3.000	3.000	3.000	2.500	2.500
ψ_{max}	degrees	35	35	0	35	0
b_{max}	inches	3.500	2.500	2.500	2.500	2.500
				OUTPUT DATA		
$(HP)_{w,\,max}$	horsepower	355.3	291.7	246.4	164.9	137.8
P_{nd}	teeth per inch	3.5	4.0	5.0	5.0	6.0
N_p		9	10	15	11	15
N_g		27	30	45	33	45
C_{nF}		4.000	3.820	3.979	4.000	4.000
$\cos\psi$		0.8571	0.8333	1.0000	0.8800	1.000
ψ	degrees	31.003	33.557	0	28.358	0
D_p	inches	3.000	3.000	3.000	2.500	2.500
D_g	inches	9.000	9.000	9.000	7.500	7.500
(C.D.)	inches	6.000	6.000	6.000	5.000	5.000
N_{np}		14.292	17.280	15.000	16.142	15.000
b	inches	3.077	2.500	2.500	2.212	2.094
m_F		1.766	1.760	0	1.672	0
m_p		0.9972	0.9663	1.3260	1.0594	1.3260

* Input data common to all problems are listed separately in text of example 8.1.

Therefore, choose $(N_p)_{min} = 8$ teeth to enter in table 8.12 for the helical gears. However, for use of the figure 8.14 (F.C.) for optimum design of spur gears we use $\psi_{max} = 0°$ as an input value along with $(N_p)_{min} = 14$ teeth as determined by equation 8.21. Similarly, for a problem of specified center distance (C.D.), we merely calculate the corresponding D_p value from equation 8.15 and enter it as input data for the values $(D_p)_{min} = (D_p)_{max} = D_p$ so calculated. Problems D and E of table 8.12 are for the case of specified (C.D.), with gear ratio $a = 3$ and $(D_p)_{min} = (D_p)_{max} = 2.500$ in corresponding to a value of (C.D.) = 5.000 in by equation 8.15a

In table 8.12 we also summarize the output data for the five problems, as calculated by the figure 8.14 (F.C.), using the digital computer program derived from it. From this table we note the gain in $(HP)_w$ obtained using helical gears compared with spur gears. Similarly, the possibilities of improved design from increased face width are illustrated. However, improvement is not always achieved in this respect, depending on whether $(C_{nF})_{max}$ or b_{max} governs, as revealed in figure 8.12.

Finally, in table 8.12 we note for the helical gears that profile contact ratio m_p is less than unity for problems A and B. This is acceptable for full control and smooth action providing that face contact ratio is large enough for good satisfaction of equation 8.32, as previously discussed. In this connection, for all five problems summarized in table 8.12 we see that the relationship between m_F and m_p is such as to assure positive control and smooth action for all the presented gearset designs, as can be seen from equation 8.32.

8.2 Optimal Design of Helical Springs

Helical springs of both the compression and the extension types are used in many places of machine design. In this section we shall present some design optimization studies for helical compression springs with realistic boundary conditions imposed. To start, we will review some basic design information which is pertinent to our discussion below [8.1–8.9, 8.17–8.20].

Basic Background Information. A typical helical compression spring is shown in figure 8.15, and for a given material its significant characteristics are uniquely defined by the wire diameter d, the mean coil diameter D, and the active number of coils N. The primary purpose of the spring is to satisfy certain functional requirements of the machine, such as those generally depicted in figure 8.16. These functional requirements are usually expressed by the specification of force versus deflection characteristics, which are required from analysis calculations for proper functioning of the machine. Hence the functional requirements of a spring are uniquely defined by the specification of the greatest force F_{max} existing at the occurrence of maximum spring deflection *and* the specification of the force gradient k. These functional requirements are determined *before designing* the spring, from either a static or a dynamic force analysis of the entire machine or mechanism. In addition, when *designing* the spring, functional requirements such as long life, the existence of space restrictions, and problems in manufacturability must be taken into account.

The maximum shearing stress τ_{max} in either an extension or a compression helical spring, because of the axial force F_{max}, occurs at the outer fibers as

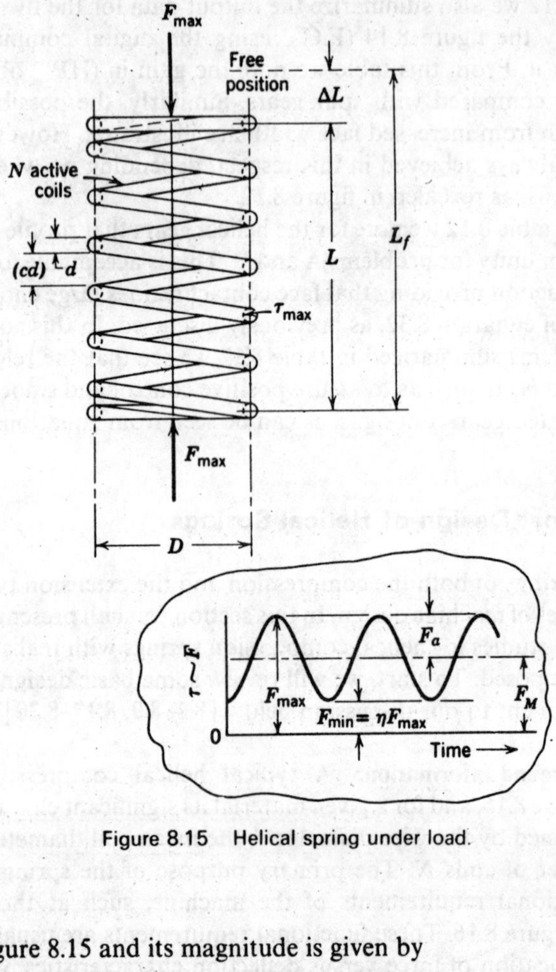

Figure 8.15 Helical spring under load.

shown in figure 8.15 and its magnitude is given by

$$\tau_{max} = \frac{8DF_{max}}{\pi d^3} K_{ts} \tag{8.33}$$

In equation 8.33, which is derived in most textbooks on machine design, τ_{max} is the maximum shearing stress in the spring wire, in pounds per square inch resulting from the axial force F_{max}, in pounds; D is the mean diameter of the coil, in inches; d is the wire diameter, in inches; and K_{ts} is the Wahl correction factor accounting for the effects of direct shear and curvature in addition to torsion. K_{ts} is given by

$$K_{ts} = \frac{4(D/d) - 1}{4(D/d) - 4} + \frac{0.615}{(D/d)} \tag{8.34a}$$

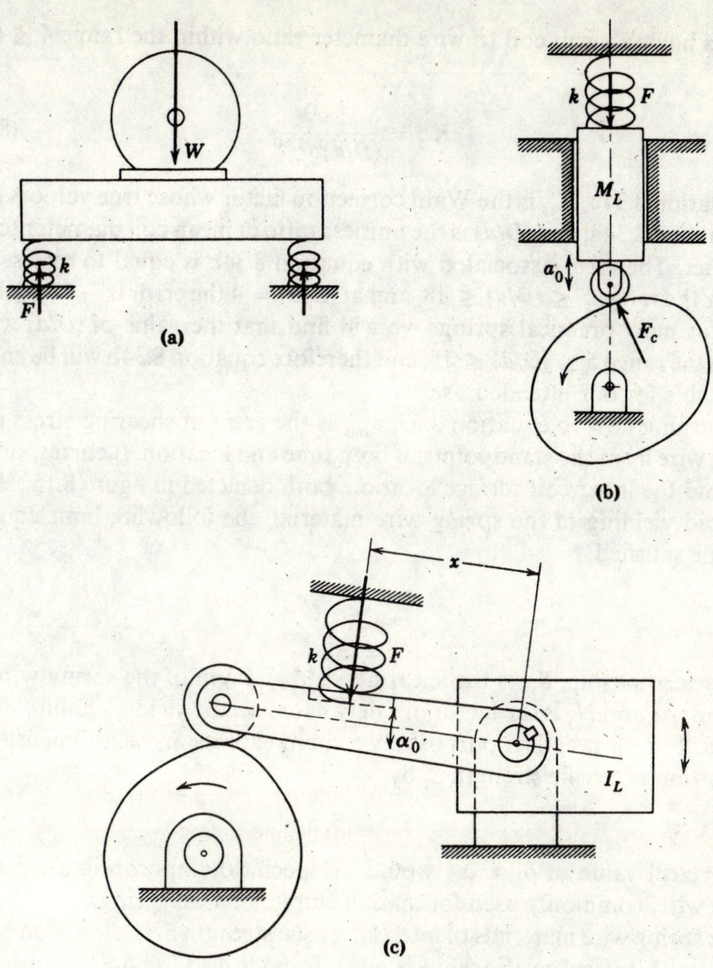

Figure 8.16 Typical uses of helical springs (similar to uses of extension springs). (*a*) Support springs; (*b*) load spring, translating body; (*c*) load spring, oscillating body.

Equation 8.34a is presented *graphically* in most machine design textbooks, and an exceptionally accurate graph is presented in figure 110 of reference 8.19.

We will facilitate the optimization studies to follow by converting equation 8.34a to simple exponential form by the techniques presented in Chapter 2. Hence, from equation 8.34a (or directly using the corresponding data of figure 110 in reference 8.19) we derive equation 8.34b. This equation is valid for design use over the practical range of interest for well-proportioned springs having mean coil to wire diameter ratio within the range $4 \leq (D/d) \leq 18$.

$$K_{ts} = \frac{1.60}{(D/d)^{0.140}} \tag{8.34b}$$

In equation 8.34b, K_{ts} is the Wahl correction factor whose true value is given by equation 8.34a, and (D/d) is the unitless ratio of mean coil diameter to wire diameter. The error associated with equation 8.34b is equal to or less than 2% for the range $5 \leq (D/d) \leq 18$, and at $(D/d) = 4$ the error is -5.4%. However, for most practical springs we will find that the value of (D/d) will lie within the range $5 \leq (D/d) \leq 18$, and therefore equation 8.34b will be entirely acceptable for our intended use.

Returning now to equation 8.33, τ_{max} is the greatest shearing stress in the spring wire from the standpoints of both time and location. It corresponds to F_{max} and the inner coil surface location, both depicted in figure 8.15. Hence, to avoid yielding of the spring wire material, the following limit equation must be satisfied.

$$\tau_{max} \leq \frac{S_{sy}}{N_y} \tag{8.35}$$

In this relationship, S_{sy} is the shearing yield strength of the spring wire material in psi, and N_y is the factor of safety based on the yielding failure phenomenon. For all practical purposes, we could express S_{sy} as a dimensionless proportion of tensile strength S_t by

$$S_{sy} \approx \beta_y S_t \tag{8.36}$$

The typical value of $\beta_y \approx 0.4$ would be specifically appropriate for A.S. 5 music wire, commonly used for manufacture of helical springs.

For spring wire materials of interest, tensile strength S_t is a function of wire diameter. A typical relationship is given in table 8.13 for A.S. 5 music wire. However, in general, from log-log plots of S_t versus d we would find that the following simple exponential type of equation is appropriate for commonly used spring wire materials.

$$S_t \approx \frac{C_t}{d^t} \tag{8.37}$$

In equation 8.37, S_t is tensile strength in psi, d is wire diameter in inches, and C_t and t are material constants determined by the curve-fitting techniques explained in Chapter 2. As a specific example, using the table 8.13 data for A.S. 5 music wire we would obtain the values $C_t = 200{,}000$ and $t = 0.145$. With these values for C_t and t, the error in equation 8.37 is less than 2% for the entire given range of $0.005 \leq d \leq 0.135$ from table 8.13.

TABLE 8.13

Minimum Tensile Strength of A.S. 5 Music Wire

Diameter d, in	S_t, psi	Diameter d, in	S_t, psi
0.005	426,000	0.028	343,000
0.006	415,000	0.032	337,000
0.007	407,000	0.036	332,000
0.008	399,000	0.042	323,000
0.009	393,000	0.048	316,000
0.010	387,000	0.055	310,000
0.012	377,000	0.063	303,000
0.014	374,000	0.072	297,000
0.016	367,000	0.080	292,000
0.018	361,000	0.092	281,000
0.020	358,000	0.105	277,000
0.024	351,000	0.120	271,000
0.026	347,000	0.135	268,000

Regarding critical stresses in helical springs, shearing stress amplitude τ_a is of importance with respect to the criterion for fatigue failure. From a location standpoint in the spring, τ_a exists at the same inner surface coil points as shown for τ_{max} in figure 8.15. With respect to time variation, τ_a corresponds to spring force amplitude F_a shown in figure 8.15. The equation for τ_a is similar to equation 8.33, except we use F_a instead of F_{max}.

$$\tau_a = \frac{8DF_a}{\pi d^3} K_{ts} \tag{8.38}$$

The terms in this equation are the same as defined after equation 8.33, except τ_a is shearing stress amplitude in psi and F_a is the corresponding force amplitude in pounds.

For avoidance of fatigue failure, the spring design should satisfy the following criterion.

$$\tau_a \leq \frac{S_{se}}{N_e} \tag{8.39}$$

In equation 8.39, S_{se} is the shearing strength endurance limit in psi, and N_e is the corresponding factor of safety estimated similar to N_y before design by the techniques of Chapter 5. In reference 8.21, Zimmerli reported a constant value of $S_{se} \approx 45{,}000$ psi for any of the commonly used unpeened spring steels in sizes under $\frac{3}{8}$ in. However, for peened steel wire the value becomes $S_{se} \approx 67{,}500$ psi.

A constraint of design significance is based on practical proportions for the helical spring. Hence we define design variable u by equation 8.40 and require satisfaction of limit equation 8.41 for an acceptable design.

$$u = \frac{D}{d} \tag{8.40}$$

$$u_{\min} \leq u \leq u_{\max} \tag{8.41}$$

Typical values for the limits are $u_{\min} = 4$ and $u_{\max} = 18$, which also correspond to the range of validity for equation 8.34b as previously explained.

The force gradient k for either an extension or a compression helical spring is given by

$$k = \frac{Gd^4}{8D^3N} \tag{8.42}$$

In equation 8.42, k is the force gradient of the spring, in pounds per inch; G is the modulus of rigidity for the material, in pounds per square inch, which is 11.5×10^6 psi for steel wire; d and D are as previously defined; and N is the active number of coils. The basic derivation for equation 8.42 is presented in most machine design textbooks.

The weight W for the active coils of a helical spring is often significant as we will see later, and it can be very closely approximated as follows on the basis of Pappus' second theorem from elementary calculus:

$$W \approx w(\pi DN)\left(\frac{\pi d^2}{4}\right)$$

$$\therefore \quad W \approx \left(\frac{\pi^2}{4}\right) wNDd^2 \tag{8.43}$$

In equation 8.43, W is the weight of the active spring coils, in pounds; w is the weight density of the material which, is 0.283 lb/in^3 for steel; and N, d and D are as previously defined.

If a helical spring is used in a high speed mechanism, longitudinal vibrations internal to the spring *can* lead to serious difficulties. If a condition of resonance exists for a significant Fourier component of the forcing function acting on the spring, large amplitude longitudinal forced vibrations in the spring will result, and because of this the *entire* mechanism may not function properly. This is so because the large amplitude internal longitudinal vibrations of the spring will appreciably affect the *force* exerted by the spring on mechanism components; also, such internal vibrations will significantly increase the maximum shearing stress in the spring, possibly resulting in a premature fatigue failure which might be disastrous. For a spring having constrained ends, the lowest natural frequency for internal longitudinal vibrations is determined from equation 8.44a, which follows. In the derivation of this equation the spring is considered as an elastic bar of weight W having a static force gradient k, as studied in vibrations textbooks such as reference 8.22. It is also derived in reference 8.17.

$$f_{n1} = \tfrac{1}{2}\sqrt{\frac{gk}{W}} \tag{8.44a}$$

In equation 8.44a, f_{n1} is the natural frequency, in cycles per second, for longitudinal internal vibrations in the lowest mode for a spring with constrained ends; g is the gravitational constant of 386 in/sec^2; k is the force gradient of the spring, in pounds per inch; and W is the weight of the active spring coils, in pounds. Another equation for f_{n1} is obtained directly from equation 8.44a by using equations 8.42 and 8.43; it is

$$f_{n1} = \left(\frac{d}{2\pi D^2 N}\right)\sqrt{\frac{gG}{2w}} \tag{8.44b}$$

In this equation, all the terms and units have been defined earlier.

In designing a helical spring for a high speed mechanism, generally the natural frequency f_{n1} from equation 8.44b should be appreciably greater than any *significant* Fourier component of the forcing function acting on the spring. Then the condition of resonance will be avoided and the spring will behave as it should, supplying the force F with a desired maximum value F_{max} and having the desired force gradient k, with no exorbitant stresses or internal vibrations. The author has found for many actual problems that a worth-while objective in designing helical springs for high speed mechanisms is the maximizing of f_{n1}.

For cases where space limitations are significant, the length of a spring, in its machine setting, measured along its axis is important. For a compression spring, the length L for the condition when F_{max} exists is expressed by equation 8.45, which is obvious from figure 8.15.

$$L \approx Nd + N(cd)$$

$$\therefore \quad L \approx Nd(1 + c) \tag{8.45}$$

In equation 8.45, L is the approximate length, in inches, of the spring for the condition when F_{max} exists, c is the decimal percentage of d which is allowed for clearance between adjacent coils, and N and d are as previously defined. Equation 8.45 is derived under the reasonable assumption that the minimum space permitted between adjacent coils must be greater for springs of large wire diameter than for springs of small wire diameter, in order that there will be no contact between adjacent coils during operation resulting from manufacturing errors in coil spacing.

Buckling is another type of failure phenomenon which we must avoid in the design of a helical compression spring. An excellent approach to the problem is derived in reference 8.23, and based on this our buckling failure criterion is derived as follows. From equation 34 in Chapter 3 of reference 8.23, conservatively assuming hinged ends for our compression spring, we must satisfy equation 8.46 for avoidance of buckling, referring to figure 8.15 for notation.

$$L \leq \frac{11.15(D/2)^2}{(F_{max}/k)}$$

$$\therefore \quad L \leq \frac{2.79D^2k}{F_{max}} \tag{8.46}$$

If we now eliminate L in the preceding criterion by combination with equation 8.45, we obtain a practical way of handling the buckling criterion as follows. We define a limited buckling parameter x by equation 8.47 and require satisfaction of limit equation 8.48 for an acceptable design.

$$L = Nd(1 + c) \leq \frac{2.79D^2k}{F_{max}}$$

$$\therefore \quad x = \frac{Nd(1 + c)F_{max}}{2.79D^2k} \tag{8.47}$$

$$x \leq 1 \tag{8.48}$$

All terms in equations 8.47 and 8.48 have been defined earlier.

Objectives for Optimum Design. There are many practical objectives for optimum design of helical compression springs, and which choice to make depends on the particular problem as related to the requirements of the total mechanical device in which the spring is to function. For example, maximization of factor of safety N_e related to fatigue breakage is an objective covered in section 9.3 of reference 8.24. For a tight space situation we might choose minimization of outside diameter D_o or minimization of length L as an appropriate objective. For high speed mechanisms or other devices we might choose to maximize lowest natural frequency f_{n1} for internal longitudinal vibrations in order to avoid resonant conditions where we have given excitation frequencies of significance. Incidentally, this has been the appropriate optimization objective for designing many load springs for high speed cam mechanisms in a number of industrial applications, such as for the cam mechanisms of figures 1.10, 1.11, and 1.12 of Chapter 1. In other situations we might choose the objective of minimizing weight W. In this section we will choose simultaneously the triple objective of maximizing f_{n1}, minimizing L, and minimizing W. We will find that the conclusions of optimum design are identical for each of these three objectives, as determined from the unique optimization study which follows, based on the MOD technique.

Boundary Conditions for the Triple Optimization Problem. For the triple optimization problem which follows, we will assume the common boundary conditions summarized in Table 8.14. In this table the limit equations for an acceptable design are summarized, as are the specified items for the optimization problem, both of which are typical for realistic design situations. Regarding avoidance of yielding, the criterion is handled as follows, using equations 8.35, 8.36, and 8.37 in the derivation.

$$\tau_{\max} \leq \frac{S_{sy}}{N_y} = \frac{\beta_y S_t}{N_y} = \frac{\beta_y C_t}{d^t N_y}$$

Hence we define a limited stress parameter y by equation 8.49, which must satisfy limit equation 8.50 for an acceptable design avoiding the yielding failure phenomenon.

$$y = \tau_{\max} d^t \tag{8.49}$$

$$y \leq \frac{\beta_y C_t}{N_y} \tag{8.50}$$

Also, we require use of standard size wire diameters d_s, such as the values of d given in table 8.13. This is a discrete value constraint on d finite in number N_{sd}, as expressed by equation 8.51.

$$d = d_s \leq d_{\max} \qquad \text{for} \quad s = 1, \ldots, N_{sd} \tag{8.51}$$

We also assume that force gradient k which the spring is to deliver must fall within the range of limit equation 8.52.

$$k_{\min} \leq k \leq k_{\max} \tag{8.52}$$

Also, we include regional constraints on N and D for a practical design, by means of limit equations 8.53 and 8.54 respectively.

$$D \leq D_{\max} \tag{8.53}$$

$$N_{\min} \leq N \leq N_{\max} \tag{8.54}$$

Finally, the limit equations 8.39, 8.41, and 8.48 have already been described, on τ_a, u, and x respectively. The handling of special cases in numerical applications for these summarized limit equations of table 8.14 are considered later for the derived flow chart.

TABLE 8.14

Boundary Conditions for Triple Optimization Problem

LIMIT EQUATIONS

$$\tau_a \leq S_{se}/N_e \qquad (8.39)$$

$$u_{min} \leq u \leq u_{max} \qquad (8.41)$$

$$x \leq 1 \qquad (8.48)$$

$$y \leq \beta_y C_t/N_y \qquad (8.50)$$

$$d = d_s \leq d_{max} \quad \text{for} \quad s = 1, \ldots, N_{sd} \qquad (8.51)$$

$$k_{min} \leq k \leq k_{max} \qquad (8.52)$$

$$D \leq D_{max} \qquad (8.53)$$

$$N_{min} \leq N \leq N_{max} \qquad (8.54)$$

SPECIFIED ITEMS

F_{max} and η (see figure 8.15); β_y, C_t, t, and N_y; S_{se} and N_e; u_{min}, u_{max}; d_s, for $s = 1, \ldots, N_{sd}$; d_{max}; k_{min}, k_{max}; D_{max}; N_{min}, N_{max}; G, w; c

Initial Formulation for Triple Optimization Problem. The (I.F.) for the triple optimization problem is summarized in table 8.15, based on the equations already discussed. Hence we have three primary design equations, 8.44b, 8.45, and 8.43, for f_{n1}, L, and W respectively. Also, note that we must include equation 8.55 as readily derived from figure 8.15, in order to complete the mathematical coordination of the variables involved.

$$F_a = \frac{(F_{max} - F_{min})}{2}$$

where $F_{min} = \eta F_{max}$.

$$\therefore \quad F_a = \frac{F_{max}(1 - \eta)}{2} \qquad (8.55)$$

Exploratory Calculations. We will now apply exploratory equations 6.1 through 6.4 of Chapter 6 to the (I.F.) of table 8.15, referring also to table 8.14 for the (L.E.)'s and the specified items. However, first we review the (I.F.) for recognition of free variables and constrained parameters.

$$n_f = 3 \quad (\text{for } K_{ts}, \tau_{max}, \text{and } F_a)$$

$$n_c = 8 \quad (\text{for } d, D, N, \tau_a, u, k, x, \text{and } y)$$

TABLE 8.15

Initial Formulation for Triple Optimization Problem

$f_{n1}\uparrow = (d/2\pi D^2 N)\sqrt{gG/2w}$	(P.D.E.) (8.44b)
$L\downarrow = Nd(1 + c)$	(P.D.E.) (8.45)
$W\downarrow = (\pi^2/4)wNDd^2$	(P.D.E.) (8.43)
$\tau_{max} = (8DF_{max}/\pi d^3)K_{ts}$	(S.D.E.) (8.33)
$K_{ts} = 1.60/(D/d)^{0.140}$	(S.D.E.) (8.34b)
$\tau_a = (8DF_a/\pi d^3)K_{ts}$	(S.D.E.) (8.38)
$u = D/d$	(S.D.E.) (8.40)
$k = Gd^4/8D^3N$	(S.D.E.) (8.42)
$x = Nd(1 + c)F_{max}/(2.79D^2k)$	(S.D.E.) (8.47)
$y = \tau_{max}d^t$	(S.D.E.) (8.49)
$F_a = F_{max}(1 - \eta)/2$	(S.D.E.) (8.55)

Limit equations and specified items are given in table 8.14

Also, we note that the number of (S.D.E.)'s in the (I.F.) is $N_s = 8$. Hence, as explained in section 6.13 of Chapter 6, we have $n_f = 3 < N_s = 8$, so the problem is of the (R.S.) type for the MOD. Thus by equations 6.1 through 6.4 of Chapter 6 we calculate the following, using $n_v = n_f + n_c = 3 + 8 = 11$.

$$D_{vs} = n_v - N_s + 1 = 11 - 8 + 1 = 4$$

$$n_e = N_s - n_f = 8 - 3 = 5$$

$$n_r = n_c - n_e = 8 - 5 = 3$$

$$A_t = \frac{n_c!}{n_e!(n_c - n_e)!} = \frac{8 \times 7 \times 6 \times 5 \times 4 \times 3 \times 2}{5 \times 4 \times 3 \times 2(3 \times 2)} = 56$$

In summary, we have very quickly learned that the optimization problem is four dimensional in character, there will be 5 eliminated parameters and 3 related parameters, and there are 56 approaches theoretically possible for derivation of final formulations (some of which may be indeterminate). As an exercise, you may wish to draw a circle diagram for summary of the 56 approaches indicating its meaning, similar to what we did for figure 8.4 in our study of helical gears.

Planning the Attack. In planning for a workable variation study, we would like to reduce D_{vs} by 1 to give $D_{vs} = 3$ instead of 4. Hence, in review of the (I.F.) constraints of table 8.14, we recognize that d has a discrete value constraint finite in number. Thus we logically choose to hold d constant for the (V.S.), from which our flow chart will be derived. Therefore, our plan for numerical problem solution is to have our digital computer execute the (F.C.) for each of the $d = d_s$ values, determining the best design for each case. The overall optimum design will be the one of most favorable optimization quantity value so determined.

Final Formulation for Triple Optimization Problem. With our described plan of attack in mind, from the (I.F.) of table 8.15 we readily recognize a relatively simple determinate approach whose general format (F.F.) equation system is given in table 8.16. It is understood that d will be held constant at a d_s value for this (F.F.). Using the table 8.16 general format as a guide, we specifically derive the (F.F.) summarized in table 8.17. Incidentally, for each equation in this table the terms held constant for the (V.S.) to follow are enclosed in either parentheses or brackets. Also, for explicit agreement with the general format of table 8.16, we could introduce phantom variables for D and N as factors of D^0 and N^0 where appropriate in the equations of table

TABLE 8.16

General Format for (F.F.) of Triple Optimization Problem, for Constant $d = d_s$

DEV. (P.D.E.)'S OF IDEAL PROBLEM

$$f_{n1}\uparrow = f_1(D, N) \tag{Ia}$$

$$L\downarrow = f_2(D, N) \tag{Ib}$$

$$W\downarrow = f_3(D, N) \tag{Ic}$$

RELATING EQUATIONS

$$k = f_4(D, N) \tag{II}$$

$$\tau_a = f_5(D, N) \tag{III}$$

$$x = f_6(D, N) \tag{IV}$$

$$u = f_7(D, N) \tag{V}$$

$$y = f_8(D, N) \tag{VI}$$

where D, N = related parameters
k, τ_a, x, u, y = eliminated parameters
(Problem constraints are summarized in table 8.14)

TABLE 8.17

Final Formulation for Triple Optimization Problem, for Constant $d = d_s$

$$f_{n1}\uparrow = \left(\frac{d_s}{2\pi}\sqrt{\frac{gG}{2w}}\right)\frac{1}{D^2N} \quad \text{(Ia)}$$

$$L\downarrow = [(1 + c)d_s]N \quad \text{(Ib)}$$

$$W\downarrow = \left(\frac{\pi^2}{4} w d_s^2\right) DN \quad \text{(Ic)}$$

$$k = \left(\frac{G}{8} d_s^4\right)\frac{1}{D^3N} \quad \text{(II)}$$

$$\tau_a = \left[\frac{6.40F_{max}(1 - \eta)}{\pi d_s^{2.86}}\right] D^{0.86} \quad \text{(III)}$$

$$x = \left[\frac{2.867(1 + c)F_{max}}{Gd_s^3}\right] DN^2 \quad \text{(IV)}$$

$$u = \left(\frac{1}{d_s}\right) D \quad \text{(V)}$$

$$y = \left[\frac{12.8F_{max}}{\pi d_s^{(2.86-t)}}\right] D^{0.86} \quad \text{(VI)}$$

where D, N = related parameters
k, τ_a, x, u, y = eliminated parameters
(Terms in brackets are held constant for the variation study to follow. Problem constraints are summarized in table 8.14)

8.17. However, for the sake of simplicity we will omit this specific detail and recognize its implication in the (V.S.) to follow.

In explanation of the table 8.17 (F.F.) equation system, the equations for f_{n1}, L, W, k, and u are unchanged from the (I.F.) of table 8.15. They did not contain any free variables and they were already organized to the general format of table 8.16. However, (S.D.E.) 8.38 for τ_a is developed by combination with (S.D.E.)'s 8.55 and 8.34b, by eliminating F_a and K_{ts} respectively in the process, giving equation III of table 8.17. Also, the (S.D.E.) 8.47 for x is developed by combination with (S.D.E.) 8.42 by eliminating k in the process, giving equation IV of table 8.17. Finally, (S.D.E.) 8.49 for y is developed by combination with (S.D.E.)'s 8.33 and 8.34b, by eliminating τ_{max} and K_{ts}

respectively in the process, giving equation VI of the (F.F.). Hence free variables have been eliminated and the (I.F.) has been properly organized to the desired format, based on equation combination of the MOD, giving the normalized (F.F.) equation system of table 8.17.

Variation Study for the Triple Optimization Problem. A three dimensional variation diagram for constant $d = d_s$ will be sketched based on the (F.F.) of table 8.17 and the constraints of table 8.14. The plan view for this diagram will have the related parameters D and N as the coordinate axes, and the feasible design space in this view will be determined by the relating equations and the constraints on both the related parameters and the eliminated parameters. From the (F.F.) of table 8.17 we note that the effects of these constraints are identical for all three optimization problems. Hence the feasible design space as shown on the variation diagram plan view will be identical for each case.

To start the three dimensional variation diagram for the triple optimization problem, we note that equations III, V, and VI for τ_a, u, and y respectively are functions of D only, being independent of N. Hence constraints imposed

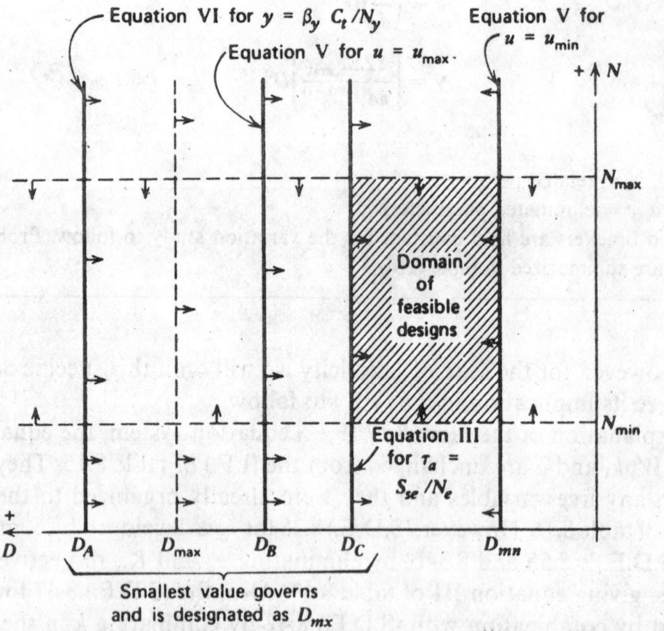

Figure 8.17 Typical plan view of 3-D diagram showing effects of constraints on N, D, τ_a, u, and y in helical spring problem. Constraints and equations referred to are in tables 8.14 and 8.17.

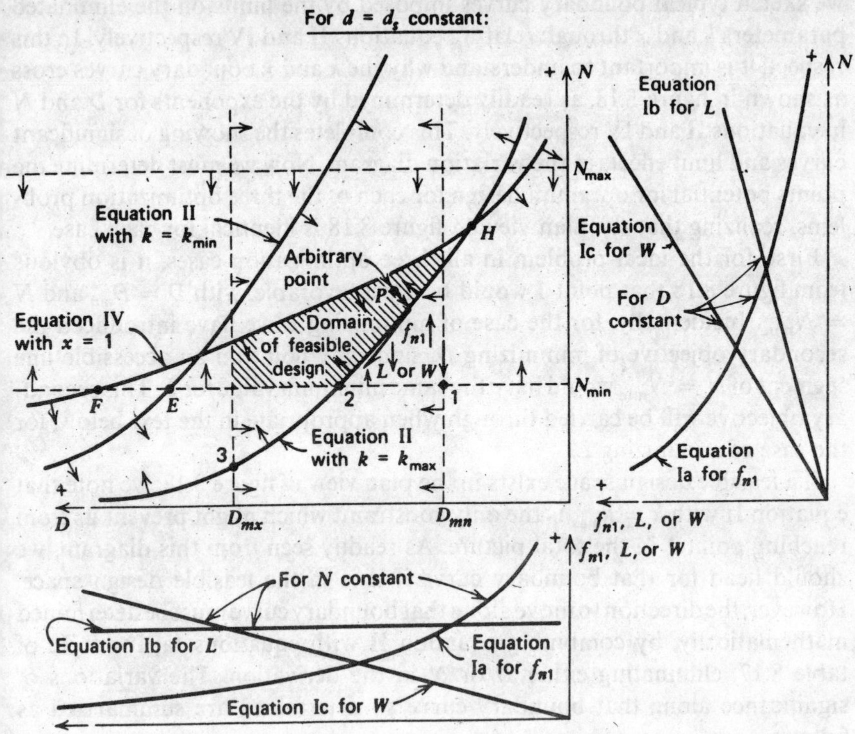

Figure 8.18 Typical three dimensional variation diagram for helical spring triple optimization problem. Constraints and equations referred to are in tables 8.14 and 8.17. D_{mx} and D_{mn} are defined in figure 8.17. Points 1, 2, and 3 are potential for optimum design for all three objectives: $f_{n1}\uparrow$, $L\downarrow$, or $W\downarrow$.

on these eliminated parameters will be reflected on the (D, N) plane as limits on D only. These effects are sketched on the (D, N) plane of figure 8.17. Thus we readily see that the greatest permissible value for D will be the smallest or governing value of D_A, D_B, D_C, and $D_{\max}$, and we designate this upper limit on D as D_{mx}. Its value will be determined in a flow chart segment discussed below. On the other hand, as shown in figure 8.17 there is only one lower limit for D as imposed by the $u_{\min}$ constraint, and we designate it as D_{mn}.

A typical three dimensional variation diagram is shown in figure 8.18 for a material of interest with $d = d_s$, incorporating the effects of all constraints in the triple optimization problem. Typical variations for the optimization quantities f_{n1}, L, and W are shown in the side and bottom cuts of figure 8.18, as sketched from equations Ia, Ib, and Ic respectively of the table 8.17 (F.F.). Next, for the plan view of the variation diagram, we sketch the limits D_{mx}, D_{mn}, $N_{\max}$, and $N_{\min}$ on the related parameters D and N respectively. Then we sketch typical boundary curves imposed by the limits on the eliminated parameters k and x through relating equations II and IV respectively. In this respect, it is important to understand why the k and x boundary curves cross as shown in figure 8.18, as readily determined by the exponents for D and N in equations II and IV respectively. This completes the showing of significant curves and limit effects in the variation diagram. Now we must determine the points potential for optimum design for each of the three optimization problems, realizing that the plan view in figure 8.18 is identical for each case.

First, for the ideal problem in all three optimization cases, it is obvious from figure 8.18 that point 1 would be most favorable, with $D = D_{mn}$ and $N = N_{\min}$. Incidentally, for the case of minimizing L we have introduced the secondary objective of minimizing D, since any point on an accessible line segment of $N = N_{\min}$ would have the same minimum value for L. This secondary objective will be carried through when appropriate in the text below, for the case of minimizing L.

If a feasible design space exists in the plan view of figure 8.18, we note that equation II with $k = k_{\max}$ is the only constraint which might prevent us from reaching point 1 in the total picture. As readily seen from this diagram, we should head for that boundary curve if it is in the feasible design space. However, the direction to move along that boundary curve must be determined mathematically, by combining equation II with equations Ia, Ib, or Ic of table 8.17, eliminating either D or N in the derivation. The variations of significance along that boundary curve so determined are summarized as follows.

1. *For maximization of* f_{n1} (combine equations II and Ia):

$$f_{n1} \sim \frac{1}{D^2 N} \sim \frac{1}{D^2(1/D^3)} = D$$

Hence f_{n1} increases as D increases along the boundary curve.

2. *For minimization of* L (combine equations II and Ib):

$$L \sim N \sim \frac{1}{D^3}$$

Hence L decreases as D increases along the boundary curve.

3. *For minimization of* W (combine equations II and Ic):

$$W \sim DN \sim D\left(\frac{1}{D^3}\right) = \frac{1}{D^2}$$

Hence W decreases as D increases along the boundary curve.

From the preceding variation study, we note the common conclusion reached for each case. That is, along the boundary curve of relating equation II with $k = k_{max}$ we should head in the direction of increasing D for improvement of the optimization quantity. Therefore, in the figure 8.18 variation diagram we readily conclude that points 2 and 3 are also potential for optimum design.

Regarding incompatible specifications, there are several possibilities as seen from the figure 8.18 variation diagram plan view. First, for the ideal problem, (I.S.) is the case if $D_{mn} > D_{mx}$. Second, for the equation II boundaries acting with the D and N constraints alone, we have (I.S.) if either $N_3 > N_{max}$ or $D_E < D_{mn}$. Next, for the equation IV boundary acting with the D and N constraints alone, we have (I.S.) if $D_F < D_{mn}$. Finally, for the boundaries of equations II and IV acting together with the D and N constraints, we have (I.S.) if either $D_H > D_{mx}$ or $N_H < N_{min}$.

In summary, from our triple optimization problem variation study of figure 8.18, we draw identical conclusions of optimum design regardless of which of the three objectives is chosen. In each case, the same three points, designated as 1, 2, and 3 in the figure, are potential for optimum design. These points of optimum design are determined by intersections of the plan view boundary curves alone in the figure, and this related parameter view is common to each of the three cases. Also, it follows that the possibilities of incompatible specifications previously discussed are common to all three cases. Thus the derived flow chart for optimal design given in figures 8.19 through 8.21 is common in all respects for the triple optimization problem of helical compression springs. Incidentally, the specific equations for the pertinent points of figure 8.18 are defined in the figure by the designated equations and constraints. These pertinent point equations are thereby derived and summarized in table 8.18. They are needed in the flow chart given in figures 8.19 through 8.21 to establish tests for (I.S.) and to determine the particular point of optimum design if a feasible design space exists.

The variation study now explained gives a common design solution for the three optimization objectives, *for any constant* $d = d_s$. However, in numerical applications we must choose specifically *one* of the three optimization objectives, since the optimum value for d_s may be different for the three cases. Hence, for generalization of the (F.C.) in figures 8.19 through 8.21, we designate the optimization quantity by Q_{mx}, which is to be maximized in each case. Q_{mx} is defined for each objective as follows, appropriately using equations Ia, Ib, or Ic of table 8.17 for establishing an equivalent maximization problem.

1. *For maximization of* f_{n1} *use*:

$$Q_{mx} = f_{n1} \tag{8.66a}$$

where f_{n1} is given by equation Ia of table 8.17.

TABLE 8.18

Equations for Pertinent Points Defined in Figures 8.17 and 8.18

$$D_A = \left[\frac{\beta_y C_t}{N_y}\frac{\pi d_s^{(2.86-t)}}{12.8F_{max}}\right]^{1.163} \tag{8.56}$$

$$D_B = u_{max} d_s \tag{8.57}$$

$$D_C = \left[\frac{S_{se}}{N_e}\frac{\pi d_s^{2.86}}{6.40F_{max}(1-\eta)}\right]^{1.163} \tag{8.58}$$

$$D_{mn} = u_{min} d_s \tag{8.59}$$

$$N_3 = \left(\frac{G}{8}d_s^{4}\right)\left(\frac{1}{D_{mx}^3 k_{max}}\right) \tag{8.60}$$

$$D_2 = \left(\frac{Gd_s^{4}}{8k_{max}N_{min}}\right)^{1/3} \tag{8.61}$$

$$D_E = \left(\frac{Gd_s^{4}}{8k_{min}N_{min}}\right)^{1/3} \tag{8.62}$$

$$D_F = \frac{Gd_s^{3}}{2.867(1+c)F_{max}N_{min}^2} \tag{8.63}$$

$$D_H = \frac{d_s}{1.861}\left[\frac{G(1+c)F_{max}}{k_{max}^2}\right]^{1/5} \tag{8.64}$$

$$N_H = \frac{Gd_s^{4}}{8k_{max}D_H^{3}} \tag{8.65}$$

2. *For minimization of* L *use*:

$$Q_{mx} = \frac{1}{L} \tag{8.66b}$$

where L is given by equation Ib of table 8.17.

3. *For minimization of* W *use*:

$$Q_{mx} = \frac{1}{W} \tag{8.66c}$$

where W is given by equation Ic of table 8.17.

Flow Chart for the Triple Optimization Problem. The main flow chart of figure 8.19 is started by reading input data, which are the specified items of table 8.14. Next, as previously explained, for numerical application we must

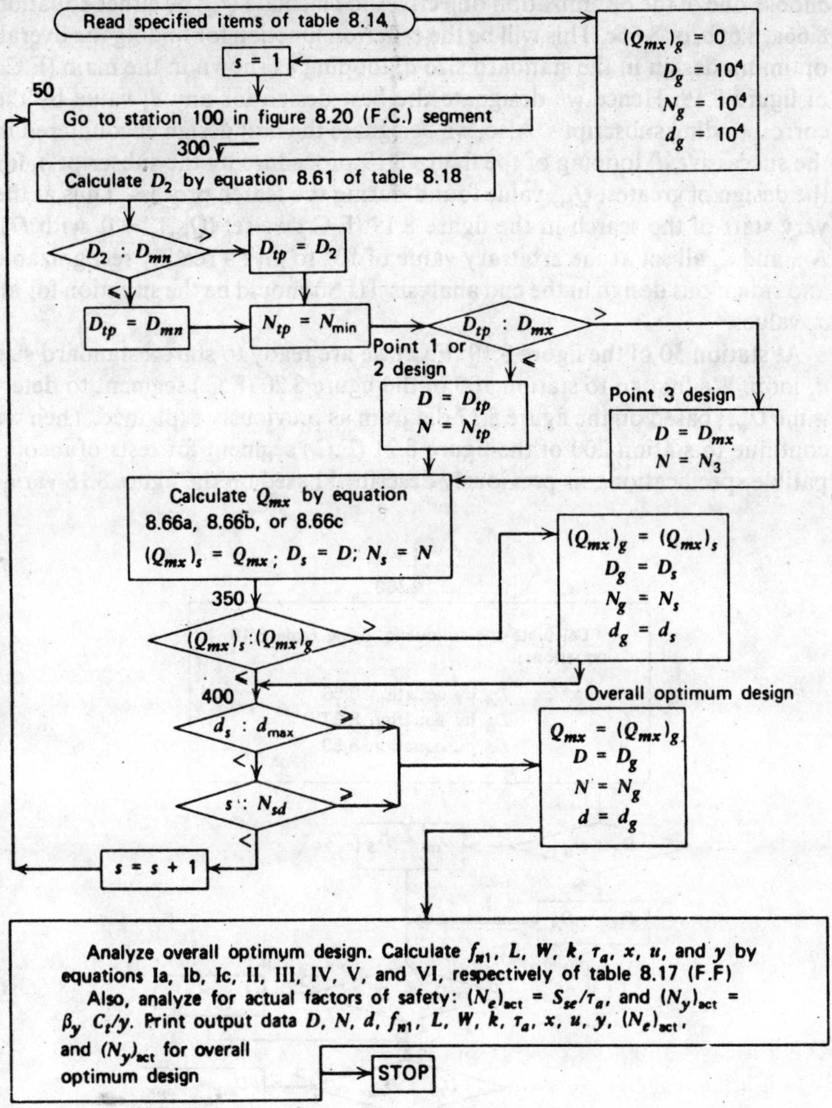

Figure 8.19 Main flowchart for helical spring triple optimization problem of section 8.2.

choose one of the optimization objectives to calculate Q_{mx} by either equation 8.66a, 8.66b, or 8.66c. This will be the criterion function for finding the overall optimum design in the standard size d_s looping as shown in the main (F.C.) of figure 8.19. Hence we designate the best design for any d_s value by the corresponding subscript s. Also, we designate the best design encountered in the successive d_s looping of the figure 8.19 procedure by the subscript g, for the design of greatest Q_{mx} value found during the search process. Thus at the very start of the search in the figure 8.19 (F.C.) we set $(Q_{mx})_g = 0$, with D_g, N_g, and d_g all set at the arbitrary value of 10^4 to give a readily recognizable and ridiculous design in the end analysis if (I.S.) should be the situation for all d_s values.

At station 50 of the figure 8.19 (F.C.) we are ready to start a standard size d_s loop. We first go to station 100 of the figure 8.20 (F.C.) segment to determine D_{mx}, based on the figure 8.17 diagram as previously explained. Then we continue to station 200 of the figure 8.21 (F.C.) segment for tests of incompatible specifications, as previously described based on the figure 8.18 varia-

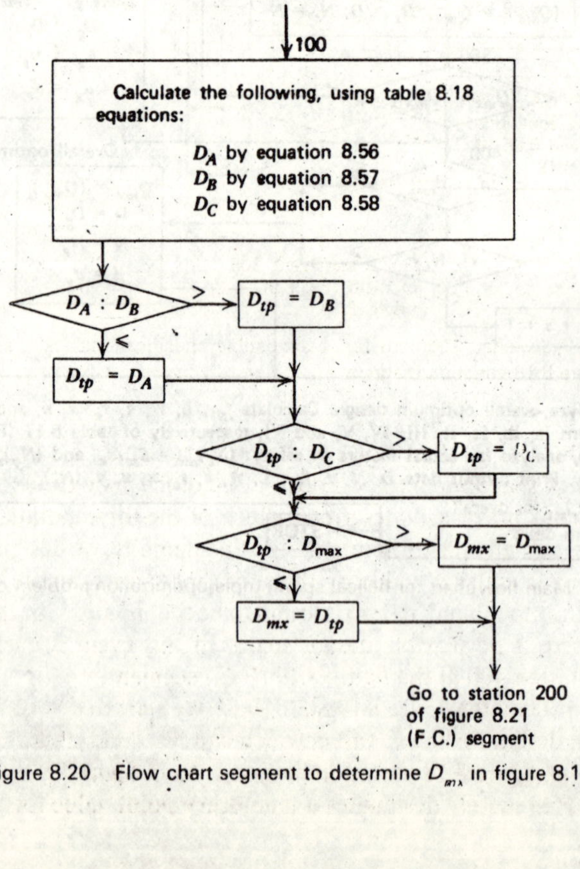

Figure 8.20 Flow chart segment to determine D_{mx} in figure 8.17.

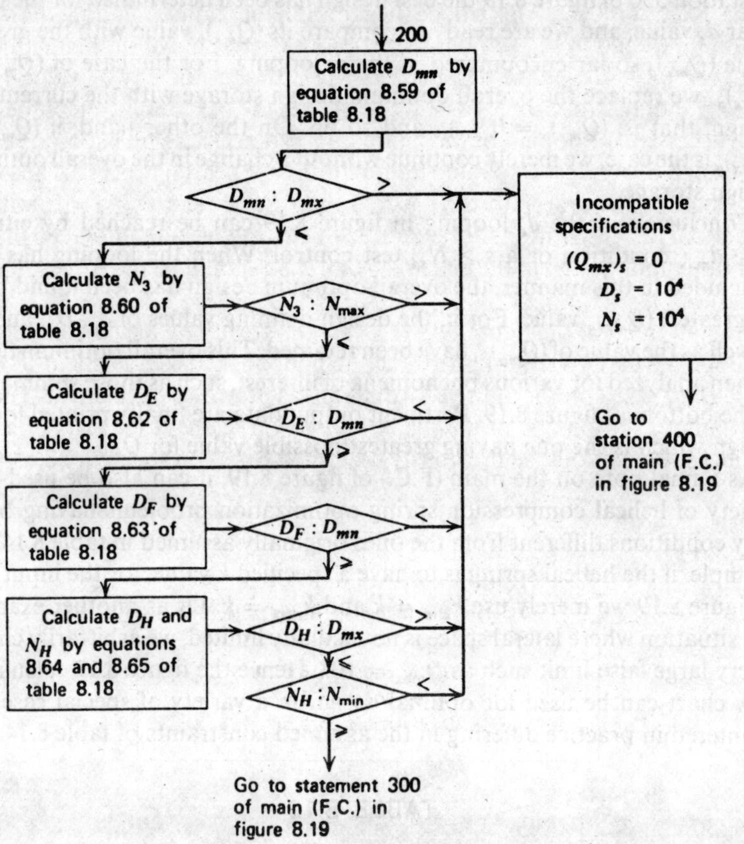

Figure 8.21 Flow chart segment for incompatible specifications tests, as described in text from figure 8.18 variation diagram.

tion diagram. If (I.S.) does not exist for the particular d_s value, we have a feasible design space and we go to station 300 of the main (F.C.) in figure 8.19 to determine as previously described which of the three points 1, 2, or 3 in figure 8.18 is actually the optimum design. Incidentally, we designate this best design for a particular d_s value by the subscript s, for example, D_s, N_s, and $(Q_{mx})_s$. On the other hand, if (I.S.) is found to be the situation for the particular d_s value, there is no feasible design space. In this event we set $(Q_{mx})_s = 0$ and go to station 400 of the figure 8.19 (F.C.) segment.

At station 300 of the figure 8.19 main (F.C.) we start to determine the point of optimum design for the d_s value, which can be found on a D comparison basis from the figure 8.18 variation diagram. As in figure 8.20, the subscript tp in figure 8.19 merely designates a temporary point value for the variable.

At station 350 of figure 8.19 the best design has been determined for the particular d_s value, and we are ready to compare its $(Q_{mx})_s$ value with the greatest value $(Q_{mx})_g$ so far encountered in the d_s looping. For the case of $(Q_{mx})_s > (Q_{mx})_g$, we replace the overall optimum design storage with the current best design, that is, $(Q_{mx})_g = (Q_{mx})_s$, and so on. On the other hand, if $(Q_{mx})_s \leq (Q_{mx})_g$ is the case, we merely continue without a change in the overall optimum design storage.

Conclusion of the d_s looping in figure 8.19 can be reached by either a $d_s \geq d_{max}$ condition or a $s \geq N_{sd}$ test control. When the looping has been concluded in this manner, the overall optimum design has been found based on greatest $(Q_{mx})_g$ value. For it, the design-defining values of d_g, D_g, and N_g, as well as the value of $(Q_{mx})_g$, have been retained. This overall optimum design is then analyzed for various phenomena of interest, such as those summarized at the bottom of figure 8.19. Pertinent output data are finally printed for this design which is the one having greatest possible value for Q_{mx}.

As a final note on the main (F.C.) of figure 8.19, it can also be used for a variety of helical compression spring optimization problems having boundary conditions different from the ones originally assumed in table 8.14. For example, if the helical spring is to have a specified k value, for the input data of figure 8.19 we merely use $k_{min} = k$ and $k_{max} = k$. Or as another example, in a situation where lateral space is not actually limited, we arbitrarily choose a very large false limit such as $D_{max} = 10^4$. Hence the figure 8.19 calculation flow chart can be used for optimal design in a variety of special cases encountered in practice differing in the assumed constraints of table 8.14.

TABLE 8.19

Brief Description of Five Numerical Problems for Example 8.2*

Problem	Objective	Input Items Not Common to All Five Problems	
		k_{max} lb/in	D_{max} in
A	$f_{n1}\uparrow$	80	0.500
B	$L\downarrow$	80	0.500
C	$W\downarrow$	80	0.500
D	$f_{n1}\uparrow$	420	0.500
E	$f_{n1}\uparrow$	80	0.300

* Common items listed in text.

EXAMPLE 8.2

Five numerical problems will be solved by the figure 8.19 (F.C.), and they are very briefly described in table 8.19. Except for the items of table 8.19, common input data for the specified items of table 8.14 for the five problems will be as follows. In each case, unpeened A.S. 5 music wire of the table 8.13 standard sizes will be assumed.

Common Input Data for Five Problems

$F_{max} = 20$ lb

$F_{min} = 10$ lb

$\therefore \quad \eta = \dfrac{F_{min}}{F_{max}} = \dfrac{10}{20} = 0.500$ (see figure 8.15)

$\beta_y = 0.4$
$C_t = 200{,}000$
$t = 0.145$
} for A.S. 5 music wire (see text, after equations 8.36 and 8.37)

$N_y = 1.344$ (from example 5.7 of Chapter 5)

$S_{se} = 45{,}000$ psi (see text, after equation 8.39)

$N_e = 1.813$ (from example 5.7 of Chapter 5)

$u_{min} = 4$ (see text, after equation 8.41)

$u_{max} = 18$ (see text, after equation 8.41)

d_s = standard diameter values in table 8.13 for $N_{sd} = 26$ standard diameters

$d_{max} = 0.135$ in largest diameter in table 8.13

$k_{min} = 50$ lb/in

$N_{min} = 3$
$N_{max} = 10$
} active coil range for (L.E.) 8.54 of table 8.14, for a practical spring

$$G = \frac{E}{2(1 + \nu)} = 29 \times \frac{10^6}{2(1.27)} = 11.4 \times 10^6 \text{ psi}$$

$w = 0.283$ lbm/in^3

} (see tables 4.1 and 4.2 of Chapter 4 for steel)

$c = 0.5$ assumed for adequate coil spacing; see text after equation 8.45, and see figure 8.15

Solutions to Optimization Problems. Based on the main (F.C.) of figure 8.19, a digital computer program was written for the triple optimization problem. With this, the overall optimum design can be determined automatically and

TABLE 8.20

Overall Optimum Design and Analysis of Helical Springs for Numerical Problems of Example 8.2*

Item	Units	A	B	C	D	E
		Numerical Problem of Table 8.19				
		DIMENSIONS FOR OPTIMUM DESIGN (see figure 8.15)				No design solution possible which satisfies all specifications
D	inches	0.346	0.500	0.346	0.220	
N		6.77	3.83	6.77	3.00	
d	inches	0.063	0.072	0.063	0.055	
		ANALYSIS OF DESIGN				
f_{n1}	cps	1091	1055	1091	5316	
L	inches	0.639	0.414	0.639	0.248	
W	lbm/10^3	6.49	6.93	6.49	1.39	
k	lb/in	80.0	80.0	80.0	408.2	
τ_a	psi	22,215	20,805	22,215	22,186	
x		0.478	0.148	0.478	0.090	
u		5.49	6.94	5.49	4.00	
y		59,514	56,826	59,514	58,276	
$(N_e)_{act}$		2.026	2.163	2.026	2.028	
$(N_y)_{act}$		1.344	1.408	1.344	1.373	
Design point of figure 8.18		3	3	3	1	(I.S.) for all d_s values

* Common input data values are listed in text; items not common are given in table 8.19.

rapidly in numerical applications. The solutions to the five numerical problems of this example are summarized in table 8.20, which is self-explanatory.

8.3 Optimal Design of Capillary-Compensated Hydrostatic Journal Bearings

Hydrostatic journal bearings [8.25–8.29] have the possible features of extremely low friction, the ability to support high load from zero shaft speed to full speed, the ability to support high load with fluids of low viscosity, smoothness of operation, no wear, no deterioration in use, and the characteristic of high mounting stiffness. However, proper design of a hydrostatic journal bearing is not a simple task. Many complicated functions must be considered in the decision-making process of design, and these functions are normally available to the design engineer in the form of multidimensional graphs and equations as illustrated in references 8.25 through 8.29.

Two types of hydrostatic journal bearings are commonly used in practice, differing by means of the compensation used for sensing changes in load as related to eccentricity. The first type is known as an orifice-compensated bearing, and its load-eccentricity characteristics are affected by changes in dynamic viscosity as will result from temperature changes in the fluid. The second type is known as a capillary-compensated bearing, and its load-eccentricity characteristics are *not* affected by changes in dynamic viscosity. The basic configuration for a capillary-compensated hydrostatic journal bearing is shown in figure 8.22, with a typical system of allied components shown in figure 8.23. Some nomenclature for the bearing design is summarized as follows, with reference to figure 8.22 where appropriate.

Some Nomenclature for Bearing Design

C = radial clearance; that is, bearing radius minus journal radius, inches
D = journal diameter, inches
d_t = capillary tube diameter, inches
e = displacement or eccentricity for journal center relative to bearing center, inches
h_0 = minimum fluid film thickness = $C - e$, inches
l = axial pocket land length, inches
L = axial bearing length, inches
l_c = circumferential distance between pockets, inches
L_p = axial pocket length, inches
L_t = capillary tube length, inches
m_{pa} = pocket aspect ratio = $(L_p/D)(l/l_c)$, dimensionless
N_j = journal speed, revolutions per minute; (for full speed operation)
p_s = manifold supply pressure for fluid, psig
Q_T = total flow rate for bearing, in^3/sec
W = load supported by bearing, lbf
ϵ = eccentricity ratio = e/C, dimensionless
μ = dynamic viscosity of fluid, reyns or lb-sec/in^2

Objective for Optimal Design. Hydrostatic journal bearings are often used in high load, smooth action, wearfree applications, where high mounting rigidity is desired for the entire range of zero to full speed. Usually the elements on the journal shaft have to be located relative to other parts of the machine, such as meshing gears or a stator element, with a required degree

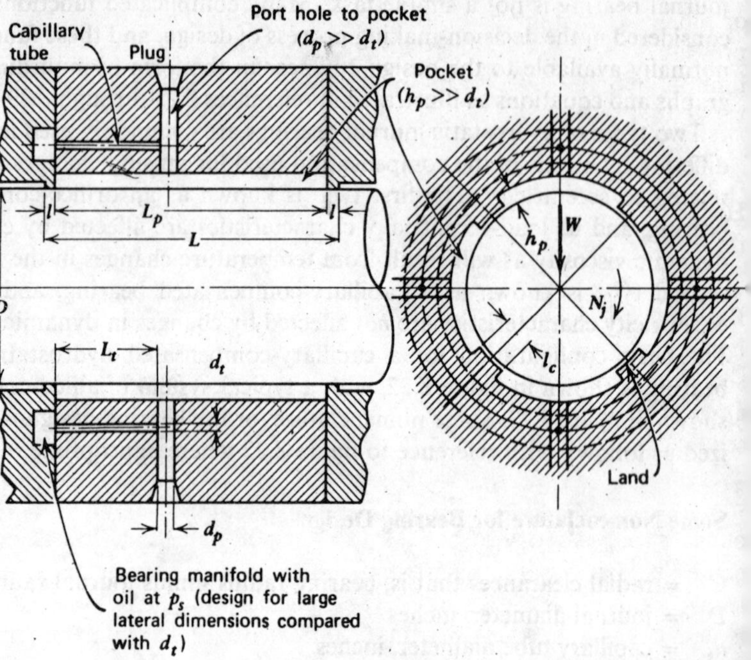

Figure 8.22 Basic configuration for capillary-compensated hydrostatic journal bearing.

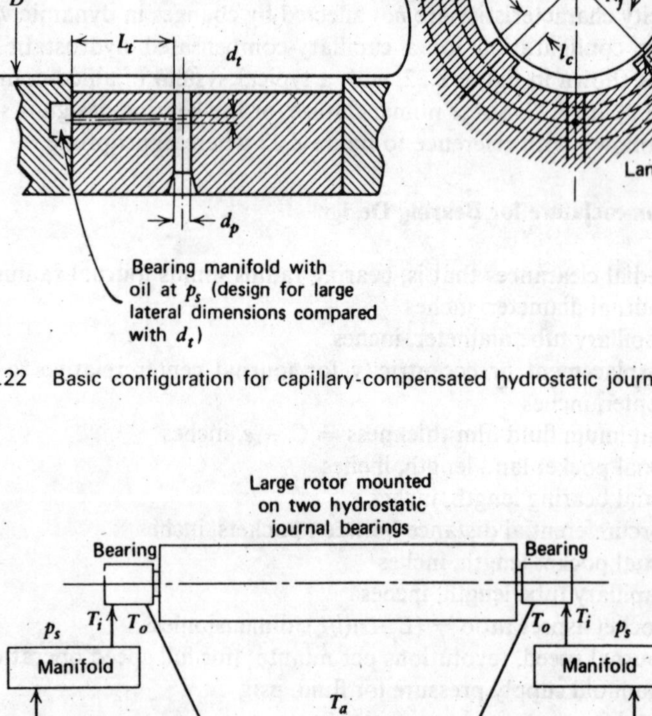

Figure 8.23 Schematic diagram of bearing system showing oil flow between system components.

of radial accuracy during operation. This design requirement can be expressed by the regional constraint of equation 8.67, where e_{max} is the specified limit for acceptable mounting accuracy.

$$e \le e_{max} \tag{8.67}$$

Hence, to maximize mounting rigidity within the limits of this constraint, the objective for optimal design will be to maximize load capacity W.

Basic Equations for Design Use. Basic equations for capillary-compensated hydrostatic journal bearings have been derived from the information presented in references 8.25 and 8.26, and they will be reviewed here before application to optimal design. To start, the load capacity relations are presented graphically in figure 6 of reference 8.25. By special multidimensional curve-fitting techniques incorporating a least squares regression analysis, similar to the techniques of Chapter 2, the basic load variable equation 8.68a of table 8.21 is derived from the test data of figure 6 in reference 8.25. It expresses load variable y_{lv} in terms of the design parameter x_{dp} and accentricity ratio ϵ. Space does not permit its derivation here. Load variable y_{lv} is defined by equation 8.68b in table 8.21. Also, design parameter x_{dp} is defined by equation 8.69 as follows.

$$x_{dp} = \left(\frac{3\pi}{16}\right)\left(\frac{x_t l}{C^3 D}\right) \tag{8.69}$$

In equation 8.69, x_t is a capillary tube design parameter defined by equation 8.70.

$$x_t = \frac{d_t^{\,4}}{L_t} \tag{8.70}$$

Similarly, fluid flow relations are presented graphically in figure 7 of reference 8.25. Again, by a least squares regression analysis curve-fitting approach, the basic flow variable equation 8.71a of table 8.22 is derived from the test data of figure 7 in reference 8.25. It expresses flow variable y_{fv} in terms of the design parameter x_{dp} and eccentricity ratio ϵ, and space does not permit its derivation here. Total flow variable y_{fv} is defined by equation 8.71b in table 8.22, and design parameter x_{dp} is defined by equation 8.69 as before.

Incidentally, as previously stated, equations 8.68a and 8.71a for y_{lv} and y_{fv} in tables 8.21 and 8.22 were derived from the test data presented in the graphs of figures 6 and 7 of reference 8.25. Hence the equations are valid only over the given ranges of test data values for the referenced graphs. Specifically,

TABLE 8.21
Load Variable Equation for Hydrostatic Bearing Problem

$$y_{lv} = \frac{\eta_{lv}}{C_1 x_{dp}^p + C_2/x_{dp}^q} \tag{8.68a}$$

where

$$\eta_{lv} = m_{lv} x_{dp} + b_{lv}$$

where

$$m_{lv} = \frac{2.7056\epsilon^2 - 3.4320\epsilon + 1.5815}{100}$$

$$b_{lv} = -0.3589\epsilon^2 + 0.4747\epsilon + 0.7780$$

and

$$p = 0.3128\epsilon^2 - 0.5015\epsilon + 0.8472$$

$$q = -1.7722\epsilon^2 + 1.3787\epsilon + 0.6024$$

$$C_1 = 0.6811\epsilon^2 - 1.4570\epsilon + 1.1101$$

$$C_2 = 14.4578\epsilon^2 - 24.9963\epsilon + 11.7636$$

$$\epsilon = \frac{e}{C}$$

x_{dp} is expressed by equation 8.69

Load variable y_{lv} is basically defined as

$$y_{lv} = \frac{\sqrt{2}W}{D(L - l)p_s} \tag{8.68b}$$

the ranges of validity for the approximations of equations 8.68a and 8.71a are

$$0.2 \lesssim \epsilon \lesssim 1.0 \tag{8.72}$$

$$0.1 \lesssim x_{dp} \lesssim 100 \tag{8.73}$$

In general, the preceding should cover the full range of practical interest for hydrostatic bearing design applications.

Pocket aspect ratio m_{pa} is defined by equation 8.74 for hydrostatic journal bearings.

$$m_{pa} = \frac{(L - 2l)l}{Dl_c} \tag{8.74}$$

TABLE 8.22

Total Flow Variable Equation for Hydrostatic Bearing Problem

$$y_{fv} = \eta_{fv} 4.213 \ln (x_{dp} + 1) \tag{8.71a}$$

where

$$\eta_{fv} = m_{fv} x_{dp} + b_{fv}$$

where

$$m_{fv} = \frac{1.6511\epsilon^2 + 1.0947\epsilon - 1.3565}{100}$$

$$b_{fv} = 0.1944\epsilon^2 - 0.1517\epsilon + 1.322$$

and

$$\epsilon = \frac{e}{C}$$

x_{dp} is expressed by equation 8.69

Total flow variable is basically defined as

$$y_{fv} = \left(\frac{24\mu l}{C^3 D p_s}\right) Q_T \tag{8.71b}$$

From figures 11 and 12 of reference 8.25 we note that load capacity of a hydrostatic journal bearing is appreciably affected by m_{pa}. However, the test data of figures 6 and 7 of reference 8.25 are specifically for a value of $m_{pa} = 0.77$, and this was the basis for derivation of equation 8.68a of table 8.21 for load variable y_{lv}. Hence the following load capacity modifying factor η_{pa} equation is derived for later application, to account for the effect of m_{pa} on y_{lv}.

$$\eta_{pa} \approx \frac{1.721}{(1.5 + m_{pa})^{0.662}} \tag{8.75}$$

Space does not permit derivation of equation 8.75 for η_{pa}, but it should be reasonable to use over the range

$$0 \leq m_{pa} \leq 3$$

This interval should cover the full range of practical interest for hydrostatic bearing design applications.

Dynamic viscosity μ is a function of average oil temperature $(T_{\text{oil}})_{\text{avg.}}$ in degrees Fahrenheit for the bearing. It can be expressed by an equation of the following form.

$$\mu = \frac{C_o}{(T_{oil})^r_{avg}} \tag{8.76}$$

Specifically, for an oil of SAE 20W grade, the constants are $C_o = 0.8754$ and $r = 2.5698$. This is derived by simple exponential equation curve fitting, as explained in Chapter 2, using graphs readily available in the machine design literature such as figure AF16 of reference 8.3. The equation should be valid over the full range of practical interest, such as $80°F \leq (T_{oil})_{avg} \leq 220°F$.

Outlet oil temperature T_o for the bearing in the figure 8.23 system is expressed by equation 8.77, with inlet oil temperature T_i expressed by equation 8.78.

$$T_o = (T_{oil})_{avg} + \frac{(\Delta T)_{oil}}{2} \tag{8.77}$$

$$T_i = T_o - (\Delta T)_{oil} \tag{8.78}$$

In these equations temperatures are expressed in degrees Fahrenheit, and $(\Delta T)_{oil}$ is the difference between inlet and outlet temperatures, that is, $(\Delta T)_{oil} = T_o - T_i$.

Increase in oil temperature through the bearing is readily calculated from the power consumption of the bearing and the oil flow through the bearing. Hence, from elementary thermodynamics we derive equation 8.79.

$$(\Delta T)_{oil} = \frac{163H_b}{c_p Q_T} \tag{8.79}$$

In this equation H_b is the horsepower loss in the bearing, c_p is the specific heat of the oil in Btu/gal-°F, and Q_T is the total oil flow in in^3/sec as before. By equation 12-27 of reference 8.26, H_b is expressed in terms of the design parameters as follows.

$$H_b = 2.61 \times 10^{-6} \frac{l D^2 N_j^2 \mu}{C} \left[\frac{D}{\sqrt{1 - \epsilon^2}} + \frac{(L - 2l)(2 - \epsilon^2)}{\pi(1 - \epsilon^2)} \right] \tag{8.80}$$

The terms and units of this equation have been defined earlier.

Horsepower required from the pump, H_p, on a per bearing basis for the figure 8.23 system, is expressed as follows from equation 12-8 of reference 8.26.

$$H_p = \frac{p_s Q_T}{6603 \eta_m} \tag{8.81}$$

In this equation all terms and units are as previously defined, except η_m is the pump efficiency, typically $\eta_m \approx 0.75$.

From a thermodynamic heat balance of the figure 8.23 system on a per bearing basis, we derive equation 8.82.

$$H_c = (1 - \eta_m)H_p + H_b + \left(\frac{h_{cr} A_e}{33{,}000}\right)\left[T_a + \frac{(\Delta T)_c}{2} - (T_{\text{oil}})_{\text{avg}}\right]. \tag{8.82}$$

In this equation, H_c is the cooling horsepower per bearing as required for the figure 8.23 system, h_{cr} is the film coefficient of heat transfer for convection and radiation, A_e is the effective surface area for such heat transfer in square inches, T_a is the ambient temperature in degrees Fahrenheit, and $(\Delta T)_c$ in degrees Fahrenheit is due to compression of the oil by the pump. Typical values to use are $h_{cr} \approx 0.18$ ft-lb/min-in^2-°F for still air in a normal environment (see page 317 of reference 8.3) and $(\Delta T)_c \approx 4$°F for $p_s = 1000$ psi (see page 349 of reference 8.26). The other terms and units for equation 8.82 have already been defined.

As previously defined, radial clearance C is expressed by equation 8.83.

$$C = \frac{e}{\epsilon} \tag{8.83}$$

Also, minimum fluid film thickness h_0 is related to eccentricity e and eccentricity ratio ϵ as follows.

$$h_0 = C - e = \frac{e}{\epsilon} - e$$

$$\therefore \quad h_0 = e\left(\frac{1}{\epsilon} - 1\right) \tag{8.84}$$

Finally, for simplification in notation, we define dimensionless parameters x_g, x_l, and x_c by equations 8.85, 8.86, and 8.87 respectively.

$$x_g = \frac{L}{D} \tag{8.85}$$

$$x_l = \frac{l}{L} \tag{8.86}$$

$$x_c = \frac{4l_c}{\pi D} \tag{8.87}$$

Terms for these three equations have already been defined. This concludes the basic equation system for our optimization problem. Because of its complexity, solution will be obtained by AOD using OPTIGO as described in Chapter 6.

Criterion Function for AOD. The objective for optimal design will be to *maximize* load capacity, W, of the hydrostatic journal bearing. However, AOD algorithms such as OPTIGO, are programmed to *minimize* a criterion function Q as explained in Chapter 6. Hence, by simple reciprocation, we define Q for our optimization problem by $Q = 1/W$. Thus we derive Q for this problem from equation 8.68b, obtaining equation 8.88, which also incorporates the effects of pocket aspect ratio m_{pa} by the η_{pa} equation 8.75 as previously described.

$$Q\downarrow = \frac{1}{W}$$

$$Q\downarrow = \frac{\sqrt{2}}{\eta_{pa}\, y_{lv} D p_s (L - l)} \tag{8.88}$$

In this criterion function for our optimization problem, η_{pa} is calculated by equation 8.75, for which we must also use equation 8.74 to calculate m_{pa}. Also, load variable y_{lv} is calculated by equation 8.68a.

Constraints and Specified Items. The constraints and specified items for the optimization problem are summarized in table 8.23. They are determined primarily from practical considerations of the design problem, including the requirements of a typical design situation as well as the necessity for achieving practical proportions for the bearing. To save on space, they will not be discussed further.

Analysis for a Design State Point. During the search procedure by OPTIGO, it is necessary to analyze hydrostatic bearings at various design state points, as explained in Chapter 6. A state of design is uniquely determined by a set of independent parameters x_i, for $i = 1, \ldots, N$. For the hydrostatic journal bearing, from a review of figure 8.22 and the basic equations now summarized, the state of design is uniquely defined by *seven geometrically independent parameters* chosen typically as follows.

$$e, \epsilon, x_t, L, D, x_l, \text{ and } x_c$$

With known values for these variables at a design state point, the hydrostatic bearing can be analyzed for the various phenomena of interest according to the procedure summarized in figure 8.24 calculated in the given order. It is assumed that the specified items of table 8.23 are known values.

TABLE 8.23

Constraints and Specified Items for Hydrostatic Bearing Problem

REGIONAL CONSTRAINTS

$$0 \leq e \leq e_{max}$$
$$0.2 \leq \epsilon \leq 1.0$$
$$(x_t)_{min} \leq x_t \leq (x_t)_{max}$$
$$L_{min} \leq L \leq L_{max}$$
$$D_{min} \leq D \leq D_{max}$$
$$l \geq l_{min}$$
$$x_l \leq (x_l)_{max}$$
$$l_c \geq (l_c)_{min}$$
$$x_c \leq (x_c)_{max}$$
$$0.1 \leq x_{dp} \leq 100$$
$$0 \leq m_{pa} \leq 3$$
$$h_o \geq (h_o)_{min}$$
$$C \leq C_{max}$$
$$x_g \leq (x_g)_{max}$$
$$Q_T \leq (Q_T)_{max}$$
$$(T_{oil})_{avg} \leq T_o \leq (T_o)_{max}$$
$$H_c \leq (H_c)_{max}$$
$$H_p \leq (H_p)_{max}$$

SPECIFIED ITEMS

p_s, $(\Delta T)_c$, c_p, η_m, h_{cr}, A_e, T_a, $(T_{oil})_{avg}$, N_j; C_o and r for oil; and all limit values in above regional constraints

Unit Based Variables. As explained in Chapter 6, for a problem of this complexity it is advisable to define unit based variables u, which will vary essentially within the range $0 \leq u \leq 1$ during the OPTIGO search. Hence we define the following unit based variables in table 8.24 for the previously chosen seven independent variables of the basic problem: e, ϵ, x_t, L, D, x_l, and x_c. In the terminology of OPTIGO, the corresponding unit based variables also will be designated by x_1 through x_7 as shown in table 8.24. As defined in the table, these unit based variables will all vary essentially within the range $0 \leq \chi_i \leq 1$ during the OPTIGO search.

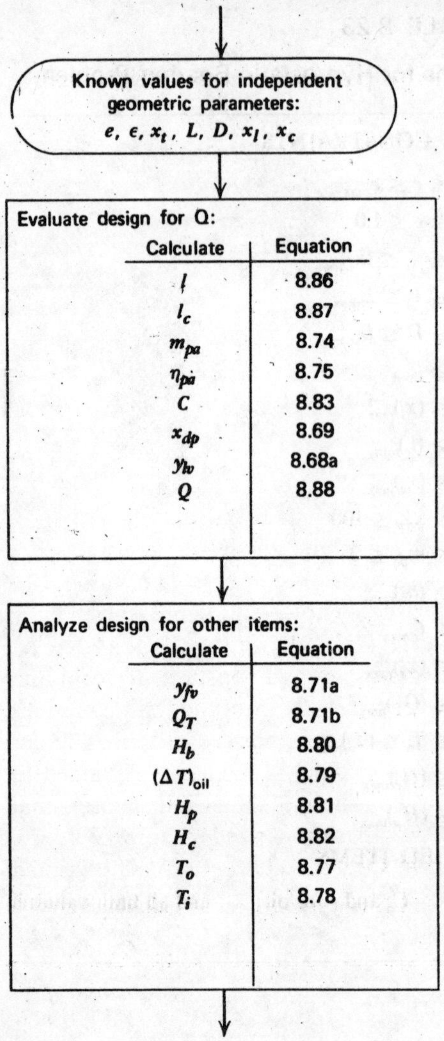

Figure 8.24 Typical procedure of analysis for a design state point in hydrostatic journal bearing problem.

AOD Equations for Use in OPTIGO. The basic equations of the AOD problem as previously summarized are transformed to the basic unit based variables by simple combination using the equations defined in table 8.24. Hence we readily obtain the unit based variable equation system summarized in table 8.25, with calculated constants C_1 through C_{15} for the total AOD problem summarized in table 8.26. As explained in Chapter 6, these C_m's for $m = 1, \ldots, N_C = 15$ remain constant throughout the search, and they are

TABLE 8.24

Definition of Unit Based Independent Variables for Hydrostatic Bearing Problem Solution by OPTIGO

$x_1 = u_e = e/e_{max}$	(8.89)
$x_2 = u_\epsilon = \epsilon$	(8.90)
$x_3 = u_{xt} = x_t/(x_t)_{max}$	(8.91)
$x_4 = u_L = L/L_{max}$	(8.92)
$x_5 = u_D = D/D_{max}$	(8.93)
$x_6 = u_{xl} = x_l/(x_l)_{max}$	(8.94)
$x_7 = u_{xc} = x_c/(x_c)_{max}$	(8.95)

programmed as explained in Chapter 6 for intial calculation in SUBROUTINE INPUT of OPTIGO. Incidentally, the given constants for OPTIGO, G_l for $l = 1, \ldots, N_G = 29$, are the specified items of table 8.23. They are summarized in table 8.27 for the G_l notation for programming in OPTIGO. The regional constraints of OPTIGO, R_k for $k = 1, \ldots, N_R = 26$, are derived in terms of the unit based variables and constants to form unit based equations as explained in Chapter 6 directly from the limit equations of table 8.23. These unit based equations are summarized in table 8.28 for programming in SUBROUTINE RK7000 of OPTIGO. As explained in Chapter 6, analysis of the optimum design is desired. Hence the chosen final items of interest F_j for $j = 1, \ldots, N_F = 23$ are summarized in table 8.29. Also, in the programming of any equation in OPTIGO, we use the subscripted variables G_l and C_m defined in tables 8.27 and 8.26 for all the specified items and calculated constants.

Programming of Equations in OPTIGO. The AOD equation system now described must be appropriately programmed in OPTIGO, as explained and illustrated in Chapter 6, for automatic solution of the hydrostatic bearing optimization problem. In this respect, it is important to remember that SUBROUTINES Q5000, EJ6000, and RK7000 are called in that order during analysis of a state point of design in the search, as seen from figure 6.19 of Chapter 6. Also, we realize that values are known for the seven independent unit variables of table 8.24 at the start of analysis for such a point.

Optimization quantity Q to be minimized must be programmed in SUBROUTINE Q5000 by equation 8.88a of table 8.25. However, we note that m_{pa}, x_{dp}, and y_{lv} are needed for the Q calculation. Hence, referring to

TABLE 8.25

Basic Equations of Hydrostatic Bearing Problem for Programming in OPTIGO*

$$Q\downarrow = C_1 \frac{(1.5 + m_{pa})^{0.662}}{y_{lv} u_D u_L [1 - (x_l)_{\max} u_{xl}]} \quad (8.88a)$$

$$x_{dp} = C_2 \frac{u_{xt} u_L u_{xl}}{u_D} \left(\frac{\epsilon}{u_e}\right)^3 \quad \text{for} \quad u_e > 0 \quad (8.69a)$$

(For $u_e \leq 0$, use $x_{dp} = 10^9$)

$y_{lv} = f(\epsilon)$ by equation 8.68a of table 8.21 for valid range $0 < x_{dp} < 100$

(For $x_{dp} \leq 0$, use $y_{lv} = 10^{-9}$, and for $x_{dp} \geq 100$, use $y_{lv} = 15\epsilon/x_{dp}$)

$$Q_T = \frac{C_4 (u_e/\epsilon)^3 y_{fv} u_D}{u_L u_{xl}} \quad (8.71c)$$

where

$y_{fv} = f(\epsilon)$ by equation 8.71a of table 8.22 for $x_{dp} > 0$

(For $x_{dp} \leq 0$, use $y_{fv} = 0$)

$$T_o = (T_{\text{oil}})_{\text{avg}} + \frac{(\Delta T)_{\text{oil}}}{2} \quad (8.77)$$

where

$$(\Delta T)_{\text{oil}} = \frac{C_5 H_b}{Q_T} \quad (8.79a)$$

where

$$H_b = C_6 \left[\frac{u_L u_{xl} u_D{}^2 \epsilon}{u_e}\right] \left\{\frac{D_{\max} u_D}{(1 - \epsilon^2)^{1/2}} + \frac{L_{\max} u_L [1 - 2(x_l)_{\max} u_{xl}](2 - \epsilon^2)}{\pi(1 - \epsilon^2)}\right\} \quad \text{for} \quad u_e > 0 \quad (8.80a)$$

(For $u_e \leq 0$, use $H_b = 10^9$)

$$m_{pa} = C_7 \left(\frac{u_L}{u_D}\right)^2 \left(\frac{u_{xl}}{u_{xc}}\right) [1 - 2(x_l)_{\max} u_{xl}] \quad (8.74a)$$

$$H_c = (1 - \eta_m) H_p + H_b + C_8 \quad (8.82a)$$

where

$$H_p = C_9 Q_T \quad (8.81a)$$

$$T_i = T_o - (\Delta T)_{\text{oil}} \quad (8.78)$$

---

* See tables 8.24 and 8.26 for unit based variables u and calculated constants C_n respectively.

TABLE 8.26

Calculated Constants C_m for Programming in OPTIGO for $m = 1, \ldots, N_c = 15$*

$$C_1 = \frac{0.8217}{D_{\max} L_{\max} p_s} \quad (8.88b)$$

$$C_2 = \frac{0.589(x_t)_{\max} L_{\max} (x_l)_{\max}}{e_{\max}^3 D_{\max}} \quad (8.69b)$$

$$C_3 = \mu = \frac{C_o}{(T_{\text{oil}})_{\text{avg}}^r} \quad (8.76a)$$

$$C_4 = \frac{e_{\max}^3 D_{\max} p_s}{24 L_{\max} (x_l)_{\max} C_3} \quad (8.71d)$$

$$C_5 = \frac{163}{c_p} \quad (8.79b)$$

$$C_6 = \frac{2.61 \times 10^{-6} L_{\max} (x_l)_{\max} D_{\max}^2 N_j^2 C_3}{e_{\max}} \quad (8.80b)$$

$$C_7 = \frac{4 L_{\max}^2 (x_l)_{\max}}{\pi D_{\max}^2 (x_c)_{\max}} \quad (8.74b)$$

$$C_8 = \frac{h_{cr} A_e}{33{,}000} \left[T_a + \frac{(\Delta T)_c}{2} - (T_{\text{oil}})_{\text{avg}} \right] \quad (8.82b)$$

$$C_9 = \frac{p_s}{(6603 \eta_m)} \quad (8.81b)$$

$$C_{10} = \frac{(x_t)_{\max}}{(x_t)_{\min}} \quad (8.96)$$

$$C_{11} = \frac{L_{\max}}{L_{\min}} \quad (8.97)$$

$$C_{12} = \frac{D_{\max}}{D_{\min}} \quad (8.98)$$

$$C_{13} = \frac{L_{\max}}{D_{\max} (x_g)_{\max}} \quad (8.99)$$

$$C_{14} = \frac{(x_l)_{\max} L_{\max}}{l_{\min}} \quad (8.100)$$

$$C_{15} = \frac{\pi D_{\max} (x_c)_{\max}}{4 (l_c)_{\min}} \quad (8.101)$$

---

* See tables 8.25 and 8.28 for use; also, program with G_l notation of table 8.27 for specified items.

TABLE 8.27

Input Data in Terms of Given Constants G_l of OPTIGO (for l = 1, . . . , N_G = 29)

G_l	Input Item of Table 8.23	Numerical Value Used for Example 8.3
G_1	e_{max}	0.0005 in
G_2	$(x_t)_{min}$	2.5×10^{-7} in^3
G_3	$(x_t)_{max}$	9.0×10^{-6} in^3
G_4	L_{min}	1.000 in
G_5	L_{max}	3.500 in
G_6	D_{min}	1.750 in
G_7	D_{max}	2.500 in
G_8	l_{min}	0.125 in
G_9	$(x_l)_{max}$	0.3
G_{10}	$(l_c)_{min}$	0.25 in
G_{11}	$(x_c)_{max}$	0.3
G_{12}	$(h_o)_{min}$	0.0005 in
G_{13}	C_{max}	0.002 in
G_{14}	$(x_g)_{max}$	2.000
G_{15}	$(Q_T)_{max}$	20.0 in^3/sec
G_{16}	$(T_{oil})_{avg}$	120°F
G_{17}	$(T_o)_{max}$	140°F
G_{18}	$(H_c)_{max}$	1.00 hp
G_{19}	$(H_p)_{max}$	0.50 hp
G_{20}	p_s	1000 psig
G_{21}	$(\Delta T)_c$	4.0°F
G_{22}	c_p	3.38 Btu/gal–°F
G_{23}	η_m	0.75
G_{24}	h_{cr}	0.18 ft-lb/min-in^2-°F
G_{25}	A_e	475 in^2
G_{26}	T_a	90°F
G_{27}	N_j	1400 rpm
G_{28}	C_o	0.8754 } For SAE 20W oil (see text after equation 8.76)
G_{29}	r	2.5698 } For SAE 20W oil (see text after equation 8.76)

TABLE 8.28

Regional Constraints R_k for Programming in SUBROUTINE RK7000 of OPTIGO; For $k = 1, \ldots, N_R = 26$*

Basic Equation for $R_k \geq 0$	R_k Value of Final Design Solution from OPTIGO in Numerical Example 8.3
$R_1 = 1 - u_e$	0.05717
$R_2 = u_e$	0.9428
$R_3 = 1 - \epsilon$	0.6335
$R_4 = \epsilon - 0.2$	0.1665
$R_5 = 1 - u_{xt}$	0.9723
$R_6 = C_{10}u_{xt} - 1$	−0.001376
$R_7 = 1 - u_L$	0.5257
$R_8 = C_{11}u_L - 1$	0.6600
$R_9 = 1 - u_D$	0.0347
$R_{10} = C_{12}u_D - 1$	0.3790
$R_{11} = 1 - C_{13}u_L/u_D$	0.6561
$R_{12} = 1 - u_{xl}$	0.4499
$R_{13} = C_{14}u_{xl}u_L - 1$	1.1916
$R_{14} = 1 - Q_T/(Q_T)_{\max}$	0.8885
$R_{15} = 1 - T_o/(T_o)_{\max}$	0.1363
$R_{16} = T_o/(T_{\text{oil}})_{\text{avg}} - 1$	0.007626
$R_{17} = 1 - m_{pa}/3$	0.8991
$R_{18} = 1 - u_{xc}$	0.2667
$R_{19} = C_{15}u_D u_{xc} - 1$	0.6678
$R_{20} = 1 - H_c/(H_c)_{\max}$	0.8754
$R_{21} = 1 - H_p/(H_p)_{\max}$	0.09958
$R_{22} = 1 - x_{dp}/100$	0.9216
$R_{23} = x_{dp}/0.1 - 1$	77.45
$R_{24} = m_{pa}$	0.3028
$R_{25} = h_o/(h_o)_{\min} - 1$	0.6296
$R_{26} = 1 - C/C_{\max}$	0.3569

* Derived directly from table 8.23 constraints; see table 8.26 for calculated constants C_m

TABLE 8.29

Final Items and Special Functions F_j for Programming in OPTIGO for Hydrostatic Bearing Problem; for $j = 1, \ldots, N_F = 23$

F_j	Final Item	Equation Number	F_j Value of Final Design Solution from OPTIGO in Numerical Example 8.3
F_1	e	8.89	0.00047 in
F_2	ϵ	8.90	0.3665
F_3	x_t	8.91	2.497×10^{-7} in^3
F_4	L	8.92	1.6603 in
F_5	D	8.93	2.4132 in
F_6	x_l	8.94	0.1650
F_7	x_c	8.95	0.2200
F_8	l	8.86	0.2740 in
F_9	l_c	8.87	0.4170 in
F_{10}	C	8.83	0.001286 in
F_{11}	h_o	8.84	0.0008148 in
F_{12}	x_{dp}	8.69a	7.845
F_{13}	m_{pa}	8.74a	0.3028
F_{14}	y_{lv}	8.68a	0.2651
F_{15}	$W\uparrow = 1/Q$	8.88a	730.6 lb
F_{16}	y_{fv}	8.71a	11.34
F_{17}	Q_T	8.71c	2.2296 in^3/sec
F_{18}	T_o	8.77	120.92°F
F_{19}	$(\Delta T)_{\text{oil}}$	8.79a	1.830°F
F_{20}	T_i	8.78	119.09°F
F_{21}	H_b	8.80a	0.08462 hp
F_{22}	H_c	8.82a	0.1246 hp
F_{23}	H_p	8.81a	0.4502 hp

table 8.29 for the equations and notation, in SUBROUTINE Q5000 we program the equations for F_{13}, F_{12}, and F_{14} in that order before equation 8.88a for Q. Programming in this subroutine is concluded by programming $F_{15} = W = 1/Q$.

Since there are no equality constraints in the AOD equation system, we have no special programming for SUBROUTINE EJ6000. However, the regional constraints R_k of table 8.28 must be programmed in SUBROUTINE RK7000. In preparation for calculation of the R_k's, we need additional values for y_{fv}, Q_T, H_b, $(\Delta T)_{\text{oil}}$, H_p, H_c, T_o, e, C, and h_o. Hence, in reference to table 8.29 for equations and notation, we start the special programming for SUBROUTINE RK7000 with the functions for F_{16}, F_{17}, F_{21}, F_{19}, F_{23},

$F_{22}, F_{18}, F_1, F_{10}$, and F_{11} in that order. Thus we are now prepared for programming of the R_k functions in the summarized order of table 8.28. To conclude the special programming in SUBROUTINE RK7000, we calculate for the solution design only the values for $\epsilon, x_t, L, D, x_1, x_c, l, l_c$, and T_i. Therefore, referring to table 8.29 for equations and notation, in this concluding segment of SUBROUTINE RK7000 we program the final items F_2 through F_9, and F_{20}. As illustrated in example 6.9b of Chapter 6 and the listing in Appendix B, such calculations can be limited only to the final solution design point by an internal control feature readily available in OPTIGO. Specifically, the F_2 through F_9 and F_{20} calculation statements are sandwiched between two Fortran IV statements included in SUBROUTINE RK7000 as shown below, which is similar to what was done for example 6.9b in SUBROUTINE Q5000 of the Appendix B listing.

```
      IF (NFC .EQ. 0) GO TO 7300
         :
      (Program equations for F2 through F9, and F20)
         :
 7300 CONTINUE
```

To conclude the special programming in OPTIGO for the hydrostatic bearing problem, we must program the equations for the calculated constants C_m in SUBROUTINE INPUT. Thus referring to table 8.26 for the equations and notation, we program the calculation statements for C_1 through C_{15} immediately after the appropriate comment statement in SUBROUTINE INPUT. These are all programmed in terms of the subscripted given constants notation, G_l for $l = 1, \ldots, N_G = 29$, referring to tables 8.27 and 8.26 for the corresponding terms.

EXAMPLE 8.3

For a numerical example of the hydrostatic bearing design optimization problem just explained, we need the input data for the READ statements listed in SUBROUTINE INPUT of OPTIGO. These are the items summarized in table 6.4 of Chapter 6. The first READ statement group is summarized in table 8.30, which gives the numerical values used for the example. The second and third READ statements of SUBROUTINE INPUT for OPTIGO are for the values of $(x_i)_{\min}$ and $(x_i)_{\max}$, $i = 1, \ldots, N = 7$. As explained in Chapter 6, these feasible range limit estimates need not be accurately determined, since they are only used for ancillary purposes, such as to calculate the start point for the shotgun C_p probe of OPTIGO. Hence, in reference to tables 8.24, 8.23, and 8.27, we estimate the x_i limit values as summarized in table 8.31 for the numerical problem. The fourth READ statement of SUBROUTINE INPUT is for the specified items or given

TABLE 8.30

First Input Data Group of SUBROUTINE INPUT of OPTIGO for Example 8.3*

Fortran IV Name	Value Used	Comments
N	7	See table 8.24
NE	0	No equality constraints in AOD equation system
NR	26	See table 8.28
NC	15	See table 8.26
NF	23	See table 8.29
NG	29	See table 8.27
RM	100	Ray limit, r_{max}
SM	100	Step limit, s_{max}
NSG	2187	Number of random design points chosen for SG search
RSD	0.3456	Arbitrary seed
NPR	1	For base point printouts during coarse search

* See table 6.4 of Chapter 6 for review of Fortran IV names.

TABLE 8.31

Feasible Range Limit Estimates for Independent Unit Variables for OPTIGO Input Data of Example 8.3*

i	Table 8.24 Variable	$(x_i)_{min}$ or XMN(I)	$(x_i)_{max}$ or XMX(I)
1	x_1	0	1.0
2	x_2	0.2	1.0
3	x_3	0.028	1.0
4	x_4	0.286	1.0
5	x_5	0.8611	1.0
6	x_6	0.20	1.0
7	x_7	0.5	1.0

* For second and third READ statements of SUBROUTINE INPUT.

constants G_l, for $l = 1, \ldots, N_G = 29$. These have already been discussed, and for the numerical problem of this example the values given in table 8.27 are used. Finally, we need not consider the fifth READ statement of SUBROUTINE INPUT for the shotgun modes of operation of OPTIGO. As seen from the listing in Appendix B, this final READ statement is automatically skipped since $N_{SG} > 0$ has already been entered from the data of table 8.30.

With the special programming and input data entered as just explained, solution to the design optimization problem for the hydrostatic bearing is fully automatic by OPTIGO. Using a Univac 90/60 digital computer, the total OPTIGO search was completed in 71.8 sec of program execution, with 18 rays required, including the coarse and final searches. The values of x_i determined for the final solution point are summarized in table 8.32. For this final design point, the values of R_k are all summarized in table 8.28 with the final items F_j summarized in table 8.29. This last table summarizes the final design specifications together with the final analysis, giving all items of interest for the optimum design. Hence, from table 8.29 we note that the optimum design could sustain a load of $W = 730.6$ lb, which is the greatest value possible for designs satisfying all constraints of the described numerical example.

TABLE 8.32

Summary of Independent Unit Variable Values as Determined by OPTIGO for Final Design Point Solution of Example 8.3

OPTIGO x_i*	Unit Variable	Final Design Value
x_1	u_e	0.9428
x_2	u_ℓ	0.3665
x_3	u_{xt}	0.02774
x_4	u_L	0.4743
x_5	u_D	0.9653
x_6	u_{xl}	0.5501
x_7	u_{xc}	0.7333

* See table 8.24 for definition of x_i.

Incidentally, from the final printout for the final design solution, the following items are also obtained, referring to table 6.3 of Chapter 6 for the basic terminology of OPTIGO.

$$P = 4.7725 \times 10^{-5}$$
$$Q = 1.3688 \times 10^{-3}$$
$$(QP) = 1.4165 \times 10^{-3}$$
$$\eta_{QP} = 1.0349$$

The latter item above is reasonably close to unity, especially in light of the fact that $\eta_p = 262{,}144$ was used by the computer on the final ray for the calculations involving equations 6.37 and 6.47 of Chapter 6. Also, from the values of R_k in table 8.28, we see that only one minor violation remained for the final design point, which is specifically for R_6 in the table. Also, the convergence pattern from the coarse search base point printouts was observed to be normal, as explained in section 6.36 of Chapter 6. Hence the design solution of table 8.29 is believed to be very close to the exact point of optimum design, for all practical purposes.

Regarding final design specifications, basic nominal dimensions are taken from the final items of table 8.29 with reference to figure 8.22 for notation. Thus for the optimum design we have the following nominal dimensions:

$$L = 1.6603 \text{ in}$$
$$D = 2.4132 \text{ in}$$
$$l = 0.2740 \text{ in}$$
$$l_c = 0.4170 \text{ in}$$
$$C = 0.001286 \text{ in}$$

Incidentally, $e = 0.00047$ in eccentricity is an operating dimension for full load $W = 730.6$ lb. Also, capillary tube specifications are determined from $x_t = 2.497 \times 10^{-7}$ in^3 of table 8.29, using equation 8.70 to determine a practical combination for d_t and L_t dimensions. Typically, for a No. 77 drill size with $d_t = 0.018$ in, we would require $L_t = d_t^4/x_t = (0.018)^4/(2.497 \times 10^{-7}) = 0.420$ in length of capillary tube in figure 8.22.

Regarding some other final design specifications, the bearing block material of figure 8.22 might be chosen as SAE 660 high leaded bronze. Practical dimensions would be chosen for the bearing pocket and manifold, in accordance with the guides recommended in figure 8.22. Also, practical tolerances and surface finishes would be appropriately assigned to control the manufacturing of the bearing. Of course, other practical items would be incorporated in the total design, such as the selection of seal rings and provisions for adequate filtering of the oil. Also, provisions would be made for temperature control of the SAE 20W grade oil to 120°F, with a cooling capacity of 0.1246 hp required from table 8.29. Pump specifications for the figure 8.23 system would be based on 0.4502 hp, delivering 2.2296 in^3/sec at 1000 psig. The analysis, of course, as stated earlier, is on a per bearing basis in the figure 8.23 system.

REFERENCES

8.1 J. E. Shigley, *Mechanical Engineering Design*, 2nd ed., McGraw-Hill, New York, 1972, 753 pp.

8.2 M. F. Spotts, *Design of Machine Elements*, 4th ed., Prentice-Hall, Englewood Cliffs, N.J., 1971, 620 pp.

8.3 V. M. Faires, *Design of Machine Elements*, 4th ed., Macmillan, New York, 1965, 624 pp.

8.4 A. D. Deutschman, W. J. Michels, and C. E. Wilson, *Machine Design Theory and Practice*, Macmillan, New York, 1975, 932 pp.

8.5 P. H. Black and D. E. Adams, Jr., *Machine Design*, 3rd ed., McGraw-Hill, New York, 1968, 678 pp.

8.6 M. J. Siegel, V. L. Maleev, and J. B. Hartman, *Mechanical Design of Machines*, 4th ed., International Textbook Co., Scranton, Pa., 1965, 576 pp.

8.7 R. M. Phelan, *Fundamentals of Mechanical Design*, 3rd ed., McGraw-Hill, 1970, 678 pp.

8.8 R. H. Creamer, *Machine Design*, Addison-Wesley, Reading, Mass., 1968, 533 pp.

8.9 V. L. Doughtie, A. Vallance, and L. F. Kreisle, *Design of Machine Members*, 4th ed., McGraw-Hill, New York, 1964, 520 pp.

8.10 D. W. Dudley, Ed., *Gear Handbook*, McGraw-Hill, New York, 1962.

8.11 D. W. Dudley, *Practical Gear Design*, McGraw-Hill, New York, 1954.

8.12 *AGMA Information Sheet for Surface Durability (Pitting) of Spur, Helical, Herringbone, and Bevel Gear Teeth*, AGMA 215.01, American Gear Manufacturers Association, Washington, D.C., Sept. 1966.

8.13 *AGMA Information Sheet for Strength of Spur, Helical, Herringbone, and Bevel Gear Teeth*, AGMA 225.01, American Gear Manufacturers Association, Washington, D.C., Dec. 1967.

8.14 C. Lipson and L. V. Colwell, *Handbook of Mechanical Wear*, The University of Michigan Press, Ann Arbor, 1961, 469 pp.

8.15 A. Y. Attia, "Dynamic Loading of Spur Gear Teeth," ASME Paper 58-SA-32, 1958.

8.16 E. Buckingham, *Analytical Mechanics of Gears*, McGraw-Hill, New York, 1949.

8.17 A. M. Wahl, *Mechanical Springs*, 2nd ed., McGraw-Hill, New York, 1963, 323 pp.

8.18 N. P. Chironis, Ed., *Spring Design and Application*, McGraw-Hill, New York, 1961, 344 pp.

8.19 R. E. Peterson, *Stress Concentration Design Factors*, Wiley, New York, 1953, 155 pp.

8.20 R. E. Peterson, *Stress Concentration Factors*, Wiley-Interscience, New York, 1974, 317 pp.

8.21 F. P. Zimmerli, "Human Failures In Spring Applications," *The Mainspring*, Associated Spring Corporation, Bristol, Connecticut, No. 17, Aug.-Sept. 1957.

8.22 J. P. Den Hartog, *Mechanical Vibrations*, 4th ed., McGraw-Hill, New York, 1956, 436 pp.

8.23 M. F. Spotts, *Mechanical Design Analysis*, Prentice-Hall, Englewood Cliffs, N.J., 1964 428 pp.

8.24 R. C. Johnson, *Mechanical Design Synthesis—Creative Design and Optimization*, 2nd ed., Krieger, Huntington, New York, 1978, 448 pp.

8.25 A. A. Raimondi and J. Boyd, "An Analysis of Orifice- and Capillary-Compensated Hydrostatic Journal Bearings," *Lubrication Engineering, Journal of the American Society of Lubrication Engineers*, Jan. 1957, pp. 28–37.

8.26 D. F. Wilcock and E. R. Booser, *Bearing Design and Application*, McGraw-Hill, New York, 1957, pp. 345–366.

8.27 J. J. O'Connor, J. Boyd, and E. A. Avallone, *Standard Handbook of Lubrication Engineering*, McGraw-Hill, New York, 1968, pp. 5–66 to 5–74.

8.28 D. D. Fuller, *Theory and Practice of Lubrication for Engineers*, Wiley, New York, 1956, 432 pp.

8.29 H. C. Rippel, "Design of Hydrostatic Bearings," *Machine Design*, Penton, Cleveland; 10 part series, starting Aug. 1, 1963, pp. 108–117.

EXERCISES

Exercises for the design optimization of some complex mechanical elements are described in the remaining part of this chapter. As appropriate for the particular exercises, the *general problem statement* given on pages 341 through 342 of Chapter 7 applies to what follows. Solve all or part of it for the particular exercise and to the extent assigned by the instructor.

Gearset Design Problems

8.1. Solve the section 8.1 gearset design optimization problem for maximization of $(HP)_w$, but for numerical values which are different from those given in example 8.1 of the text. Numerical values for the specified items of table 8.6 will be assigned by the instructor. As a check on your computer program, solve the numerical problem of example 8.1 to start and compare your results with the given solution.

8.2. Solve the section 8.1 gearset design optimization problem for maximization of $(HP)_w$, except for the following modification. Assume that the gears will be shaped and ground instead of hobbed, thus requiring standard size values for circular pitch P_c in the diametral plane instead of the P_{nd} values of table 8.5. Otherwise, solve the numerical problem of example 8.1 of the text.

8.3. A helical gearset is to transmit 200 hp at 3600 rpm for the pinion, and the angular velocity ratio is to be 2:1. Assume that 20° involute stub standard proportion hobs are available in the sizes given in table 8.5, and that normal facilities are available for grinding the gearset after hardening. To avoid an excessive amount of axial thrust, the pitch helix angle ψ should be selected equal to or less than 30°. Assuming the available materials from table 4.1, design the gearset on the basis of minimizing the center distance between the pinion and gear axes. Specify the required tooth surface hardness and all the information which would be submitted to the shop for manufacturing the gears.

8.4. A helical gearset is to transmit 200 hp at 3600 rpm for the pinion, and the angular velocity ratio is to be 2:1. Assume that shaping facilities are available for manufacturing the gears with standard 20° involute stub tooth proportions in the diametral plane. Assume that cutters are available having the standard values for diametral pitch given in table 8.5 but referred to the *diametral* plane. Normal facilities are also available for grinding the gearset after hardening. In order to avoid an excessive amount of axial thrust, the pitch helix angle ψ should be selected equal to or less than 30°. Assuming the available materials from table 4.1, design the gearset on the basis of minimizing the center distance between the pinion and gear axes. Specify the required tooth surface hardness and also specify all information which would normally be submitted to the shop for manufacturing the gears. Compare the solution with that obtained for exercise 8.3, which is the same except for the manufacturing facilities.

8.5. A helical gearset is to transmit 500 hp at 1800 rpm, and the angular velocity ratio is to be 3:1. Assume that 20° involute stub standard proportion hobs are available in the sizes given in table 8.5 and that normal facilities are available for manufacturing precision gears. To avoid an excessive amount of axial thrust, the pitch helix angle ψ should be selected equal to or less than 35°. The gearset is to be used in a machine where weight is of utmost importance, and hence optimum design is to be conducted on the basis of minimizing the gearset weight.
(*Hint*: It will be a very accurate approximation to assume that the weight W of a solid gear is given by

$$W \approx w\left(\frac{\pi D^2}{4}\right)b$$

where w is the weight density of the material, D is the pitch diameter of the gear, and b is its face width.)

8.6. The pinion of a helical gearset is to transmit power while rotating at 3600 rpm, and the angular velocity ratio is to be 2:1. The distance between the pinion and gear shaft axes cannot exceed a rigid limitation of 6 in because of space restrictions in the machine, and the pitch helix angle ψ should be selected equal to or less than 35° in order to avoid excessive axial thrust. Assume that 20° involute stub standard proportion hobs are available in the sizes given in table 8.5 and that normal facilities are available for manufacturing precision gears. Space restrictions in the machine limit permissible values of face width b to 3 in or less. Assume that all available materials are given in table 4.1. Optimum design of the gearset is to be conducted on the basis of maximizing the static torque capacity of the pinion, M_{sp}, in anticipation of momentary shock loading possibilities in the particular application. Brinelling of the teeth is to be avoided in such rare occurrences of shock loading. Under normal operation, the gearset is to transmit 200 hp.

8.7. A helical gearset is to be designed for a tight space situation where we wish to minimize face width b. Boundary conditions are the same as in table 8.6, except for the following modifications.

1. Transmitted power $(HP)_t$ is a specified value. Hence equation 8.1a becomes a (S.D.E.) instead of the (P.D.E.), and equation 8.16b can be used as the basis for deriving the (P.D.E.).
2. Remove the (L.E.) $0 \leq b \leq b_{max}$ from table 8.6, as well as the corresponding specified item b_{max}.

Helical Compression Spring Design Problems

8.8. Solve the section 8.2 spring design triple optimization problem for either maximizing f_{n1}, minimizing L, or minimizing W. However, use numerical values which are different from those given in example 8.2 of the text. Numerical values for the specified items in table 8.14 will be assigned by the instructor. As a check on your computer program, solve the numerical problem of example 8.2 to start and compare your results with the given solution.

8.9. A helical compression spring is to be designed for a tight space situation where we wish to minimize mean coil diameter D. Boundary conditions are the same as in table 8.14, except for the following modifications.

1. Remove the (L.E.) $D \leq D_{max}$ and remove the corresponding specified item D_{max}.
2. Include (L.E.) $f_{n1} \geq (f_{n1})_{min}$ and include the corresponding specified item $(f_{n1})_{min}$.

8.10. A helical compression spring is to be designed for a critical situation where we wish to minimize the chance for fatigue breakage. Hence the objective for optimal design is maximization of factor of safety N_e. Boundary conditions are the same as in table 8.14, except for the following modifications.

1. Remove the (L.E.) $\tau_a \leq S_{se}/N_e$ and remove the corresponding specified item N_e.
2. As your (P.D.E.) use the following equation

$$N_e = \frac{S_{se}}{\tau_a}$$

3. Include the (L.E.) $f_{n1} \geq (f_{n1})_{min}$ and the corresponding specified item $(f_{n1})_{min}$.

(Incidentally, a problem similar to this one is solved in section 9-3 of reference 8.24.)

Hydrostatic Journal Bearing Design Problems

8.11. Solve the hydrostatic bearing optimization problem of section 8.3 for maximization of load capacity W, but for numerical values which are different from those given in example 8.3. Numerical values for the specified items of table 8.23 will be assigned by the instructor. As a check on your computer program, solve the numerical problem of example 8.3 to start and compare your results with the given solution.

8.12. A capillary-compensated hydrostatic journal bearing is to be designed for a tight space situation where we wish to minimize its length L. Boundary conditions are the same as in table 8.23, except for the following modifications.

1. Remove the (L.E.) $L_{min} \leq L \leq L_{max}$ and the corresponding specified items L_{min} and L_{max}.
2. Include a specified item W, for maximum load to be supported by the bearing.

8.13. A capillary-compensated hydrostatic journal bearing is to be designed for a tight space situation where we wish to minimize its diameter D. Boundary conditions are the same as in table 8.23, except for the following modifications.

1. Remove the (L.E.) $D_{min} \leq D \leq D_{max}$ and the corresponding specified items D_{min} and D_{max}.
2. Include a specified item W for maximum load to be supported by the bearing.

8.14. A capillary-compensated hydrostatic journal bearing is to be designed for an application where we wish to use the smallest possible size of pump motor in the figure 8.23 system. Hence the objective is to minimize pump horsepower, H_p. Boundary conditions are the same as in table 8.23, except for the following modifications.

1. Remove the (L.E.) $H_p \leq (H_p)_{max}$ and the corresponding specified item $(H_p)_{max}$.
2. Include a specified item W for maximum load to be supported by the bearing.

Other Complex Mechanical Element Design Problems

8.15 Design of a Cylindrical Roller Bearing

Mechanical design engineers generally are confronted with the problem of *selecting* rather than *designing* ball or roller bearings. However, a more intelligent selection often can be made from the innumerable available standard bearings if the engineer is familiar with optimum design studies of bearings. The author has conducted several optimum design studies pertaining to the basic design of cyclindrical roller bearings. Similar studies could be conducted for ball bearings, but they would be somewhat more complex. Space will permit the presentation of only one of the optimum design problems for cylindrical roller bearings.

Suppose that general cylindrical surfaces in contact are transmitting a radial force P_{ab} as shown in figure 8.25. Assuming that bodies a and b are both made of elastic homogeneous materials, a maximum compressive stress σ_c would be developed at the center of the contact area between the two mating surfaces, and from equation 3.4 its value would be

$$\sigma_c = 0.591\sqrt{\frac{P_{ab}E}{2L}\left(\frac{1}{\rho_a} + \frac{1}{\rho_b}\right)} \tag{8.102}$$

In this equation, σ_c is the maximum compressive stress at the contact area, in pounds per square inch; P_{ab} is the radial force being transmitted between

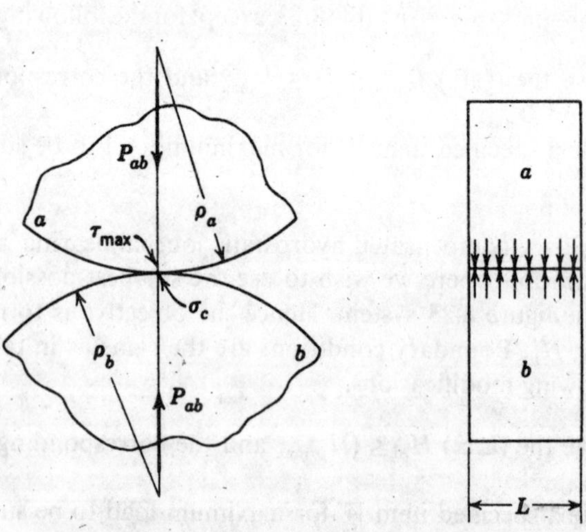

Figure 8.25 Cylindrical surfaces in contact.

surfaces a and b, in pounds; E is the modulus of elasticity, in pounds per square inch, which for the equation is assumed to be the same for both bodies; L is the length of the contact area, in inches; ρ_a and ρ_b are the radii of curvature for the free surfaces of the cylinders at the contact region, in inches, as shown in figure 8.25.

Of primary interest is the maximum shearing stress τ_{max} existing in the vicinity of contact between the two cylinders. From the theory of elasticity it is determined that the maximum shearing stress will occur at a point located a very small distance beneath the mating surfaces, and, specifically from reference 3.1 τ_{max} is related to σ_c by

$$\tau_{max} = 0.304\sigma_c \tag{8.103}$$

Incidentally, in this equation Poisson's ratio is assumed to be 0.3 for both materials a and b.

From the theory of elasticity applied to steel cylinders having typical roller and race dimensions, it is determined that the radial force P_{ab} is approximately *linearly* related to the combined deformation δ_{ab} of both bodies along the axis of the load, as given generally by

$$P_{ab} = k\delta_{ab} \tag{8.104}$$

For the *optimum design study* in this problem it will be unnecessary to know the specific relationship which expresses the force gradient k.

Let us now consider the optimum *design*, on the basis of maximizing the radial load *capacity*, of the cylindrical roller bearing depicted in figure 8.26. We assume the maximum shearing stress theory of fatigue failure with a factor of safety N_e based on the occurrence of fatigue damage in the vicinity of any surface in the bearing. We assume also that a theoretically infinite life is desired for the bearing, and therefore all critical stresses must be kept equal to or below a certain limit value determined by the theory of failure using the *endurance limit*, S_e. A similar optimum design study could be conducted by considering static loads or brief periods of overload, and exactly the same optimum design results would be obtained with respect to the bearing *geometry*. However, if a normal specified life such as 1000 hr is all that is desired, the problem becomes considerably more complex.

Assume that space restrictions in the machine limit feasible values of outside diameter D_o and feasible values of length L to the permissible ranges $D_o \leq D_{o\,max}$ and $L \leq L_{max}$. Also, from the standpoint of proportional design, express the outside wall thickness t of the bearing as a decimal percentage β of the roller diameter d as shown in figure 8.26. Hence the significant *independent geometrical* parameters for uniquely defining the bearing are D_o, d, L, α, and β, which are all shown in figure 8.26. The inside wall thickness t_i will

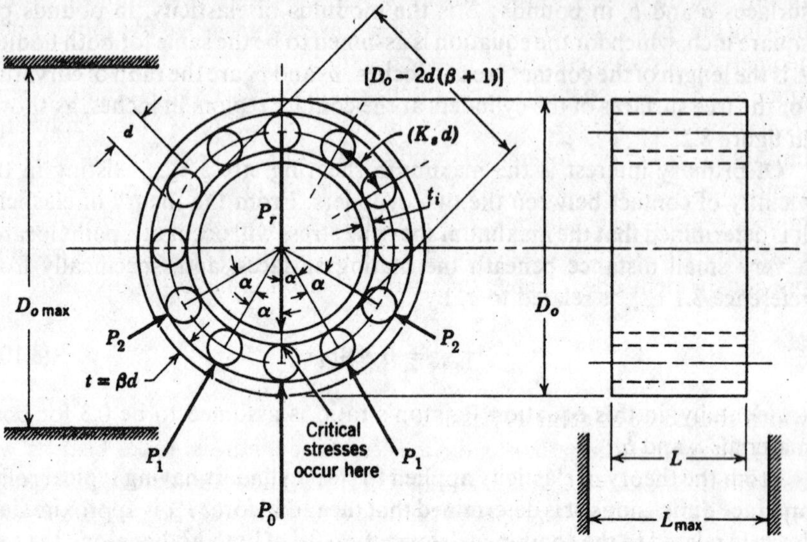

Figure 8.26 Cylindrical roller bearing for exercise 8.15.

be insignificant in this problem, if it is reasonable in size and if the shaft fits well within the inside hole of the bearing.

(a) As shown in figure 8.26, P_0 is the greatest force transmitted by a single roller because of the applied radial shaft force P_r. Derive the following approximation for expressing P_r in terms of P_0:

$$P_r \approx \left(\frac{N_R}{4}\right) P_0 \tag{8.105}$$

In this equation, N_R is the number of rollers in the entire bearing. (*Hint*: Remembering basic principles of numerical integration, the analysis of force distribution in the bearing implies that a simplification may be obtained by applying the approximating integral $\int_{-\pi/2}^{+\pi/2} \cos^2 x\, dx$, which can be readily evaluated.)

(b) Show that expression 8.105 is *exact* for even numbers of rollers equal to or greater than 4. Also, for that equation, show that the error is -4.7% for $N_R = 5$, $+1.5\%$ for $N_R = 7$, and less than 1% for all odd numbers above $N_R = 7$.

(c) Develop the primary design equation for the radial load *capacity* of the bearing:

$$(P_r)_c \approx \frac{24.3L}{N_e^2 K_s}\left[\frac{S_e^2}{(1+p)^2 E}\right][D_o - 2d(\beta + 1)] \tag{8.106}$$

In this equation, $(P_r)_c$ is the radial load *capacity*, in pounds, for an infinite bearing life; $L, D_o, d,$ and β are four of the previously selected five independent geometrical parameters; and K_s is a roller spacing factor replacing the spacing angle α, where K_s is defined by the pitch distance between roller centers, being $(K_s d)$ as shown in figure 8.26. Also, in equation 8.106, S_e is the endurance limit of the roller or race materials obtained from a reversed bending type of standard test; $p = S_e/S_y$, where S_y is the yield strength of the material determined by a simple tensile test; E is the modulus of elasticity for the roller and race materials, assumed to be the same for each; and N_e is the factor of safety previously mentioned. Hence the material selection factor is obviously

$$\frac{S_e^{\,2}}{(1 + p)^2 E}$$

which should be as great as possible.

(d) Show that the radial load *capacity* of a cylindrical roller bearing will be greatest if (1) $L = L_{\max}$; (2) $D = D_{o\,\max}$; (3) d, the roller diameter, is small; (4) $K_s \approx 1.0$, that is, rollers spaced right next to each other (i.e., angle α small); (5) β, the previously defined outside wall thickness factor, is small.

(e) For the optimum design of a practical cylindrical roller bearing, in order to assure good contact between mating surfaces over the *entire* roller length, a proportional limit equation will exist between feasible values of L and d. A reasonable limit equation to assume is $x \leq x_{\max}$, where we define $x = L/d$ and typically use $x_{\max} \approx 8$ for normal manufacturing control.

(f) Refer to the general problem statement for the case where the following constraints must be considered simultaneously.

$$L_{\min} \leq L \leq L_{\max}; \qquad D_o \leq (D_o)_{\max}; \qquad K_s \geq 1;$$
$$d_{\min} \leq d \leq d_{\max}; \qquad x \leq x_{\max}; \qquad \beta \geq \beta_{\min}$$

The optimization objective is still to maximize radial load capacity, for either momentary overload or infinite fatigue life. Assume that all limit values are known for the given regional constraints.

8.16 Hydrodynamic Journal Bearing Design

Occasionally it is necessary in the design of original machines to use a journal bearing as depicted in figure 8.27. This necessity might be imposed by a number of possible manufacturing or economic considerations, but unfortunately even a well-designed hydrodynamic journal bearing has certain inherent undesirable features. For one, the friction moment M_f for a journal bearing will always be appreciably greater than that for a ball or roller type of bearing. Second, during steady state operation there will always be an

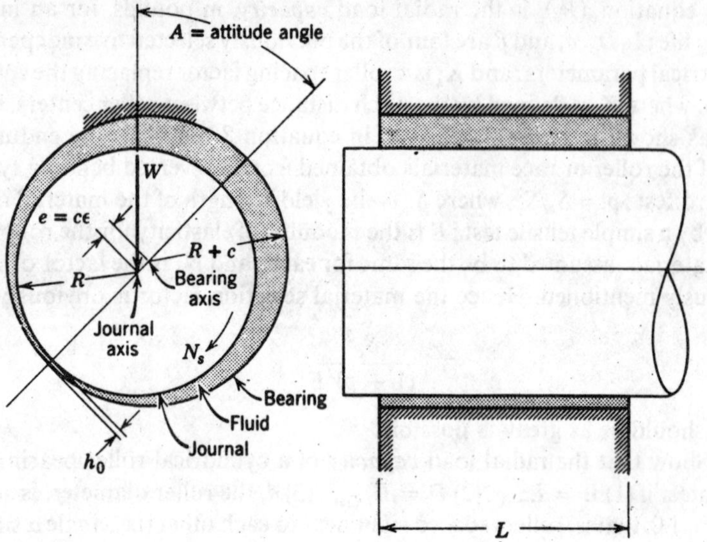

Figure 8.27 Typical hydrodynamic journal bearing in steady state operation.

eccentricity e between the journal axis and the bearing axis as shown in figure 8.27, and for precision machinery where elements such as gears have to be mounted accurately this might be a serious undesirable feature which must at least be limited to tolerable magnitudes. In this problem we shall consider the optimum design of a hydrodynamic journal bearing on the basis of minimizing the frictional moment M_f. For the sake of simplicity we shall consider the basic design shown in figure 8.27. The short bearing type will be assumed, with $L/R \lesssim 2$ for the equations used.

It is assumed that the reader is familiar with hydrodynamic journal bearings, such as that covered in reference 8.23. We shall now consider some pertinent design equations which are obtained from the study in that reference. First, the frictional moment M_f can be expressed by

$$M_f = \left(\frac{\pi^2}{15\sqrt{1-\epsilon^2}}\right)\left(\frac{\mu R^3 L N_s}{c}\right) \tag{8.107}$$

In this equation, M_f is the frictional moment exerted by the bearing on the rotating journal during steady state operation expressed, in inch-pounds; μ is the absolute fluid viscosity, in reyns or lb-sec/in^2; R is the journal radius, in inches; L is the bearing length in figure 8.27, in inches; N_s is the angular velocity of the shaft or journal in revolutions per minute; c is the radial clearance, in inches, which by definition is the difference in radii for the bearing and the journal as machined; and ϵ is the eccentricity ratio defined by

$$e = c\epsilon \tag{8.108}$$

In this equation, e is the eccentricity between the journal and bearing axes during steady state operation as shown in figure 8.27, c is the previously defined clearance, and ϵ is the unitless eccentricity ratio.

During steady state operation of a journal bearing, the smallest fluid film thickness h_0 is of significance and it is shown in figure 8.27. Hence, from the previous definitions and the geometry in the figure, h_0 is merely the difference between clearance and eccentricity as given by

$$h_0 = c - e \tag{8.109}$$

A journal bearing is generally designed to support a given radial reaction force W while the shaft is rotating at a given angular velocity N_s. Hence, from the derivation in reference 8.23, we obtain the following equation, which is based on force equilibrium:

$$\frac{Wc^2}{RL^3\mu N_s} = \left(\frac{\pi}{120}\right)\frac{\epsilon}{(1-\epsilon^2)^2}\sqrt{\pi^2(1-\epsilon^2)+16\epsilon^2} \tag{8.110}$$

In this equation, all terms and their units have been defined except W, which is the radial force being supported by the bearing, in pounds, as shown in figure 8.27.

In addition to these pertinent design equations from the study in reference 8.23, there are several significant limit equations which should be written for expressing permissible ranges of values for certain significant parameters. For one, depending on the required accuracies with respect to relative locations of other mechanical elements such as precision gears, there is a maximum permissible eccentricity e_{max}. Hence we have the following limit equation expressing the permissible range for values of e for all feasible designs:

$$0 < e \leq e_{max}$$

Also, depending on the particular application, the minimum fluid film thickness h_0 cannot be less than a specified value $h_{0\,min}$, lest very small foreign particles rub or wedge between the journal and bearing wall. Thus we have the following limit equation for the permissible range of values for h_0 for all feasible designs:

$$c > h_0 \geq (h_0)_{min}$$

Before conducting optimal design, let us consider the *independent geometrical* parameters for this problem. Hence, referring to figure 8.27, we see that the geometry for steady state operation is uniquely defined by five parameters, which can be selected, for example, as L, R, c, h_0, and the attitude angle A. However, in checking the derivation of equation 8.110, we find that the attitude angle A has already been taken into account in that equation. Thus we must conclude that there are only four *independent* geometrical parameters for this problem. Select for these parameters the journal radius R, the bearing length L, the eccentricity e, and the smallest film thickness h_0. These selections should be obvious from the problem which follows.

For the problem of optimum design, the regional constraints to be adhered to are summarized as follows.

$$0 < e \leq e_{\max}; \qquad c > h_0 \geq (h_0)_{\min}; \qquad R_{\min} \leq R \leq R_{\max};$$
$$0 < L \leq L_{\max}; \qquad x \leq x_{\max}$$

For the latter we define $x = L/R$, and require satisfaction of the constraint to be assured of acceptable practical proportions for the bearing. Typically, $x_{\max} = 2$.

The optimization objective is to minimize frictional moment M_f, which in turn will minimize heat generated by power dissipation in the bearing. Thus cooling problems will be alleviated by this approach. Assume that specified values are known for shaft speed N_s and bearing load W. Refer to the general problem statement described at the start of the exercises. Assume that all limit values are known at this stage of design for the given regional constraints.

8.17 Varying Motion Mechanisms

In the design of mechanisms having a varying type of motion, dynamic forces and dynamic stresses are generally of great significance. For high speed machinery it is usually desirable to minimize dynamic forces and to keep dynamic stresses within tolerable limits in order to obtain an optimum design. From a study of many types of varying motion mechanisms it is found that either dynamic forces or dynamic stresses can be minimized by the method of optimum design, use being made of *proportional design* when various practical limitations are involved. In the problem which follows we shall consider the optimum design of a Geneva mechanism, first considered in reference 1.3 with respect to minimizing dynamic *stresses*. However, in addition to that problem we shall consider the problem of minimizing dynamic *forces* which are felt by the machine frame.

The general principles used for solving the following problem can be applied to the minimization of dynamic stresses or dynamic forces in innumerable types of varying motion mechanisms. The author has applied the general method to many real design problems, such as those shown in the photographs of Chapter 1. Hence, in solving the following specific design problem, it is important to keep the overall consideration of *general varying motion mechanisms* in mind.

Geneva Mechanism Design.

As important background information the reader should carefully study section 9-4 of reference 8.24, and what follows is presented under the assumption that it has been read. Assume that we wish to index a given inertial load having a specified or calculated value for its mass moment of inertia of J_L, using a Geneva mechanism as typified in figure 8.28. Assume that the angle through which we wish to rotate the load per index is specified, as is the desired speed of indexing. Hence the number of stations in the Geneva mechanism is specified, as is the angular velocity ω of the drive shaft.

The typical Geneva mechanism in figure 8.28 is shown in the angular position where the drive pin force is a maximum, and the value of this force depends on the design and it is designated F_{max}. Applying the principle of motion of the mass center to a free body diagram of either the drive shaft or the Geneva wheel, we realize that the force F_{max} will be felt intermittently by the machine frame at both the Geneva wheel and the drive shaft bearings.

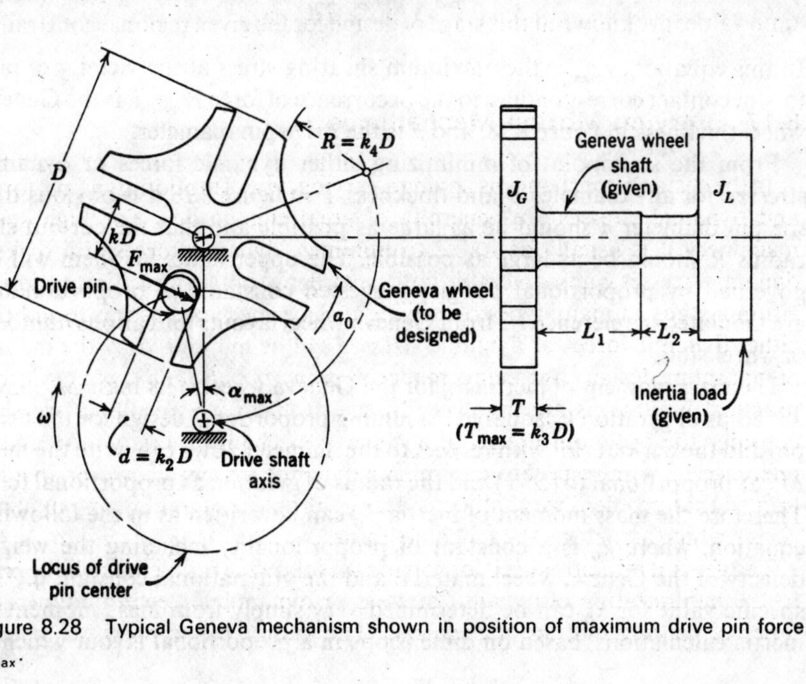

Figure 8.28 Typical Geneva mechanism shown in position of maximum drive pin force F_{max}.

The outside diameter of the Geneva wheel to be designed is designated D, and we realize that the torque arm for F_{max} on the Geneva wheel is proportional to D by a constant k which is independent of D. Thus the torque arm for F_{max} will be (kD), as shown in figure 8.28. Therefore, if we designate the *angular* acceleration of the Geneva wheel at the position shown in figure 8.28 by a_0, we can write the following equation directly from elementary dynamics:

$$F_{max}(kD) = (J_G + J_L)a_0 \tag{8.111}$$

In this equation, all terms have been defined except J_G, which is the mass moment of inertial of the Geneva wheel to be designed, about its axis of rotation.

The drive pin in figure 8.28 will have a diameter d, and it is cylindrical in shape. The drive pin force is transmitted to the Geneva wheel by means of contact between the pin and the plane faced slot of the Geneva wheel. Thus, referring to equations 8.102 and 8.103, we can write an equation for the significant maximum shearing stress τ_{max} at the contact region. Therefore, for given drive pin and wheel slot materials, we can write the following equation, where k_1 is a constant whose numerical value is determined from a consideration of equations 8.102 and 8.103:

$$\tau_{max} = k_1\sqrt{\frac{F_{max}}{Td}} \tag{8.112}$$

In this equation, τ_{max} is the maximum shearing stress at the vicinity of pin-to-slot contact corresponding to the occurrence of force F_{max}, T is the Geneva wheel thickness in figure 8.28, and d is the drive pin diameter.

From the standpoint of minimizing either dynamic forces or dynamic stresses, for any diameter D and thickness T in figure 8.28 it is obvious that the pin diameter d should be as large as possible and that the cut-out slot radius R should be as large as possible. The upper limits for them will be governed by proportional design; suggested constants of proportionality are tabulated in reference 1.3 from Geneva wheel layouts for various numbers of wheel slots.

The mass moment of inertia J_G for the Geneva wheel is its mass M_G times its radius of gyration k_G squared. Assuming proportional design for the drive pin and the cut-out slot with respect to the diameter D, we can write the mass M_G as proportional to (D^2T) and the radius of gyration as proportional to D. Therefore the mass moment of inertia J_G can be written as in the following equation, where k_6 is a constant of proportionality including the weight density of the Geneva wheel material and the gravitational constant g. The specific value for k_6 can be determined very simply from mass moment of inertia calculations, based on dimensions in a proportional layout which is made with d and R as large as reasonable, similarly to the values designated k_5 in reference 1.3.

$$J_G = k_6 T D^4 \tag{8.113}$$

All terms in this equation have been defined.

With the proportional design assumptions which have now been made, it is obvious that the only significant *independent geometrical* parameters for the Geneva wheel are the outside diameter D and the wheel thickness T. Hence, if optimum values for them can be determined, the essential geometry of the optimum Geneva wheel will be uniquely defined. Hence, we select T and D as our independent geometrical parameters, because if space restrictions should exist they will most likely affect those dimensions.

For both of the optimization problems in parts a and b which follow, assume that the following regional constraints on geometry must be adhered to for an acceptable design.

$$T_{min} \leq T \leq T_{max}$$

$$D_{min} \leq D \leq D_{max}$$

$$k_7 = x_{min} \leq x \leq x_{max} = k_3$$

For the latter constraint we define $x = T/D$ and require satisfaction of this limit equation for obtainment of a design having good practical proportions. Typically, $x_{min} = 0.05$ and $x_{max} = 0.15$ would be appropriate for steel parts. The lower limit x_{min} is imposed by such considerations as warping of the Geneva wheel in hardening, and x_{max} is indirectly imposed through the k_2 value for assurance of a good pin to slot contact condition.

For both of the optimization problems which follow, assume that J_L, index angle, and shaft speed are all specified values and that all limit values are known at this stage of design for the given regional constraints. Also, refer to the general problem statement described at the start of the exercises.

(a) Often the dynamic forces *and* the dynamic stresses will be significant in a Geneva mechanism operating at high speed. Consider the problem of designing the Geneva wheel on the basis of minimizing the dynamic force F_{max} and staying within tolerable limits for the significant shearing stress τ_{max} at the contact vicinity. Use the maximum shearing stress theory of fatigue failure presented in Chapter 4, with a factor of safety N_e based on the occurrence of fatigue damage near the pin contact region. Such fatigue damage would initiate undesirable wear of the pin, even with good lubrication.

(b) For some high speed Geneva mechanisms, dynamic *stresses* will be very high and dynamic *forces* will be relatively insignificant. Consider the problem of designing the Geneva wheel on the basis of minimizing the significant dynamic stress τ_{max}, ignoring the magnitude of the dynamic force F_{max}.

Assume the maximum shearing stress theory of fatigue failure with a factor of safety N_e based on the occurrence of fatigue damage at the pin contact vicinity.

8.18 Selection of Standard Bearing for Cam Follower

In the original design of a high speed cam mechanism a common problem is to select an appropriate follower bearing, which is usually either a standard ball bearing or a standard cylindrical roller bearing. Often the bearing is selected only on the basis of *dynamic forces* being less than the corresponding load *rating* of the bearing. However, in many cases the *contact stress* vicinity between the outside surface of the follower bearing and the surface of the cam can be a very critical region. For certain portions of the cam surface, its radius of curvature ρ_c may be small and may result in large values of significant shearing stresses according to equations 8.102 and 8.103. Hence fatigue damage near the surface might occur at such regions of the cam, eventually progressing to noticeable wear, and finally resulting in the malfunctioning of a precision high speed mechanism. Therefore this problem will be concerned with the selection of a standard available follower bearing on the basis of minimizing the maximum shearing stress τ_{max} at a critical contact region. Of course, in addition the design engineer would have to check the dynamic *forces* at the contact region in order to be sure that the internal load rating of the selected bearing is not exceeded.

Assume that it is desired to move a given mass having a weight W_L with a certain motion defined by the coordinate s in figure 8.29. Assume that values of s have already been determined for various cam angles θ at this stage of design. Hence, if it is assumed that the cam axis is fixed in space, the locus of the follower axis has already been uniquely determined. Therefore we see that the *selection* of the follower bearing will *determine* the shape of the required cam surface; grave consequences can result from making the wrong selection. Choosing the follower bearing too small can be disastrous owing to its small diameter D, which will result in a high cam contact *stress* at a critical region, and owing to the small internal load rating of the bearing itself. On the other hand, choosing the bearing too large can be disastrous because of the higher inertia *forces* and the smaller radius of curvature ρ_c resulting for the cam surface at critical regions. Thus, an optimum choice exists among a series of standard available bearings, and this problem will be concerned with the proper bearing selection. Incidentally, an analogous problem which is solved with no more difficulty exists for a swinging arm type of follower instead of the translating follower depicted in figure 8.29.

The load spring in figure 8.29 will be designed on the basis of functional

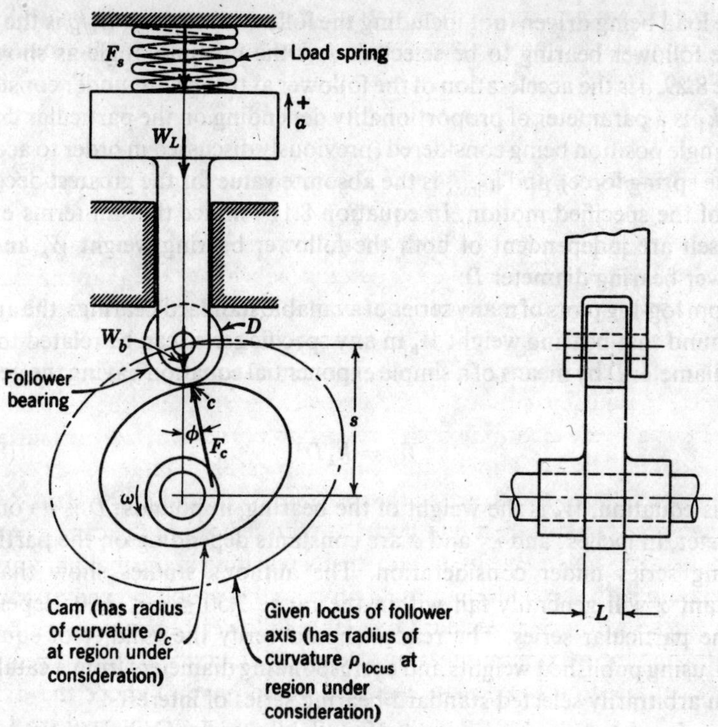

Figure 8.29 For exercise 8.18.

requirements from a dynamic force analysis, depending primarily on the follower motion and the driven inertia load. This dynamic analysis will determine the required maximum spring force $F_{s\,max}$ and the required spring force gradient k. Thus, if the greatest deceleration of the specified motion has an absolute value $|a|_{max}$ and the driven load has a mass designated $(W_L + W_b)/g$, the *required maximum* spring force $F_{s\,max}$ will be proportional to $|a|_{max}$ $(W_L + W_b)/g$. Therefore, at *any* specified angular position of the cam, it is reasonable to assume that the spring force will be proportional to $F_{s\,max}$ as linearly related to the corresponding specified displacement s. Hence, from Newton's second law, referring to figure 8.29 we can write the following equation for the contact force F_c at any specified significant cam angle position, assuming a rigid body type of follower:

$$F_c = \left(\frac{W_L + W_b}{g \cos \phi}\right)[a + k_1|a|_{max}] \tag{8.114}$$

In this equation, F_c is the contact force acting on the follower bearing as shown in figure 8.29 at the particular instant under consideration, W_L/g is the mass of the load being driven (not including the follower bearing), W_b/g is the mass of the follower bearing to be selected, ϕ is the pressure angle as shown in figure 8.29, a is the acceleration of the follower at the instant under consideration, k_1 is a parameter of proportionality depending on the particular critical cam angle position being considered (previously discussed in order to account for the spring force), and $|a|_{max}$ is the absolute value for the greatest deceleration of the specified motion. In equation 8.114 notice that all terms except W_b itself are independent of both the follower bearing weight W_b and the follower bearing diameter D.

From log-log plots of many series of available standard bearings, the author has found that bearing weight W_b in any specific series can be related to outside diameter D by means of a simple exponential equation having the general form

$$W_b = k_2 D^\alpha \tag{8.115}$$

In this equation, W_b is the weight of the bearing, in pounds; D is its outside diameter, in inches; and k_2 and α are constants depending on the particular bearing series under consideration. The author's studies show that the constant α will generally fall within the range $2.50 \leq \alpha \leq 3.00$, depending on the particular series. The reader should verify the validity of equation 8.115, using published weights and corresponding diameters from a catalogue for an arbitrarily selected standard bearing series of interest.

Also, from studies of many series of available standard bearings, the author has found that bearing width L, in figure 8.29, for any specific series can be linearly related to outside diameter D. Hence the approximation is given by the following equation, where L is the bearing width or length of cam contact area, D is the outside diameter of the bearing, and k_3 is a constant of proportionality depending upon the specific bearing series under consideration:

$$L \approx k_3 D \tag{8.116}$$

(a) Using equations 8.102, 8.103, 8.114, 8.115, and 8.116, derive the following general equation for the maximum shearing stress at the cam contact vicinity for the particular critical cam angle position of interest:

$$\tau_{max} = 0.127\sqrt{\frac{(W_L + k_2 D^\alpha)E}{gk_3 D \cos\phi}(a + k_1|a|_{max})\left(\frac{2}{D} + \frac{1}{\rho_c}\right)}$$

In this equation all terms have been defined except the modulus of elasticity E, which is assumed to be the same at the surfaces of both the cam and the follower, g, which is the gravitational constant, and ρ_c, which is the radius of curvature of the cam surface at the critical region in question.

(b) From the general equation for τ_{max} in a, derive the following equation for the optimum bearing weight. It will result in a minimum value for τ_{max} if that follower bearing is selected. (*Hint*: Consider $\partial\tau_{max}/\partial D$, noticing, in figure 8.29, that $\rho_c = \rho_{locus} - (D/2)$ and hence $\partial\rho_c/\partial D = -\frac{1}{2}$ since ρ_{locus} is *independent* of D.)

$$(W_b)_{opt} = W_L \frac{[2 + (D/2\rho_c)(1 - D/2\rho_c)]}{\{(\alpha - 2) + (D/2\rho_c)[(D/2\rho_c) + (\alpha - 1)]\}} \tag{8.117}$$

In this equation, $(W_b)_{opt}$ is the weight of the follower bearing which, if selected, will result in the smallest possible value for τ_{max} at the critical cam contact region; W_L is the weight of the driven load excluding the follower bearing weight; α is the constant for the particular series of bearings being considered, which is defined by equation 8.115 and which is determined from available catalogue data; D is the outside diameter of the optimum bearing; and ρ_c is the resulting radius of curvature for the cam surface at the critical region being considered.

It is obvious that equation 8.117 cannot be used directly to obtain an explicit selection of the optimum follower bearing, since $(W_b)_{opt}$, D, and ρ_c are all functions of the optimum bearing. Hence we shall write two more equations which can be solved simultaneously with equation 8.117 in order to determine explicitly the optimum follower bearing. First we shall write the following equation based on figure 8.29:

$$\rho_c = \rho_{locus} - \frac{D}{2} \tag{8.118}$$

At any particular critical cam angle position under consideration, the radius of curvature, ρ_{locus}, of the follower axis locus on the cam, will be known at this stage of design. Hence equation 8.118 gives us another relationship between ρ_c and D. The remaining relationship is merely equation 8.115 relating W_b with D, since for the specific bearing series of interest k_2 and α will be determined constants. Solving equations 8.117, 8.118, and 8.115 simultaneously for $(W_b)_{opt}$, D, and ρ_c will uniquely determine the optimum bearing size to select from the available series. Write a digital computer program for the iterative solution of these equations and apply your program to the solution of the following numerical problem.

(c) From catalogue data for a particular series of roller bearings, the following equation is derived from a log-log plot of bearing weight W_b versus outside diameter D:

$$W_b = 0.130D^{2.88}$$

where W_b is the bearing weight, in pounds, and D is the outside diameter of the bearing, in inches. The radius of curvature, ρ_{locus}, at a critical region of a cam, for the follower axis locus, is 0.800 in. The weight of the driven load W_L in figure 8.29 is 1.00 lb. Determine the approximate diameter of the optimum available follower bearing and its approximate weight, on the basis of minimizing τ_{max} at the critical contact region. The *closest* available standard size bearing in the series should be selected; it must *also* have an adequate internal load rating.

Answer: $D_{opt} \approx 1.00$ in, corresponding to $W_{b\,opt} \approx 0.14$ lb.

8.19 Gearbox Design

A manufacturer wishes to design a simple gearbox as depicted in figure 8.30 having a specified horsepower rating, a specified gear ratio, and a specified angular velocity for the pinion drive shaft. Because of the possible applications of the gear box, it is desirable to have it rectangular in shape, and as small in size and as light in weight as possible. Hence the helical *gearset* has been designed on the basis of minimizing the center distance, using the method of optimum design. Thus, at this stage of the design procedure, the pinion and gear pitch diameters D_p and D_g and the face width b are known numerical values.

It is realized that most of the gear box weight lies in the housing, which for economic reasons will be a casting. Therefore a final objective will be to design the housing on the basis of minimum weight. It is obvious in figure 8.30 that we will want to select the clearances c_1 and c_2 as small as practicable in order to minimize the housing weight. Incidentally, the clearance c_1 includes the addendum of either the gear or the pinion, which has been determined previously from the optimum design of the helical gearset.

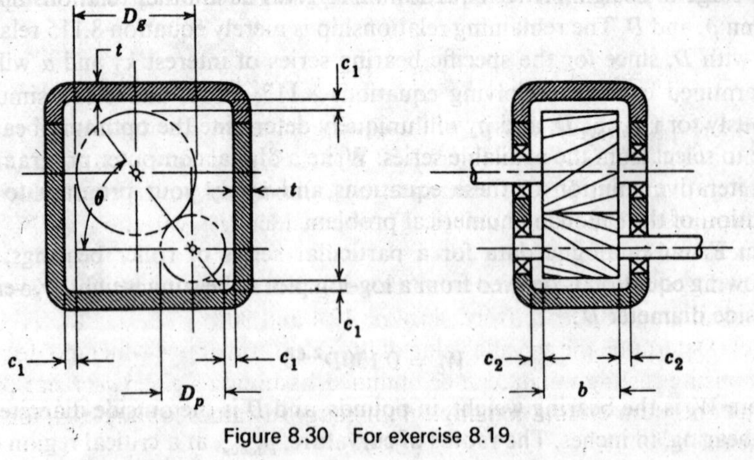

Figure 8.30 For exercise 8.19.

Therefore the smallest practicable values for the clearances c_1 and c_2 are those that will assure that no possibility exists for interference between the gears and housing walls. Hence at this time we assume that reasonable values for minimum clearances c_1 and c_2 have been selected.

It is also obvious in figure 8.30 that the housing wall thickness t should be as small as practicable in order to minimize the housing weight. The smallest feasible thickness t will be governed by practical considerations associated with the manufacturing of the casting. If the thickness t is too small, the molten metal might not flow properly when the casting is being poured into the mold. The smallest permissible value of wall thickness t will be somewhat influenced by the overall lateral dimensions of the housing, which have been roughly determined and minimized at this stage by the optimum gearset design. From past experience we know that stresses in the housing due to gear and pinion loads will be entirely insignificant, if the existence of a practicable castable wall thickness t is assumed. Thus, for this problem, we need not consider housing stresses in our design calculations. Therefore, at this stage of design, we assume that a practicable minimum wall thickness t has been selected as a numerical value.

(a) The remaining problem in designing the housing is to determine the optimum orientation of the gearset with respect to the housing walls. Assume that a rectangular parallelepiped general shape is desired for the outside of the casting. Prove that the housing weight is minimized if the orientation angle θ in figure 8.30 is equal to either 0 or 90°, for a given gear ratio of unity. Assume that the casting walls are solid, since the relatively small amounts of removed metal will be offset by the presence of mounting lugs, bosses, and bearings in the real casting.

(*Hint*: For part a prove that the housing weight equation is of the general form

$$W_h = w[A(\sin\theta + \cos\theta) + B(1 + \cos\theta)(1 + \sin\theta) + C]$$

where w is the weight density of the casting material, θ is the relative gearset orientation angle shown in figure 8.30, and A, B, C are functions of face width b, pitch diameters D_p, D_g, clearances c_1, c_2 and thickness t, all of which are determined independent values. Then consider the variation of W_h with respect to θ.)

(b) Determine the optimum values of orientation angle θ for given gear ratios of 2 and 3 respectively. *Answer*: 19.4 and 30°.

(c) Determine the specific value of the gear ratio above which no decrease in housing casting weight can be obtained, assuming a given *gear* size. What is the optimum orientation angle θ (or range of values for angles θ) for such gear ratios?

8.20 Brake Design

Brake design is a rather broad field in mechanical engineering which is closely allied with clutch design. However, in order to be specific in this book we shall consider only the design of a cam-actuated pivoted internal shoe type of brake as typically depicted in figure 8.31. Such a brake is essentially analyzed in Chapter 14 of reference 8.1; the reader should be familiar with that background or its equivalent before he solves the optimum design problems which follow.

The mechanical brake shown in figure 8.31 is in the true sense a system of mechanical elements, each of which can be designed on some optimum basis. However, having spent a considerable amount of time on optimum designs of typical mechanical elements in Chapters 7 and 8, we shall devote the following work to optimum design studies for the overall brake construction, even though the individual mechanical element designs are equally important.

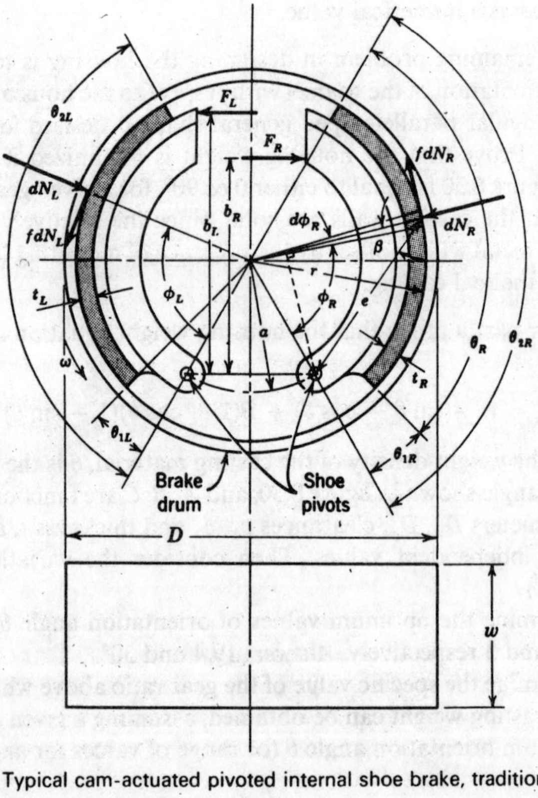

Figure 8.31 Typical cam-actuated pivoted internal shoe brake, traditional design.

Hence, as for many mechanical systems such as the Geneva mechanism in exercise 8.17, we shall employ the principle of *proportional design* whenever it is advantageous to do so.

The significant geometry of the cam-actuated brake in figure 8.31 is uniquely defined by the following independent geometrical parameters shown in that figure:

$$\theta_{1L}, \theta_{2L}, \theta_{1R}, \theta_{2R}, t_L, t_R, a_L, a_R, r, D, b_L, b_R, \text{ and } w$$

Hence, assuming proportional design, we could say that the value for drum radius r would be proportional to the value for outside diameter D, which is expressed by $r = k_1 D$, where k_1 is a proportionality parameter. Also, in order to be assured of having a good contact between the brake drum and the friction material on the brake shoe, we could say that the maximum permissible value for width w is w_{max} and this upper limit would be proportional to the radius r or the diameter D, which is expressed by $w_{max} = k_2 D$. In this case, k_2 is a proportionality parameter whose value depends on practical considerations associated with the manufacturing of the brake. Similarly, we could relate the geometrical parameters b_L and b_R by a proportionality parameter k_3, and this general procedure could be extended to all the geometrical parameters in the front view of figure 8.31.

Using the method of optimum design, it will be possible to determine the optimum values for the proportionality parameters of the basic brake design. Hence, after the conduct of such an optimum design study, which would determine certain desirabilities for proportionality parameters, we could draw a layout of the optimum brake configuration using an arbitrary value for a lateral dimension such as diameter D or drum radius r. Therefore, assuming proportional design, we could say that the brake weight W would be proportional to a lateral dimension squared times width w. We could write the following equation, where k_4 is a proportionality parameter depending on the particular proportional design of the brake elements, r is the drum radius, and w is its width as shown in figure 8.31:

$$W = k_4 w r^2 \tag{8.119}$$

Assume that weight W is the weight of the entire brake assembly, including the shoes, drum, and the actuating cam not shown in figure 8.31.

In order to be specific, for the following exercises assume that three friction materials having the physical properties given in table 8.33 are available. Assume that all three materials are obtainable in a moulded structure conforming to specifications submitted to the manufacturer. Also, assume that all three materials can be ground after bonding to the brake shoe.

TABLE 8.33

Typical Properties of Available Friction Materials

Material Number	Maximum Rubbing Speed, ft/min	Maximum Pressure p_{max}, psi	Coefficient of Friction f
1	7500	150	0.45
2	7500	150	0.35
3	7000	100	0.30

Pivoted Shoe Brake Design Problem.

A cam-actuated pivoted internal shoe brake is to have a specified value for torque *capacity* T, and the brake drum is to rotate with a specified value for maximum angular velocity ω. The brake is to be used in an application where weight W is an important consideration. Thus optimum design is to be conducted on the basis of minimizing the weight W. The available friction materials are those whose properties are given in table 8.33, the designed brake cannot be self-actuating, and overheating of the friction material must be avoided.

(a) Derive the following primary design equation:

$$W = \frac{k_4 T}{f\left(\int_{\theta_{1L}}^{\theta_{2L}} p_L \, d\phi_L + \int_{\theta_{1R}}^{\theta_{2R}} p_R \, d\phi_R\right)}$$

where W is the brake weight, k_4 is the proportionality parameter as defined by equation 8.119, T is the specified torque capacity required of the brake, f is the coefficient of friction for the brake lining material, p_L and p_R are the pressures for the left and right shoes at angles ϕ_L and ϕ_R respectively, and angles θ_{1L}, θ_{2L}, θ_{1R}, θ_{2R} are as shown in figure 8.31. From the primary design equation, prove that the following are desirable as far as minimizing the brake weight W is concerned.

1. Either drum radius r *or* drum width w can be chosen arbitrarily without appreciably affecting the brake weight W.

2. By proportional design, considering significant stresses, the brake elements should be so designed as to minimize the parameter k_4.

3. The optimum available material from table 8.33 is number 1.

4. Angles θ_{2L} and θ_{2R} should be selected as large as feasible and angles θ_{1L} and θ_{1R} should be selected as small as possible.

5. If possible, by proper design, it is desirable that $p_L = p_R$ = constant $= p_{max}$ for all angles ϕ_L and ϕ_R. See part b.

(b) Assume that the actuating cam is such that the points of application for F_L and F_R move through the same linear distance in the horizontal direction of figure 8.31.

1. For either shoe in figure 8.31 prove that pressure p is a constant for any angle ϕ *if* thickness t of the friction material is proportional to $\sin\phi$. (Use subscript L for left shoe and R for right shoe.)

2. Designate the cylindrical brake drum radius by r, and assume that the free outside radius of the shoe friction material is ground to coincide with the drum radius as shown in figure 8.32. Prove that t is proportional to $\sin\phi$ for the case where the backing surface of the friction material is cylindrical having the same radius r as the drum but eccentric to the drum center by an amount $\epsilon = t_{max}$, as shown in figure 8.32. Assume that $t \ll 2r$ and $\epsilon \ll 2r \sin\phi$ are valid. (Prove for the eccentric backing surface construction as shown in figure 8.32 that thickness $t \approx t_{max}\sin\phi$ is a close approximation. Hence, for a construction of this type, from step 1 the pressure p would be constant for any reasonable value of the angle ϕ. Therefore, this is a practical construction for obtaining the desired constant pressure distribution, rather

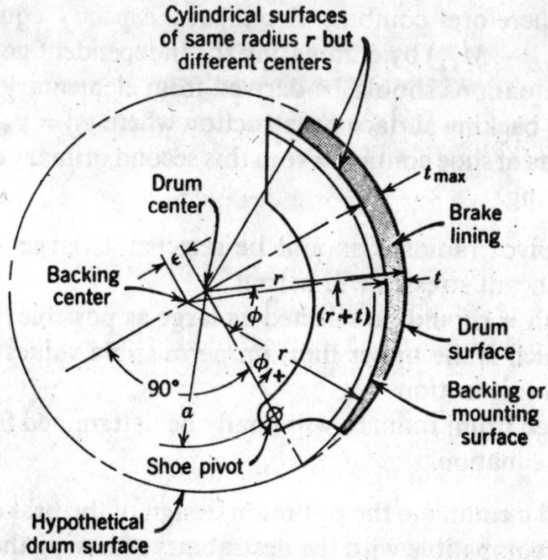

Figure 8.32 Brake shoe design for constant pressure distribution.

than the inefficient sinusoidal pressure distribution which results from a brake lining having constant thickness t.)

3. Assuming that both brake shoes have the eccentric backing surface type of construction depicted in figure 8.32, prove that $p_L = p_R =$ constant for all angles ϕ if

$$\left(\frac{a_R}{a_L}\right)\left(\frac{b_L}{b_R}\right)\left[\frac{(t_{\max})_L}{(t_{\max})_R}\right] = 1$$

The desired backing surface eccentricities $\epsilon_L = (t_{\max})_L$ and $\epsilon_R = (t_{\max})_R$ would be determined from this equation.

(c) In part a we proved that, as far as minimum weight W is concerned, either the drum radius r *or* the drum width w could be arbitrarily chosen. However, the choice of proportional values for r or w will appreciably affect the self-actuating tendencies of the brake design. For the direction of operation shown in figure 8.31, the self-actuating possibility should be considered for the left shoe. If M_{nL} is the moment of all normal drum forces on the left shoe about the shoe pivot and M_{fL} is the moment of all frictional drum forces on the left shoe about the shoe pivot, a consideration of $(M_{nL} - M_{fL})$ will indicate whether or not a self-actuating design exists. In order to avoid a self-actuating condition it is necessary to have $(M_{nL} - M_{fL})$ positive, and, because of the unpredictable nature of friction, $(M_{nL} - M_{fL})$ should be as large as feasible. Hence, conduct a second stage of optimum design on the basis of maximizing $(M_{nL} - M_{fL})$; for this stage of secondary importance adopt the *design* results already determined in parts a and b, and maximize $(M_{nL} - M_{fL})$ by the independent parameters which have virtually no effect on the value for W. Therefore, combine the torque *capacity* equation with the equation for $(M_{nL} - M_{fL})$ by eliminating the independent geometrical parameter r. These equations should be derived from elementary theory, assuming the eccentric backing surface construction where $p_L = p_R = p_{\max} =$ constant for all points of shoe contact. From this second primary design equation prove the following:

1. The shoe pivot radius a should be selected as large as proportional design and significant stresses will permit.
2. Drum width w should be selected as large as possible; that is, let $w = w_{\max} = k_2 D$, which is the upper limit on permissible values as discussed at the beginning of this section.
3. The required drum radius r will finally be determined from the derived torque *capacity* equation.

Parts a, b, and c complete the optimum design of the brake. The design so obtained is also compatible with the desirability of having the rubbing speed as low as possible.

8.21 Pivoted Shoe Brake Design Application

Design a cam-actuated pivoted internal shoe brake on the basis of minimizing its weight, where the specified torque *capacity* is 300,000 in-lb and where the brake drum will not rotate faster than 1000 rpm. Assume that the available friction materials are those given in table 8.33.

(a) Determine reasonable values for θ_1, θ_2, a, w, D, b_L, b_R, r, $(t_{max})_L$, $(t_{max})_R$.

[Typical answers from proportional design layout shown in figure 8.33: $\theta_1 = 15°$, $\theta_2 = 150°$, $a = 0.4D$, $w = 0.5D$, $b_L = \frac{8}{7}b_R$, $r = 0.47D$. Then $D_{opt} = 20.45$ in, $(t_{max})_L = 0.0437D$, and $(t_{max})_R = 0.050D$.]

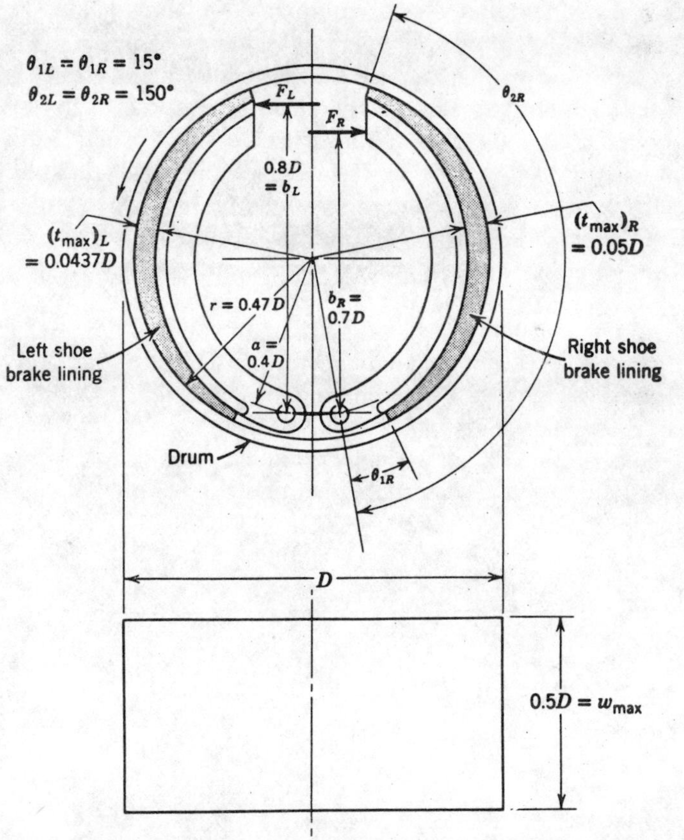

Figure 8.33 Typical proportional geometrical design of optimum brake.

(b) For the optimum design in part a calculate F_L and F_R in figure 8.33, which is a layout for the optimum design. ($F_L = 5150$ lb, $F_R = 24{,}900$ lb.)

(c) Estimate the percentage increase in weight W that would result if the optimum design had been conducted *arbitrarily assuming* a constant thickness t for the friction material, instead of the eccentric backing surface construction of the study in exercise 8.20 (37%).

APPENDIX A

Some Basic Nomenclature for MOD and AOD

A.1 Method of Optimum Design

General Notation with MOD

MOD	= method of optimum design
(P.D.E.)	= primary design equation
(S.D.E.)	= subsidiary design equation
(L.E.)	= limit "equation"
(I.F.)	= initial formulation equation system
(F.F.)	= final formulation equation system
(V.S.)	= variation study for optimization of design
(N.S.)	= case of normal specifications
(R.S.)	= case of redundant specifications
(I.S.)	= case of incompatible specifications
MSF	= material selection factor, generally only for (N.S.)
2-D	= two dimensional (V.S.) and variation diagram
3-D	= three dimensional (V.S.) and variation diagram
(I)	= developed (P.D.E.) of ideal problem in (F.F.)
(II), (III), (IV), etc.	= relating equations in (F.F.)

Exploratory Calculation Notation for MOD

A_t = number of approaches theoretically possible for derivation of (F.F.) from (I.F.)
D_{vs} = number of dimensions for (V.S.); 2-D, 3-D, ..., n-D.
n_c = number of constrained variables
n_e = number of eliminated parameters in (F.F.)
n_f = number of free variables in (I.F.)
n_r = number of related parameters in (F.F.)
n_v = number of variables in (I.F.), not counting optimization quantity
N_{ff} = number of equations in (F.F.)
N_s = number of (S.D.E.)'s in (I.F.)

A.2 Automated Optimal Design by OPTIGO

Tabulated Descriptions in Text

Item Described	See Table	Page
OPTIGO subroutine names	6.2	246
OPTIGO base point printout variables	6.3	254
OPTIGO input data items	6.4	259

Some Other Items

AOD = automated optimal design
SG = shotgun
G_l = *given constants* in AOD equation system entered as input data in OPTIGO; for $l = 1, \ldots, N_G$
C_m = *calculated constants* to reduce computational time in search process of OPTIGO, programmed in SUBROUTINE INPUT; for $m = 1, \ldots, N_C$
F_j = *final items* generally of interest in analysis of design solution, programmed in SUBROUTINES Q5000, EJ6000, and/or RK7000 generally with logical if for skip on NFC = 0; for $j = 1, \ldots, N_F$. Also, may be used to program intermediary functions as desired in these subroutines
α = finite difference size coefficient (see table 6.4)

β = starting step size coefficient (see table 6.4)

$(\beta_{sol})_s$ = solution step size coefficient for coarse search (see table 6.4)

$(\beta_{sol})_f$ = solution step size coefficient for final search (see table 6.4)

$(n_p)_{max}$ = maximum number of steps into penalty zone during self-tuning C_p probe (see table 6.4)

$(\eta_p)_s$ = starting value for penalty function wall steepening coefficient (see table 6.4)

μ = multiplier coefficient for steepening of penalty function wall during convergence to solution point (see table 6.4)

Δ = large step size along a ray, by equation 6.55a

Δ_r = general step size along a ray during search process; see equation 6.55b and figure 6.18

Δ_{sol} = solution step size along a ray for search process; see equation 6.55c and figure 6.18

APPENDIX B

OPTIGO Fortran IV Listing and Sample Printouts

The algorithm OPTIGO for automated optimal design was developed and described in the last part of Chapter 6. It is a gradient based search technique, incorporating features of shotgun start, self-tuning, interval halving, finite differences, and automatic convergence to a practical sized neighborhood of the solution point.

The Fortran IV listing of OPTIGO is given in section B.1 of this Appendix. It includes specific user programmed functions from the problem of example 6.9b in Chapter 6, and these are indicated by enclosed blocks in the listings for SUBROUTINES Q5000, EJ6000, RK7000, and INPUT. Also, a sample of the OPTIGO printout for input and output data is given in section B.2 of this Appendix for solution to the problem of example 6.9b.

Incidentally, the algorithm OPTIGO was initially developed in the author's consulting engineering office using a Monroe 1880-43 programmable calculator having 2048 program steps and 202 data storage registers. This machine calculates with an accuracy of approximately 13 significant figures, and it has magnetic card storage with hard copy printing for input and output data. However, the final development of the Fortran IV version of OPTIGO which follows was made on a Univac 90/60 digital computer in the batch mode of operation at the Worcester Area College Computation Center. The Univac 90/60 digital computer is an IBM compatible virtual machine using EBCDIC. Each user may acquire up to one megabyte of virtual storage per program on it. Calculation accuracy was approximately 7 digits for single precision variables and 15 digits for double precision variables. Hence it was necessary to use double precision accuracy for all variables affecting finite difference calculations and other calculations involving differences between numbers, such as for equation 6.31. Adaptation of the program to terminal operation with time sharing should be straightforward if this is desired. Recently the author translated the Fortran IV version of OPTIGO which follows to the Basic programming language for use on a TRS-80 microcomputer with terminal operation.

B.1 Fortran IV Listing of OPTIGO

```
C     OPTIGO - SEE OPTIMUM DESIGN OF MECHANICAL ELEMENTS SECOND EDITION
C     OPTIGO MAIN PROGRAM
C     PREPARE SUBROUTINES INPUT, Q5000, EJ6000, AND RK7000 BEFORE START
      DIMENSION C(100), DXD(25), DXR(25), E(25), F(100), G(100),
     1OPDB(25), OPDM(25), OPDP(25), R(100), X(25), XB(25), XDM(25),
     2XDP(25), XM(25), XMN(25), XMX(25), XP(25), XS(25), XSG(25)
      COMMON E, P, Q, OP, R, X, AL, B, BSS, BSF, C, CP, DRS, EM, EP,EPS,
     1EQP, F, G, N, NC, NE, NF, NFC, NPM, NPR, NR, NSG, RA, RE, RM,
     2RSD, S, SD, SM, TRA, XMN, XMX
      DOUBLE PRECISION DXD, DXM, DXR, E, P, PE, PR, Q, QP, OPDB, OPDM,
     1OPDP, R, SSQ, X, XB, XDM, XDP, XM, XP
      INTEGER RA, RM, S, SM
      CP = 0.
      CALL INPUT
      NFC = 0
      SDS = 0.
      DO 40 I = 1,N
         DXD(I) = (XMX(I) - XMN(I)) / AL
         SDS = SDS + (XMX(I) - XMN(I))**2
   40 CONTINUE
      D = SDS**.5 / B
      TRA = 1.
      BS = BSS
      EP = EPS
      IF (NSG .EQ. 0) NCP = 1
      IF (NSG .EQ. 0) GO TO 100
      DO 60 I = 1,N
         X(I) = (XMN(I) + XMX(I)) / 2.
   60 CONTINUE
      NCP = 0
      WRITE (6,70)
   70 FORMAT(48H START POINT FOR SHOTGUN CP PROBE IS AS FOLLOWS.)
  100 RA = 1
      S = 1
      SD = 0.
  104 NP = 0
C     REMOVE CP = 0. STATEMENT 105 FOR CASE OF CP SPECIFIED.
  105 CP = 0.
  106 DS = D * B / BS
  110 DO 120 I = 1,N
         XB(I) = X(I)
  120 CONTINUE
      IF (NSG .EQ. 0 .AND. CP .GT. 0.) NCP = 3
      IF (NSG .EQ. 0 .AND. CP .GT. 0.) GO TO 1200
  200 CALL OPCALC
      QB = Q
      RB = RE
      OPB = OP
      DRS = D / TRA
      IF (NPR .EQ. 0) GO TO 250
      CALL PRTOUT
  250 DO 280 I = 1,N
```

```
      XDP(I) = XB(I) + DXD(I)
      XDM(I) = XB(I) - DXD(I)
  280 CONTINUE
      DO 360 L = 1,N
         DO 320 I = 1,N
         XP(I) = XB(I)
         IF (I .EQ. L) XP(I) = XDP(I)
         X(I) = XP(I)
  320    CONTINUE
         CALL OPCALC
         I = L
         OPDP (I) = OP
         DO 340 I = 1,N
            XM(I) = XB(I)
            IF (I .EQ. L) XM(I) = XDM(I)
            X(I) = XM(I)
  340    CONTINUE
         CALL OPCALC
         I = L
         OPDM (I) = OP
  360 CONTINUE
      SSO = 0.
  400 DO 420 I = 1,N
      OPDB(I) = (OPDP(I) - OPDM(I)) / 2. / DXD(I)
      SSO = SSO + OPDB(I)**2
  420 CONTINUE
      DR = DRS
      DXM = - DR / SSO**.5
  430 DO 440 I = 1,N
      DXR(I) = DXM * OPDB(I)
  440 CONTINUE
      NHS = 0
      S = 1
      DO 460 I = 1,N
      X(I) = XB(I)
  460 CONTINUE
      OS1 = OB
      ONO = OB
      RS1 = RB
      RNO = RB
      OPS1 = OPB
  500 DO 520 I = 1,N
      XS(I) = X(I) + DXR(I)
      X(I) = XS(I)
  520 CONTINUE
      CALL OPCALC
      OS = O
      RS = RE
      OPS = OP
      IF (CP .EQ. O.) GO TO 1000
      IF (DRS .EQ. DR) GO TO 580
      NHS = NHS + 1
      IF (NHS .LT. 4) GO TO 580
      GO TO 600
  580 IF (OPS .LT. OPS1) GO TO 620
  600 IF (DR .LT. DS) GO TO 700
      DR = DR / 2.\
      NHS = 0
      DO 610 I = 1,N
         DXR(I) = - DXR(I) / 2.
  610 CONTINUE
  620 S = S + 1
      OPS1 = OPS
      IF (S .LE. SM) GO TO 500
      WRITE (6,650)
  650 FORMAT (///,30H STOP BECAUSE S LIMIT REACHED.///)
      CALL PRTOUT
      STOP
  700 SSO = 0.
      DO 720 I = 1,N
```

```
          SSD = SSD + (XS(I) - XB(I))**2
  720 CONTINUE
      SD = SSD**.5
      DO 740 I = 1,N
      XB(I) = XS(I)
  740 CONTINUE
      RA = RA + 1
      TRA = TRA + 1.
      DSA = 2. * DS
      IF (SD .LE. DSA) GO TO 760
      IF (RA .LE. RM) GO TO 200
      WRITE (6,750)
  750 FORMAT (///,32H STOP BECAUSE RAY LIMIT REACHED.///)
      CALL PRTOUT
      STOP
  760 IF (RA .EQ. 2) GO TO 800
      EP = EM * EP
      RA = 1
      GO TO 200
  800 IF (BS .EQ. BSF) GO TO 900
      BS = BSF
      DS = DS * BSS / BSF
      EP = EP  * EM
      NPR = 0
      WRITE (6,850)
  850 FORMAT(///,47H FINAL SEARCH STARTED WITH BASE POINT PRINTOUTS/
     128H SUPPRESSED AND BS, D, DS =
      WRITE (6,870) BS, D, DS
  870 FORMAT (10X, 1P3E16.6///)
      GO TO 110
  900 NFC = 1
      CALL OPCALC
      WRITE (6,920)
  920 FORMAT (///,29H SOLUTION FOUND IS AS FOLLOWS/)
      CALL PRTOUT
      GO TO 1400
 1000 IF (NP .GT. 0 .OR. RS .GT. RS1) GO TO 1100
      ONO = OS
      RS1 = RS
      RNO = RS
      GO TO 620
 1100 NP = NP + 1
      IF (NP .LT. NPM) GO TO 620
      CP = ABS((ONO - OS) / (RS - RNO))
      IF (NCP .GT. 0) GO TO 1200
      NCP = NCP + 1
      WRITE (6.1150) CP
 1150 FORMAT (///,3X,24HFOR SHOTGUN SEARCH CP = , 1PE16.6///)
      CALL SG
      IF (NSG .EQ. 0) GO TO 1170
      WRITE (6.1160)
 1160 FORMAT (3X,35HSHOTGUN SEARCH GAVE FOLLOWING POINT/)
 1170 CONTINUE
      GO TO 100
 1200 WRITE (6.1250) CP
 1250 FORMAT(///,3X,28HFOR MAIN OPTIGO SEARCH CP = , 1PE16.6///)
      WRITE (6.1300)
 1300 FORMAT (///,28H OPTIGO OUTPUT DATA FOLLOWS.///)
      WRITE (6.1350)
 1350 FORMAT (///,40H COARSE SEARCH STARTED WITH BS, D, DS = )
      WRITE (6.1370) BS, D, DS
 1370 FORMAT (10X, 1P3E16.6///)
      GO TO 200
 1400 IF (NF .EQ. 0) GO TO 1450
      WRITE (6.1410)
 1410 FORMAT (3X, 36HFINAL ITEMS F(J) FOR J = 1,NF ARE = )
      WRITE (6.1420) (F(J), J = 1,NF)
 1420 FORMAT (5X, 5G14.6///)
 1450 STOP
      END
```

```
      2000 SUBROUTINE PRTOUT
     C        PRINTS DATA FOR BASE POINTS AND SOLUTION POINT.
              DIMENSION C(100), DXD(25), DXR(25), E(25), F(100), G(100),
             1OPDB(25), OPDM(25), OPDP(25), R(100), X(25), XB(25), XDM(25),
             2XDP(25), XM(25), XMN(25), XMX(25), XP(25), XS(25), XSG(25)
              COMMON E, P, Q, QP, R, X, AL, B, BSS, BSF, C, CP, DRS, EM, EP,EPS,
             1EQP, F, G, N, NC, NE, NF, NFC, NPM, NPR, NR, NSG, RA, RE, RM,
             2RSD, S, SD, SM, TRA, XMN, XMX
              DOUBLE PRECISION DXD, DXM, DXR, E, P, PE, PR, Q, QP, QPDB, QPDM,
             1QPDP, R, SSQ, X, XB, XDM, XDP, XM, XP
              INTEGER RA, RM, S, SM
              WRITE (6,2050) RA, TRA, DRS, S, SD, EP, CP
      2050 FORMAT (///,3X,46HBASE POINT DATA RA, TRA, DRS, S, SD, EP, CP =
             13X, I8, 1P2E16.6, I8, 1P3E16.6)
              WRITE (6,2100)
      2100 FORMAT (5X,33HVALUES OF X(I) FOR I = 1,N ARE = )
              WRITE (6,2120) (X(I), I = 1,N)
      2120 FORMAT (7X, 1P5D14.6)
              IF (NE .EQ. 0) GO TO 2170
              WRITE (6,2140)
      2140 FORMAT (5X,34HVALUES OF E(J) FOR J = 1,NE ARE = )
              WRITE (6,2160) (E(J), J = 1,NE)
      2160 FORMAT (7X, 1P5D14.6)
      2170 IF (NR .EQ. 0) GO TO 2230
              WRITE (6,2190)
      2190 FORMAT (5X,34HVALUES OF R(K) FOR K = 1,NR ARE = )
              WRITE (6,2210) (R(K), K = 1,NR)
      2210 FORMAT (7X, 1P5D14.6)
      2230 WRITE (6,2250) P, Q, QP, EQP
      2250 FORMAT (5X,26HVALUES OF P, Q, QP, EQP = /8X, 1P3D16.6, 1PE16.6//)
              RETURN
              END


      3000 SUBROUTINE SG
     C        FOR SHOTGUN SEARCH IF DESIRED AT START.
              DIMENSION C(100), DXD(25), DXR(25), E(25), F(100), G(100),
             1OPDB(25), OPDM(25), OPDP(25), R(100), X(25), XB(25), XDM(25),
             2XDP(25), XM(25), XMN(25), XMX(25), XP(25), XS(25), XSG(25)
              COMMON E, P, Q, QP, R, X, AL, B, BSS, BSF, C, CP, DRS, EM, EP,EPS,
             1EQP, F, G, N, NC, NE, NF, NFC, NPM, NPR, NR, NSG, RA, RE, RM,
             2RSD, S, SD, SM, TRA, XMN, XMX
              DOUBLE PRECISION DXD, DXM, DXR, E, P, PE, PR, Q, QP, QPDB, QPDM,
             1QPDP, R, SSQ, X, XB, XDM, XDP, XM, XP
              INTEGER RA, RM, S, SM
              M = 1
              QPSG = 10.**59
              RG = RSD
      3050 DO 3100 I = 1,N
                RT = (3.14159265359 + RG)**5
                AR = AINT(RT)
                RD = RT - AR
                RG = RD
                X(I) = XMN(I) + RD * (XMX(I) - XMN(I))
      3100 CONTINUE
              CALL QPCALC
              IF (QP .GT. QPSG) GO TO 3150
      3110 QPSG = QP
              DO 3120 I = 1,N
                XSG(I) = X(I)
      3120 CONTINUE
      3150 IF (M .GE. NSG) GO TO 3170
              M = M + 1
              GO TO 3050
      3170 DO 3180 I = 1,N
                X(I) = XSG(I)
      3180 CONTINUE
              RETURN
              END
```

```
      4000 SUBROUTINE QPCALC
C     CALCULATES PENALIZED OPTIMIZATION QUANTITY QP.
           DIMENSION C(100), DXD(25), DXR(25), E(25), F(100), G(100),
          1QPDB(25), QPDM(25), QPDP(25), R(100), X(25), XB(25), XDM(25),
          2XDP(25), XM(25), XMN(25), XMX(25), XP(25), XS(25), XSG(25)
           COMMON E, P, Q, QP, R, X, AL, B, BSS, BSF, C, CP, DRS, EM, EP,EPS,
          1EQP, F, G, N, NC, NE, NF, NFC, NPM, NPR, NR, NSG, RA, RE, RM,
          2RSD, S, SD, SM, TRA, XMN, XMX
           DOUBLE PRECISION DXD, DXM, DXR, E, P, PE, PR, Q, QP, QPDB, QPDM,
          1QPDP, R, SSQ, X, XB, XDM, XDP, XM, XP
           INTEGER RA, RM, S, SM
           PE = 0.
           PR = 0.
           CALL Q5000
           IF (NE .EQ. 0) GO TO 4060
           CALL EJ6000
           DO 4040 J = 1, NE
              PE = PE + E(J)**2
      4040 CONTINUE
      4060 IF (NR .EQ. 0) GO TO 4100
           CALL RK7000
           DO 4080 K = 1,NR
              IF (R(K) .GE. 0.) GO TO 4080
              PR = PR + R(K)**2
      4080 CONTINUE
      4100 RE = PE + PR
           P = CP * EP * RE
           QP = Q + P
           EQP = QP / Q
           RETURN
           END

      5000 SUBROUTINE Q5000
C     CALCULATES OPTIMIZATION QUANTITY Q
C     MUST BE COMPLETED FOR SPECIFIC PROBLEM BY USER BEFORE START.
           DIMENSION C(100), DXD(25), DXR(25), E(25), F(100), G(100),
          1QPDB(25), QPDM(25), QPDP(25), R(100), X(25), XB(25), XDM(25),
          2XDP(25), XM(25), XMN(25), XMX(25), XP(25), XS(25), XSG(25)
           COMMON E, P, Q, QP, R, X, AL, B, BSS, BSF, C, CP, DRS, EM, EP,EPS,
          1EQP, F, G, N, NC, NE, NF, NFC, NPM, NPR, NR, NSG, RA, RE, RM,
          2RSD, S, SD, SM, TRA, XMN, XMX
           DOUBLE PRECISION DXD, DXM, DXR, E, P, PE, PR, Q, QP, QPDB, QPDM,
          1QPDP, R, SSQ, X, XB, XDM, XDP, XM, XP
           INTEGER RA, RM, S, SM
C     Q IS CALCULATED HERE FROM CRITERION FUNCTION PROGRAMMED BY USER.
C     DO NOT DUPLICATE VARIABLE NAMES LISTED IN COMMON STATEMENT
C     ALSO, APPROPRIATE F(J) FINAL ITEMS ARE PROGRAMMED HERE BY USER.
C     SKIP F(J) ITEMS NOT NEEDED UNTIL END, BY LOGICAL IF ON NFC = 0.
           Q = C(1) / (X(1) * X(2))
           IF (NFC .EQ. 0) GO TO 5100
           F(1) = G(1) * X(1)
           F(2) = G(2) * X(2)
           F(3) = F(1) * F(2)
           F(4) = 2. * (F(1) + F(2))
           F(5) = (F(1)**2 + F(2)**2)**.5
      5100 CONTINUE
           RETURN
           END

      6000 SUBROUTINE EJ6000
C     CALCULATES EQUALITY CONSTRAINTS E(J) FOR J = 1, NE.
C     MUST BE COMPLETED FOR SPECIFIC PROBLEM BY USER BEFORE START.
           DIMENSION C(100), DXD(25), DXR(25), E(25), F(100), G(100),
          1QPDB(25), QPDM(25), QPDP(25), R(100), X(25), XB(25), XDM(25),
          2XDP(25), XM(25), XMN(25), XMX(25), XP(25), XS(25), XSG(25)
           COMMON E, P, Q, QP, R, X, AL, B, BSS, BSF, C, CP, DRS, EM, EP,EPS,
          1EQP, F, G, N, NC, NE, NF, NFC, NPM, NPR, NR, NSG, RA, RE, RM,
          2RSD, S, SD, SM, TRA, XMN, XMX
           DOUBLE PRECISION DXD, DXM, DXR, E, P, PE, PR, Q, QP, QPDB, QPDM,
          1QPDP, R, SSQ, X, XB, XDM, XDP, XM, XP
```

```
281           INTEGER RA, RM, S, SM
      C       E(J)S FOR J = 1, NE ARE CALCULATED HERE AS PROGRAMMED BY USER.
      C       LEAVE BLANK FOR CASE OF NE = 0.
      C       DO NOT DUPLICATE VARIABLE NAMES LISTED IN COMMON STATEMENT
      C       ALSO, APPROPRIATE F(J) FINAL ITEMS ARE PROGRAMMED HERE BY USER.
      C       SKIP F(J) ITEMS NOT NEEDED UNTIL END, BY LOGICAL IF ON NFC = 0.
282           E(1) = 1. - C(2) * (G(1) * X(1) + G(2) * X(2))
283           RETURN
284           END

285      7000 SUBROUTINE RK7000
      C       CALCULATES REGIONAL CONSTRAINTS R(K) FOR K = 1, NR.
      C       MUST BE COMPLETED FOR SPECIFIC PROBLEM BY USER BEFORE START.
286           DIMENSION C(100), DXD(25), DXR(25), E(25), F(100), G(100),
             1QPDB(25), QPDM(25), QPDP(25), R(100), X(25), XB(25), XDM(25),
             2XDP(25), XM(25), XMN(25), XMX(25), XP(25), XS(25), XSG(25)
287           COMMON E, P, Q, QP, R, X, AL, B, BSS, BSF, C, CP, DRS, EM, EP,EPS,
             1EQP, F, G, N, NC, NE, NF, NFC, NPM, NPR, NR, NSG, RA, RE, RM,
             2RSD, S, SD, SM, TRA, XMN, XMX
288           DOUBLE PRECISION DXD, DXM, DXR, E, P, PE, PR, Q, QP, QPDB, QPDM,
             1QPDP, R, SSQ, X, XB, XDM, XDP, XM, XP
289           INTEGER RA, RM, S, SM
      C       R(K)S FOR K = 1,NR ARE CALCULATED HERE AS PROGRAMMED BY USER.
      C       LEAVE BLANK FOR CASE OF NR = 0.
      C       DO NOT DUPLICATE VARIABLE NAMES LISTED IN COMMON STATEMENT
      C       ALSO, APPROPRIATE F(J) FINAL ITEMS ARE PROGRAMMED HERE BY USER.
      C       SKIP F(J) ITEMS NOT NEEDED UNTIL END, BY LOGICAL IF ON NFC = 0.
290           R(1) = 1. - X(1)
291           R(2) = 1. - X(2)
292           R(3) = X(1)
293           R(4) = X(2)
294           RETURN
295           END

296      8000 SUBROUTINE INPUT
      C       FOR INPUT DATA OF ENTIRE OPTIGO PROGRAM.
      C       INPUT DATA CARDS MUST BE SUPPLIED BY USER BEFORE START.
297           DIMENSION C(100), DXD(25), DXR(25), E(25), F(100), G(100),
             1QPDB(25), QPDM(25), QPDP(25), R(100), X(25), XB(25), XDM(25),
             2XDP(25), XM(25), XMN(25), XMX(25), XP(25), XS(25), XSG(25)
298           COMMON E, P, Q, QP, R, X, AL, B, BSS, BSF, C, CP, DRS, EM, EP,EPS,
             1EQP, F, G, N, NC, NE, NF, NFC, NPM, NPR, NR, NSG, RA, RE, RM,
             2RSD, S, SD, SM, TRA, XMN, XMX
299           DOUBLE PRECISION DXD, DXM, DXR, E, P, PE, PR, Q, QP, QPDB, QPDM,
             1QPDP, R, SSQ, X, XB, XDM, XDP, XM, XP
300           INTEGER RA, RM, S, SM
301           WRITE (6,8050)
302      8050 FORMAT (///,27H OPTIGO INPUT DATA FOLLOWS.///)
303           AL = 1.E5
304           B = 10.
305           BSS = 2000.
306           BSF = 2.E4
307           NPM = 2
308           EPS = 1.
309           EM = 8.
      C       OPTIGO WILL DETERMINE CP AUTOMATICALLY.  HOWEVER,
      C       FOR SPECIFIED CP, ADD CP = ... HERE AND REMOVE STATEMENT 105.
310           WRITE (6,8100) AL, B, BSS, BSF, NPM, EPS, EM
311      8100 FORMAT (3X,46HSUPPLIED DATA AL, B, BSS, BSF, NPM, EPS, EM = /
             16X, 7G12.4/)
312           READ (5,8150) N, NE, NR, NC, NF, NG, RM, SM, NSG, RSD, NPR
313           READ (5,8150) (XMN(I), I = 1,N)
314           READ (5,8150) (XMX(I), I = 1,N)
315           READ (5,8150) (G(L), L = 1,NG)
      C       GENERAL FIELD DESCRIPTOR FOR ALL INPUT DATA BY STATEMENT 8150
316      8150 FORMAT (8G10.3)
317           WRITE (6,8200) N, NE, NR, NC, NF, NG, RM, SM, NSG, RSD, NPR
318      8200 FORMAT (3X,58HINPUT DATA N, NE, NR, NC, NF, NG, RM, SM, NSG, RSD,
             1NPR = /5X, 9I7, 1PE14.6, I7)
319           WRITE (6,8250)
```

```
      8250 FORMAT (3X,35HVALUES OF XMN(I) FOR I = 1,N ARE = )
           WRITE (6,8270) (XMN(I), I = 1,N)
      8270 FORMAT (5X, 1P5E14.6)
           WRITE (6,8300)
      8300 FORMAT (3X,35HVALUES OF XMX(I) FOR I = 1,N ARE = )
           WRITE (6,8320) (XMX(I), I = 1,N)
      8320 FORMAT (5X, 1P5E14.6)
           IF (NG .EQ. 0) GO TO 8410
           WRITE (6,8350)
      8350 FORMAT (3X,50HVALUES OF GIVEN CONSTANTS G(L) FOR L = 1,NG ARE = )
           WRITE (6,8400) (G(L), L = 1,NG)
      8400 FORMAT (5X, 1P5E14.6)
      8410 IF (NSG .GT. 0) GO TO 8500
C          TO SPECIFY START POINT USE STATEMENT 8420 INPUT AND INPUT NSG = 0.
      8420 READ (5,8150) (X(I), I = 1,N)
           WRITE (6, 8450)
      8450 FORMAT (3X,48HVALUES OF X(I) FOR I = 1,N OF START POINT ARE = )
           WRITE (6,8470) (X(I), I = 1,N)
      8470 FORMAT (5X, 1P5D14.6)
      8500 CONTINUE
           IF (NC .EQ. 0) GO TO 8980
C          CALCULATED CONSTANTS C(M), M = 1,NC ARE PROGRAMMED HERE BY USER.
           C(1) = 1. / (G(1) * G(2))
           C(2) = 2./G(3)
           WRITE (6,8950)
      8950 FORMAT (3X,55HVALUES OF CALCULATED CONSTANTS C(M) FOR M = 1,NC ARE
          1 = )
           WRITE (6,8970) (C(M), M = 1,NC)
      8970 FORMAT (5X, 1P5E14.6)
      8980 IF (NSG .EQ. 0 .AND. CP .GT. 0.) GO TO 8995
           WRITE (6,8990)
      8990 FORMAT (///,36H PROBE(S) FOR CP CALCULATION FOLLOW./)
      8995 CONTINUE
           RETURN
           END

   $ENTRY
```

B.2 OPTIGO Printout Samples from Example 6.9(b) of Chapter 6:

```
 OPTIGO INPUT DATA FOLLOWS.

   SUPPLIED DATA AL, B, BSS, BSF, NPM, EPS, EM =
        0.1000E 06   10.00         2000.      0.2000E 05          2   1.000       8.000

   INPUT DATA N, NE, NR, NC, NF, NG, RM, SM, NSG, RSD, NPR =
          2     1     4     2     5     3   100   100    50  3.456700E-01      1
   VALUES OF XMN(I) FOR I = 1,N ARE =
      0.000000E-01  0.000000E-01
   VALUES OF XMX(I) FOR I = 1,N ARE =
      1.000000E 00  1.000000E 00
   VALUES OF GIVEN CONSTANTS G(L) FOR L = 1,NG ARE =
      1.600000E 02  7.500000E 01  4.000000E 02
   VALUES OF CALCULATED CONSTANTS C(M) FOR M = 1,NC ARE =
      8.333333E-05  4.999999E-03


PROBE(S) FOR CP CALCULATION FOLLOW.

START POINT FOR SHOTGUN CP PROBE IS AS FOLLOWS.


  BASE POINT DATA RA, TRA, DRS, S, SO, EP, CP =
       1    1.000000E 00    1.414213E-01       1    0.000000E-01    1.000000E 00    0.000000E-01
   VALUES OF X(I) FOR I = 1,N ARE =
      5.0000000-01  5.0000000-01
   VALUES OF E(J) FOR J = 1,NE ARE =
      4.1250010-01
   VALUES OF R(K) FOR K = 1,NR ARE =
      5.0000000-01  5.0000000-01  5.0000000-01  5.0000000-01
   VALUES OF P, Q, QP, EQP =
        0.0000000-01     3.3333330-04     3.3333330-04     1.000000E 00
```

```
FOR SHOTGUN SEARCH CP =       3.326195E-04


SHOTGUN SEARCH GAVE FOLLOWING POINT


BASE POINT DATA RA, TRA, DRS, S, SD, EP, CP =
        1    1.000000E 00     1.414213E-01        1     0.000000E-01     1.000000E 00     0.000
  VALUES OF X(I) FOR I = 1,N ARE =
     7.6757810-01  8.9990230-01
  VALUES OF E(J) FOR J = 1,NE ARE =
     4.8474320-02
  VALUES OF R(K) FOR K = 1,NR ARE =
     2.3242190-01  1.0009770-01  7.6757810-01  8.9990230-01
  VALUES OF P, Q, QP, EQP =
        0.0000000-01     1.2064260-04     1.2064260-04     1.000000E 00


FOR MAIN OPTIGO SEARCH CP =        1.016890E-03


OPTIGO OUTPUT DATA FOLLOWS.


COARSE SEARCH STARTED WITH BS, D, DS =
             2.000000E 03     1.414213E-01     7.071064E-04


BASE POINT DATA RA, TRA, DRS, S, SD, EP, CP =
        1    1.000000E 00     1.414213E-01        2     0.000000E-01     1.000000E 00     1.016
  VALUES OF X(I) FOR I = 1,N ARE =
     9.8277270-01  1.0834530 00
  VALUES OF E(J) FOR J = 1,NE ARE =
    -1.9251290-01
  VALUES OF R(K) FOR K = 1,NR ARE =
     1.7227290-02 -8.3453180-02  9.8277270-01  1.0834530 00
  VALUES OF P, Q, QP, EQP =
        4.4769240-05     7.8262820-05     1.2303210-04     1.572037E 00


BASE POINT DATA RA, TRA, DRS, S, SD, EP, CP =
        2    2.000000E 00     7.071066E-02       28     1.497055E-01     1.000000E 00     1.016
  VALUES OF X(I) FOR I = 1,N ARE =
     8.8881120-01  9.8773340-01
  VALUES OF E(J) FOR J = 1,NE ARE =
    -8.1448770-02
  VALUES OF R(K) FOR K = 1,NR ARE =
     1.1118880-01  1.2266640-02  8.8881120-01  9.8773340-01
  VALUES OF P, Q, QP, EQP =
        6.7459490-06     9.4922570-05     1.0166850-04     1.071068E 00


BASE POINT DATA RA, TRA, DRS, S, SD, EP, CP =
        3    3.000000E 00     4.714044E-02       26     4.197305E-02     1.000000E 00     1.016
  VALUES OF X(I) FOR I = 1,N ARE =
     8.6341890-01  1.0211540 00
  VALUES OF E(J) FOR J = 1,NE ARE =
    -7.3567780-02
  VALUES OF R(K) FOR K = 1,NR ARE =
     1.3658110-01 -2.1154400-02  8.6341890-01  1.0211540 00
  VALUES OF P, Q, QP, EQP =
        5.9736690-06     9.4516100-05     1.0048980-04     1.063202E 00
```

```
BASE POINT DATA RA, TRA, DRS, S, SD, EP, CP =
     4   4.000000E 00   3.535533E-02      24   4.427344E-03   1.000000E 00   1.016890E-03
  VALUES OF X(I) FOR I = 1,N ARE =
     8.5970000-01  1.0187520 00
  VALUES OF E(J) FOR J = 1,NE ARE =
    -6.9791790-02
  VALUES OF R(K) FOR K = 1,NR ARE =
     1.4030000-01 -1.8752100-02  8.5970000-01  1.0187520 00
  VALUES OF P, Q, QP, EQP =
         5.3107430-06    9.5148800-05    1.0045950-04    1.055815E 00


BASE POINT DATA RA, TRA, DRS, S, SD, EP, CP =
     5   5.000000E 00   2.828426E-02      15   3.310734E-03   1.000000E 00   1.016890E-03
  VALUES OF X(I) FOR I = 1,N ARE =
     8.5698780-01  1.0206510 00
  VALUES OF E(J) FOR J = 1,NE ARE =
    -6.8334120-02
  VALUES OF R(K) FOR K = 1,NR ARE =
     1.4301220-01 -2.0650660-02  8.5698780-01  1.0206510 00
  VALUES OF P, Q, QP, EQP =
         5.1820810-06    9.5272350-05    1.0045440-04    1.054392E 00


BASE POINT DATA RA, TRA, DRS, S, SD, EP, CP =
     1   6.000000E 00   2.357022E-02      20   8.956590E-04   8.000000E 00   1.016890E-03
  VALUES OF X(I) FOR I = 1,N ARE =
     8.5697100-01  1.0197550 00
  VALUES OF E(J) FOR J = 1,NE ARE =
    -6.7984810-02
  VALUES OF R(K) FOR K = 1,NR ARE =.
     1.4302900-01 -1.9755360-02  8.5697100-01  1.0197550 00
  VALUES OF P, Q, QP, EQP =
         4.0774900-05    9.5357890-05    1.3613280-04    1.427598E 00


BASE POINT DATA RA, TRA, DRS, S, SD, EP, CP =
     2   7.000000E 00   2.020304E-02      19   6.555551E-02   8.000000E 00   1.016890E-03
  VALUES OF X(I) FOR I = 1,N ARE =
     8.0655380-01  9.7785500-01
  VALUES OF E(J) FOR J = 1,NE ARE =
    -1.1938430-02
  VALUES OF R(K) FOR K = 1,NR ARE =
     1.9344620-01  2.2145030-02  8.0655380-01  9.7785500-01
  VALUES OF P, Q, QP, EQP =
         1.1594660-06    1.0566010-04    1.0681960-04    1.010973E 00


BASE POINT DATA RA, TRA, DRS, S, SD, EP, CP =
     3   8.000000E 00   1.767766E-02      14   2.903745E-02   8.000000E 00   1.016890E-03
  VALUES OF X(I) FOR I = 1,N ARE =
     7.9003040-01  1.0017330 00
  VALUES OF E(J) FOR J = 1,NE ARE =
    -7.6739350-03
  VALUES OF R(K) FOR K = 1,NR ARE =
     2.0996960-01 -1.7328260-03  7.9003040-01  1.0017330 00
  VALUES OF P, Q, QP, EQP =
         5.0349850-07    1.0529870-04    1.0580220-04    1.004781E 00


BASE POINT DATA RA, TRA, DRS, S, SD, EP, CP =
     4   9.000000E 00   1.571348E-02      18   1.650793E-03   8.000000E 00   1.016890E-03
  VALUES OF X(I) FOR I = 1,N ARE =
     7.9126080-01  1.0028330 00
  VALUES OF E(J) FOR J = 1,NE ARE =
    -9.0709730-03
  VALUES OF R(K) FOR K = 1,NR ARE =
     2.0873920-01 -2.8333660-03  7.9126080-01  1.0028330 00
  VALUES OF P, Q, QP, EQP =
         7.3468690-07    1.0501960-04    1.0575430-04    1.006995E 00
```

```
BASE POINT DATA RA, TRA, DRS, S, SD, EP, CP =
    1   1.000000E 01   1.414213E-02    14    4.892214E-04   6.400000E 01   1.016
  VALUES OF X(I) FOR I = 1,N ARE =
     7.917397D-01  1.002934D 00
  VALUES OF E(J) FOR J = 1,NE ARE =
    -9.491615D-03
  VALUES OF R(K) FOR K = 1,NR ARE =
     2.082603D-01 -2.933502D-03  7.917397D-01  1.002934D 00
  VALUES OF P, Q, QP, EQP =
        6.423239D-06   1.049456D-04   1.113688D-04   1.061205E 00


BASE POINT DATA RA, TRA, DRS, S, SD, EP, CP =
    2   1.100000E 01   1.285648E-02    22    8.840289E-03   6.400000E 01   1.016
  VALUES OF X(I) FOR I = 1,N ARE =
     7.850550D-01  9.971485D-01
  VALUES OF E(J) FOR J = 1,NE ARE =
    -1.974517D-03
  VALUES OF R(K) FOR K = 1,NR ARE =
     2.149450D-01  2.851486D-03  7.850550D-01  9.971485D-01
  VALUES OF P, Q, QP, EQP =
        2.537320D-07   1.064532D-04   1.067070D-04   1.002383E 00


BASE POINT DATA RA, TRA, DRS, S, SD, EP, CP =
    1   1.200000E 01   1.178511E-02    14    4.020843E-04   5.120000E 02   1.016
  VALUES OF X(I) FOR I = 1,N ARE =
     7.846572D-01  9.972070D-01
  VALUES OF E(J) FOR J = 1,NE ARE =
    -1.678225D-03
  VALUES OF R(K) FOR K = 1,NR ARE =
     2.153428D-01  2.792954D-03  7.846572D-01  9.972070D-01
  VALUES OF P, Q, QP, EQP =
        1.466372D-06   1.065009D-04   1.079673D-04   1.013768E 00


BASE POINT DATA RA, TRA, DRS, S, SD, EP, CP =
    2   1.300000E 01   1.087656E-02    17    1.473822E-03   5.120000E 02   1.016
  VALUES OF X(I) FOR I = 1,N ARE =
     7.833056D-01  9.966195D-01
  VALUES OF E(J) FOR J = 1,NE ARE =
    -3.765806D-04
  VALUES OF R(K) FOR K = 1,NR ARE =
     2.166944D-01  3.380477D-03  7.833056D-01  9.966195D-01
  VALUES OF P, Q, QP, EQP =
        7.383454D-08   1.067476D-04   1.068214D-04   1.000691E 00


BASE POINT DATA RA, TRA, DRS, S, SD, EP, CP =
    1   1.400000E 01   1.010152E-02    13    6.803186E-04   4.096000E 03   1.016
  VALUES OF X(I) FOR I = 1,N ARE =
     7.826419D-01  9.964698D-01
  VALUES OF E(J) FOR J = 1,NE ARE =
     2.104774D-04
  VALUES OF R(K) FOR K = 1,NR ARE =
     2.173581D-01  3.530204D-03  7.826419D-01  9.964698D-01
  VALUES OF P, Q, QP, EQP =
        1.845206D-07   1.068542D-04   1.070387D-04   1.001726E 00


FINAL SEARCH STARTED WITH BASE POINT PRINTOUTS
SUPPRESSED AND BS, D, DS =
                  2.000000E 04   1.414213E-01   7.071061E-05
```

```
SOLUTION FOUND IS AS FOLLOWS


  BASE POINT DATA RA, TRA, DRS, S, SD, EP, CP =
       2    2.700000E 01    5.439281E-03       21    1.288703E-04    4.096000E 03    1.016890E-03
    VALUES OF X(I) FOR I = 1,N ARE =
      7.8154810-01  9.9951990-01
    VALUES OF E(J) FOR J = 1,NE ARE =
     -5.8214350-05
    VALUES OF R(K) FOR K = 1,NR ARE =
      2.1845190-01  4.8011540-04  7.8154810-01  9.9951990-01
    VALUES OF P, Q, QP, EQP =
           1.4115430-08     1.0667720-04     1.0669130-04     1.000132E 00

  FINAL ITEMS F(J) FOR J = 1,NF ARE =
       125.048         74.9640         9374.07          400.023         145.796


DIAGNOSTICS:             NO ERRORS        NO WARNINGS
COMPILATION FROM CARD INPUT:     3.777 SEC(CPU TIME) ...     5 SEC(REAL TIME).
LIBRARY SEARCH(AND COMPILE):     0.025 SEC(CPU TIME) ...     0 SEC(REAL TIME).
PROGRAM EXECUTION:               3.680 SEC(CPU TIME) ...     4 SEC(REAL TIME).
CORE STORAGE UTILIZATION:    20392 BYTES FOR OBJECT CODE: 20336 BYTES FOR ARRAY STORAGE
```

APPENDIX C*

An Acceptable-Point Algorithm for Design Optimization

G. E. JOHNSON

Department of Mechanical and Aerospace Engineering,
School of Engineering and Applied Science,
University of Virginia, Charlottesville, Virginia

M. A. TOWNSEND

Department of Mechanical Engineering and Material Sciences
School of Engineering, Vanderbilt University,
Nashville, Tennessee

C.1 Nomenclature

Most of the parameters are defined as they appear in the text. A key to the remaining operators and symbols is provided below.

* This Appendix presents an independent recently developed algorithm for AOD use. The example presented at the end correlates very well with example 7.1 of section 7.7 in Chapter 7, which was solved by the MOD.

Parameter	Definition
Vectors	
$\mathbf{x}$	Vector of independent design variables
$\mathbf{g}$	Gradient of $f(\mathbf{x})$
$\mathbf{s}$	Search direction vector
Scalars	
r_j	Penalty multiplier
λ	Step length parameter
ω_i	Constraint weighting factor
ω_k	Search direction weighting factor
Indices	
j	Penalty function subproblem number
i	Constraint number
k	Multivariate iteration number
$*$	Optimal value for variable
Operators	
$\Delta \mathbf{v}^k$	$\mathbf{v}^{k+1} - \mathbf{v}^k$
$\langle \mathbf{w}, \mathbf{v} \rangle$	Dot product of $\mathbf{w}$ and $\mathbf{v}$
$\|\mathbf{v}\|$	Euclidean norm of $\mathbf{v}$

C.2 Introduction

Problems in mechanical design synthesis can usually be posed in the form of the general nonlinear programming problem, that is,

$$\text{minimize } u(\mathbf{x}), \mathbf{x} \in R^N \tag{C.1}$$

such that

$$\psi_i(\mathbf{x}) = 0 \qquad i = 1, 2, \ldots, M \tag{C.2}$$

$$\phi_i(\mathbf{x}) \geq 0 \qquad i = M+1, M+2, \ldots, P \tag{C.3}$$

$$M < N \tag{C.4}$$

While several approaches to the solution of this problem have been suggested, the penalty function transformation, first suggested by Carroll [C.1], appears to be the most widely used and generally applicable. A suitably defined penalty function transforms the constrained problem of equations C.1 through C.4 into a single function which represents the problem, namely,

$$f(\mathbf{x}, r_j) = u(\mathbf{x}) + \Psi(\psi, r_j) + \Phi(\phi, r_j) \tag{C.5}$$

The scalar functions Ψ and Φ are combinations of the respective constraints which penalize nonsatisfaction of the constraints such that minimization of $f(\mathbf{x}, r_j)$ is equivalent to minimizing $u(\mathbf{x})$ for an optimum interior to the feasible region. Equation C.5 is a parametric penalty function due to the presence of the parameter r_j which weights the relative influence of the penalty terms and objective function; parametric penalty functions are the most common type. A variety of penalty functions are possible, although presumably certain ones would be more satisfactory. One such transformation is given in reference C.2.

$$\text{minimize } f(\mathbf{x}, r_j) = u(\mathbf{x}) + \frac{1}{r_j}\left\{\sum_{i=1}^{M} \omega_i[\psi_i(\mathbf{x})]^2 + \sum_{i=M+1}^{P} \omega_i[\bar{\phi}_i(\mathbf{x})]^2\right\} \tag{C.6}$$

where

$$\bar{\phi}_i(x) = \begin{cases} 0, \ \phi_i(\mathbf{x}) \geq 0 \\ \phi_i(\mathbf{x}), \ \phi_i(\mathbf{x}) < 0 \end{cases} \tag{C.7}$$

$$\omega_i > 0, \qquad i = 1, 2, \ldots, P \tag{C.8}$$

The solutions to the problem defined by equations C.1 through C.4 and the problem defined by equations C.6, C.7, and C.8 are equivalent in the limit as $r_j \to 0$ [C.2]. Accordingly, a solution to the constrained problem can be obtained by minimizing a sequence of unconstrained problems characterized by equations C.6 and C.7 with decreasing values for r_j. The solution vector obtained in minimizing $f(\mathbf{x}, r_j)$ is used as the starting vector for the solution of $f(\mathbf{x}, r_{j+1})$.

The penalty approach enables the application of a large body of knowledge regarding the optimization of an unconstrained function, in this case the functional (function of functions) of equations C.6 and C.7. In principle, penalty function transformations are simple and straightforward; in actual practice solutions may be difficult because of the sensitivity of $f(\mathbf{x}, r_j)$ to $\{r_j\}$ and ω_i. Most of the techniques for solving this problem are either direct search or gradient-based (descent). A Hooke-Jeeves direct search algorithm is presented elsewhere [C.3]; while this method is simple and generally reliable, it is not troublefree and may be slow. Also, it is probably a good idea to have more than one technique in one's "bag of tricks." Accordingly, we present in this appendix a recently developed general-purpose descent algorithm which has proved effective in solving problems in the form of equations

C.6, C.7, and C.8 [C.4]. In the following section, descent methods are introduced; then the new algorithm is developed in detail. Finally, an example problem (taken from Chapter 7) is solved and discussed.

In the following, the penalty function parameter r_j is suppressed to minimize notation. Accordingly, the problem considered is

$$\text{minimize } f(\mathbf{x}),\ \mathbf{x} \in R^N$$

C.3 Descent Methods for Unconstrained Function Minimization

The descent methods are motivated by a first order Taylor series expansion of $f(\mathbf{x})$ about the current point $\mathbf{x}^k$, that is,

$$f(\mathbf{x}^{k+1}) = f(\mathbf{x}^k) + \langle \mathbf{g}^k, \Delta\mathbf{x}^k \rangle \tag{C.9}$$

If $\mathbf{x}^{k+1}$ is defined as

$$\mathbf{x}^{k+1} = \mathbf{x}^k + \lambda_k \mathbf{s}^k \tag{C.10}$$

then

$$\Delta\mathbf{x}^k = \lambda_k \mathbf{s}^k \tag{C.11}$$

Substituting equation C.10 into equation C.9 and taking the limit as $\lambda_k \to 0^+$ gives

$$\lim_{\lambda_k \to 0+} \left[\frac{f(\mathbf{x}^{k+1}) - f(\mathbf{x}^k)}{\lambda_k} \right] = \langle \mathbf{g}^k, \mathbf{s}^k \rangle \tag{C.12}$$

As long as

$$\langle \mathbf{g}^k, \mathbf{s}^k \rangle < 0 \tag{C.13}$$

there exists some $\lambda_k > 0$ for which $f(\mathbf{x}^{k+1}) < f(\mathbf{x}^k)$. Values of $\mathbf{s}^k$ which satisfy equation C.13 are called *descent directions.* By successive application of equation C.10 we seek $\mathbf{x}^*$ such that $f(\mathbf{x}^*) < f(\mathbf{x})$, $\|\mathbf{x} - \mathbf{x}^*\| < \delta$ for some $\delta > 0$. Here $\mathbf{x}^*$ is a local minimizer of $f(\mathbf{x})$.

Wolfe's theorem [C.5] proves the following result for the descent methods.

Let $f(\mathbf{x})$ be a scalar valued function, continuously differentiable and bounded below. Assume $\mathbf{x}^0$ is such that $\mathbf{g}$ is uniformly continuous for all $\mathbf{x}$ satisfying $f(\mathbf{x}) \leq f(\mathbf{x}^0)$. Let $\mathbf{x}^{k+1}$ be defined by equation C.10 and $\mathbf{s}^k$ satisfy equation C.13 for all k. Then there is a sequence $\{\lambda_k\}$ which leads to

$$\lim_{k \to \infty} \left\langle \mathbf{g}^k, \frac{\mathbf{s}^k}{\|\mathbf{s}^k\|} \right\rangle = 0 \tag{C.14}$$

Wolfe observes that equation C.14 implies that $\|\mathbf{g}^k\|$ converges to zero as long as the angle between $\mathbf{s}^k$ and $\mathbf{g}^k$ stays bounded away from 90°. This is also a requirement of equation C.13; hence descent methods can be expected to converge to a point where $\|\mathbf{g}^k\| \to 0$ if λ_k are properly chosen. Such a point may not be a local minimizer of $f(\mathbf{x})$, for example, a saddle point may be found. This occurrence appears to be rare in mechanical design; however, when in doubt about the validity of an algorithmic solution, several practical tests (such as those described by Murray [C.6]) can be applied. Two problems are now identified: how to choose $\mathbf{s}^k$ and how to choose λ_k.

To the first question, an obvious direction for search is $\mathbf{s}^k = -\mathbf{g}^k$ as first suggested by Cauchy [C.7] in 1847. However, although this choice satisfies equation C.13, the resulting algorithm converges very slowly compared to alternatives [C.8, C.9, C.10]. Faster convergence can be achieved if it is assumed that $f(\mathbf{x})$ can be locally approximated by a quadratic form—even when this assumption is not globally valid. Among these algorithms are Newton's method [C.11, C.12, C.13], the quasi-Newton methods [C.8, C.9] and the conjugate direction methods [C.10, C.14, C.15, C.16]. Newton's method requires the formulation of second partial derivatives and the inversion of an $N \times N$ matrix in addition to possible stability problems [C.17]. Quasi-Newton and conjugate direction methods utilize first partial derivatives only. The quasi-Newton methods do store and manipulate an $N \times N$ matrix; however, no inversion process is required. With conjugate direction methods, no matrices are necessary to compute the search directions and operations are on the same order of magnitude as for the method of steepest descent. Due to the substantial literature extant we will omit discussion of the many methods and present a generalized conjugate direction algorithm which has proved to be reliable and efficient on several industrial problems and test functions [C.4, C.10].

Conjugate Directions. The minimum of a convex N-variable quadratic function can be found in a finite number of steps if each step locates the minimum of the function along a conjugate direction of search [C.17]. Two directions, $\mathbf{s}^k$ and $\mathbf{s}^{k+1}$, are said to be conjugate with respect to some suitable matrix A if

$$\langle \mathbf{s}^{k+1}, A\mathbf{s}^k \rangle = 0 \tag{C.15}$$

Typically A is taken as A^k, which is the current approximation of the Hessian of $f(\mathbf{x})$. In the method to be developed here, A is never explicitly evaluated and its true nature in a practical design study may be problematical. The direction for search is initially defined as

$$\mathbf{s}^0 = -\mathbf{g}^0$$

and subsequently as

$$\mathbf{s}^{k+1} = -\mathbf{g}^{k+1} + \omega_{k+1}\mathbf{s}^k \tag{C.16}$$

where ω_{k+1} is to be computed to satisfy equation C.15. For some A^k we may write

$$\mathbf{g}^{k+1} - \mathbf{g}^k = A^k(\mathbf{x}^{k+1} - \mathbf{x}^k) \tag{C.17}$$

which on substitution of equation C.11 becomes

$$(\mathbf{g}^{k+1} - \mathbf{g}^k)\frac{1}{\lambda_k} = A^k\mathbf{s}^k \tag{C.18}$$

Theoretical justification of equation C.17 is given by Townsend and Johnson [C.10]. Equation C.17 is actually the quasi-Newton equation [C.8], though the conjugate direction algorithm does not formally construct an approximation to the Hessian (A^k) or inverse Hessian as is done by the quasi-Newton methods.

Next, the right side of equation C.16 and the left side of equation C.18 are substituted into equation C.15 and the resulting expression is solved for ω_{k+1}, giving

$$\omega_{k+1} = \frac{\langle \mathbf{g}^{k+1} - \mathbf{g}^k, \mathbf{g}^{k+1} \rangle}{\langle \mathbf{g}^{k+1} - \mathbf{g}^k, \mathbf{s}^k \rangle} \tag{C.19}$$

Historically [C.16] at predetermined multivariate stages it is usual to set

$$\omega_{k+1} = 0 \tag{C.20}$$

for example, at every $N + 1$ (or more) stages. This strategy, known as reset, has been suggested to reduce the effect of round-off errors and/or nonquadratic behavior of the objective function. Whenever equation C.20 is invoked, all previous information used to compute the search directions is discarded and the algorithm continues as though from a new initial point. Although there is no theoretical justification for continuing beyond N multivariate iterations without resetting the algorithm [C.18], numerical experiments suggest that overall performance can improve if reset is delayed or avoided entirely [C.4, C.10, C.18]. Reset is actually only *required* when equation C.13 is not satisfied. Accordingly, in the computer code to be presented the algorithm resets whenever equation C.13 does not hold and at predetermined intervals which are specified by the user.

Many early convergence proofs [C.9, C.14, C.16] were based on the assumption that λ_k is equal to λ_k^* for each k where

$$f(\mathbf{x}^k + \lambda_k^*\mathbf{s}^k) = \text{local min } f(\mathbf{x}^k + \lambda\mathbf{s}^k) \qquad \lambda > 0 \tag{C.21}$$

It is actually impossible to *exactly* solve equation C.21 at each stage, and computational effort in close satisfaction may become excessive, so some compromises based on experience are always necessary.

Since it is not possible or practical to solve equation C.21 at each stage it has been shown that to construct a convergent algorithm under the assumptions of Wolfe's theorem it is only necessary to identify a value for λ_k such that [C.8]

$$f(\mathbf{x}^k + \lambda_k \mathbf{s}^k) < f(\mathbf{x}^k) \tag{I}$$

$$\|\mathbf{x}^{k+1} - \mathbf{x}^k\| > 0, \text{all } k \tag{II}$$

and

$$\langle \mathbf{s}^{k+1}, \mathbf{g}^{k+1} \rangle < 0 \tag{III}$$

that is, $\mathbf{s}^{k+1}$ as computed in equation C.16 defines a descent direction per equation C.13.

Several strategies to search for an acceptable point (i.e., a value of λ_k satisfying conditions I, II, and III above) have been proposed [C.4, C.8, C.19–C.22]. The one described here, called GSAP after "golden section acceptable point" search, is simple and is well-suited to problems where the penalty function transformation of equations C.6, C.7, and C.8 is used; it is also quite satisfactory for use with other transformations or unconstrained problems.

The GSAP search identifies an acceptable point by comparing function values along the half line originating at $\mathbf{x}^k$ in the direction $\mathbf{s}^k$. When a bracket on a point which satisfies equation C.21 is found, a golden section search [C.23] is employed to reduce the bracket. During this phase, points are tested at user-specified intervals according to conditions I, II, and III. When an acceptable point is identified, the univariate algorithm returns to the main program with the appropriate value for λ_k. If an acceptable point is not located after some user-specified maximum number of function evaluations, then the univariate search is considered to have failed—either because the search direction was a poor one or because of an overly large initial step length. The policy followed is to exit from the univariate algorithm, reset the parent algorithm, and reduce the initial step length. The search for an acceptable point in the new search direction then proceeds as before. In this way, the univariate algorithm attempts to identify opportune iterations to employ the reset option.

C.4 Operations of the GCG–GSAP Algorithm

The general conjugate gradient algorithm (GCG) and golden section acceptable point search are presented below.

GCG Algorithm

1. Initialize: Pick $\mathbf{x}^0$, set $\mathbf{s}^0 = -\mathbf{g}^0$.

At iteration k

2. Move to $\mathbf{x}^{k+1}$: Determine an acceptable value of λ_k by GSAP univariate algorithm below.
3. If $\|\mathbf{g}^{k+1}\| < \varepsilon$, terminate:

$$\mathbf{x}^* \approx \mathbf{x}^{k+1}, f^* \approx f(\mathbf{x}^{k+1})$$

4. Optional reset: If some $p > N + 1$ multivariate iterations have occurred since the last (re)initialization, define $\mathbf{x}^0 = \mathbf{x}^{k+1}$, $\mathbf{s}^0 = -\mathbf{g}^{k+1}$, set $k = 0$, and go to step 2. Otherwise go to step 5.
5. Determine next search direction: Compute $\mathbf{s}^{k+1}$ by equation C.16. If $\mathbf{s}^{k+1}$ does not satisfy equation C.13, reset as in step 4 is *required.*
6. $k = k + 1$.
7. Go to step 2.

GSAP Univariate Algorithm

1. Pick ρ_0, $\Delta_{\max}$, and $\beta > 1$ (parameters, discussed below) $K \in \{1, 2, 3, \ldots\}$, $J \in \{K, 2K, 3K, \ldots\}$
2. $\rho = \min\{\rho_0, \Delta_{\max}\}$, $n = 1$, $\lambda_0 = 0$.
3. $\bar{J} = 0$.
4. $\lambda_n = \rho$.
5. $\psi = f(\mathbf{x}^k + \lambda_n \mathbf{s}^k) - f(\mathbf{x}^k + \lambda_{n-1}\mathbf{s}^k)$.
6. If $\psi \geq 0$, go to step 9.
7. If $(\beta - 1)\rho > \Delta_{\max}$, $\rho = \rho + \Delta_{\max}$. Otherwise $\rho = \beta\rho$.
8. $n = n + 1$, go to step 4.
9. Comment—minimum is now bracketed. Make K golden section function evaluations.
10. Test best point in hand and if conditions I, II, and III obtain, return to main algorithm with $\lambda_k = \lambda$; otherwise $\bar{J} = \bar{J} + K$.
11. If $\bar{J} \geq J$ and I is satisfied, return with $\lambda_k = \lambda$ (reset may be necessary). If $\bar{J} \geq J$ and I is not satisfied, consider line search a failure, return to main algorithm, reset and reduce ρ_0; if $\bar{J} < J$ go to step 9.

Optimal choices for ρ_0, β, $\Delta_{\max}$, K, and J will depend on the topography of the objective function between $\mathbf{x}^0$ and $\mathbf{x}^*$. With large K the algorithm reverts to an "exact" line search and no advantage is taken of the acceptable point philosophy. However, if an "exact" search is required, GSAP can accommodate this need by simply assigning a large value to K (say $K = 21$). In practice $K \in \{1, 2, 3, \ldots 6\}$ and $J \in \{K, 2K, 3K, \ldots, 18\}$ have been reliable and efficient. In the computer coding to follow, β is permanently assigned the value of 2. The amount by which λ can be incremented in the bracketing phase is limited by a safety parameter, Δ_{max}. The largest interval which will ever be encountered for reduction is of length $2\Delta_{max}$ and one may wish to consider this in choosing Δ_{max}. Probably more important is the scaling of the design space, and experience is the best guide. If the multivariate algorithm fails because it has ventured beyond the constraint set, then a smaller Δ_{max} is appropriate. A reliable method to select ρ_0 is given by

$$\rho_0 = \min\left(\frac{\Delta_{max}}{\|\mathbf{s}^k\|}, \frac{\|\mathbf{x}^k - \mathbf{x}^{k-1}\|}{\|\mathbf{s}^k\|}\right)$$

It is then only necessary for the user to supply a value for ρ_0 at the first stage of the search, and any value consistent with the scaling of the design variables should work. In the example which follows, values are assigned based on extensive numerical testing [C.4], and these provide a starting point for experimentation in solving an unfamiliar problem.

C.5. Optimal Design of a Torsion Bar Spring

To illustrate the use of the algorithm consider the design synthesis problem developed in section 7.7 of the text. The design variables $\mathbf{x}$ are given as

$$x_1 = d, \qquad x_2 = l \qquad (\mathbf{x} \in R^2)$$

In standard form, the problem becomes

$$\text{minimize } M_t = \frac{x_1{}^2}{4}\left[\frac{\pi G(\text{P.E.})}{x_2}\right]^{1/2} \tag{C.22}$$

such that

$$\theta_{max} - \frac{8}{x_1^2}\left[\frac{(\text{P.E.})x_2}{\pi G}\right]^{1/2} \geq 0 \tag{C.23}$$

$$\frac{S_e}{(1+P)N} - \frac{4K_i}{x_1}\left[\frac{(\text{P.E.})G}{\pi x_2}\right]^{1/2} \geq 0 \tag{C.24}$$

$$L_{max} - x_2 \geq 0 \tag{C.25}$$

$$x_2 - L_{min} \geq 0 \tag{C.26}$$

$$d_{max} - x_1 \geq 0 \tag{C.27}$$

$$x_1 - d_{min} \geq 0 \tag{C.28}$$

As posed in section 7.7, it is necessary to determine optimal material as well as geometry. Here, the problem is formulated with material properties assumed to be known parameters. The optimization is carried out for each case separately and the best material can then be selected from the candidate materials at the end of the study. The operations are virtually identical for each material and so results are only presented for AISI 1020 CR steel (with properties and regional constraints as given in section 7.7).

With any numerical technique it is usually beneficial to scale the design variables so that their ranges are of the same order of magnitude. In this problem, the design variable x_1 is modified so that $x_1 = 10d$, giving a range comparable to that of $x_2 = l$. The problem is now transformed to a sequence of unconstrained minimizations as in equations C.6, C.7, and C.8. Guidelines for the selection of r_j and ω_i are given by Fiacco and McCormick [C.2], Himmelblau [C.17], and others. In this example, $r_1 = 0.1$, $r_j = 0.09r_{j-1}$, and $\omega_i = 1, i = 1, 2, \ldots, 6$. The computer program is given in figure C.1. The program is divided into a calling program (which controls the sequence r_j and provides the final output) and four subroutines defined below.

Figure C.1 GCG–GSAP FORTRAN listing.

```
      DIMENSION XX(20),PEN(20)
      COMMON GMAT,POTEN,THETMA,SEMAT,AKI,AP,ANS,ALMAX,ALMIN,
     1DMAX,PR1,DMIN,WP1,WP2,WP3,WP4,WP5,WP6
      READ(5,10)PR1,POTEN,ALMAX,ALMIN,DMAX,DMIN,THETMA,AKI,A
      READ(5,10)ANS,AP,GMAT,SEMAT,WP1,WP2,WP3,WP4,WP5,WP6,EP4
      READ(5,10)NN,IPR,MMAX,AADL,AADLMA,KSET,EP1,LINCAL,LINMAX
      READ(5,10)(XX(I),I=1,NN)
      READ(5,10)ROMIN1
   10 FORMAT(20G)
   22 CALL GCGSAP(NN,IPR,MMAX,AADL,AADLMA,XX,KSET,EP1,LINCAL,
     1LINMAX,ROMIN1)
      IF(PR1.LT.EP4)  GO TO 21
      PR1=A*PR1
      GO TO 22
   21 TOR=2.5E-3*(XX(1)**2)*(3.1416*GMAT*POTEN/XX(2))**.5
      PEN(1)=THETMA-800./(XX(1)**2.)*(POTEN*XX(2)/(3.1416
     1*GMAT))**.5
      PEN(2)=SEMAT/(1.+AP)/ANS-40.*AKI/XX(1)*(POTEN*GMAT)**.5/
     1(3.1416*XX(2))**.5
      PEN(3)=ALMAX-XX(2)
      PEN(4)=XX(2)-ALMIN
      PEN(5)=DMAX-XX(1)
      PEN(6)=XX(1)-DMIN
      OUTPUT PEN(1),PEN(2),PEN(3),PEN(4),PEN(5),PEN(6)
      WRITE(6,60)XX(1),XX(2),TOR
   60 FORMAT(/,'D= ',F6.3,3X,'L= ',F6.3,3X,'TORQUE= ',
     1F8.3)
      END
      SUBROUTINE FUNC(L,R,OF)
      COMMON GMAT,POTEN,THETMA,SEMAT,AKI,AP,ANS,ALMAX,ALMIN,
```

```
     1DMAX,PR1,DMIN,WP1,WP2,WP3,WP4,WP5,WP6
      DIMENSION R(L),P(6)
      TMAX=2.5E-3*(R(1)**2.)*(3.1416*GMAT*POTEN/R(2))**.5
      P(1)=WP1*(THETMA-800./(R(1)**2.)*(POTEN*R(2)/(3.1416
     1*GMAT))**.5)
      P(2)=WP2*(SEMAT/(1.+AP)/ANS-40.*AKI/R(1)*(POTEN*GMAT)**.5
     1/(3.1416*R(2))**.5)
      P(3)=WP3*(ALMAX-R(2))
      P(4)=WP4*(R(2)-ALMIN)
      P(5)=WP5*(DMAX-R(1))
      P(6)=WP6*(R(1)-DMIN)
      SUMVIO=0.0
      DO 25 I=1,6
      IF(P(I).LT.0) GO TO 10
      GO TO 25
   10 P(I)=(1./PR1)*P(I)**2.
      SUMVIO=SUMVIO+P(I)
   25 CONTINUE
      OF=TMAX+SUMVIO
      RETURN
      END
      SUBROUTINE GCGSAP(N,IPRINT,IMAX,ADL,ADLMAX,X,KRESET,
     1EPS1,INCALL,MAXCAL,ROMIN)
      DIMENSION X(20),G(20),S(20),GXN(20),SSAVE(20),DG(20),
     1XN(20)
      KFUNC=1
      KGRAD=1
      KFAIL=0
      LM=0
      ITER=0
      CALL GRAD(N,X,G)
      CALL FUNC(N,X,F)
  223 L=1
      DO 215 I=1,N
  215 S(I)=-G(I)
  220 DOTSG=0.
CCC   TEST FOR DESCENT DIRECTION
      DO 52 I=1,N
   52 DOTSG=DOTSG+G(I)*S(I)
CCC   RESET IF A NONDESCENT DIRECTION IS COMPUTED
      IF(DOTSG.GT.-1E-16) GO TO 223
      ITER=ITER+1
CCC   TEST FOR MAXIMUM MULTIVARIATE ITERATIONS
      IF(ITER.GT.IMAX)GOTO925
      GOTO926
  925 WRITE(6,930)IMAX
  930 FORMAT(/,2X,'AFTER ',I4,' STAGES, NO CONVERGENCE.')
      DO 931 I=1,N
      WRITE(6,932)I,X(I)
  932 FORMAT(/,2X,'X(',I2,')= ',E14.7)
  931 CONTINUE
      WRITE(6,934)FXN
  934 FORMAT(//,2X,'F= ',E14.7)
      WRITE(6,37)KFUNC,KGRAD
   37 FORMAT(/,2X,I6,2X,' FUNCTION EVALUATIONS AND ',I6,
     1' GRADIENT EVALUATIONS')
      GO TO 962
CCC   CHECK FOR INTERMEDIATE OUTPUT
  926 IF(MOD(ITER,IPRINT).EQ.0)GOTO901
      GO TO 902
  901 WRITE(6,950)ITER,KFUNC,KGRAD
  950 FORMAT(/,2X,'AT STAGE ',I4,/,' AFTER ',I6,
     1' FUNCTION CALLS AND ',I6,' GRADIENT CALLS')
      DO 804 I=1,N
```

```
      WRITE(6,805)I,X(I),I,S(I),I,G(I)
  805 FORMAT(/,2X,'X(',I2,')= ',E14.7,2X,'S(',I2,')= ',E14.7,
     12X,'G(',I2,')= ',E14.7)
  804 CONTINUE
      WRITE(6,806)F
  806 FORMAT(//,2X,'F= ',E14.7)
CCC   COMPUTE STEP LENGTH
  902 CALL GSAP(N,S,X,ADL,ADLMAX,F,XN,FXN,DOTSG,
     1LCOUNT,LM,DG,GXN,INCALL,MAXCAL,KGCALL,EPS1,G)
      KFUNC=KFUNC+LCOUNT
      KGRAD=KGRAD+KGCALL
CCC IF STEP SATISFIES I AND II, CONTINUE.  (RESET MAY OBTAIN.)
CCC   LM=0 SATISFIES I, II AND III.  LM=1 MAY NOT GIVE III.
      IF(LM.LE.1) KFAIL=0
      IF(KFAIL.EQ.0) ADL=AMIN1(ADL,ADLMAX)
      IF(LM.LE.1) GOTO827
CCC   IF LM=2, SEARCH FAILED.  SHRINK STEP AND RESET.
      ADL=AMAX1(((.618)**(MAXCAL-1))*ADL,ROMIN*(.618)**(MAXCAL
     1-1))
      KFAIL=KFAIL+1
CCC   IF NO PROGRESS WITH STEP OF ROMIN, QUIT.
      IF(KFAIL.GE.2.AND.ADL.LT.ROMIN)  GO TO 181
      GO TO 223
  181 WRITE(6,182) ITER,KFUNC,KGRAD,F
  182 FORMAT(/,'CONVERGENCE IS ASSUMED AFTER ',I4,' STAGES,',
     1I5,' FUNCTION CALLS',/,'AND ',I4,' GRADIENT CALLS.',/,
     1'F= ',E14.7)
      DO 183 I=1,N
      WRITE(6,184) I,X(I)
  184 FORMAT(/,'X(',I2,')=',E14.7)
  183 CONTINUE
      GO TO 962
  827 DO 221 I=1,N
      X(I)=XN(I)
  221 G(I)=GXN(I)
      F=FXN
CCC   TEST FOR CONVERGENCE BY GRADIENT TEST.
      GNORM=0.0
      DO 222 I=1,N
  222 GNORM=GNORM+G(I)**2.
      IF(GNORM.LT.EPS1) GO TO 203
CCC   RESET OPTION AT KRESET PREDETERMINED INTERVALS.
      IF(L.EQ.KRESET) GO TO 223
CCC   COMPUTE NEXT CONJUGATE DIRECTION.
      TOP=0.0
      BOTTOM=0.0
      DO 216 I=1,N
      TOP=TOP+DG(I)*G(I)
      BOTTOM=BOTTOM+DG(I)*S(I)
  216 SSAVE(I)=S(I)
CCC   DO NOT DIVIDE BY ZERO.  INSTEAD RESET ALGORITHM.
      IF(ABS(BOTTOM).LT.1E-16)  GO TO 28
  100 W=TOP/BOTTOM
      GO TO 104
   28 W=0.0
      L=1
  104 DO 219 I=1,N
  219 S(I)=-G(I)+W*SSAVE(I)
      L=L+1
      GOTO220
  203 DO 801 I=1,N
      WRITE(6,802)I,X(I)
  802 FORMAT(/,2X,'X(',I2,')= ',E14.7)
  801 CONTINUE
```

```
      WRITE(6,803)FXN
  803 FORMAT(//,2X,'F= ',E14.7)
      WRITE(6,25)ITER,KFUNC,KGRAD
   25 FORMAT(/,2X,'CONVERGED AT STAGE ',I4,' AFTER ',
     1I6,' FUNCTION CALLS AND ',/,2X,I6,' GRADIENT CALLS')
  962 RETURN
      END
      SUBROUTINE GRAD(M,Z,DF)
      DIMENSION Q(20),Z(20),DF(20)
      CALL FUNC(M,Z,F1)
      DO 128 I=1,M
      EPS3=.000001
      DO 129 J=1,M
      DEL=0.
      IF(I.EQ.J)DEL=EPS3
  129 Q(J)=Z(J)+DEL
      CALL FUNC(M,Q,F2)
  128 DF(I)=(F2-F1)/EPS3
      RETURN
      END
      SUBROUTINE GSAP(M,SSUB,XSUB,DL,DLMAX,FX,Z,FZ,DOTGS,
     1KOUNT,ML,DG1,GZ,KJ,MTEST,KKOUNT,EP,GX)
      DIMENSION SSUB(20),XSUB(20),Z(20),Z1(20),W1(20),W2(20)
     1,SUNIT(20),Y(20),DG1(20),GX(20),GZ(20),AZ(4),AF(4)
     1,Q1(20),Q2(20)
      KOUNT=0
      KKOUNT=0
CCC   BRACKET MINIMUM IN SEARCH DIRECTION.
      A=0.
      DLSUB=DL
      LSTEP=0
      FA=FX
      SMAG=0.
      DO 342 I=1,M
  342 SMAG=SMAG+SSUB(I)**2.
      SMAG=SMAG**.5
      DO 343 I=1,M
  343 SUNIT(I)=SSUB(I)/SMAG
      C=A+DLSUB
      DO 80 I=1,M
   80 Y(I)=XSUB(I)+C*SUNIT(I)
      CALL FUNC(M,Y,FC)
      KOUNT=KOUNT+1
      IF(FC-FA)20,21,21
   20 IF(LSTEP.EQ.0)GOTO41
      GO TO 44
   41 DLSUB=2.*DLSUB
CCC   LIMIT MAXIMUM INCREMENT IN BRACKET SEARCH.
      IF(ABS(DLSUB).LT.DLMAX)  GO TO 44
      LSTEP=LSTEP+1
      DLSUB=DLMAX
   44 B=C+DLSUB
      DO 81 I=1,M
   81 Z1(I)=XSUB(I)+B*SUNIT(I)
      CALL FUNC(M,Z1,FB)
      KOUNT=KOUNT+1
      IF(FB-FC)22,23,23
   22 A=C
      FA=FC
      C=B
      FC=FB
      GO TO 20
   21 B=C
      FB=FC
```

```
CCC    MINIMUM IS NOW BRACKETED.  START GOLDEN SECTION SEARCH.
    23 F1=.382
       F2=.618
       DX=B-A
       Y1=A+F1*DX
       Y2=A+F2*DX
       DO 85 I=1,M
       W1(I)=XSUB(I)+Y1*SUNIT(I)
    85 W2(I)=XSUB(I)+Y2*SUNIT(I)
       CALL FUNC(M,W1,FY1)
       CALL FUNC(M,W2,FY2)
       KOUNT=KOUNT+2
       J=2
CCC    HOPFINGER'S INTERVAL REDUCTION STRATEGY.
       AZ(1)=A
       AZ(2)=Y1
       AZ(3)=Y2
       AZ(4)=B
       MKOUNT=0
       ML=0
       AF(1)=FA
       AF(2)=FY1
       AF(3)=FY2
       AF(4)=FB
       FTK=AMIN1(AF(1),AF(2),AF(3),AF(4))
       DO 77 I=1,4
    77 IF(ABS(FTK-AF(I)).LE.1E-16)  KT=I
CCC    TEST FOR ACCEPTABILITY IF IT IS APPROPRIATE.
    14 IF(MOD(J,KJ).EQ.0) GO TO 51
    17 IF(KT.GT.2) GO TO 10
       AZ(4)=AZ(3)
       AF(4)=AF(3)
       AZ(3)=AZ(2)
       AF(3)=AF(2)
       AZ(2)=AZ(1)+F1*(AZ(4)-AZ(1))
       DO 11 I=1,M
    11 Q1(I)=XSUB(I)+AZ(2)*SUNIT(I)
       CALL FUNC(M,Q1,FQ1)
       AF(2)=FQ1
       IF(AF(2).LT.FTK) GO TO 12
       KT=2*KT-1
       FTK=AF(KT)
       KOUNT=KOUNT+1
       J=J+1
       GO TO 14
    12 KT=2
       FTK=AF(2)
       KOUNT=KOUNT+1
       J=J+1
       GO TO 14
    10 AZ(1)=AZ(2)
       AF(1)=AZ(2)
       AZ(2)=AZ(3)
       AF(2)=AF(3)
       AZ(3)=AZ(1)+F2*(AZ(4)-AZ(1))
       DO 15 I=1,M
    15 Q2(I)=XSUB(I)+AZ(3)*SUNIT(I)
       CALL FUNC(M,Q2,FQ2)
       AF(3)=FQ2
       IF(FTK.LE.AF(3)) GO TO 16
       KT=3
       FTK=AF(3)
       KOUNT=KOUNT+1
       J=J+1
```

```
      GO TO 14
   16 KT=2*KT-4
      FTK=AF(KT)
      KOUNT=KOUNT+1
      J=J+1
      GO TO 14
   51 MKOUNT=MKOUNT+KJ
CCC   IF MAXIMUM ALLOWED GOLDEN SECTION EVALUATIONS HAVE BEEN
CCC   MADE, DONT TEST FOR A DESCENT DIRECTION.
      IF(MKOUNT.GE.MTEST) GO TO 53
CCC   IF NO PROGRESS HAS BEEN MADE BUT MORE FUNCTION CALLS ARE
CCC   ALLOWED, RETURN TO INTERVAL REDUCTION STRATEGY.
      IF(AZ(KT).LT.1E-16) ML=1
      IF(AZ(KT).LT.1E-16) GO TO 17
      DO 91 I=1,M
   91 Z(I)=XSUB(I)+AZ(KT)*SUNIT(I)
      FZ=AF(KT)
      DL=AZ(KT)
      CALL GRAD(M,Z,GZ)
      KKOUNT=KKOUNT+1
      GNORMZ=0.0
      SDOTGZ=0.0
      DO 92 I=1,M
      GNORMZ=GNORMZ+GZ(I)**2.
      SDOTGZ=SDOTGZ+SSUB(I)*GZ(I)
   92 DG1(I)=GZ(I)-GX(I)
CCC   IF GRADIENT VANISHES,QUIT.
      IF(GNORMZ.LT.EP)  GO TO 3
CCC   IF LINE SEARCH IS EXACT,QUIT.
      IF(ABS(SDOTGZ).LT.1E-12)  GO TO 3
      GDOTGZ=0.0
      SDOTDG=0.0
      DO 93 I=1,M
      GDOTGZ=GDOTGZ+GX(I)*GZ(I)
   93 SDOTDG=SDOTDG+SSUB(I)*DG1(I)
      IF(ABS(SDOTDG).LT.1E-16)  GO TO 17
      CHECK=(DOTGS*GNORMZ-SDOTGZ*GDOTGZ)/SDOTDG
CCC   IF I, II AND III ARE SATISFIED, QUIT.
      IF(CHECK.LT.0.) GO TO 3
      ML=1
      GO TO 17
   53 IF(AZ(KT).LT.1E-16) ML=2
      IF(ML.EQ.2) GO TO 31
      DO 88 I=1,M
   88 Z(I)=XSUB(I)+AZ(KT)*SUNIT(I)
      FZ=AF(KT)
      DL=AZ(KT)
      CALL GRAD(M,Z,GZ)
      KKOUNT=KKOUNT+1
      DO 89 I=1,M
   89 DG1(I)=GZ(I)-GX(I)
      GO TO 3
   31 DO 90 I=1,M
      Z(I)=XSUB(I)
      GZ(I)=GX(I)
   90 DG1(I)=0.0
    3 RETURN
      END
```

Index to Computer Program of Figure C.1

Calling program	lines 1–28 (supplied by user)
Function subroutine	lines 29–51 (supplied by user)
Main program	lines 52–165 (GCG algorithm)
Gradient subroutine	lines 166–178 (Forward differences are used in figure C.1. Analytical derivatives or a different numerical scheme could be supplied by the user. In figure C.1, the differencing increment is 1×10^{-5}.)
Univariate search	lines 179–344 (GSAP algorithm)

The null tests in lines 71, 146, 249, 299, 300, 316, 322, and 328 should be assigned to be compatible with the word length on the computing machine in use. The program in figure C.1 is written for the Fortran Load and Go (FLAG) compiler available on the XDS Sigma-7 computer at Vanderbilt University. Double precision is obtained by designating the appropriate FLAG option prior to running. With other systems, double-precision and format requirements should be checked for compatibility prior to compilation.

In the input statements (lines 4 through 8) variable names are assigned as below.

Model parameters (lines 4 and 5 of figure C.1)

PR1	$\equiv r_1$ of equation C.6
POTEN	$\equiv$ (P.E.) of equations C.22, C.23, and C.24
ALMAX	$\equiv L_{max}$ of equation C.25
ALMIN	$\equiv L_{min}$ of equation C.26
DMAX	$\equiv d_{max}$ of equation C.27
DMIN	$\equiv d_{min}$ of equation C.28
THETMA	$\equiv \theta_{max}$ of equation C.23
AKI	$\equiv K_i$ of equation C.29
A	$\equiv r_j = Ar_{j-1}$ to define the sequence $\{r_j\}$ below equation C.8
ANS	$\equiv N$ of equation C.24
AP	$\equiv P$ of equation C.24
GMAT	$\equiv G$ of equations C.22, C.23, and C.24
SEMAT	$\equiv S_e$ of equation C.24
WP1-WP6	$\equiv \omega_i$ of equation C.8
EP4	$\equiv$ convergence test on r_j of equation C.6; if $r_j \leq$ EP4, quit

Algorithm Parameters (lines 6, 7, and 8 of figure C.1)

NN	≡ number of independent variables in objective function
IPR	≡ intermediate output every IPR multivariate iterations
NMAX	≡ maximum number of multivariate iterations allowed
AADL	≡ initial step length (ρ_0) for univariate search at first multivariate iteration
AADLMA	≡ maximum increment (Δ_{max}) by which bracketing phase of univariate search can increase
KSET	≡ maximum number of multivariate iterations allowed between resets
EP1	≡ convergence criterion in the gradient test, that is, convergence assumed if $\|g^{k+1}\|^2 < \text{EP1}$
LINCAL	≡ number of golden section function evaluations made before testing acceptability criteria (*K* in GSAP algorithm)
LINMAX	≡ maximum number of golden section function evaluations allowed in univariate search (*J* in GSAP algorithm)
XX(I)	≡ initial solution estimate ($\mathbf{x}^0$) provided by user
ROMIN1	≡ convergence criterion in step length test, that is, convergence assumed if $\lambda_k < \text{ROMIN1}$.

In figure C.2 the data file used to obtain a solution is presented. The actual computer output for the program of figure C.1 and the data file of figure C.2 is found in figure C.3. Referring to figure C.3, the algorithm has determined that for the AISI 1020 CR steel, the spring should be 85.0 in long and have a diameter of 1.476 in in agreement with the solution developed in Chapter 7. If optimal values for *d* and *l* are desired for other materials, it is necessary to enter the appropriate material properties in the data file and rerun the program.

Figure C.2 Data file for torsion bar spring example.

```
1.000 .1,500.,85.,20.,15.,0.,.2,1.41,.09
2.000 1.3,.62,11.417E6,37.20E3,1.,1.,1.,1.,1.,1.,1E-6
3.000 2,25,100,.1,1.,80,.0001,5,5
4.000 12.,50.
5.000 .000001
```

Figure C.3. GCG–GSAP algorithmic solution to torsion bar spring example.

```
BEGINNING EXECUTION


CONVERGENCE IS ASSUMED AFTER   18 STAGES,   176 FUNCTION CALLS
AND    10 GRADIENT CALLS.
F=    .7504421E+04
```

```
X( 1)=  .1427535E+02

X( 2)=  .9090438E+02

CONVERGENCE IS ASSUMED AFTER   21 STAGES,   165 FUNCTION CALLS
AND    15 GRADIENT CALLS.
F=    .7871123E+04

X( 1)=  .1470946E+02

X( 2)=  .8561826E+02

CONVERGENCE IS ASSUMED AFTER   14 STAGES,   110 FUNCTION CALLS
AND     9 GRADIENT CALLS.
F=    .7910786E+04

X( 1)=  .1475621E+02

X( 2)=  .8507672E+02

CONVERGENCE IS ASSUMED AFTER   11 STAGES,    86 FUNCTION CALLS
AND     8 GRADIENT CALLS.
F=    .7913874E+04

X( 1)=  .1476248E+02

X( 2)=  .8500437E+02

CONVERGENCE IS ASSUMED AFTER   11 STAGES,    81 FUNCTION CALLS
AND     7 GRADIENT CALLS.
F=    .7915720E+04

X( 1)=  .1476255E+02

X( 2)=  .8500363E+02

CONVERGENCE IS ASSUMED AFTER   12 STAGES,    90 FUNCTION CALLS
AND     9 GRADIENT CALLS.
F=    .7914284E+04

X( 1)=  .1476284E+02

X( 2)=  .8500024E+02
 PEN(1) = 7.364467123993178E-02
 PEN(2) = -5.579731805482879E-07
 PEN(3) = -2.359069566963967E-04
 PEN(4) = 65.0002359069567
 PEN(5) = .237156223611765
 PEN(6) = 14.7628437763882

D= 14.763    L= 85.000    TORQUE= 7914.189
 *STOP*
```

REFERENCES

C.1 C. W. Carroll, "The Created Response Surface Technique for Optimizing Nonlinear Restrained Systems," *Operations Research*, Vol. 9, 1961, pp. 169–184.

C.2 A. V. Fiacco and G. P. McCormick, *Nonlinear Programming: Sequential Unconstrained Minimization Techniques*, Wiley, New York, 1968.

C.3 R. C. Johnson, *Mechanical Design Synthesis—Creative Design and Optimization*, Krieger, Huntington, N.Y., 1978, pp. 428–431.

C.4 G. E. Johnson, Strategies for Optimal Design, Doctoral Dissertation, Vanderbilt University, 1978.

C.5 P. Wolfe, "Convergence Conditions for Ascent Methods," *SIAM Review*, Vol. 11, 1969, pp. 226–235.

C.6 W. Murray, "Failure, the Causes and Cures," *Numerical Methods for Unconstrained Optimization*, W. Murray, Ed., Academic Press, London, 1972, pp. 107–122.

C.7 A. L. Cauchy, "Methode Generale pour la Resolution des Systemes d'Equations Simultanees," *Comptes Rendus, Academie des Sciences, Paris*, Vol. 25, 1847, p. 536.

C.8 J. E. Dennis and J. J. More, "Quasi-Newton Methods, Motivation and Theory," *SIAM Review*, Vol. 19, No. 1, 1977, pp. 46–89.

C.9 W. C. Davidon, "Variable Metric Method for Minimization," Rep. ANL-5990 Rev, Argonne National Laboratories, Argonne, Ill., 1959.

C10. M. A. Townsend and G. E. Johnson, "In Favor of Conjugate Directions: a Generalized Acceptable-Point Algorithm for Function Minimization," *Journal of the Franklin Institute*, Vol. 306, No. 5, 1978, pp. 325–339.

C.11 W. Murray, "Second Derivative Methods," *Numerical Methods for Unconstrained Optimization*, W. Murray, Ed., Academic Press, London, 1972, pp. 57–71.

C.12 A. Mathews and D. Davies, "A Comparison of Modified Newton Methods for Unconstrained Optimization," *Computer Journal*, Vol. 14, 1971, pp. 293–294.

C.13 R. Fletcher and T. L. Freeman, "A Modified Newton Method for Minimization," *Journal of Optimization Theory & Applications*, Vol. 23, No. 3, 1977, pp. 357–372.

C.14 E. Polak and G. Ribiere, "Note sur la Convergence de Methodes Conjugees," *Revue Francaise Inf. Rech. Oper.*, 16RI, 1969, pp. 35–43.

C.15 L. Nazareth, "A Conjugate Direction Algorithm without Line Searches," *Journal of Optimization Theory & Applications*, Vol. 23, No. 3, 1977, pp. 373–387.

C.16 R. Fletcher and C. M. Reeves, "Function Minimization by Conjugate Gradients," *Computer Journal*, Vol. 7, 1964, pp. 149–154.

C.17 D. M. Himmelblau, *Applied Nonlinear Programming*, McGraw-Hill, New York, 1972.

C.18 E. M. E. Beale, "A Derivation of Conjugate Gradients," *Numerical Methods for Nonlinear Optimization*, F. A. Lootsma, Ed., Academic Press, London, 1972.

C.19 L. Armijo, "Minimization of Functions Having Lipschitz Continuous First Partial Derivatives," *Pacific Journal Mathematics*, Vol. 16, 1966, pp. 1–3.

C.20 A. A. Goldstein and J. F. Price, "An Effective Algorithm for Minimization," *Numerische Mathematik*, Vol. 10, 1967, pp. 184—189.

C.21 R. Klessig and E. Polak, "Efficient Implementations of the Polak-Ribiere Conjuga Gradient Algorithm," *SIAM Journal of Controls*, Vol. 10, No. 3, 1972, pp. 524–549.

C.22 H. Y. Huang and J. P. Chambliss, "Quadratically Convergent Algorithms and One-Dimensional Search Schemes," *Journal of Optimization Theory & Applications*, Vol. 11, No. 2, 1973, pp. 175–188.

C.23 J. Kowalik and M. R. Osborne, *Methods for Unconstrained Optimization Problems*, American Elsevier, New York, 1968.

Index